T0337922

Energy for Sustainable Society

Energy for Sustainable Society

From Resources to Users

Oguz A. Soysal
Frostburg State University
Frostburg
Maryland
USA

Hilkat S. Soysal
Independent Researcher
Frostburg
Maryland
USA

This edition first published 2020
© 2020 John Wiley & Sons Ltd

All rights reserved. No part of this publication may be reproduced, stored in a retrieval system, or transmitted, in any form or by any means, electronic, mechanical, photocopying, recording or otherwise, except as permitted by law. Advice on how to obtain permission to reuse material from this title is available at http://www.wiley.com/go/permissions.

The right of Oguz A. Soysal and Hilkat S. Soysal to be identified as the authors of this work has been asserted in accordance with law.

Registered Offices
John Wiley & Sons, Inc., 111 River Street, Hoboken, NJ 07030, USA
John Wiley & Sons Ltd, The Atrium, Southern Gate, Chichester, West Sussex, PO19 8SQ, UK

Editorial Office
The Atrium, Southern Gate, Chichester, West Sussex, PO19 8SQ, UK

For details of our global editorial offices, customer services, and more information about Wiley products visit us at www.wiley.com.

Wiley also publishes its books in a variety of electronic formats and by print-on-demand. Some content that appears in standard print versions of this book may not be available in other formats.

Limit of Liability/Disclaimer of Warranty

MATLAB® is a trademark of The MathWorks, Inc. and is used with permission. The MathWorks does not warrant the accuracy of the text or exercises in this book. This work's use or discussion of MATLAB® software or related products does not constitute endorsement or sponsorship by The MathWorks of a particular pedagogical approach or particular use of the MATLAB® software.

In view of ongoing research, equipment modifications, changes in governmental regulations, and the constant flow of information relating to the use of experimental reagents, equipment, and devices, the reader is urged to review and evaluate the information provided in the package insert or instructions for each chemical, piece of equipment, reagent, or device for, among other things, any changes in the instructions or indication of usage and for added warnings and precautions. While the publisher and authors have used their best efforts in preparing this work, they make no representations or warranties with respect to the accuracy or completeness of the contents of this work and specifically disclaim all warranties, including without limitation any implied warranties of merchantability or fitness for a particular purpose. No warranty may be created or extended by sales representatives, written sales materials or promotional statements for this work. The fact that an organization, website, or product is referred to in this work as a citation and/or potential source of further information does not mean that the publisher and authors endorse the information or services the organization, website, or product may provide or recommendations it may make. This work is sold with the understanding that the publisher is not engaged in rendering professional services. The advice and strategies contained herein may not be suitable for your situation. You should consult with a specialist where appropriate. Further, readers should be aware that websites listed in this work may have changed or disappeared between when this work was written and when it is read. Neither the publisher nor authors shall be liable for any loss of profit or any other commercial damages, including but not limited to special, incidental, consequential, or other damages.

Library of Congress Cataloging-in-Publication data applied for

HB ISBN: 9781119561309

Cover Design: Wiley
Cover Image: Images courtesy of Oguz A. Soysal and Hilkat S. Soysal

Set in 9.5/12.5pt STIXTwoText by SPi Global, Chennai, India

Printed and bound by CPI Group (UK) Ltd, Croydon, CR0 4YY

10 9 8 7 6 5 4 3 2 1

Contents

About the Authors

Oguz A. Soysal received all of his degrees in electrical engineering from Istanbul Technical University, Turkey. Over his professional experience of more than 40 years, he worked as an R&D engineer for a power transformer manufacturer; served as a professor at several universities in Turkey and the US; and participated in research projects at Ohio State University in the US and the University of Toronto in Canada. In addition to his regular employment, he often provides consulting services to numerous industrial companies and has served as an expert witness in energy-related cases. Since 1998 he has been a faculty member of the Department of Physics and Engineering at Frostburg State University in Maryland, USA. His areas of teaching include energy systems, power electronics, control systems, mechatronics, and engineering design. He has published more than 50 papers in major journals and international conference proceedings, and co-authored a textbook on "Fault Conditions in Electric Energy Systems."

   Hilkat S. Soysal received her Law degree from the University of Istanbul, Turkey. After practicing law for over 15 years as an attorney, she started teaching engineering-related law courses at the College of Engineering at Istanbul University, Turkey, where she received her MSc degree in Marine Engineering. Between 2000 and 2012 she taught engineering courses and directed several energy-related projects at the Department of Physics and Engineering of Frostburg State University in Maryland USA. She has jointly published numerous papers on renewable energy applications in technical journals and conferences.

How Was This Book Born?

We were college students in the 1970s when the energy crisis affected every aspect of life in some way. We grew accustomed to continuously increasing energy prices, scheduled electric outages, and shortages of many industrial products. At the beginning of our careers, the Three Mile Island accident weakened our faith in nuclear power. When the Chernobyl disaster spread radiation over Europe, we were living in the Black Sea area where radioactive contamination seriously affected the agriculture. In our lifetimes, coalmine accidents killed many people, and numerous environmental disasters originating from wars, deep-water oil drilling, and tanker accidents contaminated the oceans. In the meantime, we have witnessed public concerns and debates on windfarm development and hydraulic fracking for unconventional oil recovery.

We have pursued our dreams of sustainability by conducting several renewable energy projects while living in the coal-mining regions of Western Maryland, West Virginia, and Pennsylvania. With a modest funding provided by Maryland Energy Administration in 2006, we found an opportunity to install a residential-scale wind and solar powered hybrid generation system on the Frostburg State University (FSU) campus to increase the public awareness about renewable energy. As part of this project, we established the Renewable Energy Center and organized the "Renewable Energy Symposium" at FSU. We received funding from the Appalachian Regional Commission (ARC) for development of the "Wind and Solar Energy (WISE) Education Program" to train professionals and individuals who are interested in installing small-scale renewable energy systems. The Honorable Roscoe Bartlett (then 6th District Congressman) has been instrumental in helping materialize two federal appropriations for construction of a completely off-grid building on the FSU campus, which they named the "Sustainable Energy Research Facility (SERF)." Construction of SERF took longer than anticipated and the building officially opened on October 29, 2012. While the SERF building could never operate as intended and the co-authors are no longer involved in its current management, it was an extraordinary learning experience which led to the preparation of this book.

This book stems from our lifetime experiences through the evolution of energy systems from fossil-fuel dominance to cleaner and sustainable sources, which culminated in the construction of the SERF facility at Frostburg State University. The goal of this project was to create a learning environment completely energized by renewable sources where students, faculty, researchers, and the public could be involved in renewable energy research and development, while experiencing the challenges of sustainable living. SERF was intended to be a unique interdisciplinary center on the FSU campus, as well as a scientific and technical attraction for the small mountain town of Frostburg. Although the SERF project could not reach its initial goal due to non-technical matters, the lessons we learned have stimulated us to disseminate our knowledge and experience, and kept us working on our research individually.

In 2017, we drove more than 20 thousand miles to visit energy-related sites in North America. We crossed the immense corn fields of Midwestern plains and saw corn-ethanol production facilities. We visited the Experimental Breeding Reactor EBR-I, where nuclear power was used to light four lightbulbs for the first time, in 1951. We toured the hydroelectric power plants in the Columbia River Basin in Washington and Oregon States. We drove to the Great Canadian Oil Sands North of Alberta, and crossed Death Valley in Nevada to see the Crescent Dunes Concentrated Solar Power (CSP) Station, which generates solar electricity day and night.

All our love for energy-related research and lifelong experiences came together in this textbook. In this way we believe that we can reach a larger audience interested in finding means of sustainable energy supply for modern society.

Oguz A. Soysal
Hilkat S. Soysal

Preface

Energy is central for the quality of life, productivity, and prosperity of a society. On the other hand, the production, delivery, and transformation of energy sources have irreversible impacts on the ecosystem, air quality, water supply, and land use, and in the long-term leads to climate change. Sustainable economic development and progress of humanity rely on continuous availability of high-quality energy sources with minimal effects on the environment.

This text was prepared while keeping in mind an interdisciplinary group of readers. It covers technical, social, economic, and environmental aspects of energy systems. The primary goal is to supply material for a first course on fundamental energy topics at upper division undergraduate or first year graduate studies. It may be also a useful resource for technical professionals who wish to enrich and/or update their knowledge with current developments and emerging issues in the field of energy.

The book is organized in such a way that the material covered in earlier chapters provides a foundation for the subsequent chapters. The instructor or reader may change the sequence depending on the level of knowledge, area of interest, and scope of study.

While a basic mathematics and physics background is sufficient to understand most of the content, a college-level knowledge of introductory physics and calculus is needed to thoroughly absorb topics related to energy conversion. Due to the interdisciplinary nature of the subject, principles of thermodynamics, fluid mechanics, nuclear energy, electromagnetic fields, and electromechanical energy conversion are summarized as needed.

The text is arranged in 12 chapters that explain energy flow from primary sources to end-users with an interdisciplinary system approach. Topics include the current and projected future availability of primary energy sources, energy supply chain, conversions between different forms of energy, energy security, environmental implications of energy use, and the impacts of growing energy demand on global sustainability.

The sequence of the topics is organized considering the share of various sources and conversion techniques in the current global energy-supply mix. Each chapter contains a review of background information, outline of the current technologies, and potential future developments. In addition to the technical content, global, socioeconomic, ethical, and environmental issues associated with presented energy technologies are discussed throughout the text with real-world examples and case studies. Every chapter ends with a summary, self-assessment quiz, suggested discussion and research topics, and problems. A list of relevant web sites that provide reliable information and references are included in each chapter to help readers probe further the presented topics and issues.

The first chapter is a general overview of energy systems. Interactions between the major elements of the energy supply chain are described. The chapter emphasizes the interdisciplinary aspects of energy system development, decision-making, system design, and energy management tasks.

Chapter 2 introduces fundamental principles of physics underlying energy conversion and storage systems. The concepts of work, energy, and power are briefly reviewed as well as the fundamental conservation principles of physics. The chapter describes the major forms of energy and outlines various technologies used in practical energy conversion and storage systems. A more detailed coverage of thermal, mechanical, and electrical energy is left to future chapters where heat engines and electromechanical energy conversion systems are studied.

Chapter 3 presents conventional and unconventional fossil fuels. The terminology used for classification of resources and reserves is explained based on the definitions used by international organizations and professional societies. The evolution of conventional fossil fuel exploration and recovery methods are complemented with emerging unconventional technologies such as hydraulic fracturing, extra-heavy oil and sand-oil processing and are explained with contemporary examples and statistics.

Chapter 4 introduces the fundamentals of nuclear power generation and reactor types. Major benefits of nuclear power plants and concerns about nuclear power plant development, operation, radioactive waste disposal, and decommissioning of nuclear power plants are presented followed by a discussion of major nuclear accidents and nuclear reactor safety.

Chapter 5 covers major renewable energy sources such as solar, wind, hydro, geothermal, and biomass energy. The benefits and challenges of renewable energy are discussed and latest estimates of the global resource potential of each renewable source are presented. The chapter discusses the strengths and limitations of wind and solar power as well as their potential to replace fossil fuels. Hydroelectric, wind, and solar energy conversion systems are discussed in more detail in Chapters 8–10.

Electric power systems, which allow the most convenient and flexible way to transmit all kinds of primary sources to end users, are presented in Chapter 6. The chapter briefly presents the historic evolution of electric power generation and transmission systems. Fundamental concepts, laws, and methods of electricity and magnetism needed to analyze single-phase and three-phase circuits, transformers, electric machines, and transmission lines are reviewed to establish a basis for understanding various technologies discussed in the subsequent chapters.

Nearly three-quarters of electric power in the world is generated from fossil fuels, nuclear reactions, geothermal sources, biofuels, and concentrated solar power using heat engines that operate on a thermodynamic cycle. Chapter 7 presents the fundamental concepts of thermodynamics, describes the basic operation of steam turbines, gas turbines, and combined cycle heat engines. The chapter includes a discussion of the efficiency and environmental impacts of thermal power generation systems.

Chapter 8 discusses technologies currently used in hydroelectric power generation. Concepts of fluid mechanics that are necessary to understand hydro machinery are summarized. Dam structures and major turbine types used in hydroelectric generation facilities are briefly explained and compared. The chapter also discusses socioeconomic benefits and challenges of large-scale hydroelectric generation projects with real-world case studies.

Sources and properties of wind, assessment of wind energy potential, and characteristics of wind-powered generation systems are presented in Chapter 9. Wind-turbine blade aerodynamics, basic rotor design techniques, and generator types are introduced followed by output power control methods of small-scale and utility-size generation units.

Chapter 10 presents an overview of solar geometry for assessment of radiant energy captured by solar collectors and photovoltaic (PV) modules. Operation and characteristics of collector types used in solar thermal units are described. Basic solid-state electronics relevant to photovoltaic (PV) generation are introduced and various PV cell technologies are compared.

Energy security issues, which have become one of the primary concerns for industrialized countries are discussed in Chapter 11. Short-term and long-term risks that energy systems face are described with recent examples. Electric grid vulnerability to natural and intentional disasters and resilience of energy systems are explained. Economic and social impacts of long-term interruptions of energy supply are explained with real-world examples. The text also discusses the benefits and drawbacks of energy independence versus energy interdependence.

Chapter 12 discusses the role of energy systems in economic development and global sustainability. Environmental impacts of energy systems on ground-level air pollution, acid rain, and global climate change are described with statistical data and facts. Water, energy, and food nexus is one of the current debates. Recent statistical data presented in the text show use of energy for water treatment and supply in contrast to the use of water for fuel production and electric generation. Land allocated to produce biomass instead of agriculture and food production is discussed from technical, economic, and ethical perspectives. The chapter includes methods for estimation of energy return on energy invested (EROI), impacts of the on-site energy consumption on depletion of primary energy sources, and greenhouse gas emissions.

The text discusses the challenges of various energy technologies in meeting the growing energy need of the current societies *"without compromising the ability of future generations to meet their own needs"* as stated by the United Nation's World Commission on Environment and Development. The authors wish is that this textbook contributes to the readers' understanding of the responsibilities in development of energy systems by preserving the natural resources and global environment.

November 2019

Oguz A. Soysal
Hilkat S. Soysal

Acknowledgments

The authors are grateful to Maryland Energy Administration (MEA), Appalachian Regional Commission (ARC), Maryland Industrial Partnerships (MIPS) program, US Department of Energy, and the Honorable Roscoe Bartlett (US Representative for Maryland's 6th congressional district between 1993 and 2013) who was instrumental for the federal appropriation to build the Sustainable Research Energy Facility. We appreciate the warm support of our friend Professor Siddik Binboga Yarman of Istanbul University who continuously encouraged us to write this book. We thank the Wiley production team who helped during the publication of our manuscript.

1

Overview

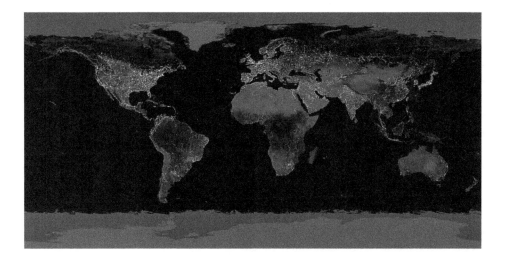

Image available at https://visibleearth.nasa.gov/view.php?id=55167/ (Accessed in August 2018).

Credit: Data courtesy Marc Imhoff of NASA GSFC and Christopher Elvidge of NOAA NGDC. Image by Craig Mayhew and Robert Simmon, NASA GSFC.

City lights show urbanized areas around the world. Although the density of lights is not necessarily proportional to the population density and degree of development, use of electricity reflects various aspects of social and economic activities. Energy intensity is higher in brighter areas of the earth such as Europe, Middle East, Southeast Asia, North America, parts of South America, and Oceania. Big metropolitan areas around the world are visible as bright light clusters. In the USA, the interstate highway network is detectable from the city lights. Dark spots in Africa, South America, Asia, and Oceania correspond to sparsely populated and less industrialized areas. Polar regions are entirely dark since they are not populated. According to the World Bank database (The World Bank Group [US] 2018), about 13% of the world population still does not have access to electricity. In darker areas of the earth, vital elements of modern society such as sanitary services, healthcare, education, transportation, water, and food supply are minimal.

Energy for Sustainable Society: From Resources to Users, First Edition. Oguz A. Soysal and Hilkat S. Soysal.
© 2020 John Wiley & Sons Ltd. Published 2020 by John Wiley & Sons Ltd.

1.1 Introduction

An energy system is a collection of elements that work together to supply the energy needs of a society. Inputs of an energy system are natural primary sources that can be economically converted to fuels, secondary energy sources, and energy carriers. Outputs are various forms of energy supplied to end-users.

Sun is the essential external energy source for life on earth. While sunlight is the natural source of heat, most primary energy sources available on earth are consequences of solar radiation heating the earth surface and atmosphere. Flowing water, wind, and firewood resulting from solar heat have energized human activities since early civilizations. Vegetation and living organisms initially developed due to the solar energy have been transformed over millions of years to coal, petroleum, and natural gas. In addition, periodic variations in gravitational attraction of the moon and other celestial bodies cause tidal motions and ocean waves.

Figure 1.1 outlines the interactions between the energy system, nature, and society. An energy system transforms primary sources into fuels and electric power to deliver various forms of energy needed for manufacturing, construction, agriculture, transportation, and public services. Commercial transactions, communication, computation, and data processing are essential economic functions that depend on energy. Economy delivers industrial products, buildings, roads, public services, food, treated water, education, recreation, and entertainment to the society.

Energy systems use water for extraction and processing of coal, petroleum, and natural gas; irrigation of crops for biofuel production; and cooling of power plants. Air is necessary for combustion of fuels and cooling of engines, motors, and generators.

On the other hand, natural resources are critical for life, productivity, and economy of the society. Obviously, all creatures need fresh water, clean air, and food to survive. Food supply depends on adequate irrigation of farmlands and drinking water for livestock. Nature offers feedstock for industrial processes and production.

Modern society cannot sustain without abundant energy, water, and food. Such commodities strongly depend on each other. Agriculture and food production rely on both water and energy supply. Energy systems use significant amount of water for fuel production and cooling purposes. Part of this water is recycled to the source, but some part is evaporated. In addition, energy systems are major sources of air, water, and land pollution. If not eliminated properly, toxic compounds released from energy facilities may be deposited in plants, seafood, and other living organisms. Air and water pollution strongly affect human health and can even cause fatal diseases. In populated areas, noise and vibration produced by mining equipment, fuel transportation trucks, freight trains, and generation units create public reactions.

Society is the receiving end of the energy system. Institutions collecting energy data categorize end-users based on their energy consumption profiles. Industrial, commercial, transportation, and residential sectors are the major groups of energy users. Each one may be expanded to subcategories for more detailed statistical evaluations of energy use.

Government offices closely watch the interactions between the energy system and society using "social and economic indicators." Such indicators reflect the welfare, living standards, and productivity of the society. Since food supply, water, and air quality affect the health and well-being of the population, social and economic indicators include pollution

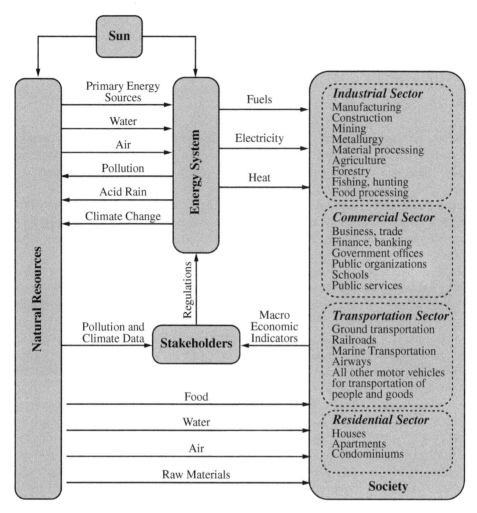

Figure 1.1 Interactions of an energy system with nature and society.

and climate change information too. Legislators, government administrators, and decision makers issue laws, policies, regulations, codes, and guidelines to ensure proper management of the energy system for the benefits of the society.

Intergovernmental organizations are also part of the feedback process. For example, the Organization for Economic Cooperation and Development (OECD) established the International Energy Agency (IEA) to help countries in a broad range of energy issues including oil, gas, and coal supply and demand, renewable energy technologies, electricity markets, energy efficiency, and access to energy. International agreements establish global dialog on energy related issues. Kyoto Protocol and Paris Agreement are examples of international movements to reduce the greenhouse effect and global warming resulting from human activities, especially from operation of energy systems.

As Figure 1.1 illustrates, an energy system is the central element of a closed loop global scheme, which also includes natural resources and society. Stakeholders are diverse groups concerned about energy production and consumption. Legislators, government offices, and local administrators regulate the management and development of the energy system on behalf of stakeholders.

1.2 Elements of an Energy System

The goal of an energy system is to transform primary sources into various forms of fuels and energy carriers to supply energy needs of the society. Figure 1.2 outlines a wide-area energy system such as a country or large geographic region where diverse types of primary sources are available and are used to supply different sectors.

Sun and earth crust offer all primary energy sources for the world. Energy systems use a mixture of renewable and non-renewable natural resources. The major components of an energy system are fuel extraction and processing facilities, energy conversion plants, and distribution networks.

Crude oil, coal, and natural gas are not suitable for practical uses in their original forms as extracted from the earth. Refineries, coal processing plants, and natural gas treatment facilities prepare fuels for diverse applications. In the Figure 1.2, fuel production facilities are grouped in a dashed box.

Energy sources reach the end users either in the form of a fuel or an energy carrier, such as ground transportation, railroads, and ships, which carry fuels to users. Pipelines are convenient means to transport natural gas and liquid fuels to consumers. Electric transmission and distribution networks deliver electricity to the point of use.

Fuel storage is imperative for continuous energy supply to end users. Massive quantities of coal, oil, and gas can be stocked for later use without significant degradation or losses. Electric power plants and transportation sectors are especially in need of uninterruptible fuel supply for reliable operation. Fuels can be stored in silos, tanks, bunkers, or piles in designated fields. Fire safety and environmental protection agencies issue codes, standards, regulations, and guidelines to ensure proper management of fuel storage facilities.

Nuclear fuel production starts with uranium mining; uranium undergoes several steps such as milling, conversion, and enrichment during fabrication of fuel rods used in nuclear reactors. After the fuel has reached the end of its useful life, it is removed from the reactor and reprocessed to make new fuel. The series of industrial processes from uranium mining to disposal of the nuclear waste is called the "nuclear fuel cycle."

Electricity is the principal energy carrier for modern society because it is convenient for long distance transmission and wide area distribution. Electric energy is easy to control, and it can be converted to thermal, mechanical, and chemical energy. Moreover, most renewable energy sources and nuclear energy can reach end users only in the form of electric power.

Centralized electric generation plants are interconnected via high voltage (HV) transmission lines, where overhead HV lines transmit electric energy to populated areas. Utility companies deliver electric energy to consumers through medium voltage (MV) and low voltage (LV) lines.

Transmission and distribution substations connect transmission lines coming from or going to various directions. In a substation, transformers change the voltage levels, and also various switchgear and automated relays protect the electric grid.

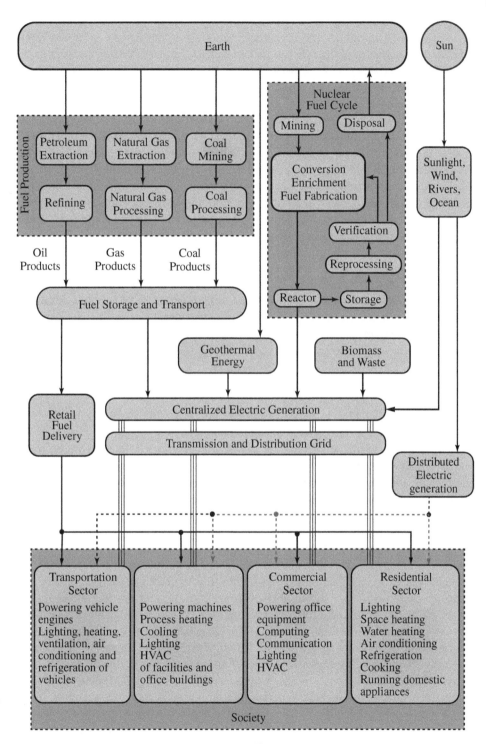

Figure 1.2 A wide area energy system with diversified sources.

Regional and national grids are interconnected to exchange electric energy. Computer controlled substations optimize electric load flow and increase reliability. Some cross-border interconnections may use high voltage DC transmission to isolate the grids with different frequency and voltage standards. DC transmission is also convenient for very long-distance transmission and connections between islands and mainland with underwater cables.

Some consumers may choose to generate part or all electric power they use. In addition, local communities, factories, public service providers, and private investors may install small or medium size power generation units. Independently owned de-centralized generation systems are called "distributed generation." As opposed to the centralized generation, distributed generation units do not need long transmission lines and expensive substations to change voltage levels. Most distributed generation systems use renewable energy and local sources. Private generation systems may be connected to the interconnected electric grid in such a way that the consumer uses grid power when local generation is insufficient or send energy to the electric grid when the generation exceeds the consumption. Net-zero energy consumers generate as much energy as they consume over a year. Self-standing off-grid consumers are not connected to an electric utility and they generate all electric energy they need. While backup generators burning oil products, natural gas, or biofuels are available for short-term electric generation, usually renewable sources such as sunlight, wind, or small streams power most of the on-site electric generation systems. Distributed generation reduces the transmission losses and increases the reliability of the electric supply.

Energy consumers are often grouped into industrial, commercial, transportation, and residential sectors. Each sector has specific energy needs and energy consumption profile. Energy supply mix, efficiency, and environmental effects are different for each sector.

The transportation sector includes all vehicles used to move people or goods from one place to another. Ground transportation vehicles include cars, busses, subways, trains, trucks, and all other ground vehicles. Passenger boats, ferries, tanker ships, cargo boats, cruise ships, military boats, submarines, aircraft carriers, and other vessels are among the marine transportation vehicles. Passenger planes, cargo planes, military aircrafts, and helicopters are examples of air transportation vehicles.

The transportation sector needs mobile energy sources; petroleum products are the most common energy sources for transportation. Electric trains, light-rail trams, trolleys, and subways use electric power generated in stationary power plants and are typically fed through a jointed arm sliding on overhead wires. Electric cars use energy stored in high capacity batteries, which are mostly charged from electric grid. In hybrid vehicles, the fuel is a petroleum product; in addition, an electric machine supports the powertrain. In hybrid locomotives and ships a fuel-burning engine drives an electric generator, which powers an electric motor. Hybrid transportation systems combine the advantages of higher energy content of the fuel and operational benefits of electric machines. Electric machines are easier to control in a wide speed range and they can allow bidirectional energy flow.

Practical application of on-site generation for transportation systems is currently limited due to high power needed for vehicles and low efficiency of wind and solar generation systems. Innovative designs for solar vehicles have been tested around the world for many years. With the contemporary PV cells, which are currently not more than 20% efficient, solar vehicles are not yet practical. Some sailing boats use wind turbines to generate electricity to power small devices on board. On the other hand, the emerging electric vehicle industry is introducing public charge stations, which are mostly supplied from electric grid. Using renewable energy to generate electricity on site for vehicle charging stations would increase the share of on-site generation for transportation sector.

The industrial sector consists of manufacturing, mining, processing, and construction facilities. Forestry, commercial fishing, and hunting are also considered part of industry since their products are used as feedstock in industrial facilities. For example, construction materials, furniture makers, and paper industries use trees; the clothing industry uses fur and leather; and the cannery industry processes fish.

The industrial sector needs energy to power heavy machinery, heat or cool industrial processes, and supply process control systems. In addition, computer and communication systems, heating, cooling, and ventilation of buildings consume energy in industrial facilities. Process industry and electric generation plants use coal products or natural gas to produce heat and steam; and coal is the major ingredient in steel production. Steelworks use electric power to melt scrap metal in arc furnaces and induction ovens to make semi-finished steel products.

The commercial sector covers shopping centers, malls, stores, offices, banks, financial companies, contractors, business, and public services. Street lighting, cleaning, trash collection, water treatment, and sewage are examples of physical public services. Government offices, hospitals, schools, religious institutions, military bases, public security, and social groups are also considered in the commercial sector because they provide numerous services to the community. Hotels, institutional living complexes such as shelters, nursing homes, assisted living facilities, correctional institutions, dormitories are also considered in the commercial sector.

The commercial sector uses energy for building lighting, heating, ventilation, air conditioning, and powering office equipment. Office equipment includes communication and computer systems, which require reliable electric supply.

The residential sector includes living areas for people. From an energy perspective, private houses and apartments are the main residential consumers. Public living facilities are excluded from the residential sector because they are generally considered in the commercial sector. The residential sector uses energy mostly for lighting, heating, appliances, communication, and entertainment.

In a modern society most activities depend on uninterrupted energy supply. The need for energy has continuously grown in modern societies. Primary sources must be processed and transferred to end users in different forms, such as fuels and electricity that can be conveniently used to produce heat, mechanical power, and light. Energy delivered to the society powers vehicles, machines, devices, tools, and appliances that make life easier and more comfortable.

1.3 Fundamental Concepts

1.3.1 Work, Energy, and Power

The common definition of energy is "the capacity to do work." In everyday language, the word "work" is used in a broader sense including intellectual, literary, artistic, social, or spiritual works. However, the term has a specific meaning in physics, which should be clearly understood to discuss topics that involve energy.

In physics, a force does mechanical work when it moves an object over a distance. The concept was first developed in classical mechanics. If a force F moves an object over a distance s, then the work it performs is equal to the scalar product of the vectors representing the force and displacement. The object can be a rigid body or a particle, but it is easier to understand and visualize mechanical work done by a force to move a rigid object. Mechanical energy is the capacity to perform mechanical work, or stated differently, the ability to move objects. On the other hand, a force that changes the shape of an elastic object also performs a work, hence deformation is the result of an energy exchange. In that case, forces holding together microscopic elements that form the body perform work during the displacement of these elements. In the international unit system (SI) the unit of work is Newton-meter (Nm). Since energy is the capacity to perform work, it has the same dimension, hence the standard unit of energy is also Newton-meter.

In 1843, James Prescott Joule demonstrated the equivalency of heat and mechanical energy. The older unit used to measure thermal energy is the calorie, which is the amount of heat that raises the temperature of 1 gram of water by 1 degree Celsius under 1 atmosphere pressure. Joule experimentally found that 1 cal is equivalent to 4.18 Nm mechanical work. Because of his contributions to the energy science, his name was adopted as unit of energy in the international unit system (SI). One Joule (J) is the energy corresponding to the work done by a force of one Newton over 1 meter distance. Hence, 1 J is equal to 1 Nm or 0.239 cal.

In the imperial unit system (IS), the unit of energy is the British thermal unit (Btu). One Btu is the amount of energy required to raise the temperature of one pound of water by 1 degree Fahrenheit. One Btu is equivalent to 1055 J. Since 1 Btu is extremely small, energy professionals express larger energy quantities in millions, billions, or quadrillions of Btus.

Diverse fields of science, engineering, and industry derived discipline-specific units to simplify accounting and interpretation of energy quantities. For example, in many energy statistics "tons of oil equivalent (toe)," "tons of coal equivalent (tce)" and "quad (10^{15} Btu)" are frequently used units. The energy of an explosive is often described as kilograms (kg) of TNT. Cooling units are also sometimes specified in kg, which is equivalent to energy needed to melt 1 kg of ice. Food industry uses "large calories," which is equal to 1000 calories, shown capitalized as Calories on the food packages to indicate the nutrition energy value. In quantum physics, energy of atomic particles is identified in electron-volts ($1\,\mathrm{eV} = 1.6\,10^{-19}$ J). In energy systems larger amounts of energy are expressed using metric prefixes such as kilo (10^3), mega (10^6), giga (10^9), tera (10^{12}), peta (10^{15}), and exa (10^{18}).

Power is the energy produced or converted in unit time. The standard unit of power in SI is a Watt, shown with the symbol W. One Watt is equal to one Joule per second (1 J/s). In the imperial unit system (IS), the unit of power is horsepower, first defined by James Watt to compare the power of a steam engine to the work done by an average horse in a unit time to turn the wheel of a mill.

Example 1.1 James Watt observed that a horse could turn a mill wheel 144 times in an hour. The wheel was 12 ft (3.7 m) in radius. Watt assumed that the horse pulled the wheel with constant 180-pound force. Determine the power of this horse in international unit system (SI).

Solution

Distance the horse travels in 1 hour is $(2\pi \times 12) \times 144 = 10,857.3$ ft. Therefore, the work done by turning the wheel 144 times with a force of 180 pound-force is

$$W = F \cdot L = 180 \times 10{,}857 = 1{,}954{,}260 \text{ pound-force} \cdot \text{feet.}$$

This is the energy delivered by the horse in 1 hour. Power is energy per unit time, therefore.

$$P = \frac{W}{t} = \frac{1{,}954{,}260}{3600} = 542.85 \text{ pound-force} \cdot \text{feet/s}$$

Using the conversion factors from imperial to metric units, we obtain:

$$1 \text{ pound-force} \cdot \text{feet} = (0.4536 \times 9.81) \times 0.3038 = 1.35 \text{ N} \cdot \text{m}$$

$$P = 542.85 \times 1.35 = 732.85 \text{ N} \cdot \text{m/s.}$$

The example above shows the method to compute metric equivalent of the imperial unit of horsepower (hp) based on James Watt's assumptions. Mr. Watt tried to compare the work that can be achieved using a steam engine instead of a horse for marketing purposes. Obviously, such comparison depends on the quality of the work horses used in a region. Later, other physicists and engineers modified the definition of hp based on their observations and assumptions.

Today, the metric equivalent of a horsepower is not unique. Mechanical horsepower (also known as imperial horsepower) is 745.7 W, while metric horsepower is 735.5 W. If the power of an electric motor is expressed in horsepower, one HP can be approximated to 736 W.

Example 1.2 An electric power plant operates at full capacity to generate 200-MW constant electric power. How much energy will this plant deliver in one day?

Solution

Energy is the product of power and time. Since the output power remains constant, the given electric generation plant will deliver in one day

$$W = P \cdot t = 200 \times 24 = 4{,}800 \text{ MWh} = 4.8 \text{ GWh.}$$

Example 1.3 Electric companies charge their customers based on the energy they consume, not the total power of appliances installed in the building. In January, the 4-kW electric heater installed in a house turns on for 20 minutes in an hour on average. Calculate total energy consumed by the heater in January.

Solution

Total time the heater is on in January is $T = 31 \times 24 \times \frac{20}{60} = 248$ hours. Therefore, total energy consumed by the heater is:

$$W = 4 \times 348 = 992 \text{ kWh.}$$

1.3.2 Energy Conservation and Transformation

Energy is a conceptual quantity. The fundamental principle of classical physics states that energy cannot be created or destroyed but can be converted from one form to another. Mechanical, thermal, chemical, nuclear, and electrical energy are the major forms of energy exchanged between the elements of an energy systems. Any process involving energy conversion or transfer must be consistent with the conservation of energy principle.

In everyday language we often use the terms "production" and "consumption" of energy, which seems to contradict with the conservation of energy principle. In reality, energy production means *conversion of energy to another form that is used for a specific purpose*. For example, when a fuel burns, the internal energy is converted to heat by chemical reactions and the fuel vanishes. Hence, fuel is consumed to produce heat. Energy is consumed when some part of it is converted to useful work and the remaining part is released to the surrounding environment as unused heat. Energy dissipated as unused heat is called "loss." Therefore, energy available to use is less than the total energy of the system since some part of it is irreversibly lost to either change the internal energy or to heat the environment.

Elements of an energy system exchange various forms of energy or transform energy from one form to another. Thermal energy is the essential part of energy systems used to heat spaces, enable industrial processes, or produce mechanical energy. More than 80% of the energy need of the world is produced by burning fuels. Combustion is a chemical energy conversion process, which transforms the internal energy of fuels to heat. In the fast oxidation reaction, the bonding energy between the molecules of the fuel and oxygen entering the reaction is more than the bonding energy of the products (mainly carbon dioxide, water vapor, and other oxides). Such chemical reactions are called *exothermic* because the change of internal energy is released as heat.

Nuclear reactions, on the other hand, convert a mass of atomic particles into thermal energy. A nuclear reaction (fission, fusion, or radioactive decay), produces new isotopes, in which the total mass of atomic particles is less than the total mass of the particles forming the atom of the element used as nuclear fuel.

Energy of the sun is also nuclear, resulting from the fusion of hydrogen and helium atoms. Heat of the sun is transferred to earth through space in the form of radiant energy. Solar radiation is the major primary source of energy heating the Earth's surface, causing wind and waves, and powering water cycles through evaporation and precipitation. In addition, solar radiation is converted to chemical energy in plants by photosynthesis. All living organisms are powered by some sort of conversion of solar radiation into biochemical energy, therefore all fossil fuels that exist in the earth crust are the result of solar energy. Solar energy is also used directly as a primary source in energy systems. Sunlight can be converted to heat using solar collectors or concentrating mirrors. Photovoltaic (PV) cells convert the radiant energy of the sun directly to electric energy.

Steam and gas turbines in power plants convert thermal energy into mechanical energy. Generators convert mechanical energy produced by turbines into electrical energy. Internal combustion engines of vehicles convert the heat obtained by burning a fuel into mechanical energy. Hydroelectric power plants convert the potential energy of the water collected in a reservoir into electric energy. Wind turbines harness the kinetic energy of the moving air during rotation, thus generating electricity.

End users of energy systems convert energy from one form to another to perform specific functions. Heat needed for various industrial processes is produced either by burning fuels or from thermoelectric devices. Buildings are heated either by burning some sort of fuel or directly by electric heaters. Household appliances use electricity to produce mechanical energy.

Mechanical and electrical energy are completely available for conversion to each other. Therefore, in the ideal situation without any heat loss, energy flow in an electromechanical system is completely reversible. In the real world, however, some part of the input energy is lost as unused heat, reducing the output energy of the conversion system.

Conversion between thermal and mechanical energy is the main focus of thermodynamics. The energy conservation principle, which is considered as the first law of thermodynamics, is insufficient to explain the energy flow when the internal energy of system components changes significantly. The second law of thermodynamics indicates that energy can only flow from higher to lower temperature. In thermodynamic cycles some part of the energy changes the internal energy of a working fluid. Efficiency of heat engines depends on the temperature difference between the heat source and the sink where part of the heat is rejected.

Operation of all main components of an energy system is based on a transformation from one form of energy to another. Operation characteristics and efficiency of energy converters affect the performance of an energy system. We will further discuss in Chapter 2 the mathematical and physical foundations of major conversion processes that have practical applications in energy systems.

1.4 Energy Statistics

Before discussing energy flow from resources to consumers, we need to understand clearly how energy is quantified and expressed by energy professionals. As we emphasized in the first section, an energy system is not merely an engineering structure. Engineers, scientists, mathematicians, economists, financial analysts, and other professionals co-operate in operation and development of energy systems. Such a diverse group of experts communicate using common terminology, statistical information, and special units to describe vast amounts of energy.

Today, an overwhelming amount of information is accessible online, mostly free of charge. However, students and professionals must be able to distinguish reliable and unbiased official resources from web pages reflecting individual opinions and speculations.

International organizations and national agencies publish continuously updated statistical data on the status of national, regional, and global energy systems. In addition, they analyze country profiles to prepare long-term energy outlooks based on various scenarios. Major independent sources of energy information are listed as "Recommended Web Sites" at the end of this chapter. Detailed statistical data on energy production, supply, and consumption are available in online databases maintained by these institutions.

National and international institutions collect energy data from companies and consumer groups through questionnaires and survey tools. The reliability of collected information depends on the transparency of governments or managements. In addition, some energy terms may have different meanings in diverse parts of the world. Therefore, statistical data must be analyzed and interpreted with caution, by keeping in mind specific definitions and unit conversion tables of each data provider.

The US Energy Information Administration (EIA) publishes a broad range of documents on energy topics. "Annual Energy Outlook" (EIA 2018a) outlines the state of the US energy system every year and delivers periodically updated statistical data. Another annual report is "International Energy Outlook" (EIA 2017), where data collected from countries are presented. Data tables presented in EIA publications are available online at the "Interactive Table Browser" (EIA 2018b).

The IEA was formed following the oil crisis in 1974 by members of the Organization for Economic Cooperation and Development (OECD). The IEA observes all energy-related issues around the world and publishes numerous reports in addition to comprehensive online data services.

Energy professionals around the world have developed non-standard units based on the energy content of various commodities. Today "ton of oil equivalent (toe)" is a common unit in international energy statistics. Energy equivalent to burning one ton of oil is assumed 41.868 gigajoules (GJ). Before petroleum was popular, energy equivalent to the heat released from burning one metric ton of coal was used as the unit in energy trade. One ton of coal equivalent (tce) energy is 28 GJ. After oil became a critical commodity worldwide, some organizations started to use toe in energy statistics.

The IEA expresses amounts of energy in million tons of oil equivalent (Mtoe). The US Department of Energy (DOE) uses quadrillion Btu or quad (10^{15} Btu) in most energy statistics.

In crude oil trade, barrel is a common unit of volume. A barrel of oil is 42 US gallons, or approximately 159 l. Units of mass in energy statistics may be slightly different from the unit of mass of the international unit system. Whereas one ton in SI is 1000 kg; the short-ton is 907.18 kg, and the long-ton is 1016.05 kg.

Conversion between different units is important in energy studies; in this book, we will mostly use metric units. Publications and web sites that provide statistical data such as EIA (2018b) and IEA (2018) include on-line calculators and conversion tables. The reader may also refer to the detailed list of conversion factors given in Appendix A.

1.5 Primary Sources

Sources of energy that are directly available in nature are called "primary sources." They must be processed to obtain fuels and other forms of energy to supply the energy needs of society conveniently and safely. Primary energy sources can be categorized as renewable and non-renewable sources. Figure 1.3 shows the share of major sources in the global total primary energy supply (TPES) based on IEA statistics (IEA 2018).

More than 80% of the world's energy is supplied from fossil fuels, which consist of coal, oil, and natural gas. Although nuclear fuel can be recycled, the uranium reserves in the Earth's crust are limited. Therefore, 86.3% of the world energy supply comes from non-renewable sources. The remaining 13.7% includes hydroelectric energy, biofuels, waste, and other renewable sources.

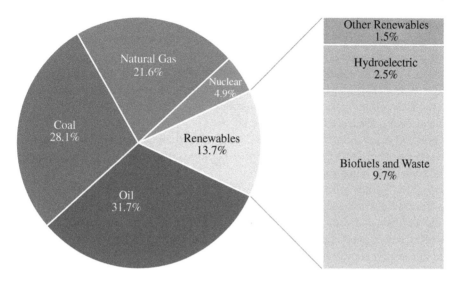

World 2015 TPES: 13,647 Mtoe (571 EJ)

Figure 1.3 Share of primary sources in the world's total primary energy supply (TPES). Source: IEA (2018).

1.5.1 Renewable Sources

Sources that do not vanish in nature when converted to other forms are renewable energy sources. Sunlight is the principal source of energy on earth. Kinetic and potential energy of wind and water are renewable energy sources resulting from the sunlight reaching the earth surface. While solar energy can be converted directly to heat or electricity, wind and ocean waves produced by temperature differences are also consequences of sunlight. Periodic tidal movements of the oceans, which are the result the earth rotation and gravitational forces of the moon and the sun, can be also harnessed as renewable energy sources.

For centuries, wind has powered sailing boats and windmills, and streams turned waterwheels. Before steam engines were invented, windmills and waterwheels were driving sawmills, grain mills, and pumps. In the pre-industrial era, sunlight was the primary source for heating. Locally available firewood, animal fat, and other forms of biofuels were used for additional heating and lighting when sunlight was unavailable. Animal and human forces were the principal energy source for construction, manufacturing, and short distance transportation.

The major challenge of wind and solar energy is their intermittence due to natural changes. Sunlight is only available during day. The incidence angle of the sunlight during the day is well defined and predictable for every instant year-round. However, the intensity of solar energy available for conversion changes significantly due to cloudiness and clarity of the air, which are uncontrollable and difficult to forecast. Wind energy may be available day and night; however, power generated by windmills changes in a wide range with wind speed.

Animal fat, hydrocarbon rich plants, and microorganisms are considered renewable primary sources if they regenerate at the same rate or faster than their consumption.

Biofuels include solid biomass, liquids, and gases derived from plants, grains, vegetable oil, or animal fat specifically for energy purposes. Biomass covers a vast group of wood-based materials generated by industrial processes or directly from forests. Firewood, wood chips, bark, sawdust, and shavings are examples of biomass. Ethanol produced from hydrocarbon rich plants such as corn, soybean, sugar cane, and sugar beet is mixed with gasoline and used in the transportation sector. Biodiesel produced by fermentation from animal fat or vegetable oil is also used as transportation fuel.

Hydraulic forces are more controllable and predictable than wind and solar radiation. In addition, the power output of hydroelectric power plants can be regulated by storing energy at a large scale using dams and artificial lakes.

According to IEA Key World Energy Statistics (IEA 2018), renewable sources produced 13.7% of the TPES worldwide in 2015. The biggest part of the renewable energy sources (9.7%) was biofuels and combustible waste collected from industrial processes and trash. Hydroelectric power plants produced 2.5% of the global energy need. Solar, wind, geothermal, and tide energy contributed to TPES only 1.5%.

1.5.2 Non-renewable Sources

Non-renewable sources diminish in quantity as they are transformed to useful energy forms. The biggest part of the world's energy need is supplied from hydrocarbons that have been produced in the Earth's crust over millions of years. All types of coal, petroleum, and natural gas are grouped as "fossil fuels." Although fossil fuels are still forming naturally in the Earth's crust from vegetation and organisms, the consumption is at a much higher rate compared to their production. Unconventional oil and gas extracted from special rock formations are also non-renewable sources.

Uranium is the main element used in nuclear reactors. Although an extremely small amount of nuclear fuel can produce huge amounts of energy, the natural reserves available to extract uranium is limited.

The convenience of "energy on-demand" in large amounts has increased the use of fossil fuels since the industrial revolution. Today, most functions of modern society rely on continuous energy supplies from non-renewable primary sources. However, combustion of fuels results in significant environmental impacts, by causing air and water pollution, producing acid rain, and leading to global climate change.

More than 90% of the world's energy is supplied by burning fossil fuels and waste. IEA estimated that 32,294 million tons of carbon dioxide was released into the atmosphere from the combustion of coal, oil, and natural gas in 2015. The energy sector produces two-thirds of all greenhouse gas emissions caused by human activities in the world.

Figure 1.4 shows primary energy consumption in the world extracted from the EIA Interactive Table Browser (EIA 2018b). Consumptions in 2010 and 2015 are based on actual data collected from countries. The following years' consumptions are projected under the reference case assumptions. The EIA World Energy Projection System Plus$^©$ model projects energy consumption for various scenarios. Reference cases estimate future consumption based on current policies, technology, and resources. Other cases consider lower or higher oil prices, economic growth, oil and gas resource technologies, nuclear costs, or a combination of multiple parameters.

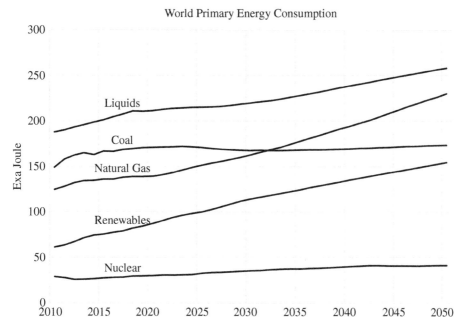

Figure 1.4 World total primary energy consumption by fuel. Source: EIA (2017).

We clearly see that fossil fuels will remain a dominant primary source in the future. However, the share of other renewables is expected to increase by 2% per year, at the biggest growth rate among all primary sources. Natural gas is the second growing energy source, and consumption of coal and liquids increase at a lower rate.

1.6 Secondary Sources

Secondary energy sources are commodities obtained by converting primary sources into more convenient forms to supply consumers. Processed fuels, electricity, and hydrogen are the most common secondary energy sources. Secondary sources contain more energy per unit mass or volume, are easier to transport, cleaner, and safer than the raw primary energy sources. For example, crude oil cannot be burned directly to run engines; coal extracted from mines must be cleaned to be used by consumers. Energy of rivers, ocean waves, wind, sun, and nuclear reactions are transferred to consumers in the form of electricity.

1.6.1 Processed Fuels

Fuels are derived by refining primary sources. Solid, liquid, and gaseous fuels are an essential part of the energy supply to end users. Fuels can be conveniently transported and stored, allowing steady energy supply to consumers. All fuels are carbon or hydrocarbon mixtures, which produce heat by combustion. Calorific value (also known as heat content) of a fuel depends on its chemical composition. Fuels are used either directly for space and water heating in buildings or power diverse types of engines. Turbines and internal combustion engines convert the heat content of a fuel into mechanical energy to power vehicles, industrial systems, and electric generators. Calorific values of common fuels are listed in Appendix B.

Before the industrial revolution, firewood, animal oil, and some forms of coal were principal substances to transport energy. With the industrial revolution, coal became the major fuel, and steam carried heat from boilers to heat engines. With the invention of internal combustion engines and development of motor vehicles, petroleum products started to be used as energy carriers. Throughout the evolution of energy systems, fuels preferred to carry energy have shifted from solid to liquid, and from liquid to gaseous forms.

1.6.1.1 Solid Fuels

Diverse types of coal such as anthracite, lignite, and bitumen are common solid fuels. Coal, as extracted from mines, is not directly suitable for use by consumers. A coal processing plant cleans coal from impurities such as soil, rocks, various metals, and minerals, particularly sulfur and mercury. A mechanical cleaning process separates solid impurities with different densities. In addition to mechanical cleaning, various chemical processes are performed to deliver cleaner, better quality, and more efficient coal to the consumers.

1.6.1.2 Liquid Fuels

Liquid fuels are derived from petroleum, hydrocarbon rich plants, animal fat, and natural gas. Emerging coal-to-liquid (CTL) technologies yield synthetic liquid fuels that can be used in many applications. CTL fuels can be used to run a variety of vehicles including cars, trucks, tanks, and jets.

Petroleum is the major source for gasoline, diesel fuel, jet fuel, and other liquid fuels. Crude oil is unusable in its raw natural form; the chemical components of the crude oil must be separated to produce usable fuels. Usually, pipelines or tanker ships carry crude oil over long distances to refineries where diverse grades of oil products are separated.

Animal fat and vegetable oil are processed with special enzymes to obtain biodiesel. Ethanol produced in bulk quantities by fermenting sugar cane, beet, corn, soybean, or other carbohydrate-rich plants is mixed with gasoline to reduce the oil-dependence of transportation sector. In most regions of the USA retail gasoline contains up to 10% ethanol. Specially adjusted engines can run on E85, which is 85% ethanol blended with 15% gasoline.

1.6.1.3 Gaseous Fuels

A mixture of combustible gases used for energy purposes is called fuel gas. While hydrogen and carbon monoxide may be used as a fuel, most fuel gases are hydrocarbons, such as methane, butane, propane, or their mixtures. Such fuels can be transported from the production facility to the consumers either through pipelines or in liquid form in pressure tanks. Many of the gas fuels are extracted from petroleum in refineries or directly from sedimentary rock formations called shales.

Natural gas extracted from earth contains contaminants, water, and combustible liquids. It must be cleaned up to meet the quality standards required for high-pressure pipelines and consumers. Natural gas processing and treatment plants are crucial for the gas distribution network.

Mixtures of different gases produced by fermentation of biodegradable organic materials including agricultural waste, manure, municipal waste, sewage, food waste, or decaying vegetables are called "biogas," which is a renewable energy source. Gaseous biofuels can be also produced by anaerobic digestion of microorganisms such as bacteria and algae.

1.6.2 Electric Power

Electricity plays a significant role in energy systems both as an energy carrier and a secondary source supplying a broad range of applications. All forms of primary sources can be converted to electricity and delivered to end-users continuously.

Electricity not only powers a broad range of equipment, tools, and appliances; it also improves the quality of life by providing comfort, security, and communication means. Computer networks, digital technology, and industrial controls can only be powered by electricity. Mission critical infrastructures, including electronic instruments in hospitals, data centers, security monitoring devices, data transmission, and defense equipment, rely on continuous electric power. The end-use of electric power is more efficient and cleaner than burning fuels.

Electric generation systems utilize a diverse fuel mix to increase energy security as well as system stability, continuity, and reliability. Nuclear, hydraulic, tidal, solar, wind, geothermal, and waste energy cannot directly power end-use applications. Such energy sources are delivered to consumers in the form of electricity.

In 2015, 19% of final energy consumption was in the form of electric power as indicated in Figure 1.5 (EIA 2017). However, it should be noted that only about one-third of primary energy sources used in generation facilities reach the end users in the form of electricity due to conversion and transmission losses. Most of these losses occur in fossil fuel burning power stations, which have around 30% overall efficiency.

Figure 1.5 Total final consumption of secondary sources worldwide in 2015. Source: IEA (2018).

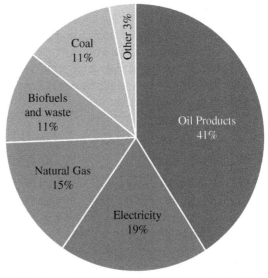

2015 World TFC by Secondary Sources: 392,884PJ

Bidirectional energy flow in electric machines is an important benefit of electric power. All electric machines can be used as a motor or generator by configuring the external circuit properly. This property makes electric power even more efficient in certain applications, especially in electric vehicles and industrial transportation systems.

The major drawback of electric power is the difficulty and cost of its direct storage. Therefore, electric power systems must deliver as much electricity as the end users demand at any time. Electric power producers continuously adjust generation to match the demand. Conventional electric generation using fossil fuels, hydropower, and nuclear reactors are more controllable than electric generation systems using intermittent renewable sources such as wind and sunlight. Integration of such renewable energy sources into the fuel mix needs battery backup or other energy storage means to keep electric grids reliable and stable.

1.7 Energy Carriers

Carrying energy from production facilities to the end users is an intermediate step in an energy supply chain. Substances or effects used to transfer energy from one place to another are called energy carriers. The arrows in Figure 1.2 represent different forms of energy carriers. Electricity, steam, water, air, and fluids are common energy carriers.

1.7.1 Electric Transmission

In the second half of the nineteenth century, discoveries in the field of electricity and magnetism revolutionized energy systems by using moving electrons as a new form of energy carrier in place of substances. Although the first electric transmission systems were in direct current (DC), today alternating current (AC) systems dominate electric energy systems because they allow more efficient energy transmission using high voltage lines. With the development of power electronics devices, high voltage DC(HVDC) systems evolved for transmission of bigger amounts of power over longer distances, interconnection of large area networks, and electric transmission through underwater cables.

Electric transmission systems are comprised of substations and high voltage power lines. In substations, step-up or step-down transformers change the voltage level; switchgear connect or disconnect transmission lines; and measurement, monitoring, and protection devices control power flow throughout the system. Electric power is delivered to end users via medium voltage or low voltage distribution systems.

1.7.2 Steam

In process industries, factories, and electric power plants steam is a preferred energy carrier for heat transfer. In large buildings, it is convenient to use steam for space heating. Many power plants produce steam for district heating by burning municipal and industrial waste.

Combined heat and power (CHP) or cogeneration plants produce both steam and electric power. They harvest either excess heat produced in industrial processes to generate electricity or transmit the heat losses of an electric power plant to an industrial process. In either case, cogeneration systems convert inevitable heat losses into useful energy.

1.7.3   Water, Air, and Heat Transfer Fluids

Water and air are convenient energy carriers for heating or cooling. Engines of vehicles are cooled by water circulation through a radiator. Heat dissipated from the engine warms the passenger sections of cars, busses, trains, and ships by hot water circulation. Central heating systems in many buildings use hot water circulation through baseboards or radiators. Water pipes buried in the floor slab create uniform and efficient space heating. Similarly, chilled water is often used for central air conditioning of buildings. Hot and chilled water can be sold and distributed to several buildings as a commodity.

Certain types of nuclear reactors have a primary loop of melted metal or salt to carry heat from the reactor core to boilers where steam is produced for the secondary loop. In solar thermal systems, hot water or a heat transfer fluid circulates between the solar collectors and water or space heaters. Concentrated solar power plants use steam to transfer heat captured from sunlight to turbines.

1.7.4   Hydrogen

Hydrogen is emerging as an energy carrier for both mobile and stationary applications. As the fuel cell technology progress, hydrogen production using electricity from water, natural gas, and biomass is becoming increasingly more practical, and the cost of hydrogen is decreasing. Hydrogen can be carried in liquid or high-pressure tanks as well as metal hydride containers. Fuel cells are bidirectional energy converters that produce electricity from hydrogen without a need for combustion and heat engines. The only byproducts from generating electricity by hydrogen fuel cells are pure water and oxygen. Hydrogen powered vehicles are potential solutions to reduce emissions and air pollution in the transportation sector. At present, due to practical and economic challenges, its use in the energy supply chain is limited. As fuel cell technologies improve, hydrogen may become a viable choice for energy transfer.

1.8   End Use of Energy

Consumers use various fuels and energy carriers to meet their energy needs. The fuel mix at the end-use depends on many factors including cost, availability, safety, reliability, continuity of supply, and convenience of use. In modern societies, electricity and natural gas have significant shares because they are safer, cleaner, and easier to control. In addition, some consumers may use alternative sources available on-site or generate electricity to offset part of their utility bills and ensure continuous energy supply in the case of extended utility interruptions. In this section, we will discuss the share of various sources supplied to consumers and the impacts of site energy consumption or production on natural resources and the environment.

1.8.1   Consumption by Sectors

Major energy sources delivered to end users are liquids, natural gas, coal, electricity, and renewables. While most liquid fuels are petroleum products, some of them are produced from natural gas and coal. Figure 1.6 shows global energy consumption by sectors and fuels (EIA 2017). Renewable fuels include a broad spectrum of combustibles such as biodiesel, ethanol, firewood, wood pellets, saw dust, municipal waste, and industrial excess materials.

The industrial sector is the dominant energy consumer worldwide. The transportation sector, including international and intercontinental transport, consumes about one-fourth of the fuels. About 95% of fuels consumed by the transportation sector are liquids. Residential and commercial sectors consume about one-fifth of delivered sources. Electricity and natural gas are dominant sources for residential and commercial sectors.

At the national level, the share of different sectors depends on the geography, climate, and economic development of a country. In colder climates, residential and commercial sectors consume more energy for heating, while in warmer climates they consume more energy for cooling. Energy consumption of the industrial sector depends on the dominant types of industry. For example, electronics, computer hardware, and software industries evidently consume less energy than metal production, material processing, and textile industries. Energy consumption of the transportation sector at national level depends on the geographic size and terrain conditions.

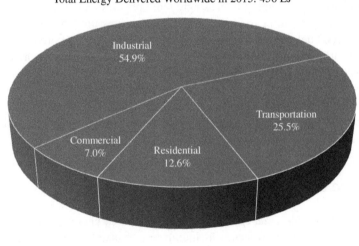

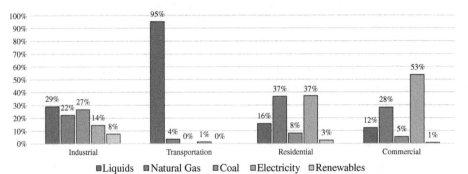

Figure 1.6 Worldwide total delivered energy by sectors and fuels 2015. Source: EIA (2018c).

In addition, energy efficiency standards and environmental protection constraints imposed by the governments affect energy consumption. Better building insulation, use of energy efficient appliances, and energy efficient lighting and heating significantly reduce residential and commercial energy consumption while maintaining the comfort level and living standards.

In developed countries, residential and commercial sectors have a bigger share in consumption of energy sources. For example, in the USA, these sectors consume nearly 40% of the total delivered energy.

Fuel mix depends on the availability and cost of energy sources. Oil and natural gas prices play key role in the choice of fuels. Before the oil crisis in 1974, it was more cost-effective to use oil products for heating and, to some extent, electric generation. Increasing oil prices led to use of electricity and natural gas. As the natural gas prices dropped in the USA over the last few decades, coal has become less dominant in electric generation and use of natural gas in industrial, commercial, and residential sectors has increased. Energy independence concerns have led the governments to rely on domestic sources rather than imported oil and natural gas.

Environmental factors also affect the fuel mix of end users. Direct use of coal as heating fuel in industrial, residential, and commercial sectors has decreased over the last decades as a natural gas network expanded in populated areas. Natural gas is cleaner and easier to control than coal. Burning coal by small users such as residences, commercial buildings, and industrial facilities causes excessive pollution particularly in big cities due to smog, soot, and disposed ash. A bigger portion of coal is used for electric generation and large industrial processes. Modern facilities reduce gas emissions by using advanced combustion and flue gas filtering techniques. Some cogeneration facilities produce carbon dioxide for food industry or other uses. Collected ash and other solid waste can be used to manufacture building materials or as landfill to minimize the environmental effects.

1.8.2 Primary Sources Consumed by End-users

Consumers are usually more concerned about their final use of energy and its direct cost when they make energy-related decisions. For example, suppose a family is trying to choose a water heater among natural gas, electric, and heat-pump type units. Beside the initial cost, the monthly gas or electric consumption is a factor. An electric heater seems to be the most cost-effective choice because it is cheaper, easier to install, would not need a flue, and is quieter since it does not have any ventilation or compressor motor. As an average consumer, many would not think about the actual effect of their new water heater on the overall energy system.

As we discussed in previous sections, primary energy sources are delivered to end-users in various forms of fuels and energy carriers. Cost, availability, reliability, and convenience of use are among the primary factors that motivate the consumers' choices. Energy preferences of end-users obviously affect the energy flow throughout the supply chain and ultimately the consumption of primary sources. To assess the actual impact of end-use of energy on natural resources, all losses and environmental effects that occur during extraction, conversion, and transmission fuels and energy carriers must be considered.

Table 1.1 Source-site energy ratios for US and Canada.

Energy type	Source-site ratio	
	USA	Canada
Electricity (grid purchase)	2.80	1.96
Electricity (on-site solar or wind) Iinstallation	1.00	1.00
Natural gas	1.05	1.02
Fuel oil (1, 2, 4, 5, 6, diesel, kerosene)	1.01	1.01
Propane and liquid propane	1.01	1.04
Steam	1.20	1.33
Hot water	1.20	1.33
Chilled water	0.91	0.57
Wood	1.00	1.00
Coal/coke	1.00	1.00
Other	1.00	1.00

Source: EPA (2018).

Electricity and natural gas are the most common energy sources for residential buildings. Coal, fuel oil, kerosene, propane, and liquefied petroleum gases (LPG) are some other fuels used for space and water heating in residential and commercial buildings. Industrial facilities, large commercial complexes, campuses, hospitals, and community services may also use district steam, hot water, and chilled water for their heating and cooling requirements. In addition, consumers may consider on-site electric generation and solar heating options.

The site-source ratio estimates the total primary energy needed to produce a fuel or energy carrier delivered to consumers. In Eq. (1.1), W_{site} is the net energy content of a secondary source consumed on site and W_{source} is the TPES used to produce that secondary source. For assessment of the energy performance of end-users, the US Environmental Protection Agency (EPA) recommends the nationwide source-site ratios listed in Table 1.1 for the US and Canada (EPA 2018).

$$R = \frac{W_{source}}{W_{site}} \tag{1.1}$$

The source-site ratio for electric power is estimated based on the fuel mix used to generate electricity. Whereas regional grids have distinct fuel mixes, national interconnections allow energy exchange between regions. Therefore, it is impossible to specify the source energy mix used to generate electricity used at a certain location. It is usually more reasonable to compare energy profile of consumers in a country using a national source-site ratio. Local source-site ratios can be obtained using the "Power Profiler" tool developed by the US EPA (EPA 2017).

From the consumer perspective, electricity is the cleanest and the most convenient energy carrier that can be easily converted to other forms of energy. Whereas electricity is convenient for space and water heating, cooking, air conditioning, and refrigeration,

Table 1.1 shows that other secondary sources which can be used for the same purposes consume less primary sources than electricity. On the other hand, electricity is indispensable to operate many systems such as light fixtures, appliances, power equipment, audio-visual sets, computers, communication devices, control equipment, and security systems. The international Energy Star® standard created by the USEPA and DOE in 1992 evaluates the energy performance of appliances, equipment, and buildings. Later, many countries including European Union, Canada, Australia, Japan, New Zealand, and Taiwan adopted similar programs.

Example 1.4 In a typical year, a single-family house purchases from utility companies 78,000 cubic feet natural gas and 12,000 kWh electric energy. Given the heat content of natural gas 1.028 kBtu per cubic feet, Calculate the total site-energy used in this house and estimate the total source energy consumed to supply this site energy.

Solution
Energy produced by burning natural gas is

$$Q_G = 1.028 \times 78,000 = 80,104 \text{ kBtu}$$

Consumed electric energy is equivalent to

$$Q_E = 12,000 \times 3.412 = 40,944 \text{ kBtu}$$

Total site energy is

$$Q_{site} = Q_G + Q_E = 121,048 \text{ kBtu}$$

Source energy is estimated using the ratios given in Table 1.1

$$Q_{source} = 1.05 Q_G + 2.8 Q_E = 84,109.20 + 114,643.20 = 198,752.40 \text{ kBtu}.\therefore$$

Note that because of conversion and transportation losses, total primary energy sources consumed to supply this house is 1.64 times the energy used on site.

Example 1.5 Suppose that the house described in Example 1.4 generates 60% of the electricity from solar modules installed on the roof. How much primary source energy is saved?

Solution
Since half of the electricity is generated by a renewable source on site, the source-site ratio for 60% of the consumed electricity is unity. The new source energy becomes:

$$Q_{source} = 1.05 \times 80,104 + (2.80 \times 0.4 + 1 \times 0.6) \times 40,944 = 154,532.88 \text{ kBtu}$$

Therefore, 198,752.40 – 154,532.88 = 44,219.52 kBtu source energy is saved by using on-site solar generation.

1.9 Energy Balance

Thus far, we have discussed the flow of energy from primary sources to end-users. Throughout the supply chain, energy is transformed from one form to another. Primary sources are converted to secondary sources, fuels, and energy carriers to be delivered to the consumers. A significant part of primary energy sources is transmitted to end-users in the form of electric power.

In energy economics, energy balance is a set of statistical data compiled to assess the energy production, consumption, and efficiency in an energy system. Energy balances help policy makers and managers to make better decisions on energy-related issues. They also serve as information sources to evaluate how energy resources are exploited in various countries, regions, or the world at different years.

Energy balance databases use a common unit to show amounts of energy exchanged between the system elements. Several governmental and international institutions such as US EIA and IEA provide online statistical data and diagrams to illustrate the yearly energy flow in specific countries or the world. Figure 1.7 is a simplified version of the USA 2017 energy flow diagram prepared by Lawrence Livermore National Laboratory (LLNL) based on DOE–EIA data. This diagram does not show energy import and export or stock changes. In addition, due to rounding errors, the numbers for separate elements may not add up to the totals shown on the diagram.

The key concept of energy balance is the conservation of energy principle. Energy is not lost or added throughout the transformation and delivery processes. However, some part of energy is lost as unused heat (W_{loss}). The efficiency of a conversion process is defined as the ratio of the useful output energy (W_{out}) to the input energy (W_{in}) as shown in Eq. (1.2).

$$\eta = \frac{W_{out}}{W_{in}} = \frac{W_{in} - W_{loss}}{W_{in}} = 1 - \frac{W_{loss}}{W_{in}} \tag{1.2}$$

According to the EIA, the US used in total 97.7 quads (quadrillion Btus) in 2017 (EIA 2018d). The fuel mix is composed of 80% fossil fuels, 8.6% nuclear energy, and 11.2% renewables, which include biofuels and combustible waste in addition to solar, wind, and geothermal energy. Nearly 40% of the primary energy sources reach the end-users through electric generation. The overall efficiency of electric power is estimated at 33%. Therefore, about two-thirds of energy sources used to generate electricity is lost while they are delivered to consumers in the form of electric energy.

In Figure 1.7 end-use efficiencies are estimated as 65% for residential and commercial, 49% for industrial, and 21% for transportation sectors. The energy balance shows that less than one-third of primary sources is consumed for energy services while two-thirds is rejected to the environment as unused heat.

Energy flow is different in every country. Government policies, incentives, market conditions, and environmental constraints affect the fuel mix and energy balance of a nation. Energy efficiency standards and programs to encourage more efficient use of energy by end-users change energy flow. Technologic advancements in building materials, vehicles, and electronic drive systems have significantly improved energy efficiency around the world over the last few decades.

1.10 Energy Indicators

Energy indicators are statistical measures that describe the energy profile of a population. They contain information about productivity, efficiency, economy, welfare, and living

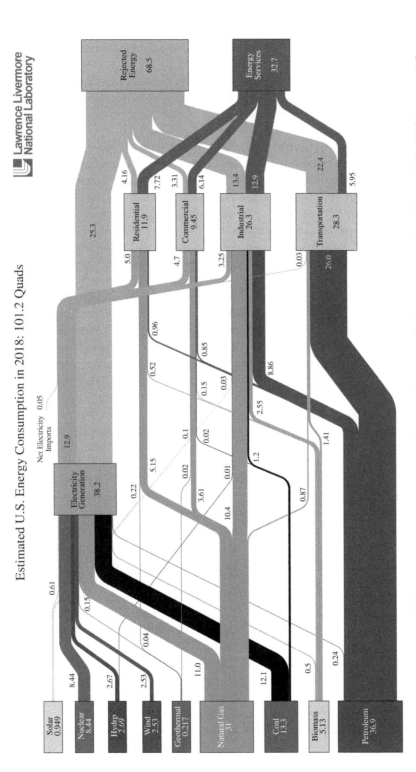

Figure 1.7 US energy flow in 2018. Source: Diagram adapted from Lawrence Livermore National Laboratory Flow Charts, LLNL, 2018, based on US Department of Energy data (EIA 2018d).

standards of a society, country, or region. Energy indicators must be interpreted carefully considering macroeconomic indicators and technical characteristics that affect the end use of energy.

Population is one of the essential economic indicators since all individuals contribute to the economy and, in return, benefit from economic development. Population shows the size of a society but does not reflect the economic productivity, wealth, or capability of individuals.

Energy consumption per person (or per-capita) reflects the overall energy use of a society. Energy consumption of a country or region includes energy used by industry, agriculture, transportation, commercial activities, services, and residential users. Therefore, energy consumption per person (or per capita) does not mean how much individuals actually consume energy on average. Energy consumption by separate sectors may give more information about the energy performance of a population.

Per-capita energy consumption of countries around the world varies in a wide range depending on the economic and social factors. Per-capita energy consumption is intricately linked to the living standards and welfare of a nation. Living in more comfortable spaces and driving private cars rather than public transportation increases the per-capita energy consumption. In wealthier districts, more energy is consumed for decorative lighting and attractions. In addition to living standards, however, geography and climate are major factors that affect energy consumption per-capita of a nation. Larger physical size and rough topography of a country increases energy used for transportation. In colder climates, buildings use more energy for heating while in warmer climates more energy is consumed for air conditioning.

Gross domestic product (GDP) is the total market value of all final goods and services produced by labor and property in a country or region. Statistical databases give GDP over a certain period; usually a quarter or a year. Since standard GDP depends on the monetary unit, nominal GDP per capita does not reflect accurately the differences of living standards since currency parity between countries or timeframes change due to inflation. GDP scaled at "purchasing power parity (PPP)" gives a more realistic view when comparing the performances of different economies. International energy statistics refer the monetary unit to US dollar value at a certain reference year, for example 2010.

Energy intensity is a ratio of delivered or consumed energy to a statistical variable, such as national GDP, population, or other metrics. Sector-specific energy intensities may use building floor space, number of households, dollar value of industrial shipments, or vehicle-miles. Consumed energy per GDP shows how efficiently energy is used for industrial and economic functions. Energy intensity in many countries has steadily decreased over the last few decades mainly because of energy efficiency improvements. However, other economic factors such as energy conservation, oil prices, recession, economic growth, or relocation of some industrial activities to other countries may also affect the energy intensity of a nation.

Earlier in this chapter we defined energy efficiency as the ratio of useful energy output to the energy input. Efficiency can be improved by using better quality fuels or more advanced technologies to increase the energy output for the same input. Energy conservation is a reduction of some services to reduce the energy consumption. Both help energy saving, but energy conservation may have a hidden cost because of reduction of services. For example,

Table 1.2 Regional indicators for 2010.

Region (in alphabetical order)	Per-capita consumption (toe/person)	Energy intensity (toe/thousand 2010 USD)	Per-capita GDP ppp (2010 USD/person)
Africa	16.1	3.50	4596.87
Asia and Oceania	56.3	5.32	10,582.67
Central/South America	60.6	4.48	13,534.09
Eurasia	153.9	8.85	17,381.68
Europe	135.8	4.06	33,413.44
Middle East	156.5	7.01	22,316.56
North America	251.0	5.97	42,045.79
World	76.7	5.31	14,440.83

Source: EIA (2018c).

oil shortages that occurred in the 1970s led many countries to develop energy conservation practices including reduction of lighting in streets, public places, and stores. Such practices have increased the crime-rate in big cities. During the energy crisis, some countries applied scheduled electric power outages up to several hours to impose energy conservation. Lacking electric power during the day affected the productivity. Restrictions on transportation for energy conservation affect commerce, tourism, and overall productivity. Daylight saving time (DST) is another typical example to energy conservation; setting the clocks to one hour earlier than the actual solar time shifts most activities to daylight, hence reduces energy consumed for lighting.

Energy intensity is also used in environmental and efficiency studies. Environmental energy indicators include the carbon dioxide emission per energy unit (for example CO_2/kWh) or per population (CO_2/capita). Distance per quantity of fuel (miles per gallon or kilometers per liter) is a common measure to compare the energy efficiency of vehicles. In the US, the Energy Star program defines the "Energy Use Intensity (EUI)" to express the energy performance of a property. The total energy used for the building divided by the gross floor area gives the EUI of the building. Energy per unit area of a production facility expresses the space used to generate certain amounts of energy in a given time interval (for example kBtu per square foot per year).

Note that the ratio of energy consumption per-capita to energy intensity gives GDP per-capita, which is one of the macroeconomic indicators that show the economic strength of a community. These three metrics can be used to assess energy performance of economies. Several databases provide detailed energy intensity and macroeconomic indicators for all nations and economic groups as well as aggregated data for the world.

Table 1.2 shows energy intensity and GDP per person in various geographic regions expressed in purchasing power parity and referred to 2010 USD (EIA, 2018c). Energy unit is chosen as tons of oil equivalent (41.9 GJ) for better visualization. Figure 1.8 illustrates the energy-economics landscape of the world based on data listed in the table.

Each one of the considered regions is dominated by a particular type of economy. For example, North America comprises the US, Canada, and Mexico which have strong

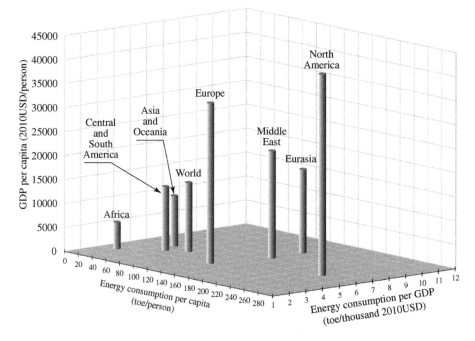

Figure 1.8 Energy intensity and GDP per capita by region for 2015. Source: EIA (2018c).

economic relations and energy trade. Eurasia is dominated by the Russian federation, which includes former Soviet Union countries. The Middle East represents the majority of oil producing countries. Europe is mainly controlled by European Union countries. Asia and Oceania contain emerging economies such as China and India. Africa covers many developing countries where a considerable part of the population has limited access to electricity, and human or animal power is still used for production.

Countries in Eurasia developed their economies during the cold war era under the Soviet Union umbrella. They went through a similar transition to a competitive free market economy after the collapse of Soviet Union in 1991.

Grouping countries by regions shows the geographic distribution of energy indicators around the world. Such distribution, however, does not display the performance of major economic groups that are spread on several continents. An alternative approach is grouping countries by their economic alliances and geographic locations. International Energy Agency (IEA) statistics divide the economies into seven groups named OECD, non-OECD Europe and Eurasia, Non-OECD Asia, Middle East, China, non-OECD Americas, and Africa. This classification combines countries across the oceans and focuses on their economic status. Most of the thirty-six members of OECD are developed and industrialized countries and a few of them have emerging economies. In all OECD countries energy policies are more transparent and social groups actively participate in energy-related decisions. Therefore, aggregated energy indicators published by IEA reveal the effects of economic and political structures on energy consumption around the world (IEA 2018).

Example 1.6 According to the IEA statistics (IEA 2018), in 2015 the world's total final consumption (TFC) was 9,384 Mtoe (Million tons of oil equivalent). The same statistics estimate the world population in 2015 as 7.335 billion. Find the average final energy consumption per capita in the world.

Solution

$$\text{TFC}_{per\ capita} = \frac{\text{TFC}}{\text{Population}} = \frac{9,384 \times 10^6}{7.335 \times 10^9} = 1.28 \text{ toe/person.} \therefore$$

Example 1.7 The IEA gives 1737 Mtoe for the 2015 final electric energy consumption in the world. The EIA estimates that residential sector used 27.1% of this energy. Assuming that the total world population of 7.335 billion lives in residences, what is the average daily residential electric energy consumption per capita in kWh/person?

Solution

In 2015 the electric energy used by the world's residential sector was

$$1,737 \times \frac{27.1}{100} = 470.7 \text{ Mtoe}$$

In Appendix A we find 1 toe = 11.630 MWh.

Using this conversion factor 1 toe = 11.630 MWh (Appendix A) we find the average daily electric consumption per person as below

$$\frac{(470.7 \cdot 10^6 \text{toe}) \times (11.630 \cdot 10^3)}{(7,335 \cdot 10^6) \times 365} = 2.042 \text{ kWh/person.} \therefore$$

1.11 Energy and Society

In earlier sections, we saw that all functions of a modern society use a certain energy mix. The availability, unavailability, and cost of energy affect the economic development, welfare, and security of a society. Utility and services offered to a society rely on energy availability. Cheaper energy attracts industries, creating jobs and stimulating economic development in a region. Shortage or extended interruptions of energy supply reduce the quality of services and productivity. Higher energy prices increase the prices of all products and the cost of living. Even crime rates increase in cities during blackouts and longer energy shortages. Energy is one of the main ingredients of the economy and social life. From local communities to multinational organizations, policymakers have developed energy-led economic development and emergency strategies.

1.11.1 Energy Sector

Elements of the energy system are highly technical structures involving all disciplines of engineering. However, we should always keep in mind that energy system is not just an engineering entity. Energy sector involves diverse groups of the society. The main function of engineers is to develop and show possible technical options. Planners, investors, policy makers, and executives take part in the operation and evolution of the energy system. Legislators set ground rules for all players in the process by making laws, rules, regulations, acts, bills, and ordinances. Information agencies provide legislators, executives, and community with reliable data and analyses. In a modern community, professional societies, media, and various citizen groups develop public opinions that influence the decision-making process. Social acceptance and preferences shape long-term policies on fuel mix and structure of the energy system. Figure 1.9 shows major groups that interact during development and operation of energy systems.

Elements of an energy system need substantial investment, technical knowledge, and advanced technology. Planning, construction, and operation of refineries, power plants, pipelines, and transmission lines usually involve multi-national companies. Certain parts of a large power plant may be even manufactured in different countries then assembled at the project site. International trade, transportation, and customs are involved in big energy projects.

In some energy projects several countries may be sharing the land and resources. Consider for example a hydroelectric power plant on an international river or a pipeline that crosses several countries. Such collaborations naturally require international agreements for financing and share of responsibilities. Sometimes development or operation of an energy facility in one country may have impacts on neighboring territories, which can even cause political conflicts.

Today, international energy markets dominate the rest of the economy. Energy sources are strategic commodities traded across borders or worldwide. Crude oil, coal, and natural gas prices continuously change based on supply-demand balance and market conditions. In addition, political tensions often cause extraordinary price increases or, in extreme cases, market instability.

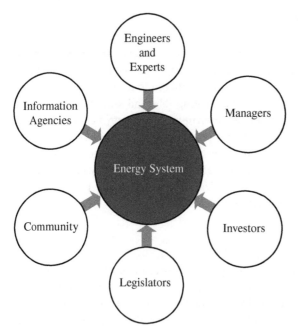

Figure 1.9 Groups involved in development and operation of an energy system.

Energy sources are strategic commodities controlled by the essential rules of economics. Production of energy sources, development of power plants, transportation, and supply coordination involve macroeconomic issues of investment, financing, planning, and resource management. Trade of energy sources follow the basic law of supply and demand. Variations of energy prices influence the cost of all industrial and agricultural products. Thus, energy sector is linked to all other sectors of the economy. Trade and use of energy sources have often resulted in international tensions escalating to conflicts and even wars.

1.11.2 Geopolitical Challenges

Because of the economic importance of energy supply, groups of countries have formed intergovernmental organizations to coordinate regional management of energy resources. On one hand, each nation has sovereign rights on energy resources available in their territory, therefore shape their energy policies according to national benefits. On the other, production, transport, trade, and use of energy sources affect cross-border economies and the global economic balance.

Organization of petroleum exporting countries (OPEC) is a multinational group initially founded by Iran, Iraq, Kuwait, Saudi Arabia, and Venezuela in 1960. OPEC's mission is "to coordinate and unify the petroleum policies of its Member Countries and ensure the stabilization of oil markets in order to secure an efficient, economic, and regular supply of petroleum to consumers, a steady income to producers and a fair return on capital for those investing in the petroleum industry." As of 2018, fourteen members of OPEC hold about three fourth of proven oil reserves and control about half of the oil production worldwide.

In several occasions, OPEC played a significant role in global economy and evolution of energy policies around the world. In 1973, OPEC countries significantly reduced oil production and applied oil embargo to industrialized western countries that supported Israel during the Arab-Israeli war. The oil embargo caused steep rise of oil prices and escalated to the historic "energy crisis" in industrialized countries and global economic recession.

Energy crisis led the industrialized nations to develop new energy policies to increase their energy independence and enhance their energy security. Immediately after the oil embargo, countries that were affected deeply from the energy crisis founded the International Energy Agency (IEA) in 1974. The initial purpose of IEA was to co-ordinate a collective response to oil supply disruptions. Over the years, IEA has evolved and expanded to ensure reliable, affordable, and clean energy for the world. Today, IEA provides unbiased and reliable information to help analysis, planning, and management of energy systems.

Following the Gulf War in 1991, a diverse group of countries formed the International Energy Forum (IEF) in Paris. The main goal of IEF is to facilitate dialog among its 72 members for mutual understanding and awareness of energy-related issues. Unlike the earlier energy organizations, IEF includes major energy producers and consumers including Brazil, China, India, Mexico, Russia, South Africa as well as OPEC and IEA members.

International conflicts and increasing dialog since 1970s have changed the view of energy production and consumption around the world. Development of energy security, efficiency, sustainability, and clean energy concepts are the most significant consequences of such interactions.

1.12   Energy Engineering

Energy engineering is an emerging interdisciplinary field that links several engineering disciplines with economic and political sciences. The main goal of professionals working in the energy field is to develop, operate, and manage energy systems in the most efficient and profitable way under ethical, social, and political constraints.

Energy engineers deal with planning, development, and operation of energy system components. Facility management, coordination of energy services, and energy efficiency are main work areas of energy engineers.

Energy related job functions require a broad knowledge in the fields of electrical, mechanical, geological, civil, and chemical engineering. Since the development of energy systems involves economic decisions, energy engineers must be able to cooperate with economists, investors, administrators, and policymakers.

Association of Energy engineers was founded in 1977 to promote the scientific and educational interests of those engaged in the energy industry and to foster action for sustainable development. More information about energy engineering, professional activities, and resources can be found at https://www.aeecenter.org (accessed in March 2020).

1.13   Chapter Review

Energy is an essential element for modern society for comfortable living in well-lighted, heated, and air-conditioned spaces equipped with modern appliances. It also saves human power and time by allowing use of power tools, electronic devices, machines, and equipment.

The basic definition of energy is "an ability to do work." According to the conservation of energy principle, energy cannot be created nor destroyed, it can only be converted from one type to another. Major types of energy exchanged in energy systems are mechanical, chemical, thermal, nuclear, and electrical.

An energy system is a collection of components that interact to supply the energy need of a society. Coal, petroleum, natural gas, and nuclear power are non-renewable primary sources. Fossil fuels are hydrocarbons extracted from the Earth's crust. Renewable primary sources include hydropower, ocean energy, sunlight, wind, biomass, and combustible waste.

The goal of an energy system is to supply the energy needs of a society. User groups are classified as residential, commercial, industrial, and transportation sectors based on their energy needs. About two-thirds of primary energy is wasted worldwide as unused heat during transformation, transmission, and end user applications.

Primary sources are processed to produce fuels and energy carriers. Heat produced by burning fuels is either used directly or converted to mechanical or electrical energy. Electricity, steam, and natural gas are major energy carriers. Electric generation worldwide heavily relies on combustion of various fuels, which are the most carbon-intensive energy sources. Energy obtained from all forms of primary sources, including fossil fuels, nuclear reactions, and renewable sources can be delivered to end users in the form of electric power.

Development and management of energy systems involve diverse groups of the society. Energy engineering is an emerging interdisciplinary field that combines professionals in various fields. Energy engineers work in multidisciplinary groups to develop, operate, and manage energy systems in a profitable and efficient way under ethical, social, and political constraints.

Among all human activities, energy production and consumption are responsible for the biggest part of air and water pollution and greenhouse gas emissions. Operation of energy systems impact the global ecology as well as water and food supply.

Review Quiz

1 The purpose of an energy system is to
   a. generate electric power using primary sources.
   b. supply the energy needs of a society.
   c. generate revenue by selling energy products.
   d. process petroleum to produce oil products.

2 Which one below is *not* a fossil fuel?
   a. Gasoline
   b. Propane
   c. Firewood
   d. Lignite

3 Which one of the following is a primary source of energy?
   a. Electricity
   b. Natural gas
   c. Kerosene
   d. Steam

4 Which one of the following is a secondary source of energy?
   a. Coal
   b. Crude oil
   c. LPG
   d. Biomass

5 Which form of energy is not directly used by end users?
   a. Electrical
   b. Mechanical
   c. Thermal
   d. Nuclear

6 Which one of the following is an energy carrier?
   a. Electricity
   b. Crude oil

c. Wind

d. Biomass

7 Which sector includes the energy consumed at the water treatment facility of a town?

a. Residential

b. Commercial

c. Industrial

d. None

8 In which sector do energy statistics consider the energy used for agriculture?

a. Residential

b. Commercial

c. Industrial

d. None

9 Energy efficiency of a device is

a. input energy divided by the output energy.

b. output energy divided by the input energy.

c. lost energy divided by the input energy.

d. lost energy divided by the output energy.

10 Which one of the following describes the energy intensity of a nation?

a. Ratio of the energy consumption to total value of goods and services produced by labor and property located in a country.

b. Product of total amount of energy produced and the total value of goods and services produced by labor and property located in a country.

c. Ratio of the gross domestic product (GDP) of a country to the population.

d. Ratio of the total energy loss to the gross domestic product (GDP) of a country.

Answers: 1-b, 2-c, 3-b, 4-c, 5-d, 6-a, 7-b, 8-c, 9-b, 10-a.

Problems

1 The Organization for Economic Cooperation and Development (OECD) is an intergovernmental group of 36 industrialized and emerging countries. Energy consumption in OPEC members represents the use of energy sources in modern societies where society is actively involved in energy-related decisions.

(a) Using EIA Interactive Table Browser[1], obtain a chart similar to Figure 1.4 for OECD countries.

(b) Compare the growth rates of consumption of each primary energy source in OECD countries and the world.

---

1 https://www.eia.gov/outlooks/aeo/data/browser/.

2 Heat content of finished motor gasoline is 121,023 Btu/gal. 10% in volume fuel ethanol with 84,762 Btu/gallon heat content is mixed to this gasoline at pump. Determine the heat content of the final fuel product delivered to customers.

3 At a gas station one gallon of regular unleaded gasoline costs $2.458 while one gallon of 10% ethanol added unleaded gasoline is priced $2.159. Assuming the heat content of gasoline and fuel ethanol are respectively 5.083 and 3.560 million Btu per barrel, calculate the cost of each fuel per unit energy.

4 A country consumes 1119 TWh electric energy for one year. What is the equivalent of this energy in quads?

5 An island consumes 1.414 quad energy in one year. This energy is supplied by imported crude oil of energy content 5.8 million BTU per barrel. How many million barrels of crude oil must be imported to meet the energy requirement of this island in six months?

6 A certain amount of fuel contains $25 \cdot 10^{12}$ Btu of energy, and it is converted into electricity in a power station having 16% overall efficiency. The average output power of the station is 7 MW. In how many days will the fuel be totally consumed?

7 In a power station $2 \cdot 10^4$ GWh of energy is to be produced in one year by burning coal. The coal used in this station can generate 900 W·years/ton. How much coal will be required in million tons?

8 In a power station $4 \cdot 10^4$ GWh of energy is to be produced in one year from natural gas. Natural gas used in this station can generate 260 Wh/CF. How much natural gas will be required in million cubic feet?

9 The global crude oil production in 2015 was 91,670 million barrels (BP, 2016). What is the coal equivalent of this production in million short tons?

10 During a 1-year period, an electric utility consumed 6 quad of coal, 2 quads of oil, 1 quad of natural gas, and 0.5 quads of hydro. If the overall efficiency of the system is 12%, how much electric energy (in GWh) could be produced by this utility?

Recommended Web Sites

- BP Statistical Review of World Energy BP Global: https://www.bp.com/
- Energy Information Administration (EIA): https://www.eia.gov/
- International Energy Agency (IEA): https://www.iea.org/
- International Energy Forum (IEF): https://www.ief.org/
- International Renewable Energy Agency (IREA): https://irena.masdar.ac.ae/gallery/
- Renewable Energy Policy Network for 21st Century (REN21): http://www.ren21.net/
- World Resources Institute (WRI): http://www.wri.org/

Further Reading

McElroy, M.B. (2010). *Energy – Perspectives, Problems, Prospects*. Oxford University Press.

Reisser, W. and Reisser, C. (2019). *Energy Sources – From Science to Society*. Oxford: University Press.

Tester, W.J., Drake, E.M., Driscoll, M.J. et al. (2005). *Sustainable Energy – Choosing Among Options*. MIT Press.

Wilson, J.R. and Burgh, G. (2008). *Energizing Our Future – Rational Choices for the 21st Century*. Wiley.

References

BP (2016). *BP Statistical Review of World Energy 2016*, London, UK: BP

EIA (2017). *International Energy Outlook 2017*. Washington DC: US Energy Information Agency.

EIA (2018a). *Annual Energy Outlook*. Washington DC: US Energy Information Administration Office of Energy Analysis, US Department of Energy.

EIA (2018b). *Interactive Table Browser*. [Online] Available at: https://www.eia.gov/outlooks/aeo/data/browser/ [Accessed August 2018].

EIA (2018c). *International Energy Statistics*. [Online] Available at: https://www.eia.gov/beta/international/data/browser [Accessed September 2018].

EIA (2018d). *Monthly Energy Review* (April 2018). Washington DC: US Energy Information Agency.

EPA (2017). *Power Profiler*. [Online] Available at: https://www.epa.gov/energy/power-profiler [Accessed 2017].

EPA (2018). *Energy Star Portfolio Manager*. [Online] Available at: https://portfoliomanager.energystar.gov/pdf/reference/Source%20Energy.pdf [Accessed 8 September 2018].

IEA (2018). *Key World Energy Statistics*. Paris: International Energy Agency.

LLNL (2018). *Lawrence Livermore National Laboratory Energy Flow Charts*. [Online] Available at: https://flowcharts.llnl.gov/content/assets/images/energy/us/Energy_US_2018.png [Accessed 22 December 2019].

The World Bank Group [US] (2018). *World Bank Open Data*. [Online] Available at: https://data.worldbank.org/indicator/EG.ELC.ACCS.ZS [Accessed August 2018].

2

Energy Conversion and Storage

(a)

(b)   (c)

Some conversion examples in energy systems. (a) Crescent Dunes Concentrated Solar Power (CSP) plant converting solar radiation into heat to generate electricity. (b) A coal-fired power plant in West Virginia, USA. (c) Generators installed in the hydroelectric power plant at Glen Canyon Dam, AZ, USA, converting mechanical energy of water into electricity.

Energy for Sustainable Society: From Resources to Users, First Edition. Oguz A. Soysal and Hilkat S. Soysal.
© 2020 John Wiley & Sons Ltd. Published 2020 by John Wiley & Sons Ltd.

2.1 Introduction

In Chapter 1 we used the word *energy* in the sense of natural sources that can be converted to useful mechanical work, heat, electricity, and other actions. In this section, we will focus on various forms of energy and conversion from one form into another.

The concept of energy was derived in sixteenth century and named using the Greek word "ergon" meaning "work." Energy can be defined in various ways leading to the dictionary definition "the capacity to do vigorous work; available power (Random House 1996)." While everybody has a certain insight of the term energy from everyday experiences, the use of the word energy in common language may lead to confusion in certain fields of science and technology. For example, you could spend the entire day reading a book or looking at a computer screen; at the end of the day, you might feel exhausted without having actually done any physical work. Where did your energy go? A person who has tried to move a big rock for hours would feel extremely tired even if the rock did not even move an inch. What is the work done by sunlight heating a surface? There are scientific explanations to all these questions that relate to some kind of physical work. In technical books, it is common to describe energy as "a capability to produce an effect" or "an ability to cause changes" (Cengel and Boles 2006).

Energy exists in many forms and can be converted from one form to another. Energy conversion is one of the oldest technologies in human history. Prehistoric humans who started fires by rubbing two sticks converted mechanical energy to heat. Although the primitive communities did not have the words to describe this breakthrough technology, they surely enjoyed the comfort of controlled heat that could be recreated and the taste of grilled meat. However, it took about one million years to develop scientific foundations of energy conversion, to produce motion from heat. In the mid-1800s steam engines converting heat to mechanical work led to the industrial revolution. By the end of the nineteenth century, converting mechanical work to electricity started the electronics era. It is ironic to see that about 85% of the energy our modern world needs still comes from burning some sort of fuel.

A modern energy system combines elements that process mechanical, chemical, nuclear, thermal, radiant, and electrical energy. Analysis and design of systems converting energy from one form to another require good understanding of fundamental concepts discussed in detail in comprehensive physics textbooks such as Young and Freedman (2016) and Giancoli (2014). In this chapter, we will focus on conversions between three essential forms of energy that are dominant in practical energy systems: thermal; mechanical; and electrical energy. Chemical, nuclear, and radiant energies support the energy system by producing heat. Thermodynamic cycles convert heat to mechanical energy. Hydrodynamics and aerodynamics interchange the properties of mechanical energy. Electromechanical systems are based on electric motors and generators that convert mechanical energy into electrical or vice versa. The goal of all types of energy conversion systems is transforming energy from one form to another to supply the energy mix needed for all activities of the modern society.

2.2 Work, Energy, and Power

Work, energy, and power are three fundamental concepts of classical physics. Although each of these words is frequently used in everyday language, they have precise definitions in physics. To establish a common terminology to discuss advanced principles of energy systems, this section will review the basic definitions and mathematical expressions of these interrelated notions.

2.2.1 Work

When a force F moves a rigid object or a particle over a distance s, the work is defined as the product of the force and distance traveled. If the force and motion are represented by vectors, the work is equal to the scalar (dot) product of these vectors. In a more general situation where the direction and magnitude of the force changes during the travel from point a to point b, the work is described by the integral expression below.

$$W = \int_a^b \vec{F} \cdot d\vec{s} \tag{2.1}$$

If the displacement is a straight line and the magnitude and angle of the force remain constant during the motion the work done can be expressed without the integral.

$$W = |\vec{F}|L \cos \alpha \tag{2.2}$$

In the expression above, L is the total distance traveled and α is the angle between the force and the displacement.

In the international unit system (SI) the unit of work is the Newton meter (Nm). In the British or Imperial unit system (IS) work is expressed in feet pound-force (ft·lb).

2.2.2 Energy

In physics, the common definition of energy is "the capability to perform a physical work." Because of the direct correlation between energy and work, they both have the same physical dimension. As noted, in the international unit system (SI), the standard unit of energy is Nm; in addition, SI defined a specific unit for energy, the Joule (J). Joule and Nm are equivalent units, therefore $1\,J = 1\,Nm$. Calorie is an older metric unit first defined to measure heat in early nineteenth century and is still in use especially in the food industry. One calorie is the energy needed to raise the temperature of 1 g of water 1 °C under normal pressure.

Energy has the dimension of foot-pound force in the imperial unit system (IS). The British thermal unit (Btu) is the standard energy unit for energy in Imperial and US customary units. One Btu is approximately equal to 1055 J. Horsepower-hour is an alternative energy unit for energy in IS derived from the power unit hp frequently used in practice.

2.2.3 Power

Power is the amount of energy converted or transferred in unit time. In mathematical sense, the function $p(t)$ that represents the power at any time is the derivative of the function $w(t)$ describing the time variation of energy. Likewise, if the time variation of power is known, total energy processed between the instants t_1 and t_2 is obtained by integrating the function representing the power in time.

$$p(t) = \frac{dw(t)}{dt} \quad \Leftrightarrow \quad W = \int_{t_1}^{t_2} p(t)dt \tag{2.3}$$

Table 2.1 Metric equivalents of imperial energy units.

Imperial units (IS)	Conversion to metric units (SI)
1 Btu	1055 J
1 Btu	0.293 Wh
1 imperial HP	745.7 W
1 metric HP	735.5 W
1 Btu/s	1.055 kW
1 Btu/h	0.293 W

If power is constant over a time interval Δt, then (2.3) simplifies to algebraic expressions.

$$P = \frac{W}{t_2 - t_1} \quad \Leftrightarrow \quad W = P(t_2 - t_1) \tag{2.4}$$

The SI unit of power is watt (W), which is equivalent to one joule per second. The only difference between energy and power is the time interval over which the energy conversion has occurred, or power is exerted. In different fields of engineering, it is common to express energy in terms of a unit of power multiplied by a unit of time.

$$1 \text{ J} = 1 \text{ W} \cdot \text{s} \quad \Leftrightarrow \quad 1 \text{ W} = 1 \text{ J/s} \tag{2.5}$$

In electrical energy systems, the processed energy is usually measured in power utilized over a one-hour period, rather than one second. Because the magnitude of energy is typically very large in an energy system, metric prefixes simplify the expressions and calculations. Appendix A includes a list of metric prefixes commonly used in the energy field. Energy consumed in residential, commercial, and small-scale industrial electric systems is usually expressed in kilowatt hours (kWh). In larger industrial systems the megawatt hour (MWh) is a more convenient energy unit. Energy generated by a power plant is in the order of gigawatt hours (GWh). Electric energy generated and used in a region or country is usually expressed in terawatt hours (TWh).

The imperial (IS) unit of power is a horsepower (hp). One imperial horsepower is 550 feet-pound per second, which is approximately 745.7 W. Metric horsepower is defined as the power required to lift an average person weighing 75 kg by 1 m in one second, which is equivalent to 735.5 W. When used for rating of an electric motor, 1 HP is equal to 746 W.

In thermal systems, power is usually expressed in terms of the energy unit as Btu per unit time, that is Btu/s or Btu/h. Relationships between essential units of energy and power in British (IS) and metric (SI) unit systems are shown in Table 2.1.

2.3 Conservation Laws

Energy systems are classified as open, closed, or isolated based on their interaction with the external environment. A system that exchanges both energy and matter with its surroundings is called an *open system*. If it exchanges only energy, not matter, it is a *closed system*. A system that does not exchange any energy and matter with its surrounding is called an *isolated system.* (Cengel and Boles 2006). Since the Earth receives energy from the Sun but does not exchange matter with outer space (neglecting the mass of meteors that may hit the Earth) can be considered a closed system.

2.3.1 Conservation of Mass

The total mass of a closed system is constant. In a system that does not exchange mass with its environment, the mass can never be created or eliminated. If the system undergoes changes through chemical reactions, its composition changes while the total mass remains the same. In an energy system, for example, while heat is produced by burning a fuel, the total mass of gas emissions, water vapor, and discarded solid compounds equals the mass of the fuel consumed.

The conservation of mass principle can be adapted to an open system that exchanges mass with its surrounding. Suppose that the mass m_{in} enters and m_{out} leaves the system. The resulting increase or decrease of mass in the system must be constant. In other words, the time rate of change of mass must be zero.

$$m_{in} - m_{out} = \frac{dm}{dt} = 0 \tag{2.6}$$

Although this seems obvious today, for centuries alchemists tried to convert various substances into gold until Antoine Lavoisier realized that it was impossible in late eighteenth century. Discovery of the conservation of mass principle allowed quantitative analysis of chemical reactions.

2.3.2 Conservation of Momentum

Newton's law of inertia states that the acceleration a of a body is proportional to the net force F acting on the body and inversely proportional to its mass m.

$$a = \frac{dV}{dt} = \frac{F}{m} \tag{2.7}$$

Since mass is constant in a closed system and acceleration is the time rate of change of velocity, the derivative of the product of mass and velocity is equal to the acting force.

$$m\frac{dV}{dt} = \frac{d}{dt}(m \cdot V) = F \tag{2.8}$$

The product of the mass and the velocity of a body is called the *linear momentum* or just the *momentum* of the body. If the total external force acting on a closed system is zero, then the momentum of the system is constant. This concept is known as the *conservation of momentum principle.*

2.3.3 Conservation of Energy

Energy is a conceptual quantity and it is independent from the system where it occurs. Physics principles and laws regarding energy interactions are the same no matter the form; mechanical, thermal, electrical, chemical, or in any other form. The conservation of energy principle can be worded in many ways, all leading to the same fundamental concept:

Total energy of an isolated system is always constant. If there is no energy exchange with the surroundings, the total energy of the system cannot increase or decrease. It can, however, be converted from one form to another, stored, or transferred from one place to another within the system.

According to the conservation of energy principle, there is no mechanism that can create energy without receiving an energy input from the external world. Similarly, no system can discard energy, but can release energy to its surroundings. The conservation of energy principle should not be confused with the term "energy conservation," which means reducing the use of energy by avoiding unnecessary and inefficient consumption. Terms like "renewable energy" and "sustainable energy" contradict with the conservation of energy principle.

The conservation of energy principle was first stated by Hermann Helmholtz. He observed the conservation of energy while he was studying mechanical power of muscles. He explained the principle of the conservation of energy principle in the article "*On the conservation of Force; A Physical Memoir*" he presented to the Physical Society of Berlin in 1847.

2.3.4 Equivalence of Energy and Mass

In classical physics, conservation of mass and conservation of energy had been two distinct concepts until Albert Einstein proved the equivalence of mass and energy in the relativity theory. Einstein's famous expression states that if an object moves at the speed of light, its mass is converted to energy equal to the product of mass m and the square of the speed of light c (in vacuum $c = 3 \cdot 10^8$ m/s).

$$E = mc^2 \tag{2.9}$$

Clearly, if a practical conversion technique were known, an extremely small amount of any matter would yield an enormous amount of energy. The conversion of mass to energy occurs in nuclear reactions when atoms of certain elements are split (fission) or two hydrogen atoms are bonded (fusion). Conversion from mass to energy is the source of nuclear energy, which is used to generate more than 20% of electric energy used in the world. A reverse process to convert energy into mass is not available in practice.

2.4 Transformation Between Energy Forms

Energy has many forms, and can be transformed from one form to another using a suitable conversion system. Most contemporary activities mainly rely on thermal, mechanical, and electric energy. Solar radiation, nuclear energy, chemical energy of fuels, hydraulic energy of flowing water, and energy of blowing wind are converted to these three essential forms to support social and economic functions.

In principle, any form of energy can be converted into any other form. Implementation of conversion techniques in energy systems depend many factors including economic viability, safety, reliability, and environmental effects. For example, plants convert carbon dioxide and water into hydrocarbons through the well-known process of photosynthesis by absorbing solar energy. However, no economically viable technology that could convert carbon dioxide back to a fuel using solar energy has been developed yet. The power source of the

Sun, fusion reactions, are known and practically applicable technology has been developed to produce huge amounts of energy from fusion. However, this technology cannot be currently applied in energy systems because extremely high temperature and pressure is required to initiate fusion reaction, which is not possible in practical energy systems. Photovoltaic energy conversion is another example, The photoelectric effect has been known since the nineteenth century, and the first photovoltaic (PV) cells were developed in the 1950s. Solar electric generation could not compete with conventional electric generation in practical energy systems until the early years of the twenty-first century because of their higher cost. The piezoelectric effect has been used in sensors for a long time to convert pressure to electric current. But commercial scale energy conversion systems based on piezoelectric effect have not been developed yet due to practical limitations.

Figure 2.1 shows interactions between various forms of energy in practical energy systems. Solar energy reaching the Earth's surface is converted to heat, which gives rise to wind, water cycle, rivers, and ocean energy. Sunlight and water yield biofuels through

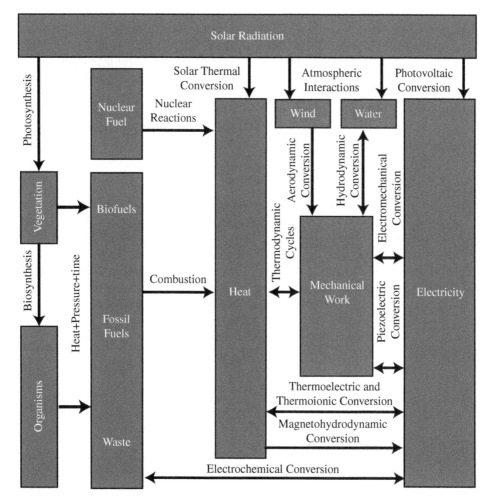

Figure 2.1 Transformations between various forms of energy.

natural processes, which eventually turn to fossil fuels by the combined effect of heat and pressure over millions of years. Photovoltaic processes convert solar energy directly to electricity. Energy interactions that occur in components of practical energy systems can be grouped in chemical, thermodynamic, and electromechanical conversion techniques.

The arrows in Figure 2.1 indicate the direction of the energy transfer. Electromechanical energy conversion is inherently bidirectional. Electric machines, for example, can operate either as a motor or generator. Hydro machines can also work as either turbine or pump. Hydrogen fuel cells generate electricity combining hydrogen and oxygen to produce water or can produce hydrogen and oxygen by electrolysis of water reversing the energy flow. Rechargeable batteries are bidirectional electrochemical converters as well.

Some conversions, however, occur in only one direction because the reverse conversion technique is either not economically viable or impractical. For example, hydrocarbons are converted to heat by combustion, but no economic method is known at present to combine carbon and hydrogen atoms to reconstruct hydrocarbon molecules that have been naturally formed over millions of years under the combined effect of pressure and temperature. Biofuel technologies use enzymes to convert vegetation or organisms into hydrocarbons, ethanol production from sugar cane, soybean, and corn, or biodiesel production from animal waste are examples to such unidirectional conversion processes. Practical thermodynamic processes are irreversible because of heat losses and non-ideal interactions. Heat produced as a result of friction or electric resistance is dissipated to the surroundings as losses unless a special combined heat and power (CHP) system is implemented to recycle this energy for a useful purpose (Fenn 1982).

Efficiency of an energy conversion system is always defined as the ratio of the output energy (or power) to the input energy (or power) as expressed below.

$$\eta = \frac{\text{Output energy}}{\text{Input energy}} = \frac{W_{in} - W_{loss}}{W_{in}} = 1 - \frac{W_{loss}}{W_{in}} \tag{2.10}$$

W_{loss} represents energy released to the surrounding space as unused heat, which is commonly called losses. Practical energy conversion systems include pumps, compressors, fans, and other ancillary equipment to support the main conversion process. Energy consumed by such equipment is also considered part of the losses.

To be implemented in a practical energy system, an energy conversion technique must be justified from cost, efficiency, reliability, and safety perspectives. Some of the known conversion techniques that are not feasible under existing economic conditions may become viable as technologies improve or market situations change. This text will focus on energy conversion techniques currently used in energy systems and emerging technologies being developed through research and development programs.

2.5 Thermal Energy

Heat is an energy interaction associated with a temperature difference. Temperature is a measure of hotness or coolness of an object or environment. Several temperature scales have been developed based on two reference temperatures, such as the temperature of an ice-water mixture and boiling water. An absolute temperature scale is defined using the three-state temperature, where solid, liquid, and gas states of water coexist (Sonmtag, Borgnakke, and Van Wylen 1998).

From an energy system standpoint, heat is the primary form of energy converted to mechanical work through thermodynamic cycles and electricity through an electrome-chanical energy conversion process. While any form of energy can be converted to heat, in energy systems heat is mostly produced by combustion of fuels.

Objects at different temperatures exchange heat. Adding or extracting heat can change the temperature of an object, occasionally leading to a change of phase between solid, liquid, and gas forms of mechanical work by expansion or compression. Heating, air conditioning, refrigeration, melting, evaporation, condensation, freezing, or production of mechanical work through thermodynamic processes are major applications of thermal energy conversions.

The physical nature of heat was not understood until the mid-nineteenth century. French scientist Antoine Lavoisier (1743–1798) made a breakthrough in physics and chemistry by discovering the principle of conservation of mass. Lavoisier defined heat as flow of an invis-ible substance, which he called *caloric*. This theory could explain heat transfer but was inadequate to explain several other properties of heat. About hundred years later, English scientist James Prescott Joule (1818–1889) discovered the relationship of heat and mechan-ical work, which established the first step leading to the conservation of energy principle and thermodynamics.

Maxwell and Boltzmann associated heat to the kinetic energy of randomly moving atoms and molecules. According to the kinetic theory, atoms of a gas move at random velocities in all directions. When the atoms collide with each other or the wall of the container, their kinetic energy transforms to heat (Young and Freedman 2016).

2.5.1 Temperature and Phase Changes

When an object is heated, the random motion of atomic particles increases, raising the tem-perature. In a solid, particles are closely tied together. When heated, the particles move farther apart causing an expansion. As the particles start to move independently, the mat-ter changes its phase from solid to liquid. Heated more, the particles can move more freely, turning the matter into gas. Under certain conditions some solid matters can directly turn into gas. This type of phase change is called *sublimation*. During a change of phase, the temperature of an object remains approximately constant.

A certain amount of energy, called *latent heat* can change the phase of a matter from solid state to liquid (melting). Similarly, a latent energy is required to transform a liquid to gas (evaporation). In the reverse processes from gas to liquid and from liquid to solid, a certain amount of energy must be taken out of a matter.

Without any change of phase, the temperature of an object changes proportionally to the heat exchanged with its surroundings. Heat capacity of a substance is the ratio of energy transferred by heat to the change of its temperature.

$$C = \frac{Q}{\Delta T} \tag{2.11}$$

Table 2.2 Specific heat of common substances.

Substance	Specific heat (J/kg °C)
Ammonia (liquid)	4700
Water, pure liquid (20 °C)	4182
Paraffin	3260
Methyl alcohol	2530
Ethyl alcohol	2440
Ice (0 °C)	2093
Water vapor (27 °C)	1864
Paper	1336
Porcelain	1085
Air (dry, sea level)	1005
Asphalt	920
Aluminum	897
Fire brick	880
Salt (NaCl)	880
Brick	840
Sand	830
Glass (pyrex)	753
Silicon	705
Iron	449
Copper	385
Brass	375
Bronze	370
Mercury	140
Lead	129

Source: https://www.engineeringtoolbox.com (accessed on 9/29/2019).

Heat capacity of a matter per unit mass is called *specific heat*. This is a physical property of a substance and corresponds to the heat needed to increase the temperature of the unit mass of that substance by one degree Celsius or Kelvin in the metric unit system. Table 2.2 shows the specific heat of selected substances. Substances with higher specific heat, such as ammonia and water require more thermal energy for unit temperature change. If c represents the specific heat and m the mass of an object, the amount of heat energy needed to change the temperature from T_1 to T_2 is

$$Q = m \cdot c \cdot (T_1 - T_2) \tag{2.12}$$

The unit of thermal energy is Joule in the international unit system (SI). The unit of thermal energy in the imperial unit system is Btu. Another unit to measure heat is the calorie (cal). James Prescott Joule demonstrated the equivalence of thermal and mechanical energy. The expression below can be used to convert calorie to joule or vice versa.

$$1 \, cal = 4.18 \, J \tag{2.13}$$

2.5.2 Production of Heat

According to the kinetic theory, heat is linked to the kinetic and potential energy of atomic particles. Therefore, all other forms of energy can be naturally converted to heat. Practical energy conversion processes produce heat that is dissipated to the surrounding region as loss. Friction between moving parts and Joule heating on a current carrying conductors are major sources of heat losses. Magnetic fields changing in time induce eddy currents heating the magnetic core of electric machines and transformers. The hysteresis property of ferromagnetic materials is also a source of heat losses in electric devices.

In this section, we will focus on the production of useful heat that enters in an energy conversion process as input energy. In energy systems, heat is used for thermodynamic processes for conversion to mechanical work, industrial processes for chemical reactions, smelting of metals, material conditioning, food processing, or space and water heating in buildings. A useful amount of heat is produced by combustion of fuels, nuclear reactions, solar radiation, or by converting electric energy through resistive, induction, or ark effects.

2.5.2.1 Combustion

Combustion is perhaps the oldest type of energy conversion process. Combustion was introduced to the human life by invention of fire. Fire and explosion are uncontrolled conversions of chemical energy into heat. Exploding dynamite converts the chemical energy of nitroglycerine to a destructive mechanical energy. Fire converts chemical energy to heat, burning surroundings.

Combustion is the controlled and accelerated oxidation of a fuel. In a chemical reaction, bonds holding atoms in molecules are broken and new bonds are established changing the composition of the reactants. In a chemical reaction, the total mass of reactants remains constant. Internal energy of the substance changes because of the change of atomic bonds, hence the reaction may absorb or release energy. A reaction that absorbs energy is called *endothermic* and a reaction that releases energy is called *exothermic*.

The oxidation of a fuel and an oxidant is an exothermic reaction. While the chemical composition of the fuel and oxidant change, heat is released. Combustion can also produce light in the form of a flame or radiance (glow).

Today, combustion of fossil fuels is the most common source of energy in the world. Fossil fuels have three essential combustible components: hydrogen, carbon, and sulfur. The simplest combustion is oxidation of hydrogen. As a hydrogen molecule is combined with oxygen, it forms water and the reaction releases heat.

$$2H_2 + O_2 \rightarrow 2(H_2O) + Heat \tag{2.14}$$

Hydrogen has the highest ignition temperature among the components of fossil fuels, but since it is a gas the combustion of hydrogen proceeds extremely fast. Hydrogen combustion is the cleanest heat production technique since pure water is part of the nature. No carbon dioxide and other oxides are emitted into the atmosphere. Hydrogen can be produced by electrolysis of water. A certain proportion of hydrogen and oxygen mixture is explosive. Hydrogen is, however, less dangerous than gasoline because it is lighter than air and a hydrogen leakage rises directly in the air without spilling on the surface. The major limitation of hydrogen is its cost of production and storage in bulk quantities.

Carbon is the essential combustible element in fossil fuels. Carbon has a lower ignition temperature, but its oxidation is slower and more difficult than hydrogen. Carbon oxidizes in two steps; it first combines with oxygen molecules to form carbon monoxide, then carbon monoxide further oxidizes to carbon dioxide.

$$2C + O_2 \rightarrow 2(CO) + \text{Heat}$$
$$2CO + O_2 \rightarrow 2(CO_2) + \text{Heat} \tag{2.15}$$

Ideal combustion of carbon emits carbon dioxide, part of which is converted back to oxygen naturally by vegetation through photosynthesis. Any remaining carbon dioxide in the atmosphere causes greenhouse effects that leads to climate change. If there is no sufficient oxygen during the combustion, the gas emission contains a certain proportion of carbon monoxide.

While many chemicals can produce heat by combustion, organic fuels that contain hydrocarbons (such as wood, coal, petroleum, and natural gas) are among the most commonly used fuels in practical energy conversions. When a hydrocarbon-based fuel burns, the molecular bonds between carbon and hydrogen atoms are broken and they are combined with oxygen atoms forming mainly carbon dioxide and water molecules. During the chemical process, energy is released to the environment. Such chemical reaction is shown below for methane (CH_4), which is the major component of natural gas.

$$CH_4 + 2O_2 \rightarrow CO_2 + 2(H_2O) + \text{Heat} \tag{2.16}$$

Note that (2.16) is a redox equation for complete combustion. If incomplete combustion occurs, such as burning with insufficient amounts of oxygen, carbon monoxide (CO) would be released from the process. Carbon monoxide is an odorless toxic gas and should be avoided to prevent health hazards and death.

All fossil fuels contain a small amount of sulfur. Combustion of pure sulfur yields sulfur dioxide.

$$S + O_2 \rightarrow SO_2 + \text{Heat} \tag{2.17}$$

As we see in Eqs. (2.14) through (2.17), all combustion processes rely on oxygen. The source of oxygen in practical burners is air. The molecular weight of air is 28.97 kg/kg·mol. By volume, dry air contains about 78% nitrogen and 21% oxygen, the remaining 1% includes argon, carbon dioxide, other gases (Culp Jr. 1979).

At exceedingly high combustion temperatures, endothermic reactions called dissociation separate oxygen and nitrogen molecules in their atoms.

$$O_2 \rightarrow 2O$$
$$N_2 \rightarrow 2N$$
$$2CO_2 \rightarrow 2CO + O_2 \tag{2.18}$$

Combustion properties of essential fuel elements are shown in Table 2.3. In energy systems, combustion occurs in specially designed combustion chambers or burners. Conditions for

Table 2.3 Combustion properties of fuel components.

Process	Ignition temperature (°C)	Required oxygen mass ratio	Lower heating value (kJ/kg,·,mol)
Oxidation of hydrogen	582	7.94	142,097
Oxidation of carbon to CO	407		110,380
		2.66	
Oxidation of CO to CO_2	609		283,180
Sulfur	243		8257

proper combustion are adequate mixes of reactants, sufficient air, sufficient temperature, sufficient time for the reaction to occur, and sufficient density to propagate the flame (Culp Jr. 1979). In practical burners, the perfect mix is never attained, and good combustion is assured by supplying excess air.

Most fuels contain small amounts of sulfur, organic matter, and minerals. Complete combustion of a fossil fuel emits a mixture of carbon dioxide, sulfur dioxide, and water vapor. In actual burners, some incomplete combustion occurs, and the exhaust gas usually contains some unburned fuel particles, carbon monoxide, sulfur oxides, and nitrogen dioxide (NO_2), nitric oxide (NO), volatile organic compounds (VOCs), particulate matter (PM), etc. The optimal air–fuel ratio is usually estimated experimentally based on analysis of exhaust gas. Modern burners are equipped with systems called scrubbers to remove sulfur from flue gas.

2.5.2.2 Nuclear Reactions

In energy systems, nuclear reactors produce steam mainly for electric generation. All nuclear reactors used in electric generation involve fission energy, which is based on splitting a large atom, such as uranium isotopes. In most of the nuclear power plants U-235 isotope is broken into one cesium-140 and one rubidium-92 atom. During this reaction three neutrons are freed up to continue the chain reaction and 3.2×10^{-11} J energy is released by splitting one U-235 atom. The complete fission of 1 kg of U-235 can produce about 63 million Btu thermal energy, which is equivalent to 3000 tons of coal burned in a typical thermoelectric power plant.

Understanding the basics of nuclear energy is important because nuclear reactors are used to generate approximately one-fourth of electricity worldwide (IEA 2018). In the US, about 20% of electric energy is currently generated by nuclear power plants (EIA 2017). Nuclear reactions and operation of nuclear reactors are explained in more detail in Chapter 4.

2.5.2.3 Electric Heating

Conversion of electric power into heat on a resistor is known as Joule heating, ohmic heating, or resistive heating. In 1840 James Prescott Joule published his experimental findings showing that an electric wire immersed in water produces heat proportional to the square of the current passing through the conductor.

Electric power dissipated on a resistor is obtained by multiplying the voltage across its terminals v and the current i flowing through it.

$$p = v \cdot i \tag{2.19}$$

Ohm's law describes the voltage and current relationship for a linear resistor R.

$$v = R \cdot i \tag{2.20}$$

$$p_{resistor} = R \cdot i^2 \tag{2.21}$$

Resistors are used to produce useful heat at the consumer end for space and water heating, cooking, and some industrial processes. Electric heating is more efficient, cleaner, and more convenient than burning a fuel at the end-user site. However, from the energy system point of view, we must consider the overall impact of electric use on the consumption of primary energy sources and the environmental impacts of electric generation. In the US interconnected system, the site-energy to source-energy ratio is estimated 2.80 (EPA 2018). That means use of one unit of electric energy at the consumer site consumes nearly three units of total primary sources. Therefore, direct use of a cleaner energy source such as natural gas, LPG, local wind and solar power is overall more efficient than using electric power from the utility grid.

Resistive heat causes significant loss in electric transmission and distribution systems. Heat produced on the wires of motors, generators, cables, transmission lines, and transformers is dissipated to the surrounding environment. Approximately 7–8% of the electric energy transmitted from generation plants to consumers is currently lost on the transmission lines.

2.5.3 Heat Transfer

Thermal energy is transferred from one place to another by three separate processes named conduction, convection, and radiation.

2.5.3.1 Conduction

Heat propagates in a solid object by changing the kinetic and potential energy of microscopic particles. If the object does not exchange heat with its environment (isolated), all points will ultimately reach the same temperature. If the object is in contact with others, heat transferred by conduction changes the temperature at different points. The conductive heat transfer is inversely proportional to the distance d between the boundaries and proportional to the cross-sectional area A and temperature gradient ΔT. The proportionality constant k is called thermal conductivity.

$$Q = \frac{k \cdot A}{d} \cdot \Delta T \tag{2.22}$$

Conductive heat transfer is particularly important in thermal insulation of energy system components. For example, pipes carrying hot water or steam must be well insulated to reduce thermal losses. Thermal insulation of a building affects the energy needed to maintain the interior at a desired temperature.

2.5.3.2 Convection

Convection is heat transfer from one point to another by motion of a fluid. If a fluid of specific mass ρ moves with a velocity u, the mass that flows per unit area in unit time is ρu. At a temperature T the fluid will transfer a thermal energy cT, where c is the specific heat of the fluid. We can, therefore, derive Eq. (2.23) to determine the rate of heat flow per unit area by convection.

$$\frac{Q}{A} = \rho u c T \tag{2.23}$$

Heat transfer by convection has practical applications in heating or cooling using fluids. For example, heat dissipated in a large machine because of losses is usually removed by circulating cooling fluids such as water or oil around the heating parts.

2.5.3.3 Radiation

Radiation is a form of heat transfer that does not rely on a surrounding fluid or contact of bodies. Visible light, infrared, and other electromagnetic waves carry heat to a distance even in vacuum. A typical example is transfer of energy from Sun to Earth.

Boltzmann's law states that the power radiated from the unit area of a black body is proportional to the fourth power of the absolute temperature of the surface. A *black body* is a surface that absorbs all incident radiation. A body that does not absorb all incident radiation is characterized by an emissivity, $\varepsilon < 1$. Emitted power density per unit area from a surface at temperature T_s is given in the expression below.

$$P_e = \varepsilon \sigma T_s^{\,4}$$
$$\sigma \simeq 5.67 \cdot 10^{-8} \; \mathrm{W} \cdot \mathrm{m}^2 \cdot \mathrm{K}^{-4}$$
$$0 \le \varepsilon \le 1 \tag{2.24}$$

The coefficient σ is known as the Boltzmann constant. Emissivity ε of a surface is a dimensionless quantity and ranges from zero to one depending on the reflective properties. Dark colored surfaces absorb radiation, whereas light colored surfaces reflect. Opaque surfaces also absorb radiation. The absorbed power density can be calculated similarly using the ambient temperature T_0. The net power density is given by Eq. (2.25).

$$P = P_e - P_s = \varepsilon \sigma (T_s^{\,4} - T_0^{\,4}) \tag{2.25}$$

2.5.4 Thermodynamics

Thermodynamics is the study of interactions between heat and mechanical work. The principle of conservation of energy is known as the first law of thermodynamics. The second law of thermodynamics deals with the interactions between heat and mechanical work and the direction of energy flow (Sonmtag, Borgnakke, and Van Wylen 1998). Fundamental concepts of thermodynamics and applications concerning the conversion of thermal energy into mechanical energy are discussed in Chapter 7.

2.6 Mechanical Energy

Mechanical energy is the capacity to do work by mechanical forces. Mechanical energy can be equally defined as *the form of energy that can be converted to mechanical work completely and directly*. Potential and kinetic energies are the forms of mechanical energy.

2.6.1 Potential Energy

Potential energy is the capacity to do work due to the position of an object. An attraction force created by Earth or deformation of an elastic element can change the potential energy of an object by performing mechanical work.

Energy associated with the vertical position of a body is called gravitational potential energy. Consider an object at a position h_1 measured in respect to a certain reference. If the object is released to fall down with gravity, the only force that acts on it is the gravitational force, in other words, its weight. As the object moves from the position h_1 to h_2, the work done by the gravitational force can be expressed as the product of the mass m, gravitational acceleration constant g (approximately 9.81), and the difference of heights measured from a reference point.

$$W_p = mg \cdot (h_1 - h_2) \tag{2.26}$$

An object that moves from a lower to a higher elevation stores potential energy. On the other hand, an object moving from a higher to a lower elevation loses potential energy. In practical energy systems the distance traveled by the elements is much smaller than the radius of earth. Since the gravitational force can be considered always in a vertical direction for such displacements, potential energy depends solely on the difference of heights of the initial and end positions, not the paths traveled.

An elastic deformation is a self-reversing temporary change of shape or dimensions of an object. A spring, rubber band, and an inflated ball are examples to elastic objects. Elastic deformations produce a force proportional to displacement. As the geometry changes, an elastic body stores or releases potential energy. Similarly, change of relative positions of objects attached by elastic elements exchanges potential energy due to the work done by elastic forces.

2.6.2 Kinetic Energy

Kinetic energy is mechanical energy associated with the motion of objects. According to Newton's law of inertia, the vector sum of all forces acting on a rigid object is equal to the mass multiplied by the acceleration of the object.

$$F = ma = m\frac{dv}{dt} \tag{2.27}$$

The work done by the force when the object moves a differential distance dx is

$$dW = F \cdot dx = m\frac{dv}{dt}(v \cdot dt) = m \cdot v \cdot dv \tag{2.28}$$

By integrating both sides of Eq. (2.28), we obtain the expression of *kinetic energy*, W_k of a moving object.

$$W_k = \int_{v_1}^{v_2} m \cdot v \cdot dv = \frac{1}{2}m(v_2{}^2 - v_1{}^2) \tag{2.29}$$

In rotational motion, kinetic energy is obtained using the moment of inertia and angular velocity of the rotating parts.

$$W_k = \frac{1}{2}J\omega^2 \tag{2.30}$$

The moment of inertia J depends on the mass and geometric shape of an object and its standard unit is kg·m². Angular velocity is measured in radians per second (rad/s). In practice the rotation speed is often expressed in revolutions per minute (rpm). The expression below converts n given in rpm value to angular velocity ω in rad/s.

$$\omega = \frac{2\pi n}{60} \tag{2.31}$$

An accelerating object absorbs kinetic energy. When it slows down, the kinetic energy is released to the outside world. An object moving at a constant velocity does not exchange kinetic energy with the external world. However, it has a stored kinetic energy due to its mass and velocity, which could be delivered if the speed decreases. To control the speed or stop the vehicle kinetic energy must be converted to some other form of energy. Mechanical breaks in vehicles convert kinetic energy to heat through friction. Electric or hybrid vehicles can convert kinetic energy to electricity and either return it to the supply system or store it in a battery bank.

2.6.3 Potential and Kinetic Energy Exchanges

Ideally, in a mechanical system where no external forces such as friction, an engine's driving force, or electromagnetic forces are present, potential and kinetic energies can fully convert to each other. For example, an ideal pendulum converts potential energy to kinetic energy and vice versa as it swings. A vehicle moving freely downhill without break force accelerates as the potential energy transforms to kinetic energy. On the contrary, when going uphill, the vehicle would slow down because the kinetic energy is converted to potential energy, if the engine does not deliver additional energy.

At free fall, potential energy is converted to kinetic energy changing the speed of the body. But when the body stops at the final point, all kinetic energy gained will be returned. If we neglect the friction forces, potential energy is completely transformed to kinetic energy, and kinetic energy is transformed back to potential energy. A force that has such property is said to be *conservative*.

In energy systems, hydroelectric power plants harness the gravitational potential energy of water stored in a reservoir. In this case, potential energy is converted to kinetic energy of turbines, and then converted to electric energy in a generator. Special types of hydroelectric power plants can be used to store a large amount of energy. Pumped hydroelectric energy storage facilities (PHES) pump water from a lower reservoir up to another reservoir at a higher elevation using excess energy. When electricity is needed, water is released from the higher reservoir through a hydroelectric turbine into the low reservoir to generate electricity. PHES is used at many locations in the United States and around the world to balance electric generation and supply. PHES power plants constitute the largest energy storage capacity in the world.

Waterwheels are ancient mechanical systems that harness kinetic energy of moving water. Similarly, windmills convert the kinetic energy of moving air (wind) into shaft rotation. Wind turbines are a modern version of windmills that generate electrical energy from kinetic energy of airflow.

2.6.4 Mechanical Power

Since power is the time derivative of energy, mechanical power is the work done by a force in unit time. In linear motion, power can be obtained by multiplying the force and velocity as shown in Eq. (2.32).

$$P = \frac{d}{dt}(F \cdot x) = F\frac{dx}{dt} = F \cdot v \tag{2.32}$$

Turbines, motors, and generators used in energy systems perform rotational motion. A point at a distance r from the center of rotation that rotates $d\theta$ radians travels the distance of $r \cdot d\theta$. The product of a tangential force and the radius of the rotational motion is defined as torque. Hence, the expression of rotational power is simplified to the product of torque T and the angular velocity ω as shown in Eq. (2.33).

$$P = \frac{d}{dt}[F_{\text{tan}} \cdot (r \cdot d\theta)] = T\frac{d\theta}{dt} = T \cdot \omega \tag{2.33}$$

The standard unit of angular velocity is radians per second. In practice, it is more convenient to express rotational speed in revolutions per minute (rpm). The relationship between an rpm value N and the angular velocity ω in radians per second is:

$$\omega = \frac{2\pi N}{60} \tag{2.34}$$

2.6.5 Mechanical Energy Balance in Incompressible Fluids

In energy systems, fluid flow can transfer thermal and mechanical energy from one point to another. When the transfer of only mechanical energy is considered, all points of the fluid are assumed to be at the same temperature, no chemical or nuclear reactions are involved, and no electrical and magnetic forces are present. The fluid is assumed to be incompressible, if the specific mass ρ remains constant. For convenience, all forms of energy are expressed for unit mass of fluid (Cengel and Cimbala 2014).

Potential energy of a fluid element at the height h per unit mass is reduced to $g \cdot h$. If the fluid element moves at a linear velocity u (m/s), its kinetic energy per unit mass is $\frac{1}{2}u^2$. In addition to potential and kinetic energies, a moving fluid has an energy associated with its pressure at any section of the flow. Pressure is the force exerted by the fluid per surface area, expressed in N/m². The fluid element with a surface area A moving over the distance d performs the work $P \cdot A \cdot d = P \cdot V$. Energy per unit mass is obtained by dividing the work with the mass of the fluid $\rho \cdot V$.

$$(P \cdot A \cdot d)\frac{1}{m} = P \cdot V\frac{1}{\rho V} = \frac{P}{\rho} \tag{2.35}$$

The combination of kinetic, potential, and flow energy per unit mass is known as Bernouilli's equation. In an incompressible fluid, total energy is constant. Therefore, Bernouilli's equation can be expressed as Eq. (2.36) at any point of the fluid.

$$\frac{P}{\rho} + \frac{u^2}{2} + gh = \text{Constant} \tag{2.36}$$

Bernoulli's equation is an approximate model of the fluid flow because frictions and laminar flow effects are neglected. It is, however, a useful tool to study most fluid flow problems in energy engineering, including hydraulic turbine performance analysis.

2.7 Electrical Energy

Electricity is a form of energy associated with the existence and displacement of charged subatomic particles. Ancient Greek philosopher and mathematician Thales of Miletus, who lived between 624 and 546 BC, noticed that a piece of amber rubbed on fur attracted lightweight objects such as hair and feather. Electricity, however, could not be used as a controllable form of energy until the eighteenth century, about 2500 years in the future. The term *electric* was derived in the mid-seventeenth century from the Greek word *elektron* and Latin word *electrum*, both meaning amber.

Electric charge is a basic physical property of matter associated with subatomic particles. Electrons are the elementary particles with negative charge, and protons have the same amount of positive charge. In a balanced atom, the total number of electrons equals the total number of protons, therefore, the net electric charge of a stable atom is zero.

Several physical effects can cause electron exchange between atoms, hence producing an excess negative or positive charge in an object. The standard unit of electric charge is Coulomb (C). Electric charge of an electron, -1.602×10^{-12}, C is the elementary charge. Total charge of an object is an integer multiple of electron charge with a positive or negative sign. Objects that have overall electric charge of opposite sign attract, while objects with an excess electric charge with the same sign repel each other. The electrostatic interaction between two charged particles, known as Coulomb force, is determined by the equation

$$F = \frac{1}{4\pi\varepsilon} \frac{Q_1 Q_2}{r^2} \tag{2.37}$$

In Eq. (2.37), r is the distance between the particles or objects carrying the charges Q_1 and Q_2. The constant ε is called permittivity of the medium. Permittivity of vacuum is $\varepsilon_0 = 8.85 \times 10^{-12}$ C/N·m². In any material, permittivity is described as $\varepsilon_0 \varepsilon_r$, where ε_r represents the relative permittivity of the material. The permittivity of air is approximately equal to unity.

Electric forces act over distance like the gravitational force. In other words, objects do not have to be in contact to each other to exert a force. Faraday developed the concept of an *electric field* that extends outward from an electric charge to every point of the surrounding space. A force is exerted on any other charge placed in such a field. Magnitude of an electric field at any point in the space can be measured by placing a test particle with a small charge q that would not have a significant effect on the field. Forces are represented by vector quantities. Hence, an electric field is also a vector with the same direction as the force it exerts on a positive charge. The electric field vector $\vec{E}$ is defined as the limit of the force vector $\vec{F}$ exerted on the test particle as the quantity of charge approaches zero (Kraus 1992).

$$\vec{E} = \lim_{q \to 0} \frac{\vec{F}}{q} \tag{2.38}$$

Similar to the gravitational potential energy in mechanics, forces acting on electric charges placed in an electric field have an electric potential energy named electromotive force (EMF), or simply potential.

2.7.1 Voltage and Current

Difference of potential energy between two points of the electric field is known as voltage. The unit for electric potential is the volt (V), named after Italian scientist Alessandro Volta, who developed the first practical battery cell in 1801. Voltage is the common word for EMF defined as the work done per unit electric charge to move charged particles from one point to another (Alexander and Sadiku 2009). In a mathematical sense, voltage is the rate at which energy changes in respect to the change of electric charge, as in the differential expression in Eq. (2.39).

$$v(t) = \frac{dw}{dq} \tag{2.39}$$

The standard unit of voltage is the volt (V), and one volt is the work done per unit electric charge. High voltage levels used in electric transmission systems are commonly expressed in kilovolt (1 kV = 10^3 V).

If an atom is in equilibrium, the number of negatively charged electrons and positively charged protons balance each other. However, physical phenomena such as an electromagnetic field, heat, or light may increase the energy of electrons forcing them to move from one atom to another. In solids, electrons are exchanged between atoms as their energy levels are increased. In liquids and gases, electrically charged molecules called ions can move with an electric or magnetic field effect.

Electric current is defined as the time rate of change of electric charge. In a mathematical sense, electric current is the time derivative of the function that describes the variation of electric charge.

$$i(t) = \frac{dq}{dt} \tag{2.40}$$

The standard unit of electric current is ampere (A). In energy systems high current values are expressed in kiloamps (1 kA = 10^3 A).

2.7.2 Electric Power and Energy

As we defined earlier, voltage is the work done to move unit electric charge. Since energy is the capacity to do work, voltage is energy per unit charge. Since power is energy per unit time, a general expression for electric power can be obtained by multiplying electric current and voltage as defined in Eqs. (2.39) and (2.40).

$$p = \frac{dw}{dq} \cdot \frac{dq}{dt} = v(t) \cdot i(t) \tag{2.41}$$

Figure 2.2 Reference voltage and current directions in passive convention.

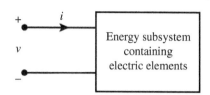

The standard international unit for electric power is the Watt (W). Metric multiples of watt such as kilowatt (1 kW = 10^3 W), megawatt (1 MW = 10^6 W), gigawatt (1 GW = 10^9 W), etc. are frequently used to signify larger amounts in electric power systems. For example, a light-bulb is rated by its power in watts, whereas a kW would be more convenient to specify the power of an electric heater. Industrial machines are typically rated in the order of kW and MW. Installed capacity of an electric generation plant is usually given in MW. Nameplate capacity of the Grand Coulee Dam power plant, which is the largest power station in the US, is 6809 MW or 6.8 GW. Mechanical output power of an electric motor is sometime specified in horsepower (hp). In electromechanical systems, 1 hp corresponds to 746 W.

In energy systems containing electrical elements energy flow may be bidirectional. Electric machines, batteries, and some solid-state elements can deliver or receive power depending on the input–output conditions. To avoid confusion, voltage across a subsystem that has electric elements is labeled with plus and minus signs to show the reference voltage direction. The reference direction of a current is indicated with an arrow as illustrated in Figure 2.2 If the voltage and current are not known, the reference directions are marked arbitrarily. Upon solving the mathematical model using the known values, direction of energy flow is determined using the algebraic signs of the voltage and current. Throughout this text, we will use the *passive convention*. In this notation if the product of voltage and current is positive, we understand that the subsystem is passive and receiving power. Otherwise, the element or subsystem is active and delivers power (Alexander and Sadiku 2009).

$$p = v \cdot i > 0 \quad \text{passive}$$
$$p = v \cdot i < 0 \quad \text{active} \tag{2.42}$$

Electric energy is the integral of electric power processed over a time interval between t_1 and t_2.

$$W = \int_{t_1}^{t_2} p(t)dt = \int_{t_1}^{t_2} v(t)i(t)dt \tag{2.43}$$

If power remains constant over a time interval, energy is simply the product of power and time. In electrical energy systems, the basic unit of energy is kWh ($3600 \cdot 10^3$ J). Electricity consumed in a house, office, or small workshop is typically measured in kWh. Larger scale energy processed in electric systems is expressed in MWh (10^3 kWh), GWh (10^6 kWh), and higher metric multiples of kWh.

Electric companies charge their customers for the energy they consume, not the installed power of their electric equipment. Therefore, no matter how much total power is installed, energy consumption will depend on how long each piece of equipment is turned on.

Example 2.1   A 100-W lightbulb is turned on for five hours every day.

(a) Find the total energy consumed over the month of January.
(b) Given the price of electricity is 0.15 \$/kWh including all taxes and fees, how much will the usage of this lightbulb cost in January?

Solution

(a) Daily energy consumption of the lightbulb is $5 \times 100 = 500$ Wh. Since January has 31 days, the total consumption will be:

$$W_{total} = 31 \times 500 = 15{,}500 \text{ Wh} = 15.5 \text{ kWh}.$$

(b) The cost of using this light bulb for five hours every day in January will be:

$$C = 0.15 \times 15.5 = \$2.30.$$

2.8   Electromechanical Energy Conversion

Conversion between mechanical and electrical energy is central in all conventional electric generation systems and production of mechanical work in all practical applications. The largest fraction of electric power is generated in thermoelectric and hydroelectric power plants using generators. Electric motors drive a wide range of equipment, tools, and appliances.

By definition, a generator is an electromechanical device that converts mechanical energy delivered by a prime mover to electrical energy. A prime mover may be a combustion engine, hydraulic turbine, wind turbine, or any system that produces mechanical work.

The opposite of a generator is an electric motor, which converts electrical input into mechanical energy to drive a mechanical load. Because electric motors are simpler and easier to control compared to combustion engines and turbines, they are preferred in industrial equipment and domestic appliances. The use of hybrid and electric vehicles in transportation is growing with advanced battery technologies that reduce the cost per stored energy while increasing the energy density.

Electromechanical energy conversion is bidirectional; that is, an electromechanical energy conversion system can operate either as a motor or generator depending on the direction of energy flow. The efficiency of an electric machine is significantly higher than a comparable heat engine. Furthermore, electromechanical conversion systems can be designed in such a way that the electric motor returns any excess energy that may arise at the load side to the source or storage element using bidirectional power electronics drives (Chapman 1991; Fitzgerald et al. 1992). For example, an electric machine that drives an elevator may be operated as a motor when the elevator goes up and as a generator when it goes down. Hybrid and electric vehicles benefit from this feature of electric machines to increase efficiency. The electric machine in an electric car operates as a motor and converts the electric energy stored in the battery to mechanical work as the car accelerates or climbs a slope. On the other hand, when the car slows down or descends from a hill, the electric machine operates as a generator to charge the battery by converting kinetic or potential energy to electrcicty.

Fundamental principles of electromechanical energy conversion are explained in Chapter 6 based on Ampere's and Faraday's laws. All rotating electric machines consist of a moving part called a rotor and a stationary part called a stator. As the shaft of the machine turns, permanent magnets or field windings mounted on either rotor or stator induce velocity voltages on coils placed on the other part. The structure of electric machines varies by the applications.

In practice, electric machines are classified as DC machines, synchronous machines, induction (asynchronous) machines, and special machines. Electric generation systems mostly rely on three-phase synchronous generators. Wind turbines generate electric power by either synchronous or asynchronous generators. The structure, operation principles, and characteristics of such machines are discussed in Chapter 6.

2.9 Photothermal Energy Conversion

Photothermal energy conversion transforms the energy of light into heat. Light is essentially an electromagnetic radiation within a certain frequency range. During the evolution of quantum physics, Albert Einstein proposed in his explanation of the photoelectric effect (1905) the existence of energy packets called photons or light quanta carried by electromagnetic radiation. Quantum physics considers light as a flux of photons that carry radiant energy proportional to the frequency of the electromagnetic wave as described in Eq. (2.44).

$$E = hf \tag{2.44}$$

In this expression, h is Planck's constant, approximately 6.63×10^{-34} Js. The propagation of light radiation is described by electromagnetic theory, while emission and absorption phenomena are explained by quantum theory.

Electromagnetic waves are usually described with the wavelength rather than the frequency. Wavelength of an electromagnetic wave is calculated by dividing the speed of light to its frequency f.

$$\lambda = \frac{c}{f} \tag{2.45}$$

Radio waves and microwaves with wavelengths in the range between hundreds of meters to millimeters transmit information at an energy level that can be produced and detected by electronic equipment. Higher frequency (or shorter wavelength) electromagnetic radiation such as X-rays, light waves, and beyond are produced by emissions from atoms, molecules, and nuclei (Kraus 1992).

Visible light emitted by an ordinary incandescent lightbulb is an electromagnetic radiation produced by electrons accelerating within a hot filament. Hot objects emit infrared (IR) radiation at a frequency just less than the frequency of visible red light. A typical human eye can see radiation between wavelengths of 380 and 740 nm (1 nm = 10^{-9} m) ranging from red to violet. A light radiation with a distinct wavelength is perceived by human eye as a particular color. Therefore, colors corresponding to the higher frequency of the spectrum, such as green, blue, or violet have higher photon energy. On the other hand, yellow, orange, and red colors have smaller photon energy (Messenger and Ventre 2004). Table 2.4 shows the characteristic values of the visible light spectrum.

Power of an electromagnetic wave per unit area is called irradiance and measured in watt per square meter (W/m^2). Solar irradiance measured on the top of atmosphere or

Table 2.4 Physical characteristics of visible light spectrum.

Color	Frequency (THz, 10^{12} Hz)	Wavelength (nm, 10^{-9} m)	Photon energy (eV, $1.602 \cdot 10^{-19}$ J)
Red	460	650	1.9
Orange	510	590	2.1
Yellow	525	570	2.2
Green	590	510	2.4
Blue	630	475	2.6
Indigo	675	445	2.8
Violet	750	400	3.1

extraterrestrial irradiance is around 1400 W/m². Solar irradiance reaching the Earth's surface is attenuated by air mass, which depends on the distance sunlight travels in the atmosphere, cloudiness, pollution, humidity, and other factors that affect clarity of the air. In general, at sea level and under clear sky, solar irradiance is approximately 900 W/m² (Markvart and Castaner 2003). This value is defined as 1 sun in solar energy conversion system. Solar modules are rated for 1 kW/m² under standard test conditions (STC).

Solar thermal panels convert solar radiation into heat. Common types of solar thermal convertors are flat plate, evacuated tube, parabolic, and concentrated solar collectors. Flat plate and evacuated tube collectors are mostly used for residential and commercial space or water heating. Concentrated solar collectors use optical elements to focus sunlight on a relatively small area to obtain high temperatures. Concentrated solar conversion systems are used for industrial applications including metal melting, steam generation, district heating, or electric generation (Kalogirou 2004). Performance and design of solar thermal energy systems will be described in more detail in Chapter 10.

2.10 Photovoltaic Energy Conversion

Photovoltaic cells convert sunlight directly into electricity without a moving part. A photovoltaic (PV) cell is essentially a semiconductor element and has many common features with diodes used in electronic circuits. Structurally, however, a PV cell is made to absorb sunlight and produce electric current. Photons entering the crystal structure of the PV cell increase the energy level of electrons creating a potential difference across the external terminals. If an external circuit is connected across the terminals, the PV cell produces a current depending on the irradiance received on its surface. The typical voltage of an illuminated PV cell when no external element is connected (open circuit) is between 0.5 and 1 V. To obtain higher voltage and current values required for practical applications, PV cells are combined in series and parallel configurations in a PV module, also called PV panel. PV cells and their interconnections to form a module are enclosed in a sturdy aluminum frame and a durable highly transparent sheet (Markvart and Castaner, eds. 2003).

Open circuit voltage of a PV module remains nearly constant for a wide range of solar irradiance, but it changes significantly with the cell temperature. On the other hand, the short circuit current is approximately proportional to the irradiance in normal direction. Solar modules are rated for Standard Test Conditions (STD) defined for 1 kW/m^2 irradiance at 25 °C ambient temperature. The voltage-current relationship of a PV module is non-linear and delivered power reaches a maximum value for a certain operating point. Electronic devices harnessing solar energy from PV arrays are designed for Maximum Power Point Tracking (MPPT) as the irradiance and temperature conditions change (Messenger and Ventre 2004).

A combination of PV modules form a PV array installed on the roof of a building, on the ground, or on poles. Output power of a PV module depends on the incident angle of the solar radiation. We will discuss the operation, performance characteristics, and design of PV generation systems in more detail in Chapter 10.

2.11 Electrochemical Energy Conversion

Electrochemical energy conversion transforms energy of bonding forces between atoms of a molecule directly into electric energy. In energy systems, electron exchanges in fuel cells, separation of elements by electrolysis, and energy storage in batteries are among the practical applications of electrochemical energy conversion.

2.11.1 Batteries

A battery consists of series and/or parallel connected cells that convert chemical energy of an electrolyte (reactant) into electric voltage. In the common terminology, a *cell* is a basic electrochemical unit that contains the electrodes, separator, and electrolyte. The term *battery*, *battery pack*, or *battery bank* signifies a group of electrically connected cells enclosed in a case.

When the electrodes of a cell are connected to an external circuit, an electric current flows through the electrolyte, changing the chemical composition of the reactant while electric energy is delivered to the external load (Andrews and Jelley 2007). In a primary cell, the reactant is consumed during the electrochemical reaction. Primary cells are designed for single use and they are discarded when all chemical energy is exhausted. Secondary cells allow reversible chemical reactions such that electrical energy supplied from an external source restores the chemical composition of the reactant for repetitive use (Hoogers 2003). Rechargeable batteries are used to store electric energy for a broad range of electric equipment from portable electronics to electric vehicles (EVs).

The first battery cell commonly called a *voltaic* or *galvanic* cell was invented by Alessandro Volta in 1799, based on Luigi Galvano's experiments showing that two different metals in contact with each other produce a potential difference that can cause reaction in the muscle of a frog leg. Volta advanced the idea by immersing two different metal electrodes (such as zinc and copper) in a saline solution. In a voltaic cell ion flow through the salt solution from anode (zinc) to cathode (copper) produces electric current (Dell and Rand 2001).

The lead-acid battery devised by French physicist Gaston Plante in 1859 is the oldest type of rechargeable cell. The battery has thin lead electrodes immersed in diluted sulfuric acid electrolyte. Because of the gas discharge during electrochemical conversion, this type is commonly known as *vented lead-acid battery*. The original design of Plante has gone through major improvements and modifications. While vented lead-acid batteries are still in use, sealed and gel batteries are more convenient and safer.

The nickel-iron (Ni-Fe) battery cell, originally developed by Thomas Edison in 1901, consists of nickel oxide-hydroxide anode and iron cathode immersed in potassium hydroxide electrolyte. Active materials are placed in nickel-plated steel tubes and the cells are robust and tolerant to over-discharge and even short circuits. Ni-Fe batteries have been preferred in applications that required higher reliability, safety, and longer life cycle. However, this type of battery has higher manufacturing cost, poor charge retention, and less energy per unit mass (J/kg) compared to other rechargeable battery types.

Nickel cadmium (Ni-Cd) and nickel metal hydride (Ni-MH) batteries are currently popular rechargeable battery types. In both cells, the positive electrode is nickel oxide hydroxide. The negative electrode of a Ni-Cd cell is cadmium, whereas Ni-MH cell uses a hydrogen-absorbing alloy.

Lithium-ion batteries (Li-ion or LIB), first developed by British chemist Stanley Whittingham in the 1970s, are rechargeable batteries commonly used in portable devices. They are becoming increasingly popular in the electric vehicle (EV) industry, military and aerospace applications, renewable energy systems, and battery banks for backup electric storage. The negative electrode of a Li-ion cell is made from carbon, typically graphite, while the positive electrode is a metal oxide. Layered lithium cobalt oxide, lithium iron phosphate, or lithium manganese oxide are among the materials used to make the positive electrode of a Li-ion cell. The electrolyte is a lithium salt dissolved in an organic solvent. Typical electrolyte solutions are ethylene or diethyl carbonate containing complexes of lithium ions. Depending on the materials used in electrodes and electrolyte, the voltage and performance parameters of a Li-ion battery change significantly.

2.11.2 Fuel Cells

The principle of the operation of fuel cells is based on oxidation of an externally supplied fuel. Whereas combustion is also an oxidation reaction, in a fuel cell the process occurs in two separate places physically isolated by an electrolyte. Supplied fuel is first oxidized on the anode by separating electrons. Electrons flow through an external circuit creating electric current. Fuel molecules lacking electrons form ions that travel through the electrolyte. As the ions reach the cathode, they react with supplied oxygen and reunite with the electrons completing their paths through the external circuit to form the oxidation products. Thus, fuel cells convert chemical energy stored in the molecules of a fuel directly into electric energy through an electrochemical energy conversion process (Hoogers 2003).

Batteries also create electric current through an electrochemical energy conversion. However, batteries consume the chemical energy of the electrolyte. The key difference between a fuel cell and a battery is that a fuel cell needs continuous supply of external fuel to sustain electric generation, whereas a battery contains a certain amount of fuel.

Several types of fuel cells have been developed to use different fuels to serve a broad range of applications from portable electronic devices to transportation vehicles and stationary power generation. Because the heat engines, electric generators, and moving parts are eliminated, fuel cells can provide higher overall efficiency and lower maintenance cost. The side products of fuel cell reactions are water and heat. Certain types of fuel cells additionally produce carbon dioxide, but it is easy to separate carbon dioxide for commercial purposes. Combined cycle systems can take advantage of the heat discharge for space, water, or process heating. Furthermore, hydrogen fuel cells allow bidirectional energy flow. Such fuel cells generate electricity when hydrogen is supplied, and they can produce hydrogen when they are supplied electricity and water. This feature is a great benefit, particularly in wind and solar powered micro grids. Fuel cells can produce hydrogen when excess power is being generated and deliver electric power from stored hydrogen when wind or solar energy is insufficient or unavailable.

A fuel cell has an electrolyte membrane inserted between one positive electrode (anode) and one negative electrode (cathode). The simplest fuel is hydrogen, but methane or other hydrocarbon gases may be used depending on the fuel cell type. The fuel is fed to the anode, and air is fed to the cathode. In an electrolyte membrane, electrons and protons or positive ions of fuel molecules are separated. Electrons flow through an external circuit, creating electric current. Protons, on the other hand, migrate through the electrolyte to the cathode, where they reunite with electrons. An oxidation occurs at the anode, while a reduction occurs at the cathode. Oxidation of the fuel produces water if the fuel is simply hydrogen. Hydrocarbon fuels produce carbon dioxide in addition to water. The chemical reaction of the fuel with oxygen or an oxidizing agent releases a significant amount of heat, which is either discharged to the surrounding area as loss or used for heating purposes (O'Hayre et al. 2016).

While the structure and basic operation principles are the same for all fuel cells, several electrode and electrolyte materials have been developed to use different fuels and serve specific purposes.

Phosphoric acid fuel cells (PAFC), developed in the 1960s, have an electrolyte of highly concentrated liquid phosphoric acid (H_3PO_4) saturated in a silicon carbide matrix (SiC). The electrodes are made of carbon paper coated with a fine platinum catalyst layer. Typical operating temperature of a PAFC is around 200 °C. A broad range of hydrocarbons can fuel this type of fuel cell. Since the operating temperature is above the boiling temperature of water, steam is discharged as a side-product. If a PAFC is used in a CHP system, the overall efficiency can be as high as 70%. PAFCs are used to power stationary power generation systems between 100 and 400 kW and large transportation vehicles.

Alkaline fuel cells (AFCs), also developed in the 1960s, use a porous material saturated with an alkaline solution such as potassium hydroxide (KOH) as the electrolyte, which transfers hydroxide ions rather than protons. Their efficiency can reach up to 70%. However, they require pure oxygen or clean air because carbon dioxide forms a deposit of potassium carbonate on the cathode, which must be cleaned by a scrubber. The need of oxygen storage increases the operation cost and limits the application field of AFC. NASA has used AFCs in Apollo series space missions and the Space Shuttle since the mid-1960s.

Polymer electrolyte membrane fuel cells, also known as proton exchange membrane fuel cells (PEM), use hydrogen as fuel. A special polymer membrane electrolyte separates the protons of hydrogen atoms. The freed electron flows through the external circuit and reaches the cathode. Hydrogen molecules regaining one electron at the cathode, combined with the oxygen molecules of the supplied air, produce water, and the reaction releases heat that must be removed out from the cell. PEM fuel cells can operate at relatively lower temperatures range from normal ambient temperature to 100 °C.

PEM fuel cells are suitable for diverse applications including stationary power production and hydrogen powered electric vehicles, since the only side-product is pure water and their output can quickly vary to meet shifting power demands. However, hydrogen has lower energy density compared to hydrocarbons, such as methane or ethanol, and methanol, which can be used if an external fuel reformer is added to the PEM fuel cell system.

Energy flow in PEM fuel cells can be reversed such that when powered with electricity and supplied pure water, they can produce hydrogen and oxygen. This feature allows storage of excess energy produced by intermittent renewable energy sources, such as wind and solar power stations in the form of hydrogen. When solar or wind generation is insufficient, stored hydrogen can be used to generate electricity in the same fuel cell.

A subcategory of PEM fuel cells known as direct-methanol fuel cells (DMFC) uses a solution of methanol as fuel. The electrolyte of DMFC is a proton conducting polymer membrane similar to PEM. Methanol is directly fed to the anode, and upon oxidation on a catalyst layer forms carbon dioxide, while protons are transferred through the membrane and electrons circulate through the external circuit producing electric current. During the redox reaction, water is consumed on the anode and reproduced on the cathode. The main benefit of DMFC is the use of liquid fuel with higher energy density. However, it has a low efficiency due to phenomena called cross-over current at higher methanol concentration, which may offset the benefit of high-energy density of methanol. DMFC is still a potential alternative energy source for mobile devices such as laptop computers and portable battery rechargers.

Molten carbonate fuel cells (MCFCs) use a porous electrolyte saturated with molten carbonate salt as the electrolyte. They operate on natural gas and biogas with high efficiency. Their higher operation temperature (approximately 600 °C) enables reforming their fuel.

Solid oxide fuel cells (SOFCs) use a thin layer of ceramic as a solid electrolyte that conducts oxide ions. They are being developed for use in a variety of stationary power applications, as well as in auxiliary power devices for heavy-duty trucks. Operating at 700–1000 °C with zirconia-based electrolytes, and as low as 500 °C with ceria-based electrolytes, these fuel cells can internally reform natural gas and biogas and can be combined with a gas turbine to produce electrical efficiencies as high as 75%.

Elements and properties of major fuel cell types are compared in Table 2.5. In addition to electricity, fuel cells produce heat. This heat can be recycled in a CHP configuration for practical uses including hot water and space heating in houses and buildings. Total efficiency of a CHP fuel cell system can reach up to 90% saving energy while reducing greenhouse gas emissions (Hoogers 2003).

Table 2.5 Comparison of major fuel cell types.

	PEMFC	DMFC	SOFC	PAFC	AFC	MCFC
Fuel	H_2	Methanol	H_2, CH_4, CO	H_2	H_2	H_2, CH_4
Change carrier	H^+	H^+	O^{2-}	H	OH^-	$CO_2{}^{2-}$
Product	H_2O	$H_2O + CO_2$	$H_2O + CO_2$	H_2O	H_2O	$H_2O + CO_2$
Electrolyte	Polymer membrane		Ceramic	Phosphoric acid	Liquid KOH	Molten carbonate
Catalyst	Platinum		Perovskites	Platinum	Platinum	Nickel
Op. temp.	50–100 °C		600–1000 °C	~200 °C	60–120 °C	~650 °C
Efficiency	35–45%		>50%	~40%	35–55%	>50%
Power	<250 kW		2 kW– megawatts	40%	<5 kW	200 kW– megawatts
Applications	Portable, automotive, stationary, CHP		Stationary CHP	Stationary CHP	Military, space	Stationary CHP

Source: Hoogers (2003).

2.12 Energy Storage

Many functions of a modern society depend on reliable and uninterrupted energy source available on demand. Today industrial production, commerce, banking, financial transactions, public services, hospital services, and many other vital supports rely on continuous electric supply. Electric generation facilities must have an adequate stock of fuel or some type of energy storage to continue delivering electric power while the load changes in a wide range. Moreover, increasing energy security concerns have enforced consumers to maintain sufficient backup energy storage for uninterruptable supply of critical loads when the grid electricity becomes unavailable.

Mobile applications rely on portable energy sources. Cars, buses, trucks, boats must carry sufficient energy source to travel a certain range. The Amount of energy per unit storage volume or mass is a dominant factor for the type of energy source used in vehicles. Mobile communication devices are also the preferred means to access information, social networking, scheduling, entertainment, and countless other applications. Mobile electronics need lightweight and high capacity battery storage.

Energy storage contributes to the main goal of an energy system in providing the society with continuous and reliable energy supply. The major benefits of using energy storage systems are summarized below.

Efficient use of renewable energy: Cost-efficient energy storage systems allow the effective use of intermittent renewable sources such as solar and wind energy, which display random variations and discontinuities in time.

Reducing energy costs: The unit cost of electricity decreases if the electric power plants operate near their rated power at all times. However, the energy generation must be always adjusted to meet the consumption. Capacity factor is defined as the ratio of the total energy generated over a time interval (usually a year) to the energy that the facility would produce continuously at its rated power. Using suitable energy storage to supply the consumers at peak demand times and restoring the consumed energy during off-peak hours reduces the unit cost of the delivered energy.

Increasing system reliability: Reliability is defined as the ability of an energy system to supply the users continuously. Energy storage systems can increase the reliability by reducing the risk of power interruptions.

Increasing thermal efficiency of buildings: Well-insulated buildings that can store energy received from sunlight during the day need less heating at night. Government incentives and building certification programs such as LEED and Passive House Certification (PHI) encourage communities to use high quality insulation materials to store energy in the buildings.

Energy can be stored in various forms. The amount and type of stored energy varies by the specific requirements and limitations of each application. Space and weight constraints, cost effectiveness, economic life, safety, and environmental interactions are among the main considerations in selection of the type of storage for a certain application. Energy storage may involve converting energy from forms difficult to store directly into other forms that can be captured in a physical enclosure more conveniently and economically.

2.12.1 Fuel Storage

A simple way to store energy is keeping a stock of some fuel. Fuel storage is essentially chemical energy storage since fuels are hydrocarbons that contain chemical energy in their molecular structures. Fuels are converted to heat by combustion. Therefore, they must be burned to extract their energy content. Liquid fuels are easier and more efficient to store and burn than solid fuels. Gas fuels are easiest and most efficient fuels, but they need to be pressurized or liquefied to increase the energy density.

The type of fuel to store depends on whether the end use is stationary or mobile. Electric or industrial power plants, buildings, and agriculture are examples of stationary end users.

Fossil fuel fired electric power stations and industrial plants usually maintain a certain amount of fuel to continue their operation when there are unexpected delays or disruptions in fuel delivery.

Coal fired power stations store fuel reserves in a large pile near the power plant. A typical coal pile contains several hundreds of thousands of tons of fuel storage. As received at the power station, coal is spread in thin layers and packed down with large earth moving machines to get as much air out of the pile as possible to reduce the risk of fire by spontaneous combustion. Some small particles of crushed coal may be washed away during rainstorms, causing a loss in the reserve and pollution of the watershed.

Fuel oil and gas-burning power plants are supplied by pipelines. Since such stations can adjust their output easier to follow hourly variations of the electricity demand, they maintain a certain amount of fuel in large tanks as short-term storage. Underground tanks and caves are convenient for long-term natural gas storage in large quantities.

The transportation sector depends on fuel that can be efficiently stored on vehicles. This is why higher energy intensity fuels such as gasoline, diesel, and jet fuel are preferred in conventional vehicles. Portable storage carried on the vehicles is a typical short-term energy storage that must allow a high-energy exchange rate when fueling the vehicle and supplying fuel to the engine. Some public transportation buses and even taxi cabs are fueled by propane that can be carried in pressurized liquid tanks on the vehicle.

Example 2.2 A car engine produces an average power of 100 HP when the vehicle moves at a speed of 60 miles per hour (mph). Calculate the volume of the fuel tank that would allow this vehicle to cover a 300-mi distance with one tank-load of gasoline with $32{,}317\,MJ/m^3$ energy intensity under the described average operating conditions.

Solution
At the speed of 60 mph, the vehicle will travel 300 mi in $t = 300/60 = 5$ hours. Energy consumed by 100 HP engine in 5 hours is

$$W = 5 \times (3600 \text{ s}) \times 100 \times (746 \text{ }^W/_{HP}) \times 10^{-6} = 1{,}342.8 \text{ MJ}$$

Energy intensity of the fuel is given as $32{,}317\,MJ/m^3$. Therefore, the amount of fuel needed to cover a 300-mi distance is

$$\frac{W}{W_D} = \left(\frac{1{,}342.8}{32{,}317} \text{ m}^3\right) \times (220 \text{ }^{gal}/_{m^3}) = 9.14 \text{ gal} \therefore$$

2.12.2 Potential Energy Storage

Energy can be stored efficiently as potential energy of a mass placed at a certain height above the point of use. As the mass is lowered, the potential energy can be converted into another form such as kinetic energy, to turn the shaft of a mechanical load or electric generator. The potential energy of water has been harnessed for centuries to drive grain-grinding mills or sawmills. Hydroelectric power stations convert the potential energy of water stored in a dam into electric energy.

Suppose that the volume of the reservoir of a hydroelectric power plant is Vm^3 and the average height of the reservoir is h meters above the turbine-generator modules. If the density of water in the reservoir is ρ, then the amount of the stored potential energy would be:

$$W_p = \rho Vgh \tag{2.46}$$

Pumped-storage hydroelectric power stations (PHP) use a pond to store energy. At times when the electric energy demand is lower, the water is pumped up to a higher pond. At times when the load is higher on the electric grid, the water stored in the pond flows through the turbines to generate electricity.

Example 2.3 Muddy Run pumped hydroelectric energy storage facility, located in Pennsylvania, has a usable storage capacity of 41.5 million cubic meters, and the average elevation of its upper reservoir is about 125 m above the turbine intake. What is the maximum storage capacity of the facility in GWh?

Solution

Assuming that the specific mass of water in the reservoir is $1 \, t/m^3$, total potential energy of the total usable water is

$$W_P = mgh = (41.5 \cdot 10^6) \times 10^3 \times 9.81 \times 125 = 5.09 \cdot 10^{13} \, J.$$

Since $1 \, J = 1 \, Ws$ and $1 \, kWh = 3600 \, Ws$, the amount of stored energy is equal to

$$W_P = \frac{5.09 \cdot 10^{13}}{3{,}600 \cdot 10^3} = 14.1 \cdot 10^6 \, kWh = 14.1 \, GWh \; \therefore.$$

Hydroelectric power generation in Niagara Falls uses large ponds to store potential energy of water. Because of the recreational and touristic importance of the waterfalls, the Niagara River is diverted to store energy in two artificial ponds that were constructed on both sides of the river in Canada and the US. The water coming from the Niagara River is pumped to the ponds at night after tourist attractions close, hence the amount of water flowing in the falls decreases. The potential energy of water stored in the ponds is used to generate electricity during the daytime, when there is higher demand for electric power.

Transportation vehicles convert the energy of fuel into potential energy as they climb to a higher elevation. Conventional vehicles powered by a fuel have no means to convert this potential energy back to a useful work. Thus, when the vehicle is going down a hill, the potential energy is converted into heat on the brakes and dissipated to the air. Hybrid or electric vehicles are designed to convert the potential energy change into electric energy and store it in batteries by regenerative braking while the vehicle moves to a lower elevation.

2.12.3 Kinetic Energy Storage

Speed variations in any mechanical system result in change of kinetic energy. When the speed increases, moving parts store kinetic energy, which can be converted to another form as the speed decreases. In linear motion, kinetic energy is proportional to mass and square of the linear velocity as described in Eq. (2.29).

A vehicle stores kinetic energy as it accelerates. To slow down on a horizontal road, the kinetic energy must be converted to another form. In vehicles powered by a conventional engine, brakes convert kinetic energy into heat by friction and heat is dissipated to the surrounding air as unused energy. Hybrid vehicles and electric cars convert kinetic energy into electric energy by regenerative braking and store it in a battery to use as needed.

Flywheels are mechanical elements commonly used to exchange rotational kinetic energy. A heavy wheel with a big moment of inertia is often attached to the shaft of a mechanical system to improve stability and provide power during fluctuations or short interruptions of the driving torque. Energy exchange can be as fast as a fraction of a revolution or the stored kinetic energy can support the rotation over a longer time.

Flywheel technology is emerging as a short-time energy storage for spacecrafts, the automotive industry, railroad locomotives, and uninterruptible power supply systems. A flywheel is also an option to reduce the speed variations of a wind turbine if the wind speed changes too fast, hence regulating the output voltage.

Table 2.6 Energy intensity of practical flywheel materials (Shape factor $k = 1$).

Material	Density ρ (kg/m^3)	Tensile stress σ (MPa)	Specific energy σ/ρ (kJ/kg)
Kevlar fiber – 50% epoxy	1400	1000	714.3
Composite carbon fiber – 40% epoxy	1550	750	483.9
Aluminum	2700	450	166.7
Titanium alloy	4500	650	144.4
High strength steel alloy	8000	900	112.5
Steel	7900	585	74.1
Cast iron	7150	140	19.5

Source: Engineering Toolbox (available online – last accessed on 3/5/2019, https://www
.engineeringtoolbox.com).

Kinetic energy stored by a flywheel is determined using the Eq. (2.30), repeated below for both angular velocity ω and rotation per minute n.

$$W_{fw} = \frac{1}{2}J\omega^2 = \frac{1}{2}J\left(\frac{2\pi n}{60}\right)^2 \tag{2.47}$$

Moment of inertia is proportional to the mass and square of the radius of the flywheel.

$$J = kmr^2 \tag{2.48}$$

Advanced flywheel storage systems have rotors spinning at speeds between 20,000 and 50,000 rpm. The shape factor k depends on the geometry of the rotor. A toroidal flywheel, where all the mass is evenly distributed on the periphery like a bicycle wheel, has the largest shape factor equal to one. For a flat solid disk of uniform thickness, the shape factor is 0.606. Other geometric forms have smaller shape factors.

The spinning speed of a flywheel is limited by the tensile strength of the rotor material. Maximal specific energy is defined as the ratio of the maximum kinetic energy of the rotor to the moment of inertia and can be obtained from Eq. (2.49), where σ represents the tensile strength and ρ is the density of the material.

$$W_{max} = \frac{W}{J} = k\left(\frac{\sigma}{\rho}\right) \tag{2.49}$$

Specific energy of selected flywheel materials is listed in Table 2.6 for shape factor $k = 1$. In the list, cast iron has the lowest specific energy. Flywheels used in conventional mechanical systems are made from steel, which provides higher specific energy. Advanced composite materials such as carbon fiber and Kevlar fiber mixed with epoxy are used in high-speed flywheels to store significant amounts of energy. Friction is the main source of loss in flywheel storage systems. Magnetic bearings may be used to minimize the friction losses.

2.12.4 Thermal Energy Storage

Energy can be stored by changing the temperature and/or phase of a thermally insulated substance. Insulated water tanks, steam, gases, chemical compounds, massive concrete

Table 2.7 Energy storage properties of selected substances.

Substance	Density at 25 °C (if not specified) (kg/m^3)	Specific heat (kJ/kg·°C)	Latent heat (melting) (kJ/kg·°C)
Ammonia (gas)	0.717[a)]	2.060	
Dry air (sea level)	1.205[a)]	1.005	
Steam (107 °C)	0.804[a)]	1.852	
Pure water (liquid, at 20°C)	1000[b)]	4.182	334
Ammonia	823.5	4.700	339
Ethyl alcohol	785.1	2.440	108
Paraffin	800	3.260	147
Salt (NaCl)		0.880	
Snow, ice	100–917	2.090	
Timber (oak)	740	2.400	
Timber (pine)	420–670	1.500	
Concrete	1300–2400	0.880	
Brick	1400–2400	0.840	
Sand, quartz	1400–1600	0.830	
Dry soil	190	0.800	
Aluminum	2700	0.897	
Steel	7820	0.490	

a) At 20 °C, 1 atm.
b) At 4 °C.
Source: Engineering Toolbox (www.engineeringtoolbox.com) accessed on 8/20/2019.

blocks, and rocks are typical thermal energy storage elements. Energy stored in a substance if mass m and specific heat c is described by Eq. (2.50).

$$Q = mc\Delta T \tag{2.50}$$

Specific heat and melting latent heat of selected substances relevant to thermal energy storage are listed in Table 2.7. Transformations from solid to liquid phase store significantly larger amounts of thermal energy than changing the sensible heat. Liquids have higher specific heat than gases and solids. Although the specific heat of liquid ammonia and other liquids are listed in the table, water is more available and convenient for energy storage. The problem with water, however, is the boiling temperature of 100 °C at normal atmospheric pressure. To increase the energy storage capacity for a certain volume, pressurized tanks can be used. At higher temperatures, liquid metals such as sodium or salt may be used as storage or heat transfer substances instead of substances that are in liquid phase at normal pressure (Dincer and Rosen 2011). For example, the Crescent Dunes concentrated solar power plant in Nevada uses liquefied salt to store solar energy during the daytime and generates electricity 24 hours a day by conventional steam turbines (see Chapter 10).

Thick building walls insulated from external temperature changes store heat during warmer hours of the day and return this heat back to heat interior spaces. Buildings heated by circulating hot water in radiators benefit from the thermal energy stored in a thermally insulated water tank. The tank size and water temperature determine the amount of energy that can be stored. Thermal energy storage is an essential part of solar space heating systems. Since solar energy is available only during a limited time of the day, a hot water tank or another thermal storage is needed to supply heat day around.

2.12.5 Compressed Air Storage

Compressed air is frequently used in industrial plants and commercial machine shops as an efficient energy storage for power tools. A compressor powered by an electric motor or gas engine fills a tank with pressurized air. The main advantage of a compressed air tanks is its being portable and small for the power delivered to lightweight hand tools.

Compressed air storage can also be used in a larger scale to regulate the power drawn from an electric grid or as a backup energy source. In such systems, air is compressed during off-peak hours and stored in large underground reservoirs. During peak hours or power interruption, stored air is released to drive a gas turbine coupled with an electric generator.

Compressed air storage is safe and environment friendly. Old oil or natural gas wells can be used as reservoirs, as well as underground structures, natural caverns, salt domes, or abandoned mine shafts. The weight of the soil above the reservoir balances the considerable pressure of the air. Significant amount of energy can be stored in large underground cavities to supply utility scale facilities. Compressed air storage can be integrated in conventional power stations to regulate power output and increase capacity factor. During off-peak hours the gas turbines compress ambient air with excess power it produces. During peak hours, the released air contributes to the output power (Dincer and Rosen 2011).

2.12.6 Hydrogen for Energy Storage

Hydrogen is a convenient material for energy storage and transfer because it is low density and does not produce any gas emissions when burned. Hydrogen is not a primary energy source since it is not available in large quantities in nature as a gas. However, it can be easily produced by electrolysis of water. Therefore, it can be used as an intermediate energy source and carrier.

Hydrogen has high energy value per unit mass of 116,300 kJ/kg, which is higher than jet fuel (46,520 kJ/kg) for example. However, its energy content per unit volume is lower than hydrocarbons. The volumetric energy density of hydrogen is 20,900 MJ/m^3, while gasoline has energy density of 34,840 MJ/m^3.

Hydrogen can be produced by simple electrolysis; however, PEM fuel cells offer the benefit of bidirectional energy flow. PEM fuel cells can produce hydrogen and oxygen from pure water when an electric potential is applied to its electrodes. On the other hand, it generates electric power by combining hydrogen and oxygen. The only byproducts of a fuel cell operation are pure water and oxygen, which makes hydrogen storage cleaner than batteries and fuels.

Hydrogen can be stored in a pressurized tank, metal hydride container, or in liquid form. Liquefaction of hydrogen is costly and pressurized tanks have a large volume per stored energy. Metal hydride containers are emerging as convenient and safe hydrogen storage at a relatively lower cost and volume per stored energy.

In practice, the major challenges of hydrogen as an energy storage medium are low energy efficiency of its production and its high storage volume.

2.12.7 Electrical Energy Storage

Direct storage of electric energy in large quantities is difficult and expensive. It is usually converted into mechanical, chemical, or thermal energy for cost-effective storage. Pumping water to a reservoir at a higher altitude using electric motors and generating electricity as needed by turning turbine-generator systems is indirect storage of electricity in the form of potential energy. Alternatively, electricity can be stored in a flywheel in the form of kinetic energy. Batteries store electric energy in chemical form. Hydrogen produced using a PEM fuel cell can be stored in a tank and reused to generate electricity as needed.

Electrochemical batteries are the most straightforward elements to store electric energy. However, they suffer from the following drawbacks compared to other storage systems (Table 2.8):

- They are heavier for the amount of energy they can store.
- Energy delivered in unit time (power) is smaller per unit mass.
- They can withstand a limited number of charge/discharge cycles.
- Their usable life significantly decreases when they are substantially depleted.
  If the battery is not used for a long time, the stored energy decays due to the chemical reactions.

Batteries used for storage purposes in energy systems have different characteristics than batteries used to start a vehicle engine. Vehicle batteries are charged over a long time period and deliver large power only until the engine starts. Storage batteries used in energy systems charge and discharge more frequently and must deliver steady power for a relatively long time, and must allow deeper discharge ratio. Some deep-cycle batteries allow a discharge level below 25%.

Table 2.8 Dominant features of selected battery types.

	Specific energy (Wh/kg)	Energy density (Wh/dm³)	Specific power (W/kg)	Durability (cycles)
Lead acid	33–42	60–110	180	500–800
Nickel cadmium	40–60	50–150	150	2000
Nickel metal hydride	60–120	140–300	250–1000	500–2000
Nickel zinc	100	280	>3000	400–1000
Lithium ion	100–265	250–620	~250 to ~340	400–1200

2.12.8 Properties of Energy Storage Systems

Several factors distinguish the suitability of energy storage systems for practical applications. Energy storage may be needed for a short time or long time. Short-time storage must be quickly accessible. That means sufficient amounts of energy should be extracted in a short time to supply the end use, and the depleted energy should be restored quickly enough. Since energy per unit time is power, short-time storage must exchange sufficient power with the supply and load. In long-term storage, the energy exchange rate is relatively less critical. On the other hand, the stored energy may decay in time due to leakages, evaporation, heat losses, or chemical decomposition. Then, long-term storage systems must be designed such that deterioration of the stored energy is minimal.

Table 2.9 Storage capacities of various technologies.

Storage technology	Notes	Typical energy densities	
		(kJ/kg)	(MJ/m³)
Thermal energy, low temperature	Water, temperature difference 100 to 40 °C	250	250
	Stone or rock, temperature difference 100 to 40 °C	40–50	100–150
	Iron, temperature difference 100 to 40 °C	30	230
Thermal energy, high temperature	Stone or rock, temperature difference 400 to 200 °C	160	430
	Iron, temperature difference 400 to 200 °C	100	800
Conventional fuels energy	Crude oil	42,000	37,000
	Coal	32,000	42,000
	Dry wood	12,500–20,000	10,000–16,000
Synthetic fuels energy	Hydrogen, gas	120,000–142,000	10
	Hydrogen, liquid	120,000–142,000	8700
	Methanol	21,000	17,000
	Ethanol	28,000	22,000
Electrochemical energy	Lead-acid batteries	40–140	100–900
	Nickel-cadmium batteries	350	350
	Lithium ion batteries	700	1400
Mechanical energy	Hydropower, 100 m head	1	1
	Compressed air		15
	Flywheel, steel	30–120	240–950
	Flywheel, composite materials	>200	>100

Source: Engineering Toolbox (www.engineeringtoolbox.com) accessed on 8/20/2019.

Energy storage density is the amount of energy captured in unit volume. Its metric unit is Joule (or multiples of Joule) per cubic meter (J/m^3). Energy per unit mass is specific energy measured in Joule per kilogram. Energy that can be delivered per second is described by power density (W/m^3). Maximum power density, which is the power per unit volume that can be drawn at the maximum possible discharge rate is an important specification of a storage element or system. Specific energy refers to the energy stored per unit mass (J/kg). Similarly, specific power is the energy that can be extracted per second from unit mass of storage.

Since stored energy decays in time due to losses, the duration storage can hold sufficient energy is another important feature. Any type of storage wears out every time it is used, therefore has a certain useful life specified in either number of charge-discharge cycles or years.

Various storage technologies are compared in Table 2.9. The estimated values are only to give an idea for comparison purpose, and they change with the development of new materials and technologies.

2.13 Chapter Review

This chapter is an overview of essential energy conversion and storage methods. Techniques used to convert primary energy sources to the forms of energy deliverable to end users in an energy system are emphasized.

Energy is the capacity to do vigorous work or cause changes in the state of a physical system. Work is the product of force and the distance traveled. The unit of energy and work are the same. In the international unit system, both are expressed in newton meter (Nm) or equivalently in Joule (J). Power is energy converted in unit time. The standard unit of power is watt (W) in the metric system and horsepower (HP) in the imperial system. In energy systems usually metric multiples of watt are used to express large amounts of power.

Conservation of mass, momentum, and energy are three essential conservation laws of classical mechanics. Quantum physics established equivalence mass end energy described by Einstein's fundamental equation $E = mc^2$.

Principal forms of energy are thermal, mechanical, chemical, radiant, nuclear, and electrical. In principle, all forms of energy can be converted into another form through a suitable energy conversion process.

In energy systems heat is the primary form of energy associated with temperature differences. Adding or withdrawing heat changes the temperature of a matter; causes transition between solid, liquid, or gas phases; or produces mechanical work due to temperature differences. Heat capacity per unit mass is specific heat; the amount of heat required to cause the unit temperature change of unit mass of a substance. Heat needed to change the temperature of a substance with specific heat c and mass m by ΔT is given by the expression $Q = mc\Delta T$. Heat can be transferred by conduction, convection, or radiation.

Heat is produced by combustion, nuclear reactions, electromagnetic radiation, friction, and flow of electric current through a resistance. Combustion is an accelerated and controlled oxidation process. The largest amount of the world's energy is produced by combustion of hydrocarbon fuels that contain high amounts of chemical energy.

All practical energy conversion processes release heat. Unused heat released to the surrounding space is called *loss*. The efficiency of a conversion process is the ratio of the output energy (or power) to the input energy (or power) expressed in the same units. The difference between the input and output energies is the total loss.

Mechanical energy is associated with the position and velocity of an object. The expression of gravitational potential energy is $E_p = mgh$, where m is the mass of an object, g represents the gravitational acceleration 9.81, and h is the vertical distance measured from a horizontal reference plane. Elastic potential energy depends on the relative position of objects or particles held together by a force proportional to their distance. Kinetic energy is related to the velocity of an object by the equation $E_k = \frac{1}{2}mv^2$.

Mechanical power in linear motion is the product of force and velocity. In rotational motion, mechanical power is obtained by multiplying the torque with angular velocity in radians per second.

In incompressible fluids, the balance of mechanical energy is described by the Bernouilli's equation $P/\rho + \frac{1}{2}u^2 + gh = $ constant.

Electrical energy is based on the motion of charged particles. Electric current is defined as the time rate of change of electric charge. Electric voltage is the work done by moving electric charge. Electric power is the product of voltage and current. Electric energy can be obtained directly or indirectly from other forms of energy. It can be also converted to practically all other forms of energy by using various conversion methods. Electromechanical energy conversion is the most widespread conversion technique to generate electricity or produce mechanical work. Electric machines allow bidirectional energy flow between mechanical and electrical energy. If the input is mechanical and the output is electrical, the machine is called an electric generator. If the input is electrical and output is mechanical, the machine is called an electric motor. Electric machines are an important part of the energy system because they are efficient, easy to control, safe, and clean.

Batteries and fuel cells convert chemical energy directly into electric energy through electrochemical energy conversion. Chemical energy is associated with the forces that hold atoms of a substance together. Batteries can convert chemical energy into electric energy. Some types of batteries can perform bidirectional energy flow between chemical and electric energy. Fuel cells convert electric energy into chemical or vice versa by separating atoms of a molecule or combining atoms to form a molecule through chemical reactions.

Solar thermal collectors convert solar radiation into heat. Common types of solar thermal convertors are flat plate, evacuated tube, parabolic, and concentrated solar collectors.

Photovoltaic (PV) cells are solid-state elements that convert solar radiation directly into electric energy. PV cells are essentially a junction of P and N type semiconductors similar to a semiconductor diode but with a large surface exposed to sunlight. Silicon crystal, amorphous silicon, gallium arsenide, cadmium sulfide, and cadmium telluride are among semiconductor materials commonly used in PV cell fabrication. PV cells are combined in PV modules and PV modules are installed in series and parallel connections to form a solar array. With decreasing cost of PV modules per watt, the installed power of PV generation systems has significantly increased during the last decades.

Energy can be stored in various forms. Fuel stocks and water storage in a higher tank or pond and thermal energy storage in insulated mass are the simplest methods. Direct storage of large amounts of electric energy is more complex and expensive. Electric energy is generally stored by converting to other forms such as potential, kinetic, and chemical energy. Pumped hydroelectric storage is cost-effective for utility scale energy storage. Batteries are convenient for mobile applications and backup power. Flywheels and hydrogen technologies are emerging for electric storage.

Review Quiz

1 In the International Unit System (SI), the unit of energy is
   a. watt.
   b. joule.
   c. horsepower.
   d. Btu.

2 In the international unit system (SI) the unit of power is
   a. watt.
   b. joule.
   c. horsepower.
   d. Btu.

3 Combustion converts
   a. mechanical energy into heat.
   b. heat into chemical energy.
   c. chemical energy into heat.
   d. heat into electric energy.

4 Two objects are in thermal equilibrium when
   a. they have the same amount of energy.
   b. they are at the same electric potential.
   c. they are at the same height from a reference plane.
   d. they are at the same temperature.

5 Heat is
   a. power needed to change the state of a matter.
   b. an energy interaction associated with a temperature difference.
   c. a measure of hotness or coolness of an object or environment.
   d. a primary energy source.

6 Which one below is an example of kinetic energy storage?
   a. A flywheel.
   b. Water stored in a pond.
   c. An elevator.
   d. A car climbing a hill.

7 Electric power is
   a. the product of voltage and current.
   b. the product of resistance and current.
   c. the ratio of voltage to current.
   d. the product of voltage, current, and elapsed time.

8 Mechanical power is obtained by
   a. multiplying the mass and velocity.
   b. multiplying the mass and the square of velocity.
   c. multiplying the force and velocity.
   d. multiplying the force and mass.

9 Generators convert
   a. mechanical energy into electric energy.
   b. electric energy into mechanical energy.
   c. thermal energy into mechanical work.
   d. thermal energy into electric energy.

10 Which energy conversion process below is irreversible?
   a. Electromechanical
   b. Piezoelectric
   c. Thermoelectric
   d. Combustion

Answers: 1-b, 2-a, 3-c, 4-d, 5-b, 6-a, 7-a, 8-c, 9-a, 10-d.

Research Topics and Problems

Research and Discussion Topics

1 Research recent advancements and current technologic challenges in Solid Oxide Fuel Cell (SOFC) technology.

2 Research recent advancements and current technologic challenges in Polymer Membrane Fuel Cell (PEMFC) technology.

3 Draft a technical report on fluidized bed combustion in coal fired electric power plants.

4 What are the benefits and drawbacks of using electromechanical energy conversion in transportation vehicles?

5 What kind of energy conversion techniques are used to harness the energy of oceans?

Problems

1   A domestic electric water heater has a 50-gal tank. How long would it take to heat 20 °C water in the tank to 45 °C using 2 kW heating element?

2   A backup generator consumes 0.4-gal gasoline per hour when it supplies a 3.5-kW load. Determine its efficiency assuming that the calorific value of gasoline is 116,090 Btu/gal.

3   A turbine receives 1150 MJ heat from steam to deliver 450 MJ mechanical work. The steam exiting the turbine is cooled by flowing water. Calculate the flow rate of water such that its temperature rises 20 °C above the ambient temperature.

4   An incline weighing 5 tons with its passengers travels uphill 2 km on a 30 degree slope. Find the potential energy it gains in kWh.

5   A coal burning power plant converts 10,000 Btu heat to generate 1 kWh electricity. Given the efficiency of the generator 90%, determine the efficiency of the turbine.

6   A 12-V 100 Amp-hour battery weighs 27 kg and has the dimensions: 22-cm, L: 30-cm, H: 17-cm. Determine its energy density and specific energy.

Recommended Web Sites

- American Journal of Physics: https://aapt.scitation.org/journal/ajp
- Engineering Toolbox: https://www.engineeringtoolbox.com
- NIST Chemistry Webbook: https://webbook.nist.gov/chemistry
- Scientific American: https://www.scientificamerican.com

References

Alexander, C. and Sadiku, M.N.O. (2009). *Fundamentals of Electric Circuits*, 4e. New York, NY: McGraw-Hill.

Andrews, J. and Jelley, N. (2007). *Energy Science*. Oxford, UK: Oxford University Press.

Cengel, A.Y. and Boles, A.M. (2006). *Thermodynamics – An Engineering Approach*, 5e. New York, NY: McGraw-Hill Higher Education.

Cengel, Y. and Cimbala, J.M. (2014). *Fluid Mechanics*. New York: McGraw-Hill.

Chapman, S.J. (1991). *Electric Machinery Fundamentals*, 2e. New York: McGraw-Hill.

Culp, A.W. Jr., (1979). *Principles of Energy Conversion. s.l.* McGraw-Hill.

Dell, R.M. and Rand, D.A.J. (2001). *Understanding Batteries*. Cambridge, UK: Royal Society of Chemistry.

Dincer, I. and Rosen, M. (2011). *Thermal Energy Storage: Systems and Applications*, 2e. West Sussex, UK: Wiley.

EIA (2017). *Monthly Energy Review*. Washington DC: DOE.

EPA, 2018. Source Energy, Technical Reference. [Online] Available at: https://portfoliomanager
.energystar.gov/pdf/reference/Source%20Energy.pdf [Accessed 5 May 2019].

Fenn, B.J. (1982). *Engines, Energy, and Entropy*. San Francisco: W. H. Freeman and Company.

Fitzgerald, A.E., Jingsley, C.J., and Umans, S.D. (1992). *Electric Machinery*. New York:
McGraw-Hill.

Giancoli, D.C. (2014). *Physics: Principles with Applications*, 7e. Upper Saddle River, NJ: Pearson
Education.

Hoogers, G. (ed.) (2003). *Fuel Cell Technology Handbook*. Boca Raton, FL: CRC.

IEA (2018). *Key World Energy Statistics*. Paris: International Energy Agency.

Kalogirou, S.A. (2004). Solar thermal collectors and applications. *Progress in Energy and
Combustion Science* vol. 30, issue 3. 231–295.

Kraus, J.D. (1992). *Electromagnetics*, 4e. New York, NY: McGraw-Hill.

Markvart, T. and Castaner, L. (eds.) (2003). *Practical Handbook of Photovoltaics*. Amsterdam,
Netherlands: Elsevier.

Messenger, R.A. and Ventre, G. (2004). *Photovoltaic Systems Engineering*, 2e. Boca Raton, FL:
CRC Press.

O'Hayre, R., Cha, S.-W., Colella, W.G., and Prinz, F.B. (2016). *Fuel Cell Fundamentals*, 3e.
Hoboken, NJ: Wiley.

Random House (1996). *Webster's Encyclopedic Unabridged Dictionary of the English Language*.
New York: Gramercy Books.

Sonmtag, E.R., Borgnakke, C., and Van Wylen, J.G. (1998). *Fundamentals of Thermodynamics*,
5e. New York: Wiley.

Young, H.D. and Freedman, R.A. (2016). *University Physics*, 14e. s.l.: Pearson Education.

3

Fossil Fuels

Strip coal mining near Frostburg, MD. Surface mining is the process of extracting coal deposits located near the surface. After the overlaying soil and rocks are removed, the exposed seam of coal is collected by excavators and bulldozers, then carried by trucks to a coal processing plant to be screened, crushed, and washed. Coal fired power plants convert the chemical energy of coal into electric energy.

According to EIA statistics, Cabin Run coalmine shown in the photo produced on average about 613 tons of bituminous coal per day in 2016 to supply electric generation plants in the region.

Energy for Sustainable Society: From Resources to Users, First Edition. Oguz A. Soysal and Hilkat S. Soysal.
© 2020 John Wiley & Sons Ltd. Published 2020 by John Wiley & Sons Ltd.

3.1 Introduction

All types of coal, petroleum, and natural gas are classified as fossil fuels. Hydrocarbons are produced in living organisms by the process of photosynthesis or biosynthesis, which convert solar radiation into chemical energy. Vegetation and dead organisms buried underground are transformed to fossil fuels by a natural process that takes millions of years.

The formation of most fossil fuels started during the Carboniferous Age about 325 million years ago. Pressure and heat have transformed in long time dead plants and organisms to hydrocarbon compounds in the absence of oxygen. The general chemical formula of hydrocarbons is C_xH_y, where x and y determine the properties of the organic compound. All fossil fuels are extracted from the Earth's crust as a mixture of various chemicals and contain high number of hydrocarbons.

In energy systems, fossil fuels are burned to produce heat. The energy output can be used directly for residential or commercial heating purposes or to produce steam for industrial processes and electric generation. Worldwide, the largest portion of fossil fuels is consumed in thermoelectric power plants to generate electrical energy.

In 2014 approximately 84% of the global energy requirement was supplied by burning fossil fuels. However, projections of International Energy Agency (IEA) based on various scenarios indicate that the use of fossil fuels for energy production will be inevitable until next century. All fossil fuels have the common limitations and drawbacks summarized below.

- *The amount of fossil fuels in the earth crust is limited.* The natural process of converting organisms into fossil fuels never ends. Therefore, some amount of coal, petroleum, and natural gas is always being formed in the Earth's crust. However, the rate at which they are consumed is much higher than the rate of their formation.
- *The cost of search and recovery process continuously increases.* Obviously, the resources easy to reach are consumed first. When these resources are exhausted, deeper and remoter resources must be explored. New techniques must be developed to reach these resources. The cost of search and recovery process continuously increases. As the remaining amount of fossil fuels decreases, new discoveries become more difficult and costlier.
- *Fossil fuel reserves are strategic commodities.* Countries control fossil fuel reserves located in their territories. International trade of these fuels may present strategic issues and be restricted by political decisions. Therefore, consumption of foreign resources may compromise energy independency.
- *Production and consumption of fossil fuels cause air, water, and land pollution.* Fossil fuel production and processing affect land use as well as water supply. Burning fossil fuels produces a significant amount of carbon dioxide, particulate matter (PM), and greenhouse gases causing ground level pollution, smog, and acid rain.
- *Use of Fossil fuels is a major cause of climate change.* Carbon dioxide, methane, and nitrogen oxides emitted to the atmosphere during combustion of fossil fuels increases the greenhouse effect, which results in global warming.

3.2   Resources and Reserves

Terms used to describe the amounts of fossil fuels that are available in the earth may some-times be confusing and even misleading because the same words are used in other areas in different contexts. Particularly the words such as resource, reserve, proved (or proven), observed, measured, discovered, and in place have specific meanings in the vocabulary used in energy statistics. If these terms are not used correctly, estimation of remaining quantities of a resource or comparison of diverse sources may be misleading. To avoid errors, one must check the definitions given in the glossary of the statistics used.

The reliability of data collected from countries is important for accurate estimation of the fossil fuel resources. To establish a unified terminology and language independent classifi-cation of fossil fuel and mineral resources, the United Nations created a framework titled UNFC-2009. UNFC classification defines resource categories based on three fundamental criteria of economic and social viability, field project status, and geological knowledge. In UFNC-2009, a fossil fuel resource is described by selecting a combination of a category or a sub-category from each of these three criteria. This language-independent numerical cod-ing system reduces ambiguity across the nations and can be visualized in three dimensions.

Energy professionals often use the term "resource" in a generic sense to signify any deposit of hydrocarbon that may be extracted from the Earth and converted to useful energy. For a certain type of fuel, resources include all discovered and undiscovered concentrations, no matter how feasible or commercially viable they are (Whitney 2010).

On the other hand, the term "reserve" is specifically used for deposits that are available for commercial extraction at some degree of confidence based on analysis of geologic and engineering data. Reserves are estimated quantities of sources that are recoverable under existing economic and operating conditions. Figure 3.1 illustrates various categories of fossil fuel resources.

Proved reserves are determined at the highest level of certainty by geological surveys, drilling wells, and using verified exploration methods. They have reasonably high con-centrations to be economically extracted at any time using available methods. Unproved

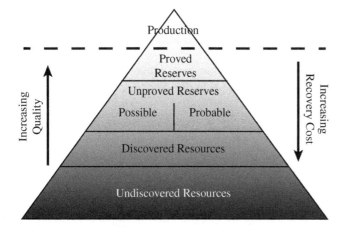

Figure 3.1   Fossil fuel resource categories.

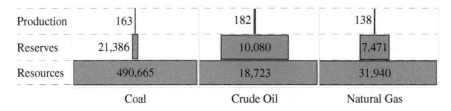

	Coal	Crude Oil	Natural Gas
Production	163	182	138
Reserves	21,386	10,080	7,471
Resources	490,665	18,723	31,940

Figure 3.2 Production, proved reserves, and estimated resources of essential fossil fuels. Unit: EJ. Source: BGR (2017).

reserves are discovered using similar techniques as proved reserves. Economic recovery of unproved reserves is, on the other hand, less feasible under current economic conditions and with existing technology.

Unproved reserves are classified as probable or possible based on the certainty level of their recoverability. Probable reserves are more likely to be recovered than not recovered (more than 50% chance of recovery). If the economic and technological circumstances change, probable reserves may be reclassified as proved. Possible reserves are less likely to be economically recovered than probable reserves.

Since the classification of reserves is based on current economic conditions and available technology, their estimated amounts may change in time as circumstances change. In fact, all estimated fossil fuel quantities have significantly increased over the last few decades, partially because of the technologic developments and partially because of new discoveries or reclassification.

Undiscovered resources are estimated based on geologic characteristics of the area that show similarities to other places where fossil fuels have been discovered. However, their concentrations, quality, and cost of recovery are not certain.

Figure 3.2 shows 2016 production, reserves, and resources of the major fossil fuel groups worldwide (BGR 2017). Fossil fuels dominate the global energy supply. Worldwide energy production from fossil fuels in 2016 was 484 EJ (EJ: exajoule = 10^{18} J), about 85% of the total energy production. Total energy content of proved fossil fuel reserves is approximately 38,937 EJ, which is about one-tenth of the estimated resources. Among all fossil fuels, coal is the most abundant energy source; estimated coal resources constitute around 90% of all fossil fuel deposits. Coal is mainly used for electric generation and industrial heat production.

Recently, crude oil reserves have increased because advanced technologies made recovery and processing of unconventional oil deposits economically viable. Crude oil, producing 30.6% of the world energy supply, has the largest share in the global energy mix. Worldwide, all types of transportation rely on oil products.

Conventional and unconventional natural gas resources are nearly four times the known reserves. Natural gas reserves also have increased significantly due to the addition of shale gas to conventional reserves. Natural gas is mainly used for heating and electric generation. Dropping gas prices led to a shift of electric generation from coal to natural gas.

Reserve to production ratio (R/P) is a statistical index used to evaluate the availability of a fossil fuel source. However, this information should be interpreted carefully. R/P is obtained by dividing the known reserves to the yearly production amount. The unit of R/P is years but it does not necessarily show how many years it takes to completely deplete a reserve. First, the production rate does not remain constant in time. Second, as we pointed out earlier, the amount of proved reserves and estimated resources may change in time because of modern technologies, improved survey methods, and resulting reclassification.

3.3 Physical Properties of Fossil Fuels

Physical properties of a fossil fuel vary by the location where it was extracted. In addition, processing, storage, and transportation of fossil fuels may affect the quality of the fuel received at the power plant. Humidity, ash, sulfur, volatile matters (VMs), nitrogen compounds, minerals, and other organic components that are present in the fuel affect its quality and the specific practical use it is suitable for. Basic properties of a fuel can be determined by a simple test, usually called "proximate test." The humidity, ash, and fixed carbon (FC) fractions can be obtained by a proximate test based on heating a fuel specimen to certain temperatures and weighing the specimen at the end of each step. A detailed test called "ultimate analysis" is necessary to accurately determine the mass fractions of carbon, oxygen, hydrogen, sulfur, nitrogen, humidity, and other components.

Heating quality of a fossil fuel is characterized by its calorific values. Gross calorific value (GCV), also called "high heating value" is the heat content of a fuel sample measured by burning it under laboratory conditions. The actual calorific value of the fuel burned in the boiler is lower, mainly due to the vaporization latent heat of the water, which is the total of the moisture content and the vapor produced during combustion. Net calorific value (NCV), also called "lower heating value," is the actual thermal energy obtained by burning the fuel in a boiler. For most solid and liquid fuels, NCV of a fuel is about 5% less than its GCV. For gaseous fuels, this difference may reach 10%. Calorific values for various common fuels are given in Appendix B.

The United Nations and IEA collect fuel data from their member countries. The collected production and consumption data for solid, liquid, or gaseous fuels are usually in mass or volume units. These physical quantities are converted to energy units using NCVs. Some statistical offices use GCV while some others use NCV. If the GCV is known from laboratory tests and ultimate analysis results are available for the moisture, hydrogen, and oxygen content of the fuel as received, the NCV can be estimated by using the equation suggested by the Intergovernmental Panel on Climate Change (IPCC) Taskforce for Greenhouse Gas Inventories (IPCC 2006).

$$NCV = GCV - 0.0245\,M - 0.212\,H_2 - 0.008\,O_2\ [MJ/kg] \tag{3.1}$$

In this equation M is the moisture content, H_2 and O_2 are the mass fractions of hydrogen and oxygen, respectively, in percent obtained by chemical analysis of the fuel as received.

Example 3.1   An electric power plant receives coal with M = 43% moisture and A = 12% ash. Dry, ash-free specimen of this coal contains $H_2 = 4.7\%$ hydrogen and $O_2 = 18.6\%$ oxygen. Given the gross calorific value of 28.922 MJ/kg, determine the net calorific value as received at the power plant.

Solution
The mass of dry, ash-free coal in the received fuel is:

$$m = 1 - (M + A)\frac{1}{100} = 1 - 0.43 - 0.12 = 0.45\ kg$$

GCV of the dry, ash-free fuel is given as 28.922 MJ/kg. Therefore, the GCV of received fuel that contains moisture and ash is

$$GCV_{rec} = GCV \cdot m = 28{,}922 \times 0.45 = 13.015 \, MJ/kg$$

The amounts of hydrogen and oxygen in the received coal are

$$(H_2)_{rec} = 4.7 \times 0.45 = 2.115\%$$
$$(O_2)_{rec} = 18.6 \times 0.45 = 8.37\%$$

The net calorific value of the received coal is calculated using Eq. (3.1).

$$NCV = 13.015 - 0.0245 \times 43 - 0.212 \times 2.115 - 0.0008 \times 8.37$$
$$= 11.506 \, MJ/kg$$

3.4 Coal

Coal is fossilized prehistoric vegetation that initially accumulated in wetlands, swamps, and marshlands during the carboniferous age between 360 and 290 million years ago. The build-up of silt and other sediments were buried in the ground with tectonic movements. The combined effect of temperature and pressure in the absence of air caused physical and chemical changes in the vegetation over millions of years. Earth's crust contains large amounts of coal that can be excavated on the surface or extracted from deep underground mine shafts.

Coal is a generic name given to solid hydrocarbon deposits. The quality of coal deposits depends on temperature, pressure, and by the length of time in formation, which is known as "organic maturity." As the organic maturity advances, the carbon content of coal increases and the moisture content decreases. In the coal formation process, compacted plants were first converted into peat, which is a very low-grade fuel. The peat was then converted gradually into "brown coal." The continuing effects of temperature and pressure converted brown coal into lignite and produced further changes progressively transforming lignite into the range classified as "sub-bituminous" coals. Further chemical and physical changes occur over time until the coal becomes harder and blacker, forming the bituminous coal. Under certain conditions, the organic maturity may increase, ultimately forming anthracite, which has the highest calorific properties.

There are several coal classifications. ASTM International (formerly American Society for Testing and Materials) classifies coal in four major ranks as anthracite, bituminous, sub-bituminous, and lignite (ASTM Subcommittee: D05.18 2018). Anthracite is at the top of the rank scale with highest carbon and energy content and a lowest level of moisture. Bituminous coal is the next in scale. Lignite and sub-bituminous coals are considered lower rank; typically softer, brownish materials that looks like soil. Lignite even shows traces of wood texture. They both have higher moisture levels and lower carbon content, and therefore their energy content is lower compared to other coal types. Higher rank coals are generally harder, stronger, darker, and shinier. Therefore, higher rank coals produce more energy per unit mass. Table 3.1 shows coal types from low to high rank and their properties.

Table 3.1 Coal ranks and properties described in ASTM D05 (ASTM Subcommittee: D05.18 2018).

Category	Heat content (MBtu/lb)	Moisture %	Carbon %	Ash %	Sulfur %
Lignite	4–8.3	30–60	25–35	10–50	0.4–1
Sub-bituminous	8.5–13	10–45	35–45	< 10	< 2
Bituminous	11–15	2–15	45–85	3–12	0.7–4
Anthracite	13–15	< 15	86–98	10–20	0.6–0.8

In statistical data, the group named "hard coal" includes sub-bituminous coal, bituminous coal, and anthracite.

3.4.1 Properties of Coal

Among the physical and chemical properties, heating value is the most important parameter of coal for energy production. Heating value or energy content of a given coal is the amount of chemical energy in its unit mass that can be converted to heat by combustion. US statistics give the higher heating values (or GCV) while IEA gives the NCV. In this text, we will use the SI unit in kilojoules per kilogram (kJ/kg). Heating value varies in a broad range depending on the rank, moisture, and impurities including sulfur, ash, and volatile matter. Heating value and coal composition affect the design of boilers and the overall efficiency of a coal power plant.

Whereas sulfur is a combustible element, the heat produced by burning it is much less than the energy produced by burning carbon. In addition its primary combustion product, sulfur dioxide (SO_2), is a major pollutant. Removing sulfur from coal before it is burned is difficult and expensive. Therefore, the sulfur content degrades the quality of coal.

Ash is a mixture of minerals deposited with the organic compounds as plants were compacted. Moisture occurs due to the exposure of coal mines to ground water or atmospheric precipitation during the storage and transportation. Ash and moisture fractions vary in a broad range for coals extracted at separate mines and even for a certain coal seam. Gases and low boiling temperature chemical compounds that evaporate when coal is heated are called volatile matter.

Fixed carbon, ash, moisture, and volatile matter fractions of coal can be determined by a simple test called "proximate analysis." Before starting the analysis, the specimen of coal is weighed. If the coal specimen is too wet, it is first air dried at 10–15°C and crushed to powder. The powdered specimen is then heated to 105–110°C (230°F) for 20 minutes and then weighed. The mass fraction of moisture (M) is obtained by dividing the mass lost during the air and oven drying by the original mass of the specimen. The remaining sample is then heated to 950°C (1750°F), in a closed pot to prevent burning, for seven minutes and weighed again. The mass loss divided by the original mass is the mass fraction of volatile matter (VM) in the sample. Volatile matter is composed of tar, lighter oils, hydrocarbon gases, hydrogen, oxides of carbon (CO and CO_2), and decomposed water products. Finally, the remaining specimen is heated to 732°C (1350°F) in an open container until it

Table 3.2 Ultimate analysis of coal extracted at major mine locations.

State/county	VM	FC	C	H$_2$	O$_2$	N$_2$	S$_2$	GCV (MJ/kg)	As received	
									M	A
PA/Schuylkill	2	98.0	93.9	2.1	2.3	0.8	0.9	34.685	2–3	8–13
MD/Allegheny	18.2	81.8	89.3	4.7	2.7	1.7	1.6	33.400	2–4	6–14
KY/Knox	40.2	59.8	83.5	5.6	7.9	1.9	1.1	34.785	2–6	2–8
OH/Jefferson	41.5	58.5	82.2	5.5	7.7	1.7	2.9	34.560	3–7	5–12
WV/Preston	31.0	69.0	87.5	5.3	4.2	1.5	1.5	36.050	2–4	4–12
ND/Stark	54.0	46.0	72,4	4.7	18.6	1.5	2.8	28.920	35–43	5–12
WY/Sheridan	45.3	54.7	74.1	5.1	18.7	1.3	0.8	29.795	18.30	5–15
TX/Houston	53.3	46.7	72.8	5.3	19.3	1.4	1.2	29.550	29–37	6–13

VM, Volatile Matter; FC, Fixed Carbon; M, Moisture; A, Ash; GCV, Gross Calorific Value.
Source: US Bureau of Mines and Culp Jr. (1979).

is completely burned. The final residue is weighed. The mass decreased during burning is divided by the original mass of the specimen to get the fixed carbon fraction (FC). The ratio of the residue to the original mass of the specimen is the ash content fraction (A) (Culp Jr. 1979; Probstein and Hicks 1982). Proximate analysis results include the mass fraction of moisture, ash, volatile matter, and fixed carbon in percent.

More detailed properties of coal are determined by a laboratory test procedure called ultimate analysis. Such analysis gives a breakdown of mass fractions of the essential elements: carbon (C), oxygen (O$_2$), nitrogen (N$_2$), and sulfur (S$_2$). Ultimate analysis results are presented in several ways. If the ultimate analysis is performed on dried samples, moisture and ash content must be considered to obtain "as-received" values. Ultimate analysis of dried, ash-free coal samples from major mine locations in the USA is presented in Table 3.2 (Culp Jr. 1979, Appendix III).

Example 3.2 Powder River Basin (PRB) underlying Wyoming and Montana contains one of the largest deposits of bituminous coal in the world. Coal mined in Sheridan County, Wyoming is representative for PRB coal deposits. As the ultimate analysis values given in Table 3.2 for this region show, the calorific value of PRB coal is lower, and its moisture and ash content are higher compared to Appalachian coal. PRB coal could not compete with midwestern and Appalachian coal because it had to be carried more than 1000 miles via railroad to the Great Lakes ports to reach populated and industrialized areas. Coal production in the PRB region increased significantly after 1988, when the Environmental Protection Agency mandated reduction of sulfur dioxide emissions from coal fired power plants.

Electric power plants use scrubbers to remove sulfur dioxide from flue gas. The average cost of a scrubber system is estimated $322 per ton of SO$_2$ removal. The estimated cost of converting an electric power plant designed to burn Appalachian coal to burn PRB coal is $113 per ton of SO$_2$. This difference made PRB coal economical in coal fired electric power generation in Midwest. The Powder River Basin in Wyoming and Montana is now producing more coal than the entire Appalachian region.

Example 3.3   An engineer working for an electric power plant in Cleveland, OH is comparing purchase options of coal mined in Sheridan County, WY at 1500 mi distance and Jefferson County, OH at 130 mi distance. Using the ultimate analysis shown in Table 3.2, compare as-received net calorific values (NCV) of these two coals.

Solution
Calculation method to obtain NCV as received is shown in Example 3.1. Using the same method for values shown in Table 3.2 we obtain:

Sheridan County, WY: Minimum 13.520 MJ/kg; maximum: 21.488 MJ/kg
Jefferson County, OH: Minimum 26.873 MJ/kg; maximum: 30.643 MJ/kg

In power generation, the ability of coal to withstand weather conditions during storage is important. Coal-fired power plants store a large amount of coal nearby to ensure continuity of production. Coal stocked in large piles should not excessively crumble due to temperature changes, rain, and wind. Another important physical property of coal is grindability. In modern power stations, coal is grinded into fine powder to be pulverized in the burner for more efficient combustion. Harder coals are more difficult to pulverize and therefore are not preferred for electric power generation. Anthracite and hard coal are mostly used for coke production, industrial processes, and direct heating.

3.4.2   Coal Reserves

Coal is the most abundant and widespread fossil fuel. According to BP Statistical Review of World Energy 2017 (BP 2018), 1,139,331 million tons (Mt) coal reserves were available worldwide, and the reserve to production ratio was 153 years by end of 2016.

While coal resources are available in more than 50 countries, USA, Russia, China, India, Australia, and Indonesia lead the coal production. Distribution of coal reserves around the world is illustrated in Figure 3.3.

China and India also import a large quantity of coal to supply electric generation plants. In Europe, coal resources are available in the UK and Germany, but they are no longer enough to supply the energy need of their industry. Many countries in the European Union import coal from Asian producers.

International projections show that coal will continue to be the major primary energy source for electric power generation. It has wide spread availability around the world, and is safe, reliable, and affordable. Another major benefit of coal is the convenience of its long-term storage as a backup energy source. However, the share of coal is expected to decrease due to environmental issues, decreasing natural gas prices, and growing use of renewable energy sources.

3.4.3   Coal Mining

Two main methods of coal mining are "surface (or opencast) mining" and "underground (or deep) mining." The choice of mining method mainly depends on the geology of the coal deposit. Worldwide about 60% of coal is extracted from deep mines, while in several countries, including the USA and Australia, surface mining is more common.

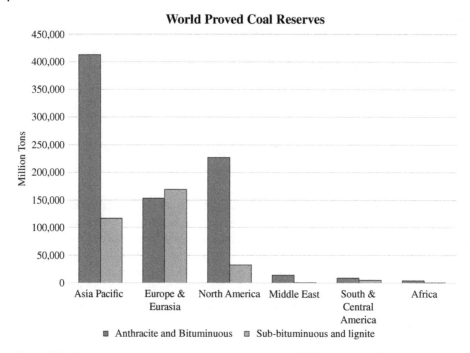

Figure 3.3 Coal proved reserves by geographic regions as of 2016. Source: BP (2018).

3.4.3.1 Underground (Deep) Mining

Deep mining is based on extracting coal from an underground coal seam. A coal shaft can be as deep as several hundred meters. Two common methods used to extract coal from underground coal deposits are named "room-and-pillar" and "longwall" mining.

Room-and-pillar mining is one of the oldest mining methods. The miners cut a network of rooms into the coal seam leaving pillars of coal to support the roof of the mine. The size of the pillars is determined empirically, and the mine progresses randomly in a convenient direction. The random layout of the mine makes ventilation difficult and there is a risk of pillar failures. A pillar failure may lead to a domino effect with collapse of overloaded neighboring pillars resulting in total collapse of a mine section. The early room-and-pillar mines were quite unsafe for workers and several mining accidents were recorded. In addition, the efficiency of this method is lower since between 20% and 40% of the coal may be left to support the overlaying rock layers. The coal that forms the pillars can be extracted by a method called "retreat mining." At the final stage of the room-and-pillar mining, the pillars are cut as the workers retreat. The mine is abandoned and left to collapse after the operation is completed.

A newer technique used in underground mining is called "longwall mining." In this method, automated mechanical shearers fully extract coal from a section of the seam. First, miners dig a hole the height of the seam and attach brackets to hold the ceiling in place. Then, they set up a mechanism that runs about 230 m (750 ft) across the seam. Self-advancing hydraulic brackets temporarily support the roof during the extraction. After the operation is complete, the roof may collapse. Longwall mining can extract up to 80%

of the coal available in the deposit. Longwall mining is more efficient but requires more expensive equipment (up to $50 million) and meticulous planning before extraction starts. Room-and-pillar method is less efficient, but coal production can start more quickly with less equipment cost (under $5 million) (WCI 2005).

3.4.3.2 Surface (Opencast) Mining

Coal deposits located near the surface, can be extracted by the surface mining method, which is also known as opencast mining, strip mining, and mountain removal. Surface mining can extract up to 90–95% of the coal using large equipment such as draglines, power shovels, bucket wheel excavators, conveyors, and large trucks.

Surface mining starts with breaking the overburden of soil and rock using explosives, which is then removed by draglines or shovel and truck. The exposed coal seam is drilled, fractured, and mined in strips. The photo on the title page of this chapter shows a strip mine in Western Maryland, near Frostburg.

The US is one of the major coal producers, owning the largest coal reserve in the world. Coal production, however, has declined significantly over the last several years. According to the Energy Information Administration (EIA) statistics, in 2016, 710 coal mines produced 728.4 million short tons of coal, marking the lowest annual production level since 1979. Surface mines produced 65% of the total coal in 2016 (EIA 2017a).

3.4.4 Preparation, Handling, and Transportation

Any kind of coal, in its raw state as extracted from ground, barely meets the consumer requirements End-users need a consistent quality that satisfies specifications for the intended application. Through a process called beneficiation, preparation plants improve the properties of the coal by removing impurities to reduce the ash and sulfur content. In a coal handling and preparation plant (CHPP) the raw material is first crushed and sorted by grain size into coarse, fine, and ultrafine grades. Rock particles, dirt, and other unwanted elements are separated in either water or a heavy liquid by a sinking or floating process. Coal processing facilities consume considerable amounts of water, which usually cannot be returned to a freshwater source. The product quality improves as the raw coal is processed more thoroughly. The product output of CHPP is the ratio of washed coal to raw coal, which is around 80% for steam coal and 65–70% for coking coal. Cleaned coal is stockpiled in separate grades and prepared for transportation. Weatherability is the property of coal to withstand exposure to environmental effects without excessive crumbling.

Coal is transported by railroad, trucks, river barges, and coastline vessels to the point of use. The largest quantity of bituminous coal is used in electric power stations and industrial plants where large earth moving machines spread the coal in layers and compact it to minimize the amount of air in the pile and reduce the risk of fire by spontaneous combustion. Fine-grain coal is temporarily stocked in large piles near electric power stations to ensure continuous electric generation. (Culp 1979).

Transportation of lignite to long distances is not cost-efficient due to its higher moisture content and lower heating value. Lignite is mostly used in electric power stations installed near the mines. Trucks and conveyor belts transport lignite to the power plant. After combustion, the ash and residues may be processed on-site to be used for production of construction materials and other byproducts. The large amount of remaining solid waste is dumped back to the extraction spots to fill the land.

While coal reserves are available in almost every region of the world and about 83% of hard coal is consumed in the country of origin (World Energy Council 2016), hard coal is still a worldwide traded energy commodity. International coal trade depends on a dedicated infrastructure and reliable transportation networks from mining to consumers. Feasible distances for economic transportation are limited by the cost.

3.4.5 Coal Production and Consumption

According to the BP Statistical Review of World Energy, in 2016, coal supplied 28.6% of the total global energy needed for all end uses (BP 2018). Figure 3.4 shows world coal production by rank.

Coal is the major source for electric power generation around the world. Nearly two-thirds of the coal extracted in the world was consumed for electric generation. A relatively smaller portion of produced coal is consumed for industrial, residential, and commercial heating purposes. Coal can also be processed to obtain synthetic fuels in gas form or liquid fuels using coal-to-liquid (CTL) technologies (Probstein and Hicks 1982).

Lignite is primarily used for electric generation in power plants installed near the extraction facility due to its lower calorific value and higher humidity. Sub-bituminous coal is also mostly used for electric generation, but it is also suitable for cement production and other industrial purposes.

Bituminous coal has higher calorific value, and lower moisture, ash, and sulfur content than lignite. This group yields higher efficiency in electric generation. Bituminous coal is also used in the metal industry for smelting iron and heat production.

Anthracite is the highest rank of coal, mostly used for commercial and domestic heating, coke, and smokeless fuel production. The steel and cement industries are major users of coal. Today, about 41% of global electric generation is fueled by coal, and 70% of the steel

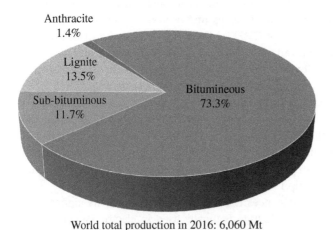

World total production in 2016: 6,060 Mt

Figure 3.4 World coal production by rank. Source: BP (2018).

produced worldwide uses coal. On average, 200 kg coal is used to produce 1 ton of cement (WCA 2018).

In addition to energy related purposes, coal and its byproducts are also used in the chemical industry. Alumina refineries, paper manufacturing, and pharmaceutical industries are some examples. Production of ammonia salts, nitric acid, naphthalene, phenol, and benzene use refined coal tar. Many products such as soap, solvents, dyes, and various fibers use coal or coal byproducts. Other specialized products using coal-based components are activated carbon filters, lightweight and strong structural material carbon fiber, silicones, lubricants, resins, cosmetics, shampoos, and toothpastes.

3.4.6 Transportation of Coal

Coal is carried from mine mouth to preparation plants and end users by a wide range of transport systems, from conveyor belts to trans-ocean bulk-cargo ships. An alternative transportation method for coal transportation to shorter distances is mixing powdered coal with water and transport the coal slurry by pipeline.

The choice of transportation mode depends on the characteristics of the coal and consumption types. In the US, electric power plants consume about 85% of the produced coal. More than 70% of the coal burned in electric power plants is delivered by rail and one-third of the coal supplied to industrial consumers is carried by trucks. Trucks are an important component of multimodal shipments that combine ground, rail, and marine transport.

Lignite is mostly consumed near the mines because of its lower calorific value and higher moisture, which makes long distance transportation uneconomical. Conveyor belts and tramways are common for transportation of lignite from mine to electric power plants. Solid boiler reside is usually sent back to fill the mine land. Lignite extracted from mines in Texas is mainly carried by coastal barges to electric generation plants installed on the Gulf Coast (NRC 2007).

Bituminous coal extracted in the Appalachian region is mostly consumed in regional electric power plants. In West Virginia and Kentucky, a significant amount of coal is moved by trucks. A typical one-way distance of coal transport by trucks is less than 100 mi (160 km), averaging 32 mi.

Western coal from the Gunpowder River area is carried by train to the Great Lakes region, and then delivered to power plants and industrial consumers by lake freighters. Western coal mined in the PRB in Wyoming and Montana travels more than 1500 mi to reach ports at the Great Lakes to be further transported by lake barges to consumers. Coal transport trains may be up to a mile (1.6 km) long and have a capacity of over 10,000 tons (see the photo in Figure 3.5).

Intercontinental transportation uses large ships ranging from 40,000 to 80,000 deadweight tonnage (DWT). According to the World Coal Association, around 700 million tons (Mt) of coal was traded internationally in 2003, and 90% of international coal transportation was seaborne. Transportation costs may reach up to 70% of the delivered cost of coal.

3.4.7 Environmental Impacts of Coal Production

Coal mining moves large amounts of materials from one place to another, and that results in landscape and vegetation change. Abandoned mine shafts may present structural risk

Figure 3.5 View of trains carrying western coal and crude oil from Golden Spike Tower, in the World's largest railyard in North Platte, Nebraska. Source: © H&O Soysal.

for buildings and roads. In mining areas, filling and grouting of the ground can increase the cost of heavier buildings. These issues must be considered in planning of coal mining sites and coal burning power plants, as well as new developments in mining areas.

Surface mining kills the vegetation on a wide area impacting the topology, hydrology, ecology, agriculture, landscape, tourism, and social life. Also called "mountaintop removal," it has been popular in the Appalachian region and several western states. Because of its destructive nature, surface mining attracts strong public reactions, but surface mining is more efficient and safer compared to underground mining. In addition, it provides jobs to local communities since trucks continuously commute between the mine and power stations carrying the produced coal. Reclamation laws require the mining companies to restore vegetation after the coal is completely extracted. Once the surface-mining operations are completed, the flat land left behind may again become convenient for agriculture and other uses. However, the change of topology usually affects permanently the flow and watershed of rivers and streams.

Figure 3.6 shows aerial photos of a mountaintop removal site in West Virginia between 1984 and 2015, published by NASA Earth Observatory (NASA 2019). Hobet mine started around 1984 in a relatively small area west of the Coal River. In general, rock debris cannot be securely graded as steeply as the original land, therefore the coal companies need to dispose the leftover rocks by filling hollows and streams. Satellite pictures show that rock and earth dams filled the valley as the mine expanded over a 31-year period, disturbing an area of more than 10,000 acres (about 40 km^2). The US Fish and Wildlife Service confirmed in a report that nearly 40% of the year-round and seasonal streams in the Mud River watershed had been filled through 1998.

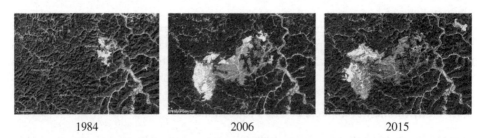

| 1984 | 2006 | 2015 |

Figure 3.6 Progress of surface mining in the Coal River area of West Virginia.

3.4.8 Coal Related Issues

Coal mining has been a major business and employment opportunity since the industrial revolution in the eighteenth century. Over the last years, the decline of coal production has resulted in a decline of employment in the coal mining industry, impacting the job market in coal producing states. According to EIA statistics, in 2016 30,005 (58%) out of 51,795 employees of all coal mines were working in underground mines. The number of coal mining employees has decreased over the last years with shrinking productive capacity of coal mining companies. At the same time, however, the average production per employee-hour rose to 6.6 short tons.

Coal deposits often release hydrogen sulfide and methane, which may accumulate in underground mines. Such poisonous gases, coal dust, and various mineral particles present health hazards for miners. Long-term exposure to such pollutants in poorly ventilated coal mines can cause chronical respiratory disorders and heart diseases.

In coal mines (particularly bituminous), a mixture of methane and other combustible gases accumulate in some cavities of the deposit forming "firedamp," which is extremely explosive. Firedamp explosions were the major cause of coal mine disasters before the flameproof lamps and gas detectors were invented. Such explosions can still occur when a metal tool strike on coal produces a spark in the presence of firedamp. In addition to flammable gases, coal dust can cause explosions that destroy mine shafts and injure or kill miners. According to the US National Institute for Occupational Safety and Health (NIOSH) statistics, 13,883 miners have died in coal mine accidents since 1839 in the US. The worst coal mine accidents were recorded at the beginning of the twentieth century. Between 1901 and 1925, 305 coal mine accidents killed 7909 miners (MSHA 2019). Improved safety standards, precautions, and advanced equipment significantly prevented coal mine accidents. Over the first decade of the twenty-first century, five coal mine accidents killed 65 miners in the US.

Many coal producing countries around the world have a history of epic coal mine disasters. On May 13, 2014, an explosion at a coal mine in Turkey (Soma, Manisa), caused an underground mine fire that continued for two days, creating the worst mine accident in the country's history. When the explosion occurred, 787 workers were underground; the official 301 deaths were confirmed (Pamuk 2014), while the local news claimed that the death toll exceeded 350.

While most coal mine accidents are related to explosions from mechanical and structural failures, and broken bones from falling, flooding, mining-triggered earthquakes, and malfunctioning equipment may also cause coal mine accidents (MSHA 2019). The coal mining process produces tons of solid waste. In addition, washing and floatation of coal to separate rock and impurities creates massive amounts of slurry liquid waste called "coal sludge." Coal sludge and waste contain toxic chemicals including mercury, lead, arsenic, beryllium, cadmium, chromium, nickel, and selenium compounds. The waste is usually stored in a dam near the coal preparation plant. In recent years, several disasters initiated by a structural breakdown or flooding killed and injured many people and contaminated the water supply. An incident known as the "Buffalo Creek flood" that occurred in 1972 in Logan County, West Virginia was a result of slurry dam burst unleashing approximately 132 million gallons (500 million liters) of black wastewater. The disaster killed 125, injured 1121, and left 4000 persons homeless. In 2008, the collapse of a retention wall at the Kingston plant in Harriman, operated by the Tennessee Valley Authority (TVA) released a combination of water and fly ash that flooded 12 homes, spilled into nearby Watts Bar Lake, contaminated the Emory River, and caused a train wreck. On January 9, 2009, a coal waste spill at the Widows Creek plant in northeast Alabama, released about 10,000 gal of toxic gypsum material, some of which spilled into Widows Creek and the nearby Tennessee River.

Disasters such as hurricanes, tsunamis, and earthquakes can cause additional disasters in areas where coal mines or coal-burning power plants operate. In 2018, the rising water during and after hurricane Florence flooded a basin of toxic coal ash at Duke Energy's Sutton Power Plant near Wilmington, North Carolina. The storm washed away more than 2000 cubic yards (1529 m^3) of coal waste. In addition, the flood contaminated the Cape Fear River and Sutton Lake (Dennis, Mufson, and Eilperin 2018).

3.4.9 Environmental Impacts of Coal Consumption

Burning coal emits large amounts of greenhouse gases including carbon dioxide, sulfur dioxide, and nitrogen oxides into the atmosphere. In addition, coal combustion products (CCPs) are formed as solid particles, "fly ash," "bottom ash," and "boiler slag." Fly ash is carried up with hot flue gases and constitutes the largest amount of coal combustion residuals by weight. Parts of the ash that are too heavy to be carried up by flue gases settles to the bottom of boilers and is called "bottom ash." When ash melts down at an elevated temperature, it forms "boiler slag."

The 1990 amendments to "The Clean Air Act (CAA)" in the US required the reduction of the toxic gases and particulate matters released to the atmosphere. In response to the regulations and National Ambient Air Quality Standards, coal-fueled electric power plants installed equipment to capture larger quantities of combustion residuals. As a result, the amount of captured fly ash increased, and massive quantities of synthetic gypsum formed in scrubbers were added to the combustion byproducts.

Several emerging control technologies reduce the amount of pollution in coal-fueled power plants. Clean coal, electrostatic precipitators, and advanced filtration systems are among them. Using selective catalytic reduction can minimize mercury emission. Flue-gas desulfurization activated carbon injection, and fabric filters are effective methods to reduce emission of metals, solid particles, and acid gases. Using fluidized bed combustor (FBC) technology, power plants can burn lower quality coal more effectively and reduce emissions. Flue gas desulfurization (FGD) systems known as scrubbers reduce sulfur dioxide (SO_2) and mercury emissions using a calcium-based reagent such as pulverized limestone. Synthetic gypsum, which is the primary feedstock for wallboard manufacturing industry is a byproduct of FGD systems.

In 2013, coal consumed in the United States resulted in 114.7 million short tons of CCP including fly ash, bottom ash, boiler slag, FGD materials, and fluidized bed combustion materials. About half of these residuals (51.6 million short tons) were utilized to make concrete components, blended cement, filling materials, road base, roofing granules, gypsum wall panel products, and snow and ice control materials. The remaining half that is not used for the manufacture of byproducts is disposed in landfills (ACAA 2015).

In addition to the greenhouse gases and acid rain, coal-burning power plants are the largest sources of mercury, arsenic, nickel, and chromium emissions in the air. Mercury emitted by power plants is transformed through natural processes to methyl mercury, which is a toxic chemical. Methyl mercury accumulates mainly in fish then transfers into land animals and humans, creating a high health risk. Other metal components create serious health risks including cancer, heart problems, asthma, and lung diseases that might result in premature deaths.

In 2011, the US EPA issued a Mercury and Air Toxics Standard (MATS) to limit the emission of toxic air pollutants produced by power plants (http://www.epa.gov/mats). Enforcement of this standard, together with other economic conditions, is leading to the retirement of many fossil fuel burning electric generation plants before 2020. While many of the existing plants already have emission control systems in place, MATS require that older coal-fired plants must also be upgraded with adequate equipment to reduce their emissions. A number of these plants have already paid off their initial costs, and the revenue obtained from sales of electric power is currently paying their operation and maintenance costs. An additional investment is needed to install new equipment to meet the MATS requirements. In addition, the competition of modern natural gas-fired plants with relatively decreasing natural gas prices result in a moderate increase of electricity prices, implying a decrease of the coal plant revenues. In the meantime, electricity demand has grown slower due to energy efficiency and use of renewable energy. A combination of these factors has made some older coal-fired plants non-profitable, motivating their owners to retire them between 2014 and 2020. According to the US EIA, the total power capacity of coal-fired plants projected to retire is about 50 GW, which is about 16% of the coal-fired electric generation capacity in 2012.

3.5 Petroleum

Petroleum is a mixture of organic hydrocarbons and small amounts of other chemicals that occurred naturally in the Earth over millions of years. Although formation of petroleum was traditionally associated with fossilized plants, today most geologists believe that it was formed from remains of aquatic organisms. Dead fish, marine creatures, algae, planktons, and other microorganisms buried under sedimentary rocks, layers of mud, and silt by tectonic movements formed petroleum over millions of years through the combined effect of temperature and pressure in a relatively oxygen-free environment.

Petroleum is a family of naturally occurring hydrocarbons in solid or liquid forms. The solid and semi-solid forms of petroleum are called *asphalt* and *tar*. Unprocessed petroleum extracted from earth in dark, viscous, liquid form is commonly called *crude oil*. The terms *oil* and *petroleum* are often used interchangeably to designate both crude oil and a broad range of petroleum products.

Crude oil is an important economic commodity. Because commercially feasible quantities are available only in certain regions of the world, oil production often creates economic and political tensions that sometimes escalate into international conflicts. In about the first three-quarters of the twentieth century, petroleum products were the preferred fuels for heating and even electric generation because of the low cost, easy installation, and convenience.

In 1973, twelve petroleum-exporting countries established a cartel named OPEC (Organization of the Petroleum Exporting Countries) to stabilize oil production and prices. The oil embargo implemented by OPEC countries during the fourth Arab-Israeli war in 1973 initiated a worldwide energy crisis. The energy crisis particularly affected countries relying on foreign oil. Skyrocketing prices of oil products, scheduled power outages up to four or five hours every day, long lines at gas stations, and the shortage of basic products and food supplies became usual during the oil crisis.

The oil crisis led to the establishment of IEA in November 1974 by member countries of the Organization for Economic Cooperation and Development (OECD). The main purpose of the IEA is to provide data and information at international levels to ensure reliable, affordable, and clean energy; economic development through energy security; environmental awareness; and worldwide engagement in energy-related issues. The IEA publishes reports, statistics, and extensive online help decision-makers in planning and management of energy resources, as well as increasing the public awareness in energy related issues.

3.5.1 Types of Petroleum Formations

Crude oil is generally classified as conventional and unconventional oil. Both types of oil have similar chemical composition, with the main difference between conventional and unconventional oil being reservoirs where they are found and the techniques used to recover and process them.

Formation of petroleum depends on many physical conditions and time. The buried organisms first build a geologic layer called *stratum*. Under the effect of temperature and pressure in the oxygen-poor environment, the organic materials transform into *kerogen*. As the temperature increases, further decomposition in time changes the structure of kerogen into oil and natural gas.

Petroleum deposits feasible for commercial extraction accumulate in specific types of fine-grained sedimentary rocks formed by the compression of sand, mud, and silt known as mudstones and shales. The rocks that have captured oil deposits are called *source rock*. Eventually, petroleum deposits migrate out from the source rock into other rock formations. Porous rocks that absorb and store petroleum are called *reservoir rocks*. A non-porous layer, called *trap rock*, keeps the collected petroleum deposit in an underground reservoir.

Oil reservoirs also contain saltwater and a mixture of combustible gases, mostly methane, called *natural gas*. Lighter types of oil float on the layer of saltwater. Natural gas accumulates on the top of the deposits. If the amount of natural gas is not commercially significant to extract separately, it is vented out and burned to prevent emission of methane, which is one of the greenhouse gases contributing to global warming. The term *flare* refers to burning unused natural gas at petroleum production facilities.

The shape of an oil reservoir depends on the rock formation and tectonic movements that took place around the oil deposit. The geologic term for a layer of rock that creates a shape like an arch is *anticline* (Figure 3.7). A conventional oil reservoir forms when the oil deposit accumulates on the top of an anticline and is sealed with an impermeable layer of rock. Oil deposits coexist with saltwater and gas. Lighter oil floats on saltwater, but even lighter gas collects on the top of the layer of oil. Synclines, which are the opposite of anticlines are not ideal formations to collect oil deposits.

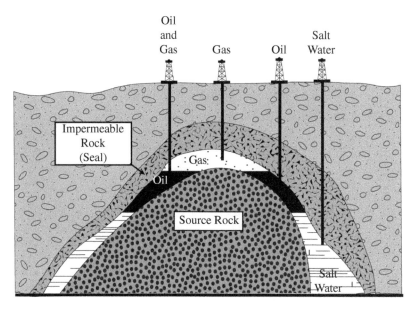

Figure 3.7 Anticline type petroleum and natural gas formation.

A pool of oil may also form when a fault moves porous reservoir rock against an impermeable seal, such as shale or salt. If the fault does not leak, the upward migration of oil to the surface is blocked, and an oil deposit accumulates against the fault. If a porous reservoir rock is enclosed in an impermeable layer of shale or salt, then a trap may form a reservoir, with thickness decreasing to zero. In the oil industry, such formation is called a *pinch-out*.

A petroleum reservoir must contain enough hydrocarbons to be commercially viable. Many of the reservoirs contain more of a salty solution called formation water, than hydrocarbons. A reservoir is usually considered productive if at least 40% or more of the fluids contain hydrocarbons. If the water content is greater than 60%, the reservoir practically produces saltwater. The oil industry term used for such reservoirs is "wet." Economic value of an oil well is determined by the rate of production and the length of economic production. A well that does not produce oil or gas to make enough profit is abandoned. Such wells are called *dry holes* or *dusters*.

3.5.2 Properties of Crude Oil

Both conventional and unconventional crude oil is a mixture of liquid hydrocarbons with a small concentration of mineral compounds. The carbon mass fraction changes between 84% and 87%, the hydrogen mass varies around 11–16%, and the sum of the oxygen and nitrogen content is up to 7%. Crude oil also contains tiny amounts of non-hydrocarbons such as sulfur, various metals, and minerals. The quality of a crude oil extracted in a region is specified by its sulfur content. Crude oil that has a lower mass fraction of sulfur is specified as sweet oil, while crude oil with higher sulfur content is referred to as sour.

Density of the deposit is one of the major properties affecting recovery and feasibility of a petroleum resource. The American Petroleum Institute (API) and the National Bureau of Standards developed a scale relative to the density of water. API gravity is calculated using

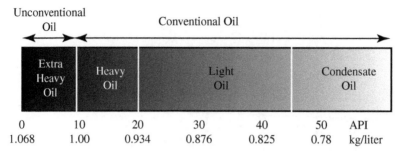

Figure 3.8 Oil classification by density.

Eq. (3.2), where σ is the specific gravity of degassed oil at the surface at 60 °F temperature.

$$G_{API} = \frac{141.5}{\sigma} - 131.5 \tag{3.2}$$

If the API gravity is greater than 10, the oil floats on water; if less than 10, it is heavier than water and does not float. The API gravity of most crude oil sources extracted and refined by conventional methods is in the range between 30–40 °API. Based on API gravity, crude oil is classified as light (higher than 31.1), medium (22.3–31.1), and heavy (below 22.3). Oil having 10 °API gravity or less is called "extra heavy." Crude oil classification based on density is illustrated in Figure 3.8.

Example 3.4   Calculate the weight of 1 bbl of crude oil (in kg) with 40 °API gravity

Solution
We can extract the specific mass of oil from (3.2):

$$\sigma = \frac{141.5}{G_{API} + 131.5} = \frac{141.5}{40 + 131.5} = 0.825 \text{ kg/liter}$$

One standard oil barrel contains 42 gal, which is equivalent to $42 \times 3.785 = 537.5$ liters. Therefore, 1 bbl of 40 °API crude oil weighs 443.5 kg.

In international oil markets, crude oil is characterized by its API gravity and sulfur content. Crude oils that are lighter (higher API gravity) and sweeter (lower sulfur content) are priced higher. One of the lighter and sweeter oils traded internationally is Brent Blend, a blend of oil extracted from 15 oil fields in the North Sea, which has API gravity of 38.3 and contains 0.37% sulfur. Another benchmark crude oil referenced in the oil trade is West Texas Intermediate (WTI) with an API gravity of about 39.6 and 0.24% sulfur content. Other major markers are Dubai Crude, Russian Export Blend, and OPEC blend.

Another important property that affects the value of an oil resource is its viscosity, which is informally known as "thickness." The international (SI) unit of viscosity is "Pascal second," (Pa·s). In the oil industry, the most frequently used unit is centi-Pascal (1 cP = 0.01 Pa s). Viscosity of heavy and extra heavy oils varies in a wide range from 20-cP to more than 1,000,000-cP. Bitumen (also called asphalt) is the most viscous hydrocarbon and it is solid at room temperature. Bitumen must be heated to become a fluid.

The more viscous the oil, the more difficult it is to process. Although there is no direct relationship between density and viscosity, the terms "heavy" and "viscous" are commonly used for unconventional oils because oil that has higher density also has higher viscosity. Increased viscosity and density make recovery of especially unconventional crude oil more challenging and costlier. Because viscosity changes significantly with temperature, heavy oil recovery methods are divided into two categories named "cold" and "warm" production.

Since the beginning of the oil industry, as easily accessible oil reserves were depleted, new search, extraction, and process technologies have been developed to exploit oil deposits located deeper in Earth, under oceans, and in arctic regions. In addition, tar sands near the surface and extra heavy oil in shale formations became increasingly important sources of unconventional oil.

3.5.3 World Oil Resources

Conventional crude oil is a strategic commodity and its production has always presented political issues. Crude oil prices depend on supply–demand balance, rather than remaining resources. Market conditions, political decisions, and consumer preferences are major factors that affect the supply–demand balance.

OPEC is a major intergovernmental organization that regulates about 42% of the world's oil production. OPEC members, especially the Middle East, have the biggest conventional oil reserves (71%).

According to a BGR Energy Study 2017, conventional oil reserves account for 38% of the world's total proved oil reserves (BGR 2017). However, conventional oil production rates dominate global crude oil markets. When crude oil prices decrease, investments on more expensive unconventional drilling and processing activities decrease. In addition, conventional oil production affects development of alternative energy sources.

Petroleum that is too heavy and viscous to be recovered by pumping out from conventional vertical wells is categorized as "unconventional oil." Essentially, the final products obtained from conventional and unconventional resources are the same, but extraction methods are different. The final product obtained by unconventional methods is sometimes called "synthetic crude oil."

Data sources give the crude oil quantities in barrels, tons, or energy units. A US barrel (bbl) is 42 gal (approximately 159 l). "Tons of oil equivalent (toe)" is an energy unit used to compare quantities of primary energy sources to crude oil. One ton of oil equivalent is considered as $41.8 \cdot 10^9$ J. Statistical data about oil resources and reserves may change in time based on reclassification of resources and development of advanced recovery technologies. Advanced exploration methods allow discovery of additional resources. Also, some of the resources that were not viable with older technologies may become feasible with emerging technologies. Changes in classification of oil types also affect the reported quantities, hence the overall estimations. Figure 3.9 shows the trend of global crude oil reserves over two decades (BP 2018).

Unconventional oil deposits have been known for a long time. For example, the unconventional oil production in Alberta, Canada started in 1967. At that time, extraction costs were remarkably high compared to the conventional oil, which are easier to recover and cheaper. Since 1990 the production of synthetic crude oil (SCO) has grown approximately five times in Alberta's Athabasca oil fields.

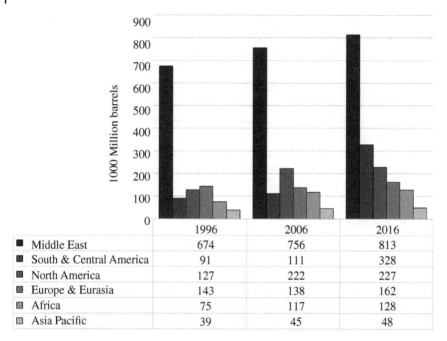

	1996	2006	2016
■ Middle East	674	756	813
■ South & Central America	91	111	328
■ North America	127	222	227
▣ Europe & Eurasia	143	138	162
▣ Africa	75	117	128
▢ Asia Pacific	39	45	48

Figure 3.9 Trend of proved oil reserves over two decades BP (2018).

Declining conventional oil reserves, rising oil prices, and growing energy demand are among the major motivations for exploration of alternative oil resources that were previously uneconomical.

The main sources of unconventional oil are shale oil (also known as tight oil), bitumen oil sands, and extra-heavy crude oil. Natural gas plant liquids (NGPLs); biofuels, including biomass-to-liquids (BTLs) and gas-to-liquids (GTLs); coal-to-liquids (CTL); and kerogen are emerging as other unconventional liquid fuels.

Unconventional oil deposits constitute about 70% of the known petroleum resources (Figure 3.10). Unconventional deposits are named extra heavy oil, tight oil, shale oil, natural bitumen (or asphalt) and oil sands (also known as tar sands).

According to the *2010 Survey of Energy Resources* conducted by the World Energy Council (WEC), natural bitumen deposits are found in 598 sites located in 23 countries. The largest deposits are in Canada, Kazakhstan, and Russian Federation.

Oil shale deposits exist in many parts of the world. Worldwide, 162 deposits of extra-heavy oil have been recorded in 21 countries. Smaller deposits have little or no economic value. Only larger deposits that are concentrated in a large area and contain a significant amount of shale oil may be feasible. Total world resources of shale oil are conservatively estimated at 4.8 trillion barrels.

At the end of 2008, the world's total conventional crude oil proven reserves were estimated at 163,038 million tons. According to the 2013 surveys, conventional oil reserves increased to 179,682 million tons by the end of 2011, with new discoveries. This distribution, however, must be interpreted with caution because the quantities are obtained through questionnaires and surveys sent to national agencies. First, the reserves reported

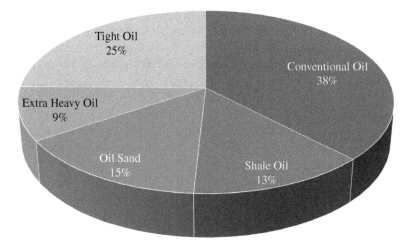

Figure 3.10 Global oil resources by type. Source: BGR (2017).

by different countries depend on their understanding of the classification and definitions. Sometimes the resources are overestimated or inflated for several reasons. Some countries may not even disclose their actual resources for political reasons. The percentages change due to reclassification of some resources, new exploration methods, and development of more efficient recovery technologies. Some resources that were considered uneconomical may become feasible with advanced technologies. Also, exploitation of unconventional petroleum resources depends on social acceptance and environmental restrictions in addition to economic feasibility.

3.5.4 Oil Exploration

Commercially feasible amounts of oil deposits are relatively rare compared to coal because their formation requires a specific combination of natural circumstances developing in a certain amount of time, and petroleum can deposit only in specific types of geologic formations. Exploration of oil resources is challenging and requires specific techniques. In earlier years of the oil industry, search methods were based on guessing, trial, and error. Modern oil geologists and geophysicists use seismic waves, sensitive sensors, and computer software to determine underground rock formations.

Seismic waves are vibrations that travel through the earth like waves produced by an earthquake or volcanic eruption. Artificially shock waves are generated by either dynamite explosions or thumper trucks. Dynamite explosions have environmental drawbacks; while thumper trucks produce more environment-friendly vibrations. Boundaries between diverse types of rock formations reflect seismic waves as a mirror reflects light or a solid surface reflects sound waves. Specialists can estimate the underground profile of the terrain by analyzing the reflection data.

Recording devices called geophones are placed on the surface at various distances and directions from the wave source. Travel time of the seismic wave to hit the formation boundary, reflect at an angle, and reach the geophone give information about the depth of the underground formations. A simple technique is processing such data to obtain a two-dimensional mapping of the underground formations. Today, computer software can generate three-dimensional models of the formations by processing data collected as an intersecting grid of seismic lines. By using such advanced techniques, specialists can estimate the depth, size, and type of an oil reservoir.

Other methods are based on using sensors to measure slight changes in the Earth's gravitational and magnetic fields. Gravity meters measure the changes in the Earth's gravitational field, and sensitive magnetometers detect small variations of the magnetic field of the Earth. Mathematical interpretation of gravity and magnetic field variations provide information on the size and location of underground formations. Devices known as sniffers are also available to sense the presence of hydrocarbons.

Undersea explorations use compressed guns to produce a pulse of air into the water. The shock waves traveling in the water enter the ground and are reflected to the surface. Travel time is calculated by collecting data from hydrophones placed on the water surface. Computer software then produces three-dimensional maps of the underwater formations.

No matter which techniques are used to estimate the formations, the most accurate information about the oil deposits is eventually obtained by actual drilling.

3.5.5 Well Drilling Techniques

After the potential field is adequately surveyed, the only way to verify the existence and value of the reservoir is actual drilling. Oil well drilling process requires substantial investment. Drilling a modern oil well on the land can cost several million dollars. The construction of an offshore platform costs up to three to four billion dollars (Janardhan and Fesmire 2011). Therefore, sophisticated methods have evolved to make exploration and recovery of petroleum more reliable, cost-efficient, and productive.

3.5.5.1 Planning

Drilling the first well in a potential oil field is a comprehensive project that requires thorough planning. The drilling work has significant impacts on the environment and community. Therefore, possible effects on the vegetation, wildlife, land, and water must be studied.

Drilling operations may require relocation of the population living on and around the site. Some locations contain historic places, archeological spots, cemeteries, and memorials important for cultural heritage. Information and discussion sessions with focus groups are part of the preparation process. The approval process usually requires board or committee decisions based on community hearings, which sometimes lead to intense public reactions. Especially if the project requires horizontal drilling and fracturing, public reactions might expand to larger groups. Expansions of existing oil fields also need a permit from government authorities.

After resolving all legal issues and obtaining the official permit, the company must lease or purchase the land. Access to the drilling site may require right-of-way agreements or contracts. Offshore and deep-water drilling requires additional international agreements.

After resolving the legal and administrative issues, the company clears and secures the drilling site. If the site is not easily accessible from existing roads, additional access roads must be constructed to transport equipment and personnel. If the field is in a remote location, temporary accommodation must be prepared for employees. Off-shore deep-water oil platforms are self-sufficient and have living facilities for workers.

Drilling operations need electric power, fuel storage, water supply, and disposal of toxic waste. In general, utility power is not available or expensive to access at remote locations. Portable diesel-powered generators are convenient for continuous and reliable electric power. A reserve pit is prepared to dump the rock, dirt, and mud taken out of the ground during the drilling process. The pit is usually lined with plastic sheet to prevent contamination of the ground water from toxic chemicals. If the site is in an ecologically sensitive area like marsh lands and wilderness protection zones, the rock and mud must be hauled away for off-site disposal.

3.5.5.2 Vertical Drilling

Conventional crude oil is extracted by drilling a vertical well. Exploitation of unconventional oil deposits is more difficult and obtained crude oil is thicker. Diluting the product to a thickness convenient for pumping in a pipeline and processing in refineries needs expensive initial processing.

Vertical drilling starts by setting an oil rig and digging a starter hole to install equipment. The essential parts of a typical oil rig are derrick, turntable, and blowout preventer (BOP). The derrick supports the pipes that will be bored into the ground. Diesel engines and electric generators power the turntable and drill bit. Depending on the type of ground, diverse types of drill bits made from steel, hard metal alloys, and even diamond are used. The drill bit, collar, and drill pipe are assembled and inserted in the starter hole. The actual string of pipes bored into the earth is typically between 5 and 36 in. of diameter. A drill bit attached to the lower end of the first pipe advances in the ground by cutting rocks. A drilling fluid that contains a mixture of water, chemicals, and mud is pumped down in the pipes to cool the drill bit while transporting the debris to the surface. As the drill bit goes deeper, new sections of pipes called casing are attached to the top. Casing is cemented to protect freshwater sources and isolate the hydrocarbon deposits and formations with different pressure gradients.

At various stages of drilling, electric sensors and gas detectors lowered into the hole measure the pressure, porosity, and permeability of the rocks, and detect the presence of combustible gases. Samples collected at certain depths are also analyzed to determine the characteristics of the geologic formations. When the samples reveal the oil sand from the reservoir rock, wellbore may have reached the final depth.

When the drilling stream reaches the reservoir rock, the drill pipe, collar, and bit are removed. To allow controlled flow of oil into the well, a perforating gun with charges of explosives is lowered into the wellbore to open holes through the section of the casing in the reservoir. Finally, a smaller diameter pipe inserted in the wellbore serves as a conduit to allow oil (or natural gas) flow from the reservoir to the surface. When the drilling operations are completed, the crew removes the drilling equipment and the rig, connects valves to the tubing, and cements the area around the well.

3.5.5.3 Directional Drilling

Directional drilling refers to boring an oil well in directions other than vertical, including horizontal drilling. The first directional well in the US was drilled in 1933 while fighting the fire in Conroe oil field in Texas. George Failing had patented a portable digging truck in 1931 and was able to control the direction of the well bore. Using his rig-mounted truck, he drilled about one dozen 600-ft relief wells in a short time, enabling the firefighters to extinguish the fire. After the fire was out, the growing lake of oil continued to supply other wells. The growing dimensions of the oil-filled lake required digging a relief well 400 ft away and deviating the borehole underground to reach the crater's source. The joint effort of George Failing and John Eastman, owner of the Eastman Oil Well Survey Co. in Long Beach, California, allowed them to successfully reach the source of the oil flow. Then, steam-powered pumps forcing thousands of tons of high-pressure water into the well could finally stop the erupting oil flow (AOGHS 2005).

Today, using advanced techniques, the wellbore can be drilled precisely in any direction. The position of the wellbore tip is determined by surveying the depth, inclination, and magnetic azimuth at the drill bit location. The progress and location of a wellbore is continuously tracked by consecutive surveys.

Directional drilling has many benefits. For example, several reservoirs can be accessed from a single rig to reduce the footprint of the oil or gas development field. Also, deposits beneath a city, park, environmentally sensitive land, historic site, or strategic area can be tapped without destroying the surface.

While a horizontal well is more expensive to drill compared to conventional vertical wells, it can be a cost-effective solution to reach an underwater reservoir that can be reached from the shore.

Directional wells can increase the "pay zone" of oil within a relatively narrow deposit. Suppose, for example, a reservoir rock of only 40 or 50 ft deep; a vertical well would be underproductive in such a deposit. However, a several-thousand-foot-long horizontal well can increase the productivity. When combined with "hydraulic fracturing," directional drilling can convert unproductive shales into productive reservoir rocks. Horizontal drilling is particularly used to extract crude oil from shales and unconventional tight oil formations.

3.5.5.4 Hydraulic Fracturing

Hydraulic fracturing of a rock formation, also known as "fracking," is based on cracking a rock formation using pressurized fluid, usually containing a mixture of chemicals. The fracking fluid, which is typically high-pressure water containing sand and special solid materials called proppants is injected into the wellbore. Fracturing proppants contain sand, aluminum oxide, and other ceramic materials that hold the fractures open after the hydraulic pressure is removed from the well (Mader 1989). Fracturing stimulates the flow of oil or gas in an oil deposit that has not escaped from the source rock and accumulated in a permeable reservoir rock.

Some rock formations were fractured by explosives (mainly dynamite) to stimulate oil recovery as early as the 1860s. Hydraulic fracturing, however, began experimentally in 1947 and was commercially applied after 1950. Today, hydraulic fracturing is a common procedure to extract light tight oil (LTO), shale gas, and tight gas around the world.

Hydraulic fracturing is, however, highly controversial because of the environmental risks it creates. Groups supporting hydraulic fracturing advocate its economic benefits and exploitation of shale oil reserves available in a broader area than conventional oil. The opponents react to the potential contamination of water supply from fracturing chemicals, pollution, methane leakage, and risk of triggering earthquakes.

3.5.5.5 Offshore and Deep Water Drilling

Drilling the ocean floor or arctic ice is more challenging than drilling on land and requires special techniques. Although the idea of exploiting undersea oil reserves is not new, deep-water drilling was not economically viable due to the higher cost of exploration and recovery when oil prices were lower. Today, only in the Gulf of Mexico, there are more than 3400 deep-water wells producing about 18% of total US crude oil and 5% of the natural gas (EIA 2018a).

If the water is not more than 1600 ft (500 m) deep, fixed platforms can be used to drill the sea floor. A fixed platform is directly anchored to the sea floor and supports a deck above the surface. All drilling equipment and crew quarters are on the deck. Fixed platforms are stable, steady, and safer since they are less susceptible to waves and weather disturbances. On the other hand, it is more expensive to build permanent structures. Therefore, fixed platforms are justified when relatively larger oil discovery is expected. For smaller oil deposits jack-up rigs present a more cost-effective solution.

Jack-up rigs are floating units that are mostly used for drilling exploratory wells or recovering oil or gas under shallower water. They are towed into position by barges, and when the drilling location is finalized, the support legs are lowered to the sea floor and fixed. Then the platform is raised up above the water surface, where the height of the platform can be adjusted.

Compliant tower rigs are also anchored to the seabed and hold all drilling modules above the surface like fixed platforms, but they can slightly move with water and wind action because their platform structure (jacket) is broken into two or more sections. The lower section serves as a base to the upper platform above the sea surface. Compliant towers are taller and narrower than the fixed platforms and they can operate at depths reaching 3000 ft.

A "tension-leg platform," is a floating structure secured in place by vertical tendons attached to the sea floor. These rigs are cost-effective to recover smaller deposits in narrower areas.

Semi-submersible platforms have some parts floating but the wellhead is typically located on the sea floor. These rigs can operate between 660 and 6560 ft (200 and 2000 m) below the surface.

While most deep-water rigs are installed above the water surface, there are subsea systems too. Subsea systems have wellheads sitting on the sea floor. Oil is pumped up to the sea surface through pipes. Since subsea systems can feed nearby platforms, local production hubs, or an onshore site, they are a versatile choice for oil companies.

In all types of deep-water platforms, precautions must be taken to prevent oil leaking into the sea. The tragic accident of the Deepwater Horizon oil rig that happened in the Gulf of Mexico in 2010 is an example for the magnitude of disasters that underwater oil spills can cause. On April 20, 2010 an oil gusher exploded drilling equipment, killing 11 workers. Based on scientific estimations, between 35,000 and 60,000 bbl of crude oil a day spilled into the Gulf of Mexico until BP, owner of the rig, was able to shut down the rig on July 15 (Robertson and Krauss 2010).

Figure 3.11 An oil well and pipeline construction in North Dakota. Source: Photo © H&O Soysal.

3.5.6 Recovery of Conventional Oil Deposits

Conventional oil deposits can be extracted by drilling vertical wells. Wells are the only way to confirm the existence and quality of the oil deposit. Therefore, once a well is drilled, well logs are interpreted to determine the types of rock formations and fluids these rocks contain. This information is used to decide whether a well should be completed and used for oil or gas production or filled with cement and abandoned.

Before regular oil extraction from a new well begins, the pressure of the reservoir is released by flowing some oil, water, and gas out of the well, so crude oil is brought to the surface. Depending on the physical properties of the formation, crude oil may flow naturally to the surface or can be stimulated using chemicals and artificial lifting by using pumps and enhanced recovery methods.

A rod pump is the most common artificial lift system used in oil fields on land. A motor turns a beam-and-crank assembly, which produces a reciprocating motion in a sucker-rod string (see Figure 3.11). The pump consists of a plunger and valve assembly to convert the reciprocating motion of the rod string to create an upward flow of crude oil. Oil extraction continues until the production is no longer economically sustainable. Then, the equipment is removed, the well plugged, and abandoned.

The economic life of a well depends on the crude oil extraction rate and proportion of oil to saltwater in the fluid coming out of the well. If the bigger part of the extracted fluid is saltwater, the well is called "wet." A well that does not produce enough oil or gas to make a profit is called a *dry hole* or *duster*.

3.5.6.1 Light Tight Oil Recovery

LTO is a term to designate any type of oil produced from shales, tight sandstones, or other low permeability formations using hydraulic fracturing in horizontal wells. While shale oil is also often used to designate LTO, it has the risk of confusion with oil produced from shales containing kerogen, which must be further processed by heating to obtain crude oil. Bakken Shale and Eagle Ford are major tight oil formations in North America.

Tight oil deposits have been known since the beginning of the oil industry. Extraction methods were, however, too expensive to compete with the extraction of conventional oil. With increasing oil prices, growing energy security concerns, and advanced extraction technologies, tight oil deposits have become competitive to conventional reserves.

Recovery of most unconventional oil deposits requires drilling a horizontal well and fracturing the rock formation. Extra heavy, tight, and shale oil deposits are examples of unconventional oil reserves. Extra heavy crude oil is further processed with hot water, steam, naphtha, and other chemicals to be adjusted to a viscosity that can flow in the pipeline and meet the refinery specifications.

Case Study 3.1

Tight Oil Production in North Dakota

One of the major light tight oil reserves in North America is Bakken formation covering around 200,000 mi^2 (520,000 km^2) below North Dakota, South Dakota, and Montana in the US, and the adjacent Canadian provinces of Saskatchewan and Manitoba. The center of the basin is roughly 16,000 ft (4800 m) below Williston, ND. The US Geological Survey (USGS) estimated in 2013 that at least 7.4 billion barrels of crude oil, 6.7 trillion cubic feet of dissolved natural gas, and 0.53 billion barrels of natural gas liquids (NGL) can be recovered in the US section of the formation (Gaswirth, Marra, Cook et al. 2013).

The Bakken formation consists of four geologic "members." The deepest member named "Pronghorn" (formerly known as "Spanish Sand") is composed of sandstones and siltstones. Three members of the formation sitting above Pronghorn member form the basin. The upper and lower shales are marine deposits rich in organic matters, which are the principal source rocks. The middle dolomite layer is the main oil reservoir situated roughly 2 mi (3 km) below the surface.

Light tight oil was discovered in 1951 in the land originally owned by Henry Bakken. In earlier years, extraction of unconventional oil was not cost-effective and large-scale commercial production did not start until the beginning of the twenty-first century. With the advanced horizontal drilling and hydraulic fracturing technologies, oil production boomed in North Dakota. Crude oil production in North Dakota reached more than a million barrels per day in 2015, making the state second oil producer in the US after Texas.

The authors of this text visited Williston basin in September 2017. In the area, it is usual to see many small oil wells operating in the middle of corn fields partially leased to oil companies. Oil production in North Dakota created several boom towns such as Williston, Watford City, Stanley, and Minot, where populations tripled after 2010 census.

While oil production brought the unemployment rate in the area to 3.8%, the lowest level in the country, it also had social challenges. The oil rush brought concerns

---

(Continued)

Case Study 3.1 (Continued)

---

among residents of the rural communities about rising crime, including prostitution and illegal drugs. Another complaint was traffic jams from all the pick-up and tanker trucks traveling back and forth to the thousands of new wells that were drilled.

Fast development of oil fields in the area created transportation issues. Produced crude oil was transported to refineries mainly by rail until the Dakota Access Pipeline (DAPL) was constructed. The project attracted opposition (The Guardian 2016) because of its possible impact on the environment and to sites sacred to Native American tribes. Construction started in June 2016 and the pipeline delivered its first oil in May 2017. The 1172-mi long, roughly 30 in. diameter pipeline transports about half a million barrels of crude oil per day from the Bakken oil field to a refinery in Patoka, Illinois, near Chicago.

3.5.6.2 Sand Oil Recovery

Oil sands, also known as tar sands, are permeable and porous sedimentary rocks that contain oil deposits in the form of bitumen, which is a thick and heavy liquid or black solid that melts at a low temperature. Commercially exploited deposits typically contain 5–15% of bitumen. Crude oil production from such deposits must be processed with water and chemicals. Sand oil is also called "synthetic crude oil."

Deposits near the surface can be collected by surface mining and transported to a processing plant to separate oil. Bitumen is extracted from deeper deposits by drilling wells. Either a "cold" or "warm" method is used to extract crude oil from sands that contain oil deposits. Cold production methods are based on surface or subsurface mining of deposits containing heavy hydrocarbons. For example, open-pit mining is used in Athabasca Oil Sands located in Alberta, Canada. The estimated 28 billion m^3 (equivalent to 176 billion barrels) of oil sands of Athabasca can be economically accessed from the surface (ERCB 2013). This mining method is like strip mining used to recover coal. Excavated oil sands are transported to processing plants, where hot water is used to separate bitumen from sand. Then, bitumen is thinned with lighter hydrocarbons such as naphtha to produce upgraded synthetic crude oil (SCO).

Another cold production method uses horizontal wells and injected naphtha or other diluents to decrease viscosity. Electric submersible pumps lift the diluted fluid to the surface to be transported to processing plants. In some areas that are too deep for surface mining bitumen can be produced cold from some wells by utilizing enhanced recovery methods. "Cold heavy-oil production with sand" and "water flooding" may be applied in such fields.

Thermal methods to extract bitumen directly in the reservoir are commonly called "in-situ," which means "in place." These methods use heat to reduce the viscosity. "Cyclic Steam Stimulation (CSS)" is also known as "steam soak" or "huff and puff" method. In this method, injected steam heats oil, then oil is separated from water (Alboudwarej, Felix, Taylor et al. 2006). Another in-situ method is "Steam-Assisted Gravity Drainage (SAGD)."

Thermal technologies for recovering oil sands and extra-heavy oil deposits are mostly energy intensive and have significant environmental impacts. Some part of the natural gas needed for thermal processes may be produced on site. To produce 1 bbl of bitumen at the reservoir (in-situ), about 1.05 GJ energy is needed. This energy is produced by burning roughly 1000 cubic feet (28.3 m³) of natural gas.

In addition to natural gas requirements, in-situ plants require a significant amount of electric energy. Plants without cogeneration must purchase this electricity. Cogeneration plants generate electricity using natural gas and sell the excess electricity after supplying their own processes.

Mining of oil sands uses about four units of water to produce one unit of crude oil at a viscosity suitable for refining. In-situ processes consume about two units of water to one unit of oil. The crude oil separated from oil sands is very thick and is typically diluted by adding naphtha before sending to a refinery via pipeline. The energy efficiency of upgrading oil sand is around 75%; in other words, one unit of energy is used to produce the amount of crude oil that yields three units of energy (Alboudwarej, Felix, Taylor et al. 2006).

Case Study 3.2

Athabasca Oil Sands

The Athabasca Oil Sands is one of the world's largest crude bitumen deposits, located in Northastern Alberta, Canada, near Fort McMurray. The formation covering about 40,000 km² has a zone thickness averaging about 45 m. Deposits are close enough to the surface to allow recovery of 20% of the bitumen by open-pit (surface) mining. The rest of the bitumen can be recovered by currently available underground and in-situ techniques. The total amount of oil that can be economically recovered from Athabasca oil sands with current technology is about 164 billion barrels (CAPP 2018).

Figure 3.12(a) shows an overview of the oil sand production site near McMurray operated by Syncrude Canada Ltd. Figure 3.12(b) The "Cyrus" Bucketwheel Excavator donated by Suncor Energy in 1988 is on display at the Oil Sands Discovery Center. It was manufactured by Bucyrus-Erie Company of South Milwaukee, Wisconsin, in 1963. After being used in a dam construction in California, it was used by Great Canadian Oil Sands between 1971 and 1984 for overburden removal and mining. The operating weight of Cyrus is 7850 tons – the weight of over 500 mid-sized cars. The wheel diameter is 9.15 m and has 10 buckets, each with 1913 l capacity. The bucketwheel can cut 230 m/min and discharge 5371 m³/h. The mobile generator, seen in the photograph background, generates 18-MW at 460-V and travels with the bucketwheel connected by a flexible cable. Figure 3.12(c) shows a demonstration of bitumen recovery in the Athabasca Oil Sands Discovery Center.

When the Rocky Mountains were formed, deformations of the earth crust trapped foreland basins that contained marine sediments, which later became source rock for hydrocarbons through biodegradation of organisms. Hydrocarbons migrated up and eroded from the newly built mountains. Over time, the actions of water and bacteria transformed the lighter hydrocarbons into bitumen, a much heavier, carbon rich, and extremely viscous crude oil.

Oil sands are a bitumen saturated mixture of silica (silicon dioxide), clay, and water. Unlike the kerosene in light oil shale, bitumen dissolves in organic solvents and its viscosity decreases when heated. At room temperature, bitumen is semi-solid, hence it is sometimes called "tar." Above 150 °C, it becomes a thick fluid, and starts to flow at

(Continued)

Case Study 3.2 (Continued)

Figure 3.12 Athabasca oil sands. (a) An overview of the Syncrude sand oil facility. (b) "Cyrus" Bucketwheel Excavator. (c) Demonstration of bitumen recovery at the Discovery Center. Source: © H&O Soysal.

higher temperatures. In the Athabasca deposits, the bitumen is captured in a porous sand matrix, with porosity of typically 25–35%. Sand consists of predominantly quartz grains of between 75 and 250 μm in size, with a small amount of clay particles attached to the surface of sand grains. A thin layer of water between the sand grains in contact with each other forms a continuous sheath. The cavities of the structure are filled with bitumen, which is separated from the sand grains by the layer of water. In contrast to the relatively impermeable oil shales, fluids can flow through the mineral matrix of oil sands. Bitumen constitutes about 18% in mass of the Athabasca oil sands (Probstein and Hicks 1982).

Aboriginal tribes of the region were using bitumen to seal their canoes. European immigrants were aware of the tar sands since the early eighteenth century. Crude oil

(Continued)

Case Study 3.2 (Continued)

extraction started in the 1930s by an American investor, Max Ball, who established Abasand Oils Ltd. Ball installed a separation plant that could process 250 tons of oil sand per day with a federal grant he obtained through the province of Alberta. The initial procedure was to drill holes and blast explosives to loosen the ground. Then, dump trucks would carry the oil sand to a separation plant. During World War II, the federal government overtook the control of Abasand Oils. Production continued until 1946 and stopped because at that time the cost of oil produced from bitumen could not compete with conventional oil prices.

Bitumen is a viscous mixture of hydrocarbons with extremely long chain of molecules that contain larger numbers of carbon atoms than natural crude oil components. A typical ultimate analysis of the chemical composition of the Athabasca bitumen shown in Table 3.3 is relatively constant over a large geographic area. The sulfur content is significantly higher than kerogen. The hydrogen to carbon (H/C) atom ratio of Athabasca bitumen is 1.49, higher than conventional crude oil. Because of its higher viscosity and density (8–14 °API) bitumen cannot be recovered through conventional oil wells. Moreover, bitumen cannot be refined in its natural state into common petroleum products used in practical applications

Table 3.3 Composition of Athabasca bitumen.

	Carbon	Hydrogen	Nitrogen	Sulfur	Oxygen
Mass %	83.1	10.3	0.4	4.6	1.6

Source: Probstein and Hicks (1982).

In the Athabasca River valley, a 40–60 m (130–200 ft) thick layer of oil sand is easily accessible from the surface. Oil sand is surface-mined using extra-large bucket wheels and hauled by heavy trucks (as large as 400 tons) to crushers. Hot water is added to the oil sand and the slurry mixture is conveyed to the extraction plant through pipelines by hydro transport. Separation vessels separate the mixture into layers of sand, water, and bitumen. Bitumen, skimmed from the top layer, is cleaned and further processed to adjust the pH level. Bitumen can be also upgraded on site or diluted by adding naphtha or other solvents before sending to refineries.

Bitumen is upgraded in a four-step process. The first step of the upgrading process is coking to reduce the carbon content and break large bitumen molecules into smaller parts. Then, a mixture of hydrocarbon molecules is sorted into its components by distillation, and catalytic conversion transforms hydrocarbons into more valuable forms. Finally, hydrotreating removes sulfur and nitrogen while adding hydrogen molecules. The end-product of this treatment process is commonly called *synthetic crude oil*. About 2 tons of oil sand is needed to produce 1 bbl of upgraded synthetic crude oil. Produced synthetic crude oil is transported to refineries at an average speed of 5 km/h. Crude oil leaving the Athabasca oil sands arrives to refineries in Edmonton via pipeline in about three days.

3.5.7 Crude Oil Production

According to IEA statistics, 4365 million tons of crude oil was produced worldwide in 2017 (IEA 2018). This corresponds to approximately 88 million barrels per day (1 metric ton ~7.33 US bbl). The US EIA estimates that in the reference case scenario, global production of petroleum and other liquid fuels would increase to 122 million barrels per day by 2050 (EIA 2017b). The reference case model assumes that the current technologies, regulations, and economic conditions will remain unchanged through the projection horizon. Oil prices and economic growth are key factors in estimating future production.

OPEC members produce about 42% of the world's oil from conventional oil resources. Figure 3.13 shows global petroleum production in 2017 by regions (BP 2018). The Commonwealth of Independent States (CIS), which includes the Russian Federation and former Soviet Union countries in Central Asia are major oil exporters after the Middle East. The increase of oil production in the world is mostly due to the increased consumption of developing countries and emerging economies in Asia. In most developed countries, oil consumption is gradually decreasing because of improvements in energy efficiency that in turn reduces energy consumption. Increasing use of other fuels, such as natural gas and renewable energy, and improved energy efficiency in many areas, especially in transportation and buildings, have reduced the share of oil in the global fuel mix. North America has reduced its dependence on foreign oil by increasing production from unconventional oil

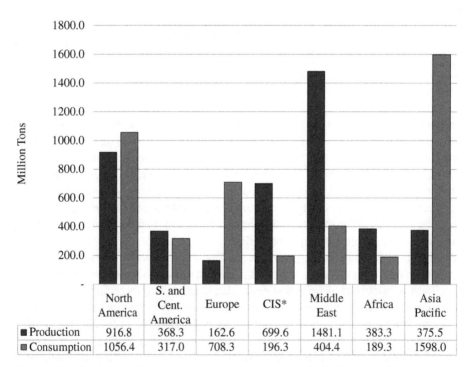

	North America	S. and Cent. America	Europe	CIS*	Middle East	Africa	Asia Pacific
■ Production	916.8	368.3	162.6	699.6	1481.1	383.3	375.5
▣ Consumption	1056.4	317.0	708.3	196.3	404.4	189.3	1598.0

Figure 3.13 Total oil production and consumption by regions in 2017. Source: BP (2018). *CIS: Commonwealth of Independent States. Includes Russian Federation and former Soviet Union countries.

and natural gas resources, and by increasing the share of renewable energy sources in the fuel mix and improving energy efficiency in all sectors.

3.5.8 Fuel Conversions

Crude oil extracted from a reservoir can hardly meet the fuel requirement for any end-use. It has to be processed to obtain diverse fuels and products for specific applications. Refineries process crude oil to produce secondary energy sources and many petroleum products used in various sections of the industry.

Petroleum processes are based on three main steps, which can be applied sequentially or combined; they are separation, cracking, and purification. Crude oil is a mixture of various hydrocarbons with different density, viscosity, and vaporization temperature. Fractional distillation separates these components according to their vaporization points. Crude oil is heated in a chamber until it vaporizes. As the oil vapor is cooled down its components, which condense back to the liquid form at different temperatures, are collected in separate compartments stacked at different heights in the distillation tower. Heavier fractions condense first, then lighter ones, and finally hydrocarbons that stay in gas form at normal ambient temperatures accumulate on the top of all liquids. Figure 3.14 is a schematic illustration of oil products separated through fractional distillation by their vaporization temperatures. Note that fractional distillation of 42-gal of crude oil yields roughly 44.6 gal of oil products (Janardhan and Fesmire 2011). This is because end-products with lower density take up more space than the original mixture.

Economic production of gasoline requires additional processing called *cracking*. A combination of heat, low pressure, and chemical additives help extract gasoline from heavier fractions. The opposite of cracking, called *unification*, and chemical reconfiguration of molecules termed *alteration* can extract gasoline by combining lighter fractions. One barrel (42-gal) of crude oil typically yield about 20-gal of gasoline. Use of cracking and additional chemical processes can even allow a refinery to convert up to 70% of the crude oil input into gasoline (Janardhan and Fesmire 2011).

Fuel oil is an important group of oil products. Commercial grades of fuel oil are identified by numbers one through six. Fuel oil No. 1 is the lightest fuel oil with lowest viscosity, typically used in vaporizing burners. Fuel oil No. 2 is a general-purpose heating oil mostly delivered to residential consumers. Fuel oil No. 3 is no longer commercially available. Fuel

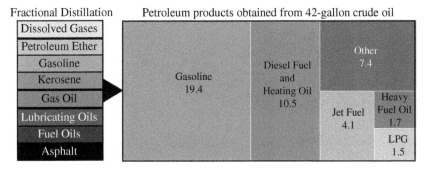

Figure 3.14 Fractional distillation of crude oil.

oil No. 4 is relatively light commercial-grade heating oil. Fuel oils No. 5 and 6 are thicker (more viscous) and must be heated before they can be pumped. Both are used for heating purposes in commercial and industrial sectors (Culp Jr. 1979).

Diesel fuel is a light oil with a higher density than gasoline, which can be used in special types of internal combustion engines that are ignited by injection rather than spark. Diesel fuel is primarily used as a fuel in freight trucks, trains, construction and farm equipment, and small-scale portable or backup electric generation units. Diesel powered cars are popular in Europe because of their efficiency and performance characteristics.

Jet fuel is less dense than diesel but denser than the gasoline used to power aircrafts. Liquefied petroleum gases (LPGs) are lightweight fuels containing propane, ethane, and butane. LPG is mainly a refinery product, but it can be also produced from natural gas. LPG is used as heating and cooking fuel in residences and small commercial businesses, where natural gas is not available. It is mostly liquified and delivered to consumers in pressurized tanks. Although LPG is the lightest grade of crude oil product, it is heavier than air. Density of LPG creates serious risk of fire or explosion in its transportation and consumption.

Figure 3.15 shows 2016 consumption of major petroleum products. While about 80% of refined petroleum products is used as fuel, crude oil also provides feedstock for the petrochemical industry, with about 20% of refinery products used to produce a wide spectrum of materials including lubricants, insecticides, fertilizers, cosmetics, medicines, detergents, clothing materials, plastics, resins, polyesters, and many products we use in our everyday life(EIA 2018b).

Since the development of internal combustion engines, petroleum has been the main fuel for transportation. Oil products are the principal fuel, supplying about one-third of the world's energy consumption. Figure 3.16 shows the share of petroleum consumed by end user categories in 2014 (IEA 2018). Fuels supplied for transportation constitutes the biggest portion of the oil consumption.

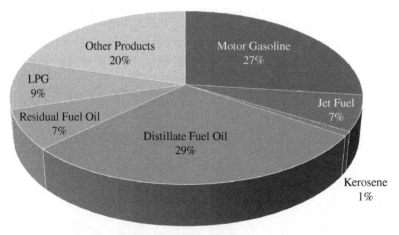

Refined Petroleum Products
2016 total consumption: 96.917 million barrels per day
(Data sourcs: EIA (2018b))

Figure 3.15 Petroleum end products by type.

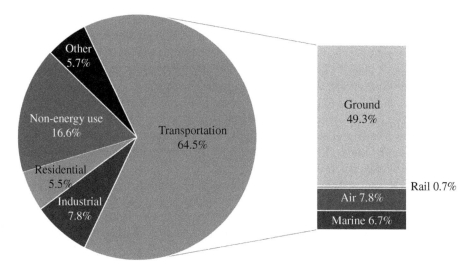

Figure 3.16 Consumption of oil products by sectors. Source: IEA (2018).

3.5.9 Oil Transportation and Distribution

Due to the higher energy density, crude oil and refinery products have several advantages over coal and natural gas in transportation and distribution. According to BP Statistical Review of World Energy 2107.8 million tons of crude oil and 1105.2 million tons of oil products were traded between countries worldwide in 2017 (BP 2018).

Bulk quantities of oil are traded between continents via marine transportation. Pipelines are the most efficient and safest means to transmit crude oil in land. Pipelines allow continuous flow of crude oil from oilfields to refineries. Long distance cross-border pipelines constitute major trade routes between countries.

On land, rail is a common way to transport massive quantities of oil over long distances where pipelines are not available. The retail oil trade mainly depends on tanker trucks to deliver oil products to local dealers.

3.5.10 Challenges of the Petroleum Industry

While oil derivatives are essential fuels that mobilize the modern world and supply feedstock for the petrochemical industry, petroleum production has a toll.

3.5.10.1 Oil Well Tragedies

Sometimes oil wells can discharge significant amounts of crude oil and destroy surroundings, damage the ecology, and take lives. An uncontrolled flow of oil or natural gas from a well is called a blow out, gusher, or spouter (British term). Until the first successful blowout preventer (BOP) became available in 1924, blowouts were common during the oil well drilling. When a high-pressure zone is breached, oil would rise at a high rate creating a gusher. A BOP is a valve fixed to the wellhead that can be closed in the event of drilling into a high-pressure zone. Today, advanced well-control techniques are used to prevent gushers.

Explorers in the first oil field in the US in northwestern Pennsylvania encountered several gushers. On April 17, 1861, the Little & Merrick well in Oil City, PA started gushing oil. The oil flowed out at a rate of about 3000 bbl (480 m³) per day, drawing 150 people to watch the blowout. Later the oil gusher burst into flames and killed 30 people with a rain of fire.

The Lakeview Gusher was one of the most historic blowouts in the US. While drilling was happening on March 15, 1910, oil roared out of the 2,225-ft (667-m) deep well at an initial flow rate of 125,000 bbl/d. A 20-ft diameter column of sand and oil reached 200 ft (60 m) in height. The gusher remained uncapped for 18 months and spilled over 9,000,000 bbl (1,400,000 m³) of crude oil. Despite a pipeline installed in an amazingly short time, only half of this oil could be recovered.

A more recent oil gusher happened in the Gulf of Mexico on April 20, 2010. A semi-submersible drilling rig near Rigel gas field named *Deepwater Horizon* caught fire, burned for two days, and sank in 4992 ft of water. This rig was owned by Transocean Ltd. and was on contract to British Petroleum to drill the Macondo prospect. Eleven persons died, and another 115 men and women were evacuated, among which 17 were injured. The cost of this accident to BP was around $1 000 000 per day. The spill was finally controlled on July 15, 2010 by lowering a 125-ton steel containment dome over the spot where most of the oil was leaking from the sea floor. This succeeded in capturing most, but not all, of the oil (Robertson and Krauss, 2010).

3.5.10.2 Oil Transport Hazards

Oversea transport of petroleum often leads to significant marine pollution that may result in intergovernmental disputes to resolve environmental and commercial consequences (Soysal 1997). In recent history, tragic oil spills resulted from oil-carrying tanker accidents. A sinking oil tanker can release thousands to millions of barrels of oil into oceans, seas, rivers, or lakes. Oil is extremely toxic and harmful for marine life. Oil spills damage the marine ecosystem and have long-term effects on the food chain. Oil tanker accidents can also ignite severe fires, which may reduce the oil spill but cause severe ground-level air pollution and emission of enormous amount of toxic gases to the atmosphere.

On November 15, 1979 the Romanian tanker Independenta carrying 714,760 bbl of crude oil collided with the Greek ship Evrialy in Istanbul, Turkey, at the southern entrance of Bosporus. The Independenta exploded immediately and both ships began to burn. The fire lasted for about a month and caused significant air pollution over the largest metropolitan city in the region, with a population of nearly 10 million at the time of the accident. Bosporus, which is the only marine trade route out of the Black Sea, divides the European and Asian sides of Istanbul. Although the fire prevented an oil spill, it caused severe air pollution and affected the marine traffic in the busy harbor for months.

The Exxon Valdez accident is one of the most dramatic examples of marine oil spill disasters. On March 24, 1989, the oil tanker Exxon Valdez, full of crude oil, struck a reef in Prince William Sound on the east side of the Kenai Peninsula in Alaska. The port of Valdez is at the southern end of the Trans-Alaska Pipeline System. Following the strike, 11 million gallons of oil spilled into the gulf, and a storm blowing soon after spread the oil widely. Eventually, more than 1000 mi of coastline were contaminated, and hundreds of thousands of animals died.

In November 2002, the oil tanker Prestige carrying heavy fuel oil sank 30 mi off the coast of Spain, and was one of the biggest ecological disaster in Europe. Four thousand tons of fuel leaked from the tanker and impacted more than 200 km of the Galician coast, in northern Spain. After the tanker sank, it contained between 60,000 and 70,000 tons of oil within its compartments. This amount of oil would eventually leak to the ocean where the water is over 3000 m deep, creating long-term damage to the marine life.

On January 6, 2018, the oil tanker Sanchi carrying 136,000 tons of crude oil from Iran to South Korea collided with the cargo ship CF Crystal 160 nautical miles off the coast of Shanghai, China. The collision triggered a fire, and the oil tanker eventually sank eight days later. All 30 Iranian and two Bangladeshi seafarers died in the accident.

Box 3.1 Humble Beginnings of the Oil Industry

Before becoming the essential fuel for transportation, petroleum had been known for centuries and used for many other purposes. According to historic findings, tar was used as building material in Babylon (Tower 1909). Around 347 AD, oil wells were drilled in China using bits attached to bamboo poles. During the siege of Persia in 331 BC, Alexander the Great's tent was lit by containers filled with oil brought from shores of the Caspian Sea (Aliyev 1994). Herodotus reported that oil-burning lamps were lighting ritual scenes in ancient Egypt. In the thirteenth century, Marco Polo witnessed mining of seep oil on the shores of the Caspian Sea during his travels through Baku in medieval Persia (today Azerbaijan) (Tower 1909). Oil was one of the elements of the "Greek Fire" used to defend cities and fortresses in medieval times.

A thick, black liquid seeping from the Carpathian Mountains in Poland was known back in the seventeenth century. Lamps burning this sticky liquid extracted from hand-dug pits were lighting the streets of the Polish town Krosno. This fuel burned with more smoke and soot compared to other lamp oils made from animal fat. In the early 1850s, a Polish chemist Ignacy Lukasiewicz successfully distilled seep oil to obtain clear burning kerosene as a cheaper alternative to whale oil, which was a popular lamp fuel at the time. Lukasiewicz registered his distillation process on December 31, 1853 in Vienna. Several other entrepreneurs hand drilled 30–50 m deep wells to extract crude oil in the Carpathian Mountains.

In North America, Seneca Indians in northwestern Pennsylvania were using petroleum as insect repellent, laxative, and stimulant. First settlers used this fluid as a cure-all medicine sold as *Seneca Oil*. The black greasy fluid with a strong smell contaminating brine wells around Allegany river, near Pittsburgh, PA was annoying salt producers. In 1818, a salt producer who hit oil in Kentucky, dumped this disgusting byproduct into Cumberland River. The oil turned a section of the river to black, then caught fire destroying surroundings. Local community started calling this fluid "devil's tar."

Around 1854, Benjamin Silliman successfully distilled the Seneca oil into several fractions, including an illuminating oil already known as kerosene. Edwin Drake, who has perhaps heard the use of similar fuel in Europe, collected the oil by blocking the stream now named Oil Creek near Titusville in northwestern Pennsylvania. At the beginning Drake was skimming the top of the pond to produce a lubricant for mill machinery. To increase the production, he tried drilling a well to locate the origin of the seep. In 1859, he hired Billy Smith, who had

(Continued)

Box 3.1 (Continued)

experience in drilling brine wells using steam-powered equipment. Drilling was progressing slowly without significant increase in oil production. When the well reached about 70 ft, they realized oil rising in the hole. After installing a hand-operated lever pump they could produce up to 40 bbl a day. Daily production soon dropped off to a steady 10 bbl, and the well continued at around that rate for a year or more (Giddens 1948). Although this venture could not produce a significant amount of oil, it was the beginning of the oil industry. During the US Civil War, steam powered mechanical drills were being used to drill oil wells in Pennsylvania. Soon, the oil industry boomed, producing 5 million barrels in 1870, and 26 million in 1880. Other oil fields were discovered all over the country to the West Coast. In 1920, 203,400 wells were producing 1.2 million barrels of crude oil per day. G. O. Smith estimated in his article published in the February 1920 issue of the National Geographic Magazine that oil supplied to the US in 1918 (including imports) was equivalent to the flow of water in Niagara Falls for about three hours. At that time, the total length of crude oil pipeline network in the east coast between New York and the Gulf of Mexico was about 30,000 mi (Smith 1920). Oil production in the US peaked in 1970 at 9.637 million barrels per day, then started to decline (EIA 2018b).

The oil industry was well-established during the first few decades of the twentieth century. However, coal was the world's principal source of energy. During World War I, the British Royal Navy started to switch from coal burners to internal combustion engines running on fuel oil. Use of petroleum allowed higher speeds and permitted faster refueling at sea with less labor force. Development of the automobile industry and airplanes increased the use of petroleum products. Particularly during World War II, the army, navy, and air force of all nations relied on petroleum. Hence, oil became a strategic commodity. Today transportation sector relies on oil derivatives and apparently petroleum will remain the essential primary source for all transportation modes at least till the next century.

3.6 Natural Gas

Natural gas is a mixture of hydrocarbons that occur naturally in sediments that contain deposits of ancient plants and organisms. Dead organisms buried under sedimentary rocks have released a mixture of combustible gases over thousands of years with the effect of heat and pressure. Natural gas coming out of the well is primarily methane mixed with other hydrocarbons such as ethane, propane, butane, and pentane. The mixture also contains small fractions of pollutants including hydrogen sulfide, carbon dioxide, water vapor, and other gases (Table 3.4). The composition of natural gas depends on the depth and type of the sedimentary rocks where it was formed.

Natural gas has been known since the beginning of the petroleum industry. Gas discharge occurs naturally as oil wells are drilled. Many petroleum reserves also contain a certain amount of natural gas accumulated above or dissolved in the crude oil. If the gas deposit has commercial value, it may be extracted separately, processed, and marketed. Natural gas extracted from a crude oil well is called *associated gas*. During the early years of the oil industry, natural gas was either released into the atmosphere or just burned to eliminate. In the oil industry, the term *flaring* means burning natural gas coming out of the well without any energy-related or commercial purpose. Still, in many oil fields where commercial collection

Table 3.4 Content of raw natural gas and classification of components.

Constituents	%	Boiling temp. °C at 1 bar	Classification		
Methane (CH$_4$)	70–90	−161	Dry gas		
Ethane (C$_2$H$_6$)		−88			
Propane (C$_3$H$_8$)	0–20	−42	LPG	Natural gas liquids (NGL)	
Butanes (C$_4$H$_{10}$)		~0			
Pentanes		30–36			
Naphta		35	Heavier fractions		
Natural gasoline		−			
Condensate		−			
Nitrogen (N$_2$)	0–0.5				
Hydrogen sulfide (H$_2$S)	0–0.5		Non-hydrocarbons	Acid gases	
Carbon dioxide (CO$_2$)	0–8				
Oxygen (O$_2$)	0–0.2				
Rare gases (He, A, Ne, Xe)	Trace				

and transmission facilities are not available, natural gas is flared to prevent emission of methane, which is one of the potential greenhouse gases causing climate change.

Natural gas reservoirs are not limited to petroleum formations. Many natural gas reserves do not contain crude oil at all, or the amount of petroleum is not suitable for commercial extraction. The term *non-associated* denotes natural gas that comes from gas wells and condensate wells, which produce additional liquid hydrocarbons called *natural-gas concentrates*. Another source of natural gas is methane deposits in the pores of coal seams, usually named *coal bed methane* (CBM). Several shale formations in the US produce non-associated natural gas.

Natural gas obtained from underground deposits is a non-renewable fuel because its formation rate is much smaller than its consumption. However, sewage, animal waste, and trash collection sites also release methane, which is the main constituent of natural gas.

Natural gas market professionals call natural gas that is practically free of hydrogen sulfide (H$_2$S) *sweet gas*. Natural gas that contains a relatively small amount of hydrogen sulfide is *sour gas*. A gas mixture that contains significant amounts of hydrogen sulfide, carbon dioxide, and other acidic impurities is called *acid gas*. Natural gas is considered dry when it is almost pure methane. "Wet" natural gas contains other hydrocarbons in addition to methane (not water vapor).

3.6.1 Purification and Processing of Natural-Gas

While the primary component of natural gas is methane, when extracted it contains several other combustible gases that are more valuable for a variety of applications. In addition, raw natural gas at the wellhead contains small amounts of carbon dioxide (CO$_2$), hydrogen sulfide (H$_2$S), methanethiol (CH$_4$S), mercury, nitrogen, helium, water vapor, and other impurities that must be removed before it is delivered to consumers. Purification and separation of natural gas components is relatively simpler than refining crude oil.

After water vapor, particulate matter, acidic gases, and other impurities are removed, natural gas is processed to separate hydrocarbons and produce pipeline quality *dry natural gas*. To be delivered to consumers via pipelines, natural gas must meet quality standards. While these standards slightly vary based on the pipeline specifications and customer needs, transmitted natural gas must be free of any particulate matter and water vapor, and contain insignificant amounts of mercury and acidic gases (H_2S, CO_2, NO_x, etc.). Delivered natural gas must have a specific heating value. For example, in the US the energy content of natural gas delivered to end-users must be about 1035 Btu/ft^3 at 60 °F at normal atmospheric pressure.

While purification and some part of the separation can be done at the wellhead, complete production occurs in a natural gas processing plant. If the natural gas is associated with crude oil, gas and condensates are separated from oil at or near the wellhead. Most of the water is also removed at the field but complete removal of water requires more complex treatment at the processing plant, either by absorption or adsorption. Absorption is based on dehydrating natural gas using a chemical agent such as glycol or solid desiccant. Adsorption is condensing and collecting water on the surface.

Sulfur exists in natural gas in the form of hydrogen sulfide. In practice, the common term for removing hydrogen sulfide is called "sweetening" the natural gas. About 95% of US gas sweetening plants use amine solutions to remove hydrogen sulfide. Monoethanolamine (MEA) and diethanolamine (DEA) are two principle amine solutions that effectively absorb sulfur compounds as natural gas streams pass through. Sulfur is separated to be sold as a side product. Hence, the regenerated amine solution can be reused in the process.

Hydrocarbons separated from methane are an important group of byproducts named "natural gas liquids" (NGL). Ethane, propane, butane, isobutane, and natural gasoline that form NGLs are sold separately. Because each of these hydrocarbons are more valuable than methane, they are usually recovered from purified natural gas and marketed separately. This is usually done in two steps; first the NGLs are extracted all together from the natural gas stream, then the base components are separated from the NGL mixture. The absorption method and cryogenic expander process are the major techniques to extract NGL from raw natural gas. The absorption method is similar to the method used to remove water. As opposed to using glycol or dehydration agent, NGL is extracted using an absorbing oil that has affinity to these gases. As the natural gas stream passes through the absorption tower most of the NGL is captured in the absorbing oil. "Rich" oil that contains NGL leaves the tower from the bottom. The mixture is heated to a temperature higher than the boiling point of the NGLs but below the boiling temperature of the absorbing oil. This process can extract about 40% of ethane and nearly all the other NGLs.

The cryogenic expansion process is based on cooling the gas stream down to about −120 °F. One of the methods to effectively drop the temperature is to use an external refrigerant gas, then rapidly expand the mixture to further cool down. Cryogenic expansion methods can recover up to 95% of ethane.

Practical uses of NGL are summarized in Table 3.5. Propane and butane are mainly used as fuel. Pentane is used as an additive to vehicle fuels and bitumen thinner in oil sands. Ethane and isobutane are important feedstock in petrochemical industry.

Table 3.5 Principal applications and end-uses of natural gas liquids.

NGL	Applications	End-use products
Ethane (C_2H_6)	Ethylene for plastics production; petrochemical feedstock	Plastics, anti-freeze, detergents
Propane (C_2H_8)	Residential and commercial heating; cooking fuel petrochemical feedstock	LPG tanks for home heating, small stoves, barbecues, etc.
Butane (C_4H_{10})	Blending with propane or gasoline; petrochemical feedstock	Synthetic rubber, tires, LPG tanks, lighter fuel
Isobutane (C_4H_{10})	Refinery and petrochemical feedstock	Alkylate for gasoline, aerosols, refrigerant
Pentane (C_5H_{12}) – natural gasoline	Blending with vehicle fuel, bitumen production in oil sands, blowing agent for polystrene foam	Transportation, insulation, and packing materials

Source: DOE (2017).

3.6.2 Natural Gas Resources and Reserves

As soon as the commercial value of natural gas was realized, the oil industry focused on the gas collected in reservoirs above or around conventional petroleum deposits. Natural gas is available in many regions of the world, especially where commercial scale petroleum deposits are present. Worldwide natural gas resources are more abundant than petroleum, but natural gas reserves and production is currently less than crude oil (see Figure 3.1). Survey technologies based on ultrasound and seismic shock waves developed for petroleum exploration are also used to locate natural gas reserves. Advanced software can produce a mapping of the underground formations. However, locating a natural gas reserve is more challenging than crude oil because some underground formations, especially salt deposits, distort or block the seismic waves.

At the end of 1996, the total proved natural gas reserves were estimated at 123.5 trillion cubic meters. In two decades, at the end of 2016, global natural gas reserves rose 50% reaching 186.6 trillion cubic meters (see Figure 3.17) (BP 2018). This increase is mainly due to the addition of unconventional gas reserves that were considered uneconomical in the past. As advanced technologies made unconventional resources feasible, consumption also increased. The world ended the year 2016 with the reserve to production ratio (R/P) of 52.5 years, which means that if the reserves and production remain the same, the natural gas reserves would be completely exhausted in about half a century.

3.6.3 Unconventional Natural Gas

Natural gas is more environmentally friendly and efficient than both coal and oil. Increased use of natural gas motivated the petroleum industry to explore alternative resources that

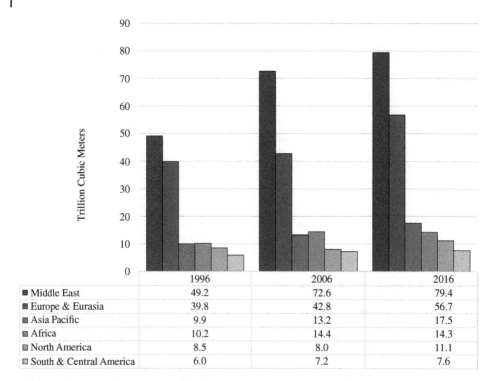

	1996	2006	2016
■ Middle East	49.2	72.6	79.4
■ Europe & Eurasia	39.8	42.8	56.7
■ Asia Pacific	9.9	13.2	17.5
▣ Africa	10.2	14.4	14.3
▣ North America	8.5	8.0	11.1
▢ South & Central America	6.0	7.2	7.6

Figure 3.17 Natural gas proved reserves over two decades. Source: BP (2018).

require more challenging methods than conventional gas production. Natural gas absorbed in relatively thin layers of sedimentary rock formations classified as "tight gas," and "shale gas" are among such unconventional gas resources. Other emerging sources of combustible hydrocarbons are "coal bed methane" (CBM), and "gas hydrates." Tight gas is extracted from low-permeability sandstones or carbonates. Shale rock formations are another important natural gas source. CBM is a highly explosive gas naturally developed in coalmines. In the early years of coal mining, CBM was being removed as a safety measure to reduce the risk of explosions in mine shafts. Commercial production of CBM started in 1970s as coal mine degasification and safety techniques were developed in the US. Today CBM is captured in many countries as a source of energy.

Unconventional gas recovery methods are relatively new and usually require drilling horizontal or directional wells, combined with hydraulic fracturing of the rock formations. To fracture a rock sediment a large quantity of pressurized water containing sand and a mixture of special chemicals is pumped into the well. These chemicals include acid solutions, corrosion inhibitors, friction reducing agents, and substances to form gel. The composition of additives depends on the properties of the reservoir and are usually proprietary to the drilling company. Fractures artificially produced in a rock formation stimulate the flow of natural gas and oil for more cost-effective extraction. The fractures may extend several hundred feet away from the wellbore. As a final stage, a mixture of sand, ceramic pellets, or other small incompressible particles called "proppants" are injected to keep the fractures open (Figure 3.18).

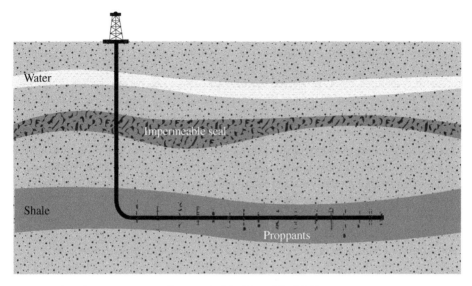

Figure 3.18 Tight gas recovery. Source: Modified from IEA (2012).

Unless natural fractures are present, almost all tight sand reservoirs require hydraulic fracturing to release gas. Hydraulic fracturing technology is usually required to stimulate gas flow in deep rock formations. Unconventional gas production has become economical with the development of advanced technologies for horizontal drilling and hydraulic fracturing.

Shale gas is widely available in more than 38 countries, covering almost all regions of the world. In 2012, shale gas was a significant fraction of the gas supply in the US (39%) and Canada (15%).

It is important to note that many of the major shale formations in many countries are in densely populated areas. In the US, for example, Marcellus Shale, which is one of the major natural gas resources, lies in a large region in the eastern US, beneath several eastern states including New York, Pennsylvania, Maryland, West Virginia, Ohio, and Kentucky. On one hand, this is an advantage from an economic perspective since resources are closer to the consumers. On the other hand, development and production unconventional gas infrastructure depends highly on social acceptance and regulations due to the environmental impacts. Fracking Marcellus shale to recover unconventional gas has attracted significant public reactions.

3.6.4 Natural Gas Transportation

Natural gas is transported in two ways: via pipeline or in liquefied natural gas (LNG) tankers. Because of the cost of liquefaction and degasification, LNG transport is more suitable for long distance marine transport or land transport to storage facilities.

Pipeline network is more efficient for land transportation of natural gas. The route from wells to consumers consists of three distinct types of pipelines named "gathering," "interstate," and "distribution" systems.

Gathering systems carry natural gas from wells to the processing plants. Raw natural gas contains corrosive gases and impurities. Therefore, small diameter, low-pressure steel pipes resistant to corrosion of hydrogen sulfide and carbon dioxide are installed in gathering pipelines.

Interstate pipelines carry processed natural gas from processing plants to distribution hubs near populated regions where large amounts of natural gas is consumed. Long-distance pipelines with 16–48 in. diameter transmit natural gas at 200–1500 psi (15–100 bar) pressure. High pressure increases the amount of transmitted energy and helps efficient gas flow over long distances. Lateral branches of the mainline transmission pipes are typically 6–16 in. diameter. Most pipes are made of strong carbon steel material meeting API or similar international and national standards. Compressor stations installed at intervals of 40–100 mi (65–160 km) along the pipeline maintain the pressure. Metering stations, control centers (supervisory control and data acquisition; SCADA), and valves are other essential parts of the long-distance pipelines.

Distribution systems deliver natural gas from distribution hubs to consumers. While most distribution lines are smaller diameter made from carbon steel, advanced plastic pipes may be used in some distribution lines for flexibility, versatility, and ease of replacement. More information about pipeline networks is available at http://naturalgas.org.

LNG is mostly used for overseas transportation of natural gas. Particularly for natural gas extracted from deep-sea wells or remote locations that are not convenient for pipeline construction, LNG is the only way to transport vast quantities of natural gas. Methane is liquefied at around −260 °F (−162 °C) at atmospheric pressure. Liquefaction of natural gas is an expensive process that can be achieved safely at specialized industrial facilities. LNG is shipped from production facilities to receiving ports by special LNG carriers at atmospheric pressure and −260 °F temperature.

Natural gas is not explosive at liquid state and can burn when it is mixed with air at 5–15% concentration. The narrow range of flammability makes natural gas safe for transportation. In addition, in the case of a spill, NGL evaporates quickly preventing the risk of long-term sea or land contamination. On the other hand, methane has high global warming potential (GWP). Release of large amounts of methane to the atmosphere during processing, transportation, and storage contributes to climate change.

3.6.5 Storage of Natural Gas

Consumers' need for natural gas varies in time. Therefore, it is more economical to store natural gas to supply short-term or long-term fluctuations of consumption. Because natural gas is a major heating source, residential consumers use significantly more natural gas in winter than in summer. Industrial consumption is more stable, but as the fuel for electric power plants is shifting from coal to natural gas, variations of electric consumption began to reflect on natural gas consumption. Electric consumption fluctuates both seasonally and daily. In summer electric demand is higher because of air conditioning needs. Electric power fluctuates during the day depending on the activities of consumers. Electric consumption is typically higher in evening hours because of lighting. In peak hours when electric consumption reaches a maximum, natural gas fired power plants must generate more electricity.

In addition, natural gas storage increases reliability and security of an energy system by providing a backup source for emergency situations created by unexpected supply shortages caused by natural disasters or political tensions.

Natural gas can be stored efficiently in massive quantities for either short or long term. Short-term storage is usually needed to compensate peak loads. Short-term storage facilities are designed for faster input and output to respond to sudden changes of consumption. Such facilities can also supply base load variations by delivering smaller amounts of gas and being refilled quickly. Long-term storage is needed for seasonal variations or energy security. Larger storage spaces with minimal leakage are required for long-term storage.

Underground reservoirs are convenient and cost-effective for natural gas storage. Three main types of underground reservoirs are *depleted reservoirs*, *aquifers*, and *salt caverns*. To recover stored gas back from the reservoir, a certain pressure is needed. Therefore, some part of the stored gas cannot be recovered from larger reservoirs. The amount of gas that can be recovered is the *working capacity* of the storage facility.

Depleted gas reservoirs are the formations where all the economical amount of natural gas has already been extracted. The space emptied of natural gas can be used as storage depending on the geologic and geographic characteristics of the reservoir. The benefit of such reservoirs beside their larger capacity is the leftover equipment from the productive operation period of the oil well. Converting depleted gas formations into storage is a cost-effective way to create an immense storage capacity.

Aquifers are underground permeable rock formations where large amounts of ground water has collected naturally. Some of the aquifers may be reconditioned to be used as gas storage. However, converting an aquifer into a gas reserve requires more investment than a depleted gas reservoir.

Salt caverns are underground salt formations that may be suitable for natural gas storage. The walls of a salt cavern provide enough structural strength and the amount of gas leakage is small. Salt caverns may be as large as a mile (1.6 km) in diameter and 30,000 ft (9 km) high. Most of the salt domes used as natural gas storage are between 6000 and 1500 ft (1800–450 m) below the surface.

3.6.6 Natural Gas Consumption

Consumption of natural gas has increased during the last several decades as a fuel for heating, electric generation, and internal combustion engines because of recently discovered unconventional reserves and smaller carbon dioxide emission compared to coal and petroleum. Natural gas is a popular and desirable heating fuel for commercial and residential consumers in urban areas where pipeline infrastructure is has been developed.

As we discussed earlier, NGLs are valuable feedstock for production of ammonia, fertilizers, plastics, rubber, aerosols, and many other petrochemical products. Propane and butane can be economically liquefied and sold in pressurized tanks as heating fuel in less developed regions or rural areas where pipeline network is unavailable. Today, about one-fifth of the world energy is supplied from natural gas. Figure 3.19 illustrates the share of natural gas use by consumer groups.

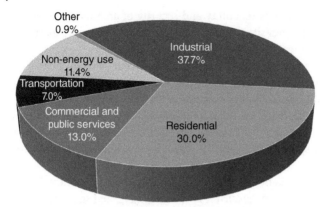

2015 World total natural gas consumption: 1401 Mtoe
(Data source: IEA Key World Energy Statistics, 2017)

Figure 3.19 Global share of natural gas consumption by end-users. Source: IEA (2018).

3.6.7 Environmental Impacts of Natural Gas Consumption

The estimated GWP of methane 25 times greater than carbon dioxide (IPCC 2006). During the construction of wells and related infrastructure, significant amounts of methane is released to the atmosphere until natural gas is collected and transported for commercial use. In the production phase some part of the natural gas also escapes to the atmosphere because of leakages, tests, maintenance, and repairs. Such release of natural gas is called "venting." Venting emits additional chemicals called "volatile organic compound" (VOC) and "particulate matter" (PM). Reactions between VOC and nitrogen oxides produce ground level ozone, which is one of the components of smog. PM and smog cause heart and respiratory health issues.

Natural gas production, transport, distribution, and usage directly or indirectly affect air quality with gas emissions such as nitrogen oxides (NO_x), carbon monoxide (CO), and sulfur dioxide (SO_2). Diesel engines used in the drilling process, driving pumps and compressors in pipelines, and natural gas fueled vehicles produce such gas emissions. Natural gas is increasingly used in electric generation to reduce the toxic gas emissions, solid waste, and water contamination. However, natural gas fueled electric power plants produce significant amounts of carbon dioxide emission.

Recovery of unconventional natural gas requires a significant amount of water for hydraulic fracturing. Water use depends on the depth, horizontal length, and the geologic structure of the formations through which the hole is drilled. Drilling and hydraulic fracturing to recover shale gas at different reservoirs in the US consumed between 2.7 and 6.85 million gallons of water for each well (NETL 2014; CRS 2009). If the water use is expressed in terms of energy produced, the estimated average water-intensity of shale gas production is about 1.3 gal per million Btu (Mielke et al. 2010). This amount translates to 17 l/MWh.

The water use for natural gas extraction is lower than all other energy sources except unconventional natural gas. However, water availability and pollution are more crucial for the future of unconventional natural gas extraction and processing. In the western

United States water availability is limited near some shale formations and distances to water sources are greater. Water availability causes challenges for drilling and production, especially in warmer summer months. In the eastern US shale formations are in densely populated areas. Whereas sufficient water is available in these areas, pollution of fresh water because of additives used in hydraulic fracturing is a bigger concern.

Gas production from CBM formations may reduce the water available for other uses as water in the coal bed is removed to lower reservoir pressures (GAO 2012).

Shale gas drilling presents potential risks to water quality from spills or releases of chemicals and wastes. Many incidents such as tank ruptures, blowouts, equipment or impoundment failures, overfills, vandalism, accidents, ground fires, operational errors, or storm water runoff may cause pollution of freshwater sources (GAO 2012).

3.7 Chapter Review

Fossil fuels are hydrocarbon deposits formed over millions of years with the combined effect of heat and pressure on dead organisms buried under sedimentary rock formations. While fossil fuels are typically used to produce heat by combustion, they also provide valuable feedstocks for industry.

The term "resource" signifies any type of hydrocarbon deposit available in the Earth's crust. Recoverable and proved "reserves" are deposits that can be extracted economically with known technologies. Reserve to production ratio (R/P) shows how many years the reserves would last if the production remained constant at the current amount.

The main constituents of fossil fuels are carbon, hydrogen, oxygen, sulfur, and nitrogen. Trace amounts of mercury, cobalt, manganese, and uranium compounds, as well as other minerals, may be present in their composition. The amount of humidity, ash, sulfur, and volatile organic matter can be determined by a simple proximate test. An ultimate analysis in a specialized laboratory determines the mass fractions of all components.

Thermal properties of fossil fuels are characterized by "gross calorific value" (GCV) and "net calorific value" (NCV). The difference between these values is the latent heat of the water that vaporizes during combustion.

Coal is a common name for solid hydrocarbon deposits ranked by their organic maturity. Peat and brown coal are the lowest rank, lignite and bituminous coal are higher rank. Anthracite is the highest rank coal with the highest carbon content. Lignite and bituminous coal mainly fuel electric generation plants. Anthracite is mostly used for heating and as feedstock for the steel industry and coke production.

Open pit (surface) mining or underground (deep) mining are the common methods for coal extraction. Coal handling and preparation plants separate coal from rocks, dirt, sulfur, and other unwanted elements. Coal mining, processing, and consumption have considerable environmental and societal impacts.

Crude oil is a mixture of liquid hydrocarbons formed from marine organisms buried under sedimentary rocks. Kerogen is an intermediate deposit in oil formation. Conventional petroleum deposits are accumulated around source rock and sealed by an impermeable layer called shale.

While natural gas deposits may be associated with oil formations, a considerable number of formations contain particularly natural gas. Oil and natural gas exploration techniques are based on sending shock waves into ground and computer analysis of collected data.

Density, viscosity, and sulfur content are the main specifications of crude oil. The American Petroleum Institute (API) gravity is a measure of the oil density. Crude oil, having an API gravity greater than 10, floats on water. Light crude oil has an API density higher than 31.1. Heavy and extra heavy crude oil is typically recovered from unconventional deposits such as oil sands, tight oil, and extra heavy oil deposits. About 60% of global oil resources are found in unconventional formations. Crude oil recovery from unconventional deposits requires horizontal (or directional) drilling, hydraulic fracturing (fracking), and additional processing.

Crude oil is an important commodity that affects the global economy and international relations. Oil products are the main fuels for the transportation sector and feedstock for the petrochemical industry. Refineries produce specific oil products by fractional distillation of crude oil.

Natural gas, as extracted from the ground, is a mixture of hydrocarbons in a gaseous state at normal temperature and atmospheric pressure. Processing plants separate methane from NGLs, which are more valuable fuels and industrial feedstock. Several NGLs are marketed separately in pressurized tanks (LPG). Some LNGs are important raw materials for chemical industry.

Natural gas commercially delivered to end users is nearly pure methane. The biggest challenges of natural gas are transportation and distribution. Delivery of natural gas to end users depends on the construction of a reliable pipeline infrastructure since liquefaction of methane is extremely costly. Methane turns into liquid form at −260 °F (−120 °C) under atmospheric pressure. LNG is mainly produced for overseas transportation by special tanker ships.

Natural gas is the cleanest fossil fuel. Advanced extraction techniques including horizontal drilling and hydraulic fracturing made unconventional natural gas deposits commercially feasible. Environmental benefits over coal and energy security concerns regarding petroleum availability increased natural gas production significantly over the last several decades.

Review Quiz

1 About what percentage of the worldwide energy is supplied from fossil fuels?
   a. Less than 20%
   b. About 25%
   c. Between 30% and 50%
   d. More than 75%

2 In the field of energy, the term "resource" means
   a. a deposit of fuel that is available at some degree of confidence for commercial extraction.
   b. a deposit of fuel that can be extracted from earth and converted to useful energy.

c. a deposit of fuel that is readily accessible.

d. a deposit of fuel that can be economically produced.

3 In the field of energy, the term "reserve" means

a. a deposit of fuel that is available at some degree of confidence for commercial extraction.

b. a deposit of fuel that can be extracted from earth and converted to useful energy.

c. a deposit of fuel that is readily accessible.

d. a deposit of fuel that can be economically produced.

4 Which one of the fossil fuel resources listed below has the biggest energy potential worldwide?

a. Coal

b. Conventional crude oil

c. Natural gas

d. Shale gas

5 Gross calorific value of a fossil fuel is also called

a. net calorific value.

b. high heating value.

c. low heating value.

d. None of the given answers.

6 Which type of coal has the highest calorific value?

a. Lignite

b. Bituminous

c. Peat

d. Anthracite

7 Which region of the world has the biggest proved reserves of anthracite and bituminous coal?

a. Asia and Pacific

b. Europe and Eurasia

c. North America

d. Africa

8 In the oil industry what does the term "flaring" mean?

a. Glare of petroleum extracted from a well at dark

b. Burning gasses that come out of an oil well to prevent emission

c. Explosion of an oil well

d. Burst of the first crude oil coming out of the ground

9 Which consumption sector is the biggest consumer of lignite?

a. Transportation

b. Residential

    c. Commercial

    d. Electric generation

10 Which sector has the biggest share in the consumption of oil products?

    a. Transportation

    b. Residential

    c. Commercial

    d. Electric Generation

11 About what percentage of world's oil reserves contains conventional crude oil deposits?

    a. Less than 20%

    b. Between 25% and 35%

    c. Between 35% and 45%

    d. More than 50%

12 The most efficient way to transport natural gas is

    a. rail.

    b. pipeline.

    c. trucks.

    d. tanker ships.

13 The biggest component of natural gas is

    a. propane.

    b. butane.

    c. methane.

    d. ethane.

14 Natural gas liquid (NGL) is

    a. liquefied natural gas.

    b. a liquid fuel that give natural gas when evaporated.

    c. a group of fuels obtained by processing natural gas.

    d. any liquid obtained by condensing natural gas.

15 The major end use of ethane is for

    a. petrochemical industry feedstock.

    b. residential and commercial heating.

    c. industrial heating.

    d. blending with vehicle fuel.

16 The major end use of propane is for

    a. petrochemical industry feedstock.

    b. residential and commercial heating.

    c. coking.

    d. blending with vehicle fuel.

17 Liquefied natural gas can be transported by
  a. long distance pipelines.
  b. regular tanker trucks.
  c. special LNG vessels.
  d. river barges.

18 Which sector consumes the biggest part of natural gas production?
  a. Transportation
  b. Commercial and public services
  c. Residential
  d. Industrial

19 Liquefied natural gas (LNG) is not used as a fuel for vehicles because
  a. turning natural gas into liquid is too expensive.
  b. it is explosive.
  c. burning methane causes pollution.
  d. it has less energy per mass than gasoline.

20 Which one below is the biggest challenge of natural gas industry?
  a. Difficulty of combustion
  b. Elimination of toxic pollutants
  c. Building a transportation network
  d. Separation of sulfur and nitrogen compounds

Answers: 1-d, 2-b, 3-a, 4-a, 5-b, 6-d, 7-a, 8-b, 9-d, 10-a, 11-c, 12-b, 13-c, 14-c, 15-a, 16-b, 17-c, 18-d, 19-a, 20-c.

Research Topics and Problems

Research and Discussion Topics

1 In a short essay, discuss the benefits and drawbacks of coal fueled electric generation.

2 Statistics show an increase of crude oil reserves in the world between 1996 and 2006. Discuss in a short essay the reasons for this increase.

3 Statistics show an increase of natural gas reserves in the world between 1996 and 2006. Discuss in a short essay the reasons for this increase.

4 What are the major factors that affect crude oil prices?

5 What are the major factors that drive market prices of oil products?

6   Discuss the benefits and challenges of unconventional gas recovery.

7   What is hydraulic fracking and where is it used?

8   What are the justifications of the groups supporting shale gas recovery?

9   What are the arguments of social groups that are against shale gas recovery?

10 Are there alternative techniques to hydraulic fracturing for unconventional oil and gas extraction?

11 In your opinion how will the Paris Agreement affect fossil fuel consumption in the world?

12 Compare the unit price of gasoline and bottled drinking water on the market where you live.

13 How would the food industry be affected from an increase of the oil price?

14 Which countries are major exporters of liquefied natural gas (LNG)?

15 Which countries are major importers of liquefied natural gas (LNG)?

16 What is the estimated natural gas storage capacity in the USA?

Problems

1   A coal burning power plant converts 10,000 Btu heat to generate 1 kWh electricity. Estimate the amount of bituminous coal this power plant burns per day given the net calorific value of the received coal is 26.5 MJ/kg and the average electric output of this generation plant is 80 MW.

2   A natural gas fired power plant converts 10,000 Btu heat to generate 1 kWh electricity. Estimate the amount of natural gas this power plant burns per day given the net calorific value of the natural gas is 1035 Btu/ft^3 and the average electric output of this generation plant is 100 MW.

3   Quantities of crude oil are measured in either barrels or metric tons. One standard oil barrel is 42 US gallons. How many barrels are there in one metric ton of West Texas Intermediate (WTI) crude oil with 39.6 API gravity?

4   The average calorific value of a fuel oil is given as 138,324 Btu/gal. Determine the calorific value in kJ/kg if the API gravity of this fuel oil is 20.

5 A household has burned 154 Ccf (1 Ccf = 100 cubic feet) of natural gas in one month. The gas company indicates that the heat content of delivered natural gas is 1.035 Therm/Ccf. How much energy did this household use in kWh? (1 Therm = 100,000 Btu)

6 The household paid $149.49 for 154 Ccf natural gas consumption including all delivery and supply charges, taxes, and fees. Given the heat content of the delivered natural gas 1.035 Therm/Ccf, calculate the overall cost of delivered energy in dollars per kWh.

7 How much anthracite with net calorific value of 30 MJ/kg would be needed to heat a house that consumes 1000 Therm of natural gas in a year for heating? Assume that the coal and gas heaters have the same efficiency.

Recommended Web Sites

- BGR (Federal Institute for Geosciences and Natural Resources-Germany): https://www.bgr.bund.de/EN/Home
- BP Statistical Review of World Energy: https://www.bp.com/en/global/corporate/energy-economics
- Canadian Energy Research Institute (CERI): https://ceri.ca
- Center for Liquefied Natural Gas: https://lngfacts.org
- Geoscience News and Information: https://geology.com/articles
- Intergovernmental Panel on Climate Change (IPCC): http://ipcc.ch
- NASA Visible Earth: https://visibleearth.nasa.gov
- Natural gas blog: http://naturalgas.org
- Oil and Gas Historical Society: https://aoghs.org
- San Joaquin Valley Geology (SJVG) The History of the Oil Industry: http://www.sjv-geology.org/history/index.html
- The World Bank Open Data: https://data.worldbank.org/indicator/EG.ELC.ACCS.ZS
- United Nations Sustainable Development (UN): https://www.un.org/sustainabledevelopment/energy
- USGS Library: https://library.usgs.gov
- World Coal Association (WCA): https://www.worldcoal.org
- World Energy Council (WEC): https://www.worldenergy.org

References

ACAA (2015). *Key Findings 2015: Coal Combustion Products Utilization.* s.l.: American Coal Ash Association (ACAA).

Alboudwarej, H., Felix, J., Taylor, S., and Badry, R. (2006). Highlighting heavy oil. *Oilfield Review*, Summer Issue 18: 34–53.

Aliyev, N. (1994). *The History of Oil in Azerbaijan*, 22–23. Azerbaijan International, Summer.

AOGHS, (2005). *Technology and the "Conroe Crater."* [Online] Available at: https://aoghs.org/technology/directional-drilling [Accessed 18 November 2018], American Oil and Gas Historical Society.

BGR, (2017). *BGR Energy Study 2017 – Data and Developments of German and Global Energy Supplies*. Hannover: s.n.

BP (2018). *BP Statistical Review of World Energy 2018*. London, UK: BP.

CAPP, (2018). *Canada's Oil Sands*. [Online] Available at: www.canadasoilsands.ca [Accessed 11 December 2018].

CRS (2009). *Unconventional Gas Shales: Development, Technology, and Policy Issues*. Washington, DC: Congressional Research Service.

Culp, A.W. Jr., (1979). *Principles of Energy Conversion*. s.l.: McGraw-Hill.

Dennis, B., Mufson, S. and Eilperin, J., (2018). Dam breach sends toxic coal ash flowing into a major North Carolina river. The Washington Post; 22 September.

DOE (2017). *Natural Gas Liquids Primer with a Focus on the Appalachian Region*. Washington, DC: US Department of Energy.

EIA (2017a). *Annual Coal Report 2016*. Washington, DC: US Energy Information Administration.

EIA (2017b). *International Energy Outlook 2017*. Washington, DC: EIA.

EIA, (2018a). *Gulf of Mexico Fact Sheet*. [Online] Available at: https://www.eia.gov/special/gulf_of_mexico [Accessed 16 November 2018].

EIA, (2018b). *Interactive Table Viewer*. [Online] Available at: https://www.eia.gov/outlooks/aeo/data/browser [Accessed August 2018].

ERCB (2013). *Alberta's Energy Reserves 2012 and Supply/Demand Outlook 2013–2022*. Calgary, Alberta, CA: Energy Resources Conservation Board.

GAO (2012). *Oil and Gas: Information on Shale Resources, Development, and Environmental and Public Health Risks*. Washington, DC: US Government Accountability Office.

Gaswirth, S.B., Marra, K.R., Cook, T.A. et al. (2013). *Assessment of Undiscovered Oil Resources in the Bakken and Three Forks Formations, Williston Basin Province, Montana, North Dakota, and South Dakota*. s.l.: USGS.

Giddens, P.H. (1948). *Early Days of Oil – A Pictorial History of the Beginnings of Industry in Pennsylvania*. Princeton, NJ: Princeton University Press.

IEA (2012). *World Energy Outlook Special Report on Unconventional Gas – Golden Rules for a Golden Age of Gas*. Paris: International Energy Agency.

IEA (2018). *Key World Energy Statistics*. Paris: International Energy Agency.

IPCC, (2006). *2006 IPCC Guidelines for National Greenhouse Gas Inventories*. s.l.: IGES, Japan. http://www.ipcc-nggip.iges.or.jp.

Janardhan, V. and Fesmire, B. (2011). *Energy Explained – Conventional Energy*. New York, NY: Rowman & Littlefield Publishers, Inc.

Mader, D. (1989). *Hydraulic Proppant Fracturing and Gravel Packing*. Amsterdam: Elsevier Science Publishers.

Mielke, W., Anadon, L.D., and Narayanamurti, V. (2010). *Water Consumption of Energy Resource Extraction, Processing, and Conversion*. Cambridge, MA: Harward Kennedy School, Belfer Center for Science and International Affairs.

MSHA, (2019). *USA Mine Safety and Health Administration*. [Online] Available at: www.msha.gov [Accessed 30 September 2019].

NASA, (2019). *Earth Observatory - World of Change*. [Online] Available at: https://earthobservatory.nasa.gov/WorldOfChange/hobet.php [Accessed 2019].

NETL (2014). *Environmental Impacts of Unconventional Natural Gas Development and Production*. Washington, DC: US Department of Energy, National Energy Technology Laboratory.

NRC (2007). *Coal: Research and Development to Support National Energy Policy*.National Research Council. Washington, DC: The National Academies Press.

Pamuk, H., (2014). *Turkish mine disaster town under lockdown as death toll rises to 301*. s.l.: s.n.

Probstein, R.F. and Hicks, E.R. (1982). *Synthetic Fuels*. Mineola, NY: Dover Books on Aeronautica.

Robertson, C. J. and Krauss, C., (2010). Gulf Spill is the Largest of Its Kind, Scientists Say. *New York Times*, 2 August.

Smith, G. O., (1920). Where the World Gets Its Oil - But Where Will Our Children Get It When American Wells Cease to Flow? *National Geographic Magazine*, February, pp. 181–202.

Soysal, H.S. (1997). *Sharing Responsibility in Marine Pollution* (in Turkish). Istanbul: Istanbul University.

The Guardian, (2016). *Dakota Access Pipeline: The Who, What and Why of the Standing Rock Protests*. [Online] Available at: https://www.theguardian.com/us-news/2016/nov/03/north-dakota-access-oil-pipeline-protests-explainer [Accessed 26 November 2018].

Tower, W. S., (1909). The Story of Oil. s.l. :(Reprinted by Bibliobazaar, Charleston, SC in 2008).

WCA, (2018). *Resources*. [Online] Available at: https://www.worldcoal.org/resources. [Accessed 15 October 2018], World Coal Association.

WCI (2005). *The Coal Resource: A Comprehensive Overview of Coal*. London: World Coal Institute www.worldcoal.org.

WEC (2016). *World Energy Resources*. s.l.: World Energy Council.

Whitney, G. (2010). *U.S. Fossil Fuel Resources: Terminology, Reporting, and Summary*. Washington, DC: CRS.

4

Nuclear Energy

Aerial view of Three Mile Island nuclear power plant on March 28, 1979, after the accident. (Photo: DOE, Public Domain.)

The Three Mile Island Unit 2 (TMI-2) reactor, near Middletown, PA, partially melted down on March 28, 1979. This was the most serious accident in US commercial nuclear power plant operating history.

The facility was closed on September 20, 2019. Decommissioning is expected to take about 60 years.

Energy for Sustainable Society: From Resources to Users, First Edition. Oguz A. Soysal and Hilkat S. Soysal.
© 2020 John Wiley & Sons Ltd. Published 2020 by John Wiley & Sons Ltd.

4.1 Introduction

Unlike other forms of energy, nuclear energy is not a consequence of solar radiation reaching the Earth. Uranium and other radioactive elements used in nuclear reactors were created by nuclear reactions when the solar system was formed, not as a consequence of degradation of vegetation and living organisms like fossil fuels.

Nuclear energy was discovered while scientists were trying to understand the power source of the sun. Speculations started around the middle of the nineteenth century by several scientists including Kelvin and Helmholtz. Eddington proposed in 1920 that the fusion of hydrogen atoms could be the source of energy in the sun. Nuclear fusion was, however, not fully understood until quantum theory was developed in the late 1920s. Today, it is known that fusion reactions occur in the core of the sun and other stars where extremely high temperature hydrogen plasma is held by extremely large gravitational forces. Fusion reactions were realized experimentally after 1930s. Using the fusion of hydrogen atoms, the US and Soviet Union tested hydrogen bombs (known as H-bombs).

Following the discovery of neutrons by James Chadwick in 1932, several scientists carried out experiments to bombard different nuclei with neutrons. In 1934 Fermi and collaborators found that uranium atoms bombarded by neutrons produced several radioactive nuclei. Meitner and Frisch discovered in 1934 that a uranium nucleus absorbing a neutron oscillates and splits into two smaller nuclei, through the process named "fission." The total mass of the particles that arise after fission is smaller than the mass of the original nucleus. The total mass decreased during the fission process is converted into energy according to the mass-energy equivalency principle initially derived by Albert Einstein (Ferguson 2011).

On August 2, 1939, Einstein sent a letter to US President Franklin D. Roosevelt bringing up the possibility of a nuclear weapon using the energy released from fission. This letter helped to secure the governmental support to start the Manhattan Project, in which a group of scientists and engineers developed the atomic bomb. Einstein, however, has never been part of the Manhattan Project because of his pacifist views. Although the first two atomic bombs ended World War II, nuclear industry has always suffered from such a destructive beginning.

Practical application of nuclear energy is newer than all other primary sources. The first demonstration of nuclear energy was the two atomic bombs detonated over Hiroshima and Nagasaki in 1945. Peaceful applications of nuclear energy to produce electric power, propel submarines, and medical uses were developed in the early years of the Cold War era, following World War II.

Development of the first nuclear power plant in the US started after World War II in Argon Laboratory, in Chicago, IL. Because of the high risk of a nuclear explosion, the first reactor designed to generate electricity, named Experimental Breeding Reactor I (EBR I), was built in a remote location in Idaho. The reactor started to operate in 1952 by powering four light bulbs, each 100 W. These light bulbs signaled the most important peaceful application of nuclear power.

Starting from the mid-1950s, nuclear power plants were built in many countries, allowing them to generate electricity much cheaper than oil and coal. The oil crisis in 1973 was a strong justification to build more than one hundred nuclear power plants in the US to avoid dependence on foreign oil. The steep rise of nuclear power in the US continued until the Three Mile Island nuclear accident on March 28, 1979. Although its small radioactive releases had no detectable health effects on plant workers or the public, its aftermath brought about sweeping changes involving emergency response planning, reactor operator training, human factors engineering, radiation protection, and many other areas of nuclear power plant operations. With the lessons learned from the TMI accident, the US Nuclear Regulatory Commission (NRC) strengthened safety standards and has been extremely strict in issuing permits for new nuclear power plants or capacity increases of the existing plants. Seven years later in 1986, the Chernobyl nuclear power plant near the city of Kiev in Ukraine exploded mainly due to human factors, causing the death of several thousands of persons from radiation-related diseases. In 2011 a tsunami damaging the Fukushima nuclear reactor on the coast of Japan marked the third significant nuclear disaster of the recent history.

Nuclear power generation is virtually carbon-free. According to the Nuclear Energy Institute (https://www.nei.org), as of April 2017, 30 countries around the world are operating 449 commercial nuclear reactors to generate electricity and 60 new nuclear plants are under construction in 15 countries. Today nuclear power plants generate approximately one-fourth of total electric energy worldwide. In the US, about 20% of electric energy is currently generated by nuclear power plants (IEA 2017). Several countries in Europe including France, Belgium, Slovakia, Ukraine, and Hungary generate more than half of their electric power from nuclear energy. In recent years, climate change, and energy security concerns, spurred the nuclear power plant development especially in eastern Asia.

Nuclear energy relies on uranium and thorium deposits in the Earth's crust, which are non-renewable resources. Although uranium ore is abundant in the Earth, economically recoverable reserves are in limited quantity. Currently, the known uranium reserves are estimated to deliver about 612 EJ, about 1.5% of the total 39,530 EJ that all non-renewable reserves can produce (BGR 2017). Moreover, nuclear energy is delivered to end-users in the form of electricity, with the average efficiency of nuclear power plants being around 40%. Therefore, once expected to be the solution of the energy problem, nuclear energy is far from replacing fossil fuels in long term.

4.2 Basic Concepts of Nuclear Physics

Nuclear physics deals with the forces and interactions between atomic particles. In this section, we will review the essential concepts needed to understand the operation of nuclear power plants. For better understanding of the subject, the reader should refer to textbooks such as Jelly (1990), Close (2015), and other references

An atom consists of a nucleus with electrons spinning around it. The nucleus contains a certain number of protons and neutrons. Protons and electrons have the same electric charge and, therefore their numbers are also equal.

In a chemical reaction, atoms that form molecules are combined or separated, but their identities remain unchanged. The total number of atoms in the reactants is equal to the atoms in the resulting products. The combustion of carbon shown in the equation below is a typical example for a chemical reaction.

$$C + O_2 \rightarrow CO_2 + 4.2eV \tag{4.1}$$

Oxidation of one mole of carbon (12 g) and one mole of oxygen (16 g) produces one mole of carbon dioxide (28 g). Chemical reactions change the chemical composition of reactants but the total mass of compounds before and after the reaction remains the same. Chemical reactions can be endothermic or exothermic, which means that energy is absorbed or released, respectively. The reaction described in Eq. (4.1) is exothermic and releases 4.2 Ev $(6.4 \times 10^{-19}$ J) of energy. Combustion converts the chemical energy stored in the molecules into heat.

Nuclear reactions, however, change the identity of the atoms involved. While the total number of protons and neutrons in a nuclear reaction remains constant, new nuclei that contain different numbers of protons and neutrons are formed throughout the reaction.

Nuclear properties of an unstable atom may change in time if it loses a neutron, proton, or electron. Isotopes are different forms of the same element with the same atomic number but different numbers of neutrons. Therefore, all isotopes of an element have different masses. For example, $_{92}U^{233}$, $_{92}U^{234}$, $_{92}U^{235}$, $_{92}U^{236}$, and $_{92}U^{238}$ are different isotopes of uranium.

4.2.1 Basic Definitions

In expressions describing nuclear reactions, number of protons and the atomic mass number are indicated beside the chemical symbol of the element. In a notation like $_ZX^A$ or $X(Z,A)$, X is the chemical symbol and Z is the atomic number of the element, which is the number of protons in the nucleus. The superscript A is the sum of the protons and neutrons that form the nucleus, known as the atomic mass number. The term *nucleon* signifies subatomic particles that form the nucleus, mainly protons and neutrons. In a non-ionized atom, the number of electrons that orbit the nucleus is equal to the number of protons.

The number of protons characterizes the chemical behavior of an element. Therefore, elements with the same atomic number are basically the same chemical element. Isotopes of an element have the same number of protons but a different number of neutrons. Isotopes of the same element have similar chemical properties but different nuclear behaviors.

The mass of atomic particles is too small to be quantified in kg for calculations. Instead, they are expressed in atomic mass unit (a.m.u.).

1 atomic mass unit (a.m.u) $= 1.660539040 \cdot 10^{-27}$kg

Mass of a proton 1.007277 (a.m.u)

Mass of a neutron 1.008665 (a.m.u)

Mass of an electron 0.0005486 (a.m.u) (4.2)

The mass of an electron is usually negligible compared to the mass of protons and neutrons. Electrons stay on an orbit around the nucleus because of the electrostatic attraction of the protons.

$$\text{Charge of an electron} = 1.602 \times 10^{-19} \text{ Coulomb} \tag{4.3}$$

In nuclear energy calculations it is more convenient to use the unit electron-volt represented as eV. One electron volt is the energy gained by a particle carrying the charge of one electron, when it is accelerated by a potential of one volt.

$$1 \text{ eV} = 1.602 \times 10^{-19} \text{J} \tag{4.4}$$

The energy equivalent of 1 a.m.u. is approximately 931 MeV.

4.2.2 Binding Energy and Mass Defect

The actual mass of the nucleus of all atoms, except hydrogen isotope $_1H^1$, is less than the sum of the mass of particles forming the nucleus (nucleons). This difference is known as "mass defect." The energy equivalent to the mass defect, termed "binding energy," holds the particles together in the nucleus. The protons, all having positive charge, would otherwise repel each other with the Coulomb force tearing apart the nucleus.

The binding energy is different for all atoms and increases with the number of nucleons Z. Total binding energy of an atom can be calculated by subtracting the mass of the nucleus from the total mass of protons and neutrons.

$$\text{Mass defect} = Zm_p + (A - Z)m_n - \text{Nucleus mass} \tag{4.5}$$

Measuring the mass of a nucleus is more difficult than measuring atomic mass, which includes the mass of orbiting electrons. Atomic mass of all elements and their isotopes has been measured precisely using mass spectroscopy techniques. Lists of experimentally obtained atomic masses are available in several publications (Audi and Wapstra 1995; Audi et al. 2003). The equation of mass deficiency can be modified by using the atomic mass of the light hydrogen atom $(_1H^1)$ $m_H = 1.007825$ a. m. u., instead of the mass of a proton. Since the light hydrogen atom contains only one proton and one electron, the mass defect is found by subtracting the measured atomic mass of an atom, which includes the total mass of electrons, neutrons, and protons, from $Z \times m_H$. Hence, the total mass of electrons is canceled, and the expression below gives the mass defect in terms of the atomic mass, rather than the mass of the nucleus.

$$\text{Mass defect} = Zm_H + (A - Z)m_n - \text{atomic mass} \tag{4.6}$$

Substituting the masses of a proton and a neutron given in (4.2), we can rewrite the expression for mass defect in the form below.

$$\text{Mass defect} = 1.007825Z + 1.008665(A - Z) - \text{atomic mass} \tag{4.7}$$

Binding energy of a nucleus can be obtained by multiplying the mass defect and 931 MeV, the energy equivalent of 1 a.m.u.

$$E_b = 931 \times (\text{Mass defect}) \tag{4.8}$$

The average binding energy per nucleon is $b = E_b/A$.

Example 4.1 Calculate the mass defect and average binding energy per nucleon for the uranium isotope $_{92}U^{235}$ (atomic mass: 235.0439 a.m.u.).

Solution

Mass defect $= 1.007825 \times 92 + 1.008665(235 - 92) - 235.0439 = 1.9151$ a.m.u

Total binding energy $= 931 \times 1.9151 = 1782.95$ MeV

Average binding energy per nucleon $= \dfrac{1782.95}{235} = 7.587$ MeV

When two lightweight nuclei are combined (fusion reaction) or a heavy nucleus splits into two lighter nuclei (fission reaction), the total number of subatomic particles remain constant but the total mass decreases because of the increasing mass defect. The mass difference is converted to energy according to Einstein's well-known expression in Eq. (4.9).

$$E = \Delta m \cdot c^2 \qquad (4.9)$$

In this equation c is the speed of light, $(299{,}792{,}458 \cong 3 \cdot 10^8$ m/s). Because the speed of light is a large number, a small change of the mass releases a huge energy.

Example 4.2 Find the energy equivalent of the mass equal to 1 a.m.u. given in Eq. (4.2).

Solution
Using the mass-energy Eq. (4.9), we obtain:

$$E = 1.660539040 \cdot 10^{-27} \times (299{,}792{,}458)^2 = 1.49 \cdot 10^{-10} \text{ J}$$

or

$$\frac{1.49 \cdot 10^{-10}}{1.602 \times 10^{-19}} = 931 \cdot 10^{-6} \text{ eV} = 931 \text{ MeV} \,\therefore$$

The average binding energy per nucleon is approximately the minimum energy required to remove a proton or neutron from the nucleus. It is also the excitation energy of the nucleus if it absorbs a proton or neutron. The excitation energy is usually released in the form of radiation and it may be sufficient to initiate splitting of some heavy-mass nuclei.

Lightweight nuclei like hydrogen, helium, and lithium have smaller binding energy than the elements located around the middle of the periodic table. Elements with an atomic mass between 40 and 100 (such as iron and nickel) have an average binding energy per nucleon between 8 and 9 MeV, therefore they have most stable nuclei. The average binding energy per nucleon decreases for elements with heavier nuclei such as uranium, thorium, and plutonium. In any nuclear reaction where a heavier nucleus is broken into two intermediate-mass nuclei, an excess binding energy is released. On the other hand, if two lightweight atoms such as hydrogen and helium isotopes are combined to form a heavier atom, again an excess binding energy is released. Binding energy is an indication of the stability of a nucleus. Figure 4.1 shows variation of the binding energy per nucleon with the atomic mass number calculated using standard atomic masses given in IUPAC (1997).

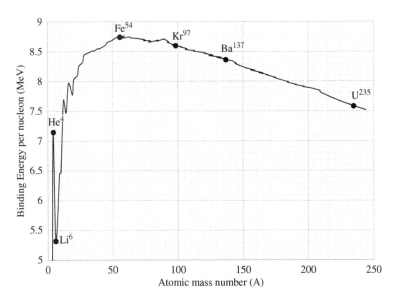

Figure 4.1 Binding energy per nucleon as a function of the atomic mass number.

4.3 Nuclear Reactions

Three types of nuclear reactions release energy; fusion, fission, and radioactive decay. Fusion is the process of combining two lightweight nuclei to form a heavier one. Fission is splitting a heavy nucleus by bombarding with a nucleon, typically a neutron. Radioactivity is a spontaneous release of subatomic particles from an unstable isotope.

All nuclear reactions result in a change of the mass deficiency in the nuclei, which is converted into energy according to Eq. (4.9). In all three types of nuclear reactions the total number of nucleons (A) and the total charge (Z) are conserved. Therefore, no sub-atomic particle is destroyed; only excess binding energy that results from mass deficiency is released.

4.3.1 Fusion Reaction

In a fusion reaction, two or more lightweight nuclei are combined to form a heavier nucleus. For example, when two hydrogen atoms are combined to build a helium atom, energy is released. Nuclear fusion is the source of energy produced by stars, including the Sun (Andrews and Jelley 2007). The five most probable fusion reactions, along with the energy and temperature required to initiate them, are listed in Table 4.1 (Culp, Jr. 1979).

As two lightweight atoms approach each other, there is first an attraction force and then a repulsion force develops. In order to fuse together, the nuclei must approach each other at a high velocity such that their kinetic energy can overcome their mutual electrostatic repulsion force. Such kinetic energy can be reached at extremely high temperatures. Therefore, temperatures in the order of millions of degrees are required to initiate and sustain a fusion reaction. Fusion reactions can sustain in the Sun and stars where such high temperatures are naturally available.

Table 4.1 Probable fusion reactions.

Reaction	Ignition energy (keV)	Ignition temperature (K)
$_1H^2 + _1H^2 \rightarrow _2He^3 + _0n^1 + 3.26$ MeV	50	$5.8 \cdot 10^8$
$_1H^2 + _1H^2 \rightarrow _1He^3 + _1H^1 + 4.03$ MeV	50	$5.8 \cdot 10^8$
$_1H^3 + _1H^2 \rightarrow _2He^4 + _0n^1 + 17.4$ MeV	10	$1.2 \cdot 10^8$
$_2He^3 + _1H^2 \rightarrow _2He^4 + _1H^1 + 18.3$ MeV	100	$1.2 \cdot 10^9$
$_3Li^6 + _1H^1 \rightarrow _2He^3 + _2He^4 + 4.0$ MeV	200	$2.3 \cdot 10^9$

Source: Culp Jr. (1979).

The theory of nuclear fusion process is well established and tested by a few countries in thermonuclear weapons, commonly known as an H-bomb. The high temperature to initiate fusion can be obtained on Earth only by nuclear fission reaction. At present, a safe technique to initiate and control a sustained fusion reaction is not available. Therefore, nuclear fusion is not currently used for practical conversion in energy systems.

4.3.2 Fission Reaction

In a fission reaction, heavier atoms are bombarded by a projectile nucleon, typically a neutron. If the target nucleus is split in two smaller isotopes, their total mass is smaller than the original atom as seen in Figure 4.1. The change of mass is converted into energy according to the Einstein's Eq. (4.9). Today, all nuclear power plants around the world generate about 20% of electricity from energy released by fission of uranium, plutonium, or thorium atoms, and subsequent radioactive decay processes.

To start a fission reaction, a heavy nucleus is bombarded by a projectile particle, usually a neutron. The impact may result in several situations depending on the change of energy balance. The strike on the target nucleus may be elastic and according to the law of conservation of momentum, the particle scatters changing its direction. If the neutron has enough energy, it can penetrate in the target nucleus. The change of energy balance causes instability in the modified nucleus, which either breaks apart or undergoes radioactive changes by emitting particles to reach a stable state (Ferguson 2011).

A fission reaction occurs when the nucleon has sufficient energy to split the target nucleus. Heavy elements with odd atomic numbers such as $_{92}U^{235}$ and $_{94}Pu^{239}$ are called *fissile* because they can be split by neutron bombardment. The equation below shows a common fission reaction of $_{92}U^{235}$. When a $_{92}U^{235}$ nucleus captures a neutron, it becomes a new radioactive isotope $_{92}U^{236}$ and emits gamma particles. If the excitation energy of the new nucleus is large enough, it breaks into barium 56 and krypton 36 isotopes.

$$_0n^1 + _{92}U^{235} \rightarrow _{92}U^{236} + \gamma \rightarrow _{56}Ba^{137} + _{36}Kr^{97} + 2_0n^1 \tag{4.10}$$

If the excitation energy is not large enough, the nucleus may not undergo a fission reaction.

The two or three neutrons released by splitting a U^{235} nucleus may collide with other U^{235} atoms if they have sufficient energy. Consecutive collisions of neutrons with U^{235} atoms that

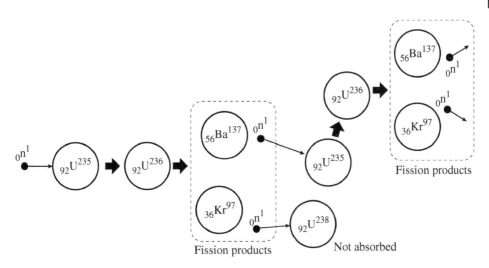

Figure 4.2 A chain reaction example.

result in new fission products is a *chain reaction*. Figure 4.2 illustrates an example of chain reaction in a reactor containing a mixture of U^{235} and U^{238} elements as fuel.

Suppose that a heavy nucleus that contains A_1 total nucleons with binding energy b_1 is split into two lighter nuclei, each having A_2 and A_3 total nucleons with average binding energy per nucleon b_2 and b_3 respectively. Energy released in such a reaction is approximately equal to the change of the total binding energy of nuclei before and after fission.

$$E = A_2b_2 + A_3b_3 - b_1(A_2 + A_3) \tag{4.11}$$

Binding energy per nucleon can be determined using Eq. (4.8) and the actual mass of the nucleus given in Audi et al. (2003), or estimated from the plot in Figure 4.1.

Example 4.3 Find the energy released by fission of U-235 into Ba-137 and Kr-97 described in Eq. (4.10). The actual mass of U^{235} nucleus is 235.0439 a.m.u. The masses of Ba^{137} and Kr^{97} nuclei are 136.9061 and 96.9212 a.m.u., respectively.

Solution
Energy can be found by calculating the mass defect. The masses on the right and left sides of Eq. (4.10) are:

$$235.0439 + 1.00867 \rightarrow 136.9061 + 96.9212 + 2 \times 1.00867$$

The left side of the equation yields 236.05257 but the right side is 235.84464; so the difference is $236.05257 - 235.84464 = 0.20793$ a. m. u. The energy equivalent of 1 a.m.u. is 931 MeV. Thus,

$$E = 0.20793 \times 931 = 193.6 \text{ MeV}.$$

Alternatively, using the metric units and the equation $E = mc^2$, we can find

$$(0.20793 \times 1.66 \times 10^{-27}) \times (3 \times 10^8)^2 = 3.10 \times 10^{-11} \text{J} \therefore$$

Example 4.4   Estimate the energy released when U-235 is broken into Ba-137 and Kr-97 using average binding energy per nucleon.

Solution

Average binding energy per nucleon for the isotopes in the reaction can be approximately found in Figure 4.1.

U-235: $b_1 \cong 7.6\,\text{MeV}$
Ba-137: $b_2 \cong 8.4\,\text{MeV}$
Kr-97: $b_3 \cong 8.6\,\text{MeV}$

Substituting these values in (4.11), we obtain:

$$E = 137 \times 8.4 + 97 \times 8.6 - 235 \times 7.6 = 199\,\text{MeV}.$$

Note: This result is close to the energy obtained in the previous example. The difference is due to the estimation of the average binding energy per nucleon from a plot.

Fission of a massive nucleus like U-235 yields two lighter nuclei and releases two or three neutrons. The fission fragments and freed neutrons release more than 80% of the fission energy. At the same time, the excited nucleus emits radiant energy as it breaks. Following the fission, radioactive decay of gamma rays, beta particles, and neutrinos release additional energy. The distribution of energy release in a typical fission of one U-235 nucleus is approximately 200 MeV or 3.2×10^{-11} J total energy, as shown in Table 4.2. This energy is the difference in total binding energies of the two nuclei from that of the original nucleus.

Two or three neutrons (in average 2.43 in a large number of fissions) released from splitting the original nucleus collide with other nuclei to continue the chain reaction, provided that they have suitable kinetic energy. Slower neutrons that have smaller kinetic energy have more chance to cause fission. Such slow neutrons are called thermal neutrons.

Table 4.2   Energy release in a typical fission of a U-235 nucleus.

Source	Released energy (MeV)
Prompt release	
Kinetic energy of fission fragments	167
Kinetic energy of fission neutrons	8
Instantaneous gamma rays	7
Delayed release	
Gamma rays	7
Beta particles	7
Neutrinos	11
Total	204

Example 4.5   How much energy is released by complete fission of 1 kg U-235 isotope?

Solution

The number of atoms in 1 g U-235 can be calculated using the Avogadro number 0.60225×10^{24}.

$$\frac{\text{Avogadro Number}}{\text{Atomic mass of U}^{235}} = \frac{0.60225 \times 10^{24}}{235.0439} = 2.56 \times 10^{21} \text{ atoms}$$

Fission of each atom releases on average 3.1×10^{-11} J (see Example 4.3). Total energy released by splitting 1 kg U-235 isotope is

$$1{,}000 \times 2.56 \times 10^{21} \times 3.1 \times 10^{-11} = 7.9 \times 10^{13} \text{J}$$

Converted to MWh, this energy is equal to

$$\frac{7.9 \times 10^{13}}{3600} = 21.9 \times 10^{9} \text{Wh} = 21.9 \text{GWh}$$

Energy produced by complete fission of 1-kg of U-235 is equivalent to approximately 3000 t of coal burned in a typical thermoelectric power plant.

4.3.3 Radioactive Decay

Radioactivity is emission of subatomic particles as an unstable atom undergoes nuclear changes to reach a more stable configuration. All elements with an atomic number higher than 83 (bismuth and heavier elements) spontaneously release particles from their nucleus until a more stable nucleus is formed.

Radioactive decay means a spontaneous disintegration of a radioactive element. The unit of radioactivity is Becquerel (Bq) in the international system of units (SI). One Becquerel corresponds to the decay of one nucleus per second, therefore, Becquerel is a derived unit equal to s^{-1}. Radiation is associated with three forms of particles named alpha, beta, and gamma particles.

- *Alpha particles* consist of two protons and two neutrons, which corresponds to a helium ($_2\text{He}^4$) nucleus. Alpha particles are rather massive and relatively slow. A nucleus that emits alpha particles becomes the nucleus of an element with a lower atomic number by two and a lower mass number by four. The energy released in alpha decay is a few MeV. However, the protons in an alpha particle ionize the air.

  Alpha particles are the least penetrating particles and typically can be blocked by a thin sheet of paper. But if digested, they cause severe illness due to ionization of the tissues. An example of alpha decay is transformation of radium $_{88}\text{Ra}^{226}$ to radon $_{86}\text{Rn}^{222}$. Radon occurs naturally in underground mineral deposits, especially coal. Radon causes significant health issues if it accumulates in poorly ventilated basements and closed spaces.

- A *beta particle* is a single electron or positron emitted by an atom. They are much less massive than alpha particles, but have more energy. Beta decay occurs if a neutron spontaneously transforms to a proton increasing the net charge of the nucleus, which repels an electron. The atomic number of a nucleus that emits a beta particle increases by one, but the mass number does not change. For example, the hydrogen isotope $_1\text{H}^3$ becomes helium $_2\text{He}^3$ by releasing a beta particle.

Beta particles are not as ionizing as alpha particles, but they penetrate matter more easily. Beta particles can be stopped by a thin sheet of most metals or a thickness of 5–10 cm of water or plastic.

- *Gamma particles* occur when a nucleus in an excited state emits electromagnetic radiation as it falls to a lower energy state. Gamma decay may follow alpha or beta decays because they are released from an excited nucleus. Gamma particles, also known as *gamma radiation*, propagate as a short wavelength electromagnetic wave. Gamma radiation is the most penetrating radioactive radiation form that can be stopped by a massive block of matter such as a lead layer or thick concrete wall.

Gamma radiation propagates in all directions around its source. Therefore, at a distance of r meters from the source of A Becquerel gamma radiation, the flux becomes

$$F = \frac{A}{4\pi r^2}, \tag{4.12}$$

therefore, its effect on living organisms decreases inversely proportional to the square of the distance from the source.

The rate of radioactive decay varies from one element to another. In a short time dt the number of disintegrating nuclei is proportional to the total number N of remaining nuclei that has not disintegrated. In other words, the rate of disintegration per second is proportional to the number of nuclei that are present in the matter.

$$\frac{dN}{dt} = -\lambda N \tag{4.13}$$

The radioactive decay constant λ has different values for different isotopes and indicates how fast the decay occurs. The negative sign shows that N decreases in time. By separating the variables on each side of Eq. (4.13) and integrating, we obtain

$$\int \frac{dN}{N} = \int -\lambda dt \Rightarrow \ln N = -\lambda t \tag{4.14}$$

Time variation of the number of nuclei that have not disintegrated is found in Eq. (4.15).

$$N = N_0 e^{-\lambda t} \tag{4.15}$$

Time that is required for half of the remaining radioactive material to disintegrate is termed *half-life time* and can be obtained as

$$\lambda T = \ln 2 \quad \text{or} \quad T = \frac{\ln 2}{\lambda} = \frac{0.693}{\lambda} \tag{4.16}$$

Half-life time indicates how fast radioactive decay occurs; this value ranges from fractions of a second to millions of years. Another parameter to measure the disintegration speed is the mean time T_m, which is the average time elapsing from the time the nucleus is formed to the moment it disintegrates. The mathematical expression of the mean lifetime is:

$$T_m = \frac{-\int_0^\infty N dt}{N_0} = \frac{-\int_0^\infty N_0 e^{-\lambda t} dt}{N_0} = \frac{1}{\lambda} \tag{4.17}$$

As seen in Table 4.3, half-life of the uranium isotopes commonly used in nuclear power plants is in the order of thousands of years.

Radioactive decay is an important part of the energy production process in nuclear reactors. For example, in a fission reaction of natural uranium that contains U-238 and U-235

Table 4.3 Half-life of some common isotopes.

Element	Isotope	Half-life	Abundance (%)
Carbon	$_6C^{12}$	Stable	98.93
	$_6C^{14}$	$5.73 \cdot 10^3$ yr	Trace
Thorium	$_{90}Th^{230}$	$14 \cdot 10^9$ yr	0.02
	$_{90}Th^{232}$	22.1 min	99.98
Plutonium	$_{94}Pu^{238}$	87.7 yr	Trace
	$_{94}Pu^{239}$	$24.1 \cdot 10^3$ yr	Trace
	$_{94}Pu^{240}$	$6.50 \cdot 10^3$ yr	Trace
Uranium	$_{92}U^{234}$	$245 \cdot 10^3$ yr	0.005
	$_{92}U^{235}$	$704 \cdot 10^6$ yr	0.720
	$_{92}U^{238}$	$4.47 \cdot 10^9$ yr	99.27

isotopes, there is a probability that a neutron hits a U^{238} nucleus instead of a U^{235} atom. If the energy of the projectile particle is large enough to enter the target nucleus without splitting, it causes instability. The excited nucleus undergoes transmutation and becomes a new element with the trapped particle and radioactive decay starts following the collision. The nucleus then emits gamma radiation until it reaches another stable state. A typical nuclear reaction where $_{92}U^{238}$ is transformed into $_{94}Pu^{239}$ is shown in Eq. (4.18).

$$_{92}U^{238} + {_0}n^1 \rightarrow {_{92}}U^{239} + \gamma \rightarrow {_{93}}Np^{239} + {_{-1}}e^0 \rightarrow {_{94}}Pu^{239} + {_{-1}}e^0 \qquad (4.18)$$

Thorium isotopes can be also used to produce a fissile uranium isotope as shown in the equation below.

$$_{90}Th^{232} + {_0}n^1 \rightarrow {_{90}}Th^{233} \rightarrow {_{91}}Pa^{233} + {_{-1}}e^0 \rightarrow {_{92}}U^{233} + {_{-1}}e^0 \qquad (4.19)$$

The half-life of Th-233 is 22.1 minutes and transforms in a short time into a protactinium isotope Pa-233 by emitting a β-particle. Pa-233 has a lifetime of 27 days and converts into U-233 by radioactive decay emitting a β-particle.

U-233 and Pu-239 are fissile isotopes and can be used in nuclear reactions. Isotopes like U-238 and Th-232 that can be transformed by neutron bombardment into a *fissile* element are classified as *fertile*. Fissile isotopes can be produced in reactors fueled by a combination of fertile and fissile materials such as Th-232 and U-233 or U-238 and Pu-239. If a reactor produces more of a fissile than it consumes, in other words if the breeding rate is greater than unity, it is called a *breeder reactor*.

4.3.4 Health Effects of Nuclear Radiation

Particles emitted during radioactive decay carry high energy sufficient to form ions by moving electrons away from atoms of the substance they penetrate. Emission of all radioactive particles causes ionizing radiation. When body tissue absorbs too many radioactive particles, ionization of the cells can cause cancer over time. The probability of developing cancer depends on the amount and duration of radiation received. There is uncertainty

Table 4.4 Dose conversion coefficients (Sv/Bq) for selected radionuclides.

Radionuclide	Inhalation	Ingestion
U^{238}	$2.9 \cdot 10^{-6}$	$4.5 \cdot 10^{-8}$
U^{234}	$3.5 \cdot 10^{-6}$	$4.9 \cdot 10^{-8}$
U^{235}	$3.1 \cdot 10^{-6}$	$4.7 \cdot 10^{-8}$
Th^{230}	$1.4 \cdot 10^{-6}$	$2.1 \cdot 10^{-7}$
Th^{228}	$4.0 \cdot 10^{-5}$	$7.2 \cdot 10^{-8}$

Source: UNSCEAR (2008).

and controversy among researchers about the cancerogen effects of the radiation. A low dose of radiation absorbed in a short duration is usually not harmful. The immune system of human body can often develop sufficient protection against cancer cells produced by low levels of radiation.

The dose of the absorbed radiation is measured in Joule per kilogram (J/kg) of absorbing matter. The effective dose, which represents the probability of health effects of low levels of ionizing radiation on human body, is expressed in sievert (Sv). Sievert is a derived unit of SI in the dimension of J/kg. Since Sv considers the biological effectiveness of the radiation based on the radiation type and energy, conversion tables are available to obtain the Becquerel value for specific radionucleide (UNSCEAR 2008). Dose conversion coefficients for selected radionuclides are listed in Table 4.4. The older cgs unit rem (Roentgen equivalent man) is still commonly used in practice.

A low level of ionizing radiation is always present in nature. Natural sources of radiation include cosmic radiation coming from space, and terrestrial radiation also known as background radiation, produced by the decay of radioactive minerals that exist in the Earth's crust. Coal mines, oil deposits, and rare earth materials often contain radioactive minerals. Radon gas accumulated in poorly ventilated basements of buildings is radioactive. Agricultural products also contain a certain amount of radiation coming from the ground water. Nuclear power plants, transport of nuclear materials, nuclear waste, and nuclear tests are among the man-made radiation sources. National and international standards limit the amount of radioactive materials emitted to the environment from nuclear activities (UNSCEAR 2008). The annual average dose of natural sources of ionizing radiation worldwide is 2.4 mSv and typical individual doses range between 1 and 13. Public exposure to a nuclear fuel cycle can be up to 0.02 mg at 1 km from some nuclear reactor sites. During the Chernobyl accident in 1986 more than 300,000 recovery workers were exposed to nearly 150 mSv, and 350,000 other individuals received doses greater than 10 mSv (UNSCEAR 2008).

Although assessment of possible health hazards from low-level ionizing radiation depend on many uncertainties and often based on controversial models, public health policies tend to minimize the exposure to excess radiation. Statistical studies can correlate the probability of cancer cases to the level of radiation, however the individual effects cannot be predicted.

High doses of ionizing radiation can cause immediate health issues and lead to death. Such effects are proportional to the dose a body receives. Nausea and vomiting are typically the first noticeable radiation sickness symptoms. Higher doses can cause diarrhea and hair loss. Excessive doses can damage the immune system and cause internal bleeding. Nuclear explosions, power plant accidents, and powerful unshielded radioactive sources can create such radiation hazards.

4.4 Nuclear Fuels

4.4.1 Resources, Reserves, Production, and Consumption

The initial fuel for most nuclear reactors is obtained from various types of mineral deposits that contain uranium oxides such as pitchblende (U_3O_8) and uraninite (UO_2). Nearly 40% of uranium is produced by underground mining and 25% by open pit (surface) mining. The rest of the uranium is extracted by in-situ leaching or other methods, including coproduct or byproduct of various manufacturing processes (IAEA 2009).

Uranium is a metallic element used in many engineering applications including ballast for airplane tails and armor-penetrating ammunition, in addition to nuclear weapons and reactors. Uranium resources worldwide are approximately 11.6 Mt, however, because of its strategic importance, some countries are reluctant to disclose the actual estimated uranium resources.

Whereas uranium is quite common in the Earth's crust in low concentrations in soil, rocks, and water, high concentration is needed for commercial mining. Economically recoverable uranium deposits are available in the western US, Australia, Canada, Central Asia, Africa, and South America.

Thorium is considered to be an alternative to uranium since Th-232 can be converted into U-233 in the reaction described in Eq. (4.19) or used in breeder reactors to produce fissile isotopes. Although no commercial nuclear reactor currently uses it for power generation, significant thorium resources are available in the Earth's crust. Figure 4.3 shows available uranium and thorium resources in the world by geographic regions.

Natural uranium ore contains about 99.27% of U-238, 0.72% of U-235, and a trace of U-234. All uranium isotopes are weakly radioactive with extremely slow decay rate. Half-life of the most abundant isotope U-238 is 4.47 billion years. The only naturally available fissile isotope U-235 has 704 million years half-life (Table 4.3). Uranium-238 is a fertile isotope that can be used to produce fissile isotopes with a reaction as described in Eq. (4.18).

Nuclear energy production and consumption are not directly correlated to the availability of uranium reserves in a region. Nuclear fuel production requires a high level of technology. Development of nuclear power plants depends on political decisions, public acceptance, and significant investment.

Figure 4.4 shows the uranium reserves, nuclear energy production, and uranium mineral consumption by geographic regions region. Uranium reserves in the Middle East are insignificant. Many European countries depend on nuclear power for their electric generation. France generates 70% of its electricity in nuclear power plants. Germany, Sweden, Finland, Switzerland, Belgium, and the UK also generate significant fraction of their electricity in nuclear power plants (see Table 4.6). However, western Europe possesses smaller uranium reserves compared to other regions. Countries in South America, and Africa, on the other hand, have not developed many nuclear power facilities despite significantly larger reserves.

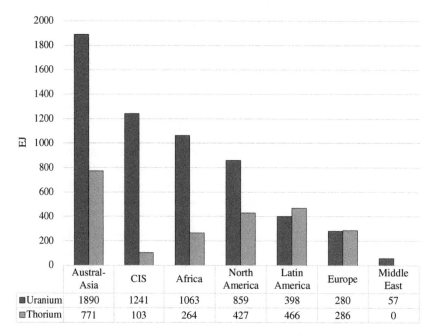

	Austral-Asia	CIS	Africa	North America	Latin America	Europe	Middle East
■Uranium	1890	1241	1063	859	398	280	57
■Thorium	771	103	264	427	466	286	0

Figure 4.3 Uranium and thorium resources by regions. Source: BGR (2017).

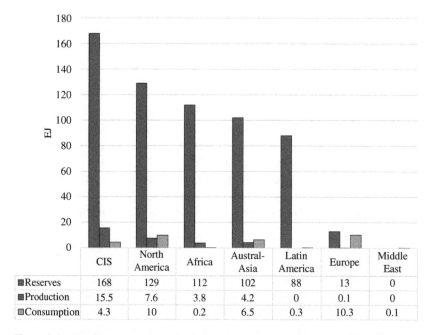

	CIS	North America	Africa	Austral-Asia	Latin America	Europe	Middle East
■Reserves	168	129	112	102	88	13	0
■Production	15.5	7.6	3.8	4.2	0	0.1	0
□Consumption	4.3	10	0.2	6.5	0.3	10.3	0.1

Figure 4.4 Uranium reserves, production, and consumption by geographic regions.

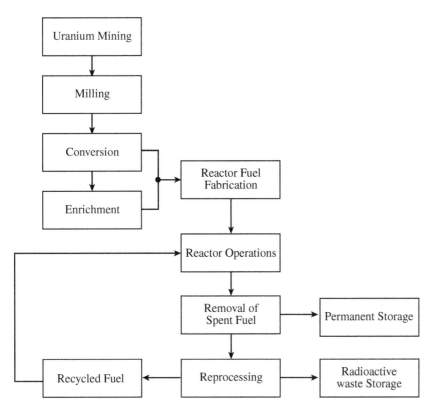

Figure 4.5 Nuclear fuel cycle.

4.4.2 Nuclear Fuel Cycle

The process of preparation, transformation, handling, and disposal of nuclear fuel is known as the *nuclear fuel cycle*. Most nuclear reactors in use today operate on nuclear fuel made from uranium. Therefore, the fuel cycle includes uranium mining, milling, reactor fuel fabrication, consumption of the fuel to obtain heat, used fuel reprocessing, and disposal of nuclear waste material. Figure 4.5 illustrates the major steps of the nuclear fuel cycle.

4.4.2.1 Fuel Preparation
The first step in fuel fabrication is uranium mining. The recovered material is separated from rocks, soil, and physical impurities by milling. The crushed mineral is then chemically processed to produce a coarse powder which contains a high concentration of mixed uranium oxides, known as *yellowcake* because of its color. Yellowcake is then converted to chemical forms suitable for producing natural uranium fuel or transported to uranium enrichment facilities as feedstock.

4.4.2.2 Uranium Enrichment
Natural uranium contains only 0.7% of the fissile isotope U-235, but many reactor designs require a higher concentration of fissile material to increase the probability of fission. The purpose of the enrichment process is to increase the U-235 concentration in the uranium fuel.

Enrichment starts with converting impure uranium oxide into uranium hexafluoride (UF_6) by a series of chemical reactions. UF_6 is solid at room temperature but evaporates (sublimes) around 56 °C. Uranium hexafluoride is extremely corrosive and reacts with water; it is stored and transported in solid form. Several methods have been developed throughout the evolution of nuclear technology to separate U-235 and U-238 isotopes that the UF_6 vapor contains.

The first enriched uranium was produced as part of the Manhattan Project to make the first atomic bombs. A combination of electromagnetic isotope separation (EMIS), thermal diffusion, and gaseous diffusion methods were used to obtain the highly enriched uranium needed for the bombs. In the EMIS method, particle accelerators are used to separate U-235 and U-238 isotopes, and this method is quite inefficient and time consuming. The thermal diffusion method is based on the fact that lightweight substances rise faster than heavier substances when heated. This method was used for a limited time at Oak Ridge National Laboratory in Tennessee but abandoned because it could not produce a significant amount of U-235. The gaseous diffusion method, in which vaporized hexafluoride compounds of U-235 and U-238 isotopes pass through the holes of a barrier at different speeds, is a technique that consumes huge amounts of energy.

Among several methods that can successfully increase the concentration of U-235, the centrifugal technique is currently preferred in commercial facilities. In this method, heavier molecules of uranium hexafluoride (UF_6) gas containing U-238 are separated from lighter molecules containing U-238 by centrifugal forces. Several centrifuge units are cascaded in an enrichment plant to increase the concentration of U-235.

Uranium enrichment plants can produce fuel for nuclear reactors as well as material for nuclear weapons. While a concentration of U-235 between 3% and 5% is sufficient for nuclear reactor fuel, nuclear weapons require a concentration as high as 90% or greater. Plutonium obtained from fuel spent in commercial reactors is lower grade than needed for weapons but can still be used for nuclear detonation. Thus, high security measures are applied in nuclear fuel processing plants. International agencies inspect the nuclear fuel reprocessing plants regularly and, as necessary, scrutinize the possibility of producing weapon rate nuclear material.

4.4.2.3 Nuclear Fuel Assembly

Enriched uranium is converted into UO_2 powder and pressed into fuel pellets. The pellets are stacked into long, thin rods. A typical fuel rod is around 10 mm in diameter and several meters in length. Fuel rods are bundled together by means of metallic brackets to form a fuel assembly. The rod bundles are spaced by a few millimeters to allow coolant flow between them.

4.4.2.4 Critical Mass for Sustained Chain Reaction

During the operation of a nuclear reactor, U-235 that the fuel rods contain undergoes fission reactions. Fission of every 10 U-235 nuclei releases 24 neutrons on average. Out of two or three neutrons released from one fission, only one must be absorbed by a U-235 nucleus to sustain the chain reaction. The other neutrons are absorbed by other materials in the fuel rods, typically U-238 nuclei. By neutron absorption, $_{92}U^{238}$ transforms to neptunium-239 ($_{93}Np^{239}$), then plutonium-239 ($_{94}Pu^{239}$) through a series of radioactive decay processes. Plutonium-239 is a fissile element and its fission contributes to the energy released by the reactor. The fuel rods must allow sufficient paths for released neutrons to ensure the necessary number of collisions with a fissile nucleus. Critical mass is the amount of nuclear material mass that guarantees that enough collisions of neutrons with fissile nuclei occur.

Power available from fission of uranium is approximately 1 MW/g of uranium. A typical nuclear power plant with 1000 MW electric generation capacity consumes about 27 t of UO_2 each year, considering that only 3–5% of uranium is actually converted to energy (Gupta 2012). About 80% of the energy available from the uranium remains in the spent fuel, therefore, reprocessing of used fuel can save a significant amount of energy.

As we showed in Eq. (4.18), by neutron absorption U-238 may produce Pu-239, which is a fissile isotope. Plutonium produced in earlier reactors was mainly used for nuclear weapons.

4.4.2.5 Disposal of Used Nuclear Material

A nuclear fuel assembly can power the reactor from several months to years. After a certain amount of nuclear fuel is spent, the remaining radioactive material is removed from the reactor core. The spent fuel is extremely hot and has to be cooled. Most of the heat is produced by the radioactive decay of medium-mass fission products, remaining uranium, and plutonium. The radioactive material is placed in deep pools of water to cool down and also to block the radiating particles to protect the workers from radioactivity.

The depleted fuel assembly may stay in the pool as temporary nuclear waste storage. After the decay heat has sufficiently decreased, the radioactive material may be removed from the water and placed in dry casks for long-term storage in a disposal facility. The spent fuel can be also transported to a reprocessing facility or be reprocessed and recycled to produce new nuclear fuel.

The spent fuel may stay for years in interim storage at the power plants in pools or dry casks. Reprocessing the nuclear fuel includes extraction of the unused uranium and the plutonium produced during the reactor operation. Decay of the radioactive waste materials is typically hundreds of years. Reprocessing nuclear fuel does not eliminate the need for disposal or long-term storage of the remaining radioactive waste materials.

4.5 Nuclear Reactors

All commercial nuclear reactors in operation today produce heat through controlled fission reactions of a nuclear fuel. Heat is used to produce steam, which turns a turbine. Most nuclear reactors are designed to generate electric power by driving large generators with a steam turbine. Reactor, heat engine (typically a steam turbine), and generator are the main parts of a nuclear power plant. In principle, the only difference between a fuel fired generation and nuclear generation is the method of heat production. Instead of combustion of a hydrocarbon fuel, nuclear power plants produce heat by nuclear reaction. Figure 4.6 illustrates the layout of a typical nuclear power plant.

The primary function of nuclear reactors is to generate heat by controlled chain reaction. "Burner reactors" consume the fissile material in the fuel without producing significant

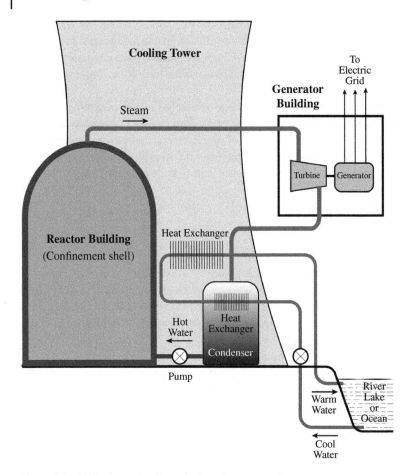

Figure 4.6 Main elements of a typical nuclear power plant.

amount of new fissile isotopes. Some reactors are designed to breed new fissile material as the excess neutrons are absorbed in fertile nuclei during the course of the reaction. For example, Th-232 and U-238 can be converted into U-233 and Pu-239, respectively. Such reactors are named "converter reactors" or "breeding reactors" depending on the breeding ratio.

If the neutrons are not slowed down in a moderator, the reactor is called a "fast reactor." Such reactors must use enriched fuel, but during the chain reaction more fertile material is converted into fissile material. Fast reactors are more compact compared to a thermal reactor with the same power output and they are easier to shield to prevent radiation. However, control of the chain reaction is more difficult because fast neutrons have a short life.

Neutrons causing fission may be slowed down by passing through a moderator medium for easier control of the chain reaction. Almost all the power reactors used today are this type and they are called "thermal reactors." Less enriched fuel or natural uranium may be used in such reactors. Distilled water (light water), heavy water (deuterium oxide, D_2O), graphite, molten salt, liquid metal, and organic materials can be used as moderators. The majority of commercial reactors use light water as moderator.

4.5.1 Reactor Core

The fuel assembly, moderator, control rods, and coolant form the reactor core, which is enclosed in a pressure vessel with thick concrete walls. A reactor core has an upright cylindrical shape with a diameter between 0.5 and 15 m, depending on the power of the reactor. Figure 4.7 outlines a typical light water reactor core. Fuel rods are tubular elements that contain reactor fuel. Moderator sections are placed between fuel rods. Control rods can move up and down to change the chain reaction rate. A reflector layer is placed around the core to send neutrons that were not absorbed back to increase the efficiency of the reaction. Radiation shielding protects the walls of the reactor from radiation damage and operating personnel from radiation exposure. A coolant fluid transfers the enormous heat produced in the reactor core to the external heat exchangers to produce steam.

The core structure described above is called heterogeneous because the fuel and moderator are in separate sections. Free neutrons move in different environments of the fuel elements and the moderator substance. Such reactors are more common worldwide because they offer more design options and are easier to build.

In a homogeneous reactor core, the fuel is homogeneously mixed with the moderator. Chain reaction occurs in a slurry solution of fuel and moderator, which is usually heavy water (deuterium). The reactor core has two sections. The central part contains a solution of uranyl sulfate in heavy water, while the peripheral section contains a slurry of thorium in heavy water. Both the solution and the slurry mix are circulated through separate heat exchangers to produce steam. Such reactors have the benefit of better heat transfer and the possibility of adding nuclear fuel during the reactor operation.

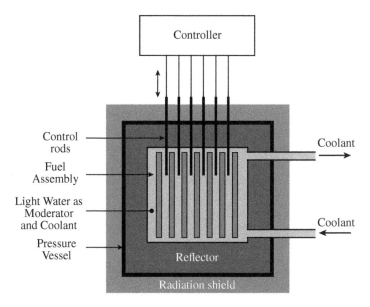

Figure 4.7 A light water reactor core.

4.5.2 Fuel Assembly

The basic fuel used in nuclear reactors is U-235, which is the only fissile isotope that is found in natural uranium deposits. While reactor grade purified natural uranium can be used as fuel, most reactors rely on enriched uranium oxide, uranium carbide, or uranium sulfate. Fast breeder reactors (FBRs) use a combination of uranium and plutonium or uranium and thorium.

Cost of nuclear fuel impacts the cost of electricity generated in nuclear power plants. In addition to being cost-effective, nuclear fuel must be resistant to radiation damage and should be suitable to operate at high temperatures.

Nuclear fuels are stored in tubular containers made from stainless steel, beryllium, magnesium, zirconium, and aluminum alloys. Cladding protects the fuel from oxidation and corrosion as well as preventing contamination of the coolant fluid by fission products. Cladding materials must be able to withstand high temperatures that arise in the reactor core, have high thermal conductivity, and be resistant to the effects of radiation and corrosion; Zircaloy (zirconium and aluminum), and Magnox (alloy of magnesium) are among preferred cladding materials. Fuel rods are bundled together to form a fuel assembly. The fuel assembly of a typical reactor comprises 200–300 fuel rods and between 100 and 200 fuel assemblies are installed to power the reactor.

4.5.3 Moderator

Moderator is basically a medium used to reduce the velocity of neutrons to a range suitable for a controlled chain reaction. Neutrons released during the fission reaction have high kinetic energy due to their speed, about 1.5×10^7 m/s. Their probability of being absorbed by the U-238 nuclei is higher than the probability of causing fission of the U-235 nuclei. In addition, faster neutrons may escape from the reactor core without encountering a fissile nucleus. The probability of fission can be increased either by increasing the concentration of fissile U-235 atoms or reducing the velocity of the released neutrons. This is why fast reactors do not need a moderator but require costly enriched fuel. Thermal reactors have to reduce the neutron velocity by a moderator to use less enriched fuel.

Nuclei of elements with a smaller mass number, such as hydrogen, beryllium, and carbon, scatter the colliding neutrons at lower speed due to the conservation of momentum principle. Hydrogen and helium gases are not suitable as moderators because of the lower atom density, thus lower probability of collision with neutrons. However, light water (H_2O), heavy water (deuterium oxide, D_2O), and graphite are common moderator substances.

Light water (H_2O) is the lowest cost and the most effective moderator; in addition, it is abundant and can be used as the cooling fluid at the same time. However, light water can capture free neutrons and turn the reactor to a subcritical state if the fuel is not sufficiently enriched. Therefore, natural uranium cannot be used in light water reactors.

Heavy water (D_2O) is also an effective moderator because of the high concentration of deuterium ($_1H^2$) and its lower tendency to capture free neutrons compared to light water. A small amount of heavy water is present in regular light water. During the electrolysis, light water decomposes faster than deuterium oxide. Heavy water can be obtained by repetitive electrolysis of light water. This is obviously an energy intensive, and therefore expensive, process. However, heavy water reactors can use natural uranium with low U-235 concentration. Canadian Deuterium Uranium (CANDU) reactors use pressurized heavy water as moderator and natural uranium as fuel.

4.5.4 Control Rods

The purpose of the control rods is to adjust the chain reaction rate at a level suitable for the desired heat output. In a chain reaction, the neutrons released from previous fission cause new neutrons to be released as they collide with other fissile nuclei. If the number of new electrons increases, the reactor is said to be supercritical and it may turn to an atomic bomb. If the number of new neutrons decreases, the reactor becomes subcritical and the chain reaction rate decreases until the reactor stops operating. Therefore, the number of new neutrons must be controlled to maintain the ratio K given by Eq. (4.20) around unity.

$$K = \frac{\text{Number of neutrons causing fission in any generation}}{\text{Number of neutrons released from fission in preceding generation}} \qquad (4.20)$$

Control rods are made of an absorber material such as boron, cadmium, silver, and indium. As the other reactor elements, control rods must be heat, radiation, and corrosion resistant. Control rods are moved up or down in the space between fuel rods. In regular reactor operation, a control system adjusts the position of control rods. If needed, they can also be moved manually. There are three types of control rods; shut off rods, coarse regulation rods, and fine regulation rods. Shut off rods are used in emergency situations, while coarse and fine regulation rods are continuously adjusted during the normal operation of a reactor.

4.5.5 Cooling System

Coolant is a fluid that transfers heat produced in the reactor core to the heat exchangers where steam is produced. It may be a gas or liquid that has high heat transfer properties, low neutron absorption, and low viscosity. In addition, the coolant must be non-corrosive, and easy to pump. Carbon dioxide, hydrogen, helium, and air can be used as coolant. Light water, heavy water, and liquid sodium, or a mixture of liquid sodium, potassium, and organic fluids are common liquid coolants. Air and light water are the most abundant and cost-effective coolants.

Cooling fluid is circulated through heat exchangers to produce steam for turbines. Most steam turbines operate based on the Rankine cycle discussed in more detail in Chapter 9. The efficiency of a heat engine directly depends on the higher and lower temperatures. The remaining heat in the steam-water mixture is rejected to a river, lake, or air by a condenser. Hyperbolic cooling towers as depicted in Figure 4.6 have become the symbol of nuclear power plants in popular imagination. Similar cooling towers are also used in many fossil fuel burning power plants.

Light (regular) water is a good heat transfer fluid because it is abundant and inexpensive. However, because it boils at 100 °C at normal pressure, it must be kept under pressure to increase the boiling point. Light and heavy water can serve both as a coolant and moderator.

In FBRs liquid sodium is used as coolant because of its low neutron absorption. Sodium is solid at room temperature but melts at 98 °C. Its boiling point is 883 °C at normal pressure, which removes the need to keep it under high pressure. The heat transfer properties of sodium are superior to water, carbon dioxide, and air. However, sodium must be kept above 98 °C at all points of the cooling system to avoid the risk of clogging. Upon neutron capture, sodium becomes a radioactive isotope Na-24 that emits gamma rays as it decays with a half-life of 14.8 hours. To prevent the radiation hazard, heat is transferred in a well-shielded heat exchanger. Sodium cooled reactors must have a secondary coolant circulation to produce steam.

4.5.6 Reactor Types

Nuclear reactors are usually categorized by the type of coolant and moderator materials. Six types of reactors described next are currently used in the world, mainly to generate electric power. The first two reactor types called a pressurized water reactor (PWR) and a boiling water reactor (BWR) are dominant in commercial nuclear power plants.

4.5.6.1 Pressurized Water Reactor (PWR)

Pressurized water reactors are the most common type among the commercial nuclear power reactors operating around the world. As of December 2017, out of 450 operational nuclear power reactors worldwide, 298 were this type. PWR uses ordinary water both as moderator and coolant. Since light water absorbs some of the neutrons released in chain reaction, these reactors must consume enriched uranium. The reactor core contains between 80 and 100 t of enriched uranium dioxide fuel pellets. The cladding material of the fuel rods is stainless steel or Zircaloy™ (zirconium alloy).

High pressure cooling water circulates between the core and pressure vessel as illustrated in Figure 4.8. Water is pressurized to increase the boiling temperature up to 374 °C.

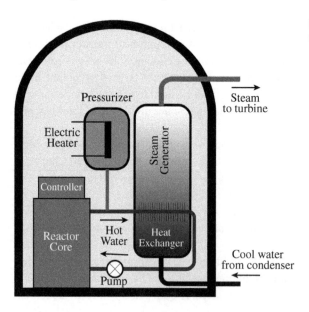

Figure 4.8 Pressurized water reactor.

The pressure must be high enough to maintain water in liquid phase at this temperature. For safe operation of a typical PWR, the water pressure and temperature are usually set around 155 bar and 315 °C, respectively. The pressurizer is essentially a boiler in which the temperature is maintained constant by an electric heater to compensate the fluctuations due to reactor activity and load variations. The steam generator, which is basically a heat exchanger, also isolates the radioactive coolant from the steam circulating in a secondary loop through the turbine that turns the electric generator.

Steam quality is defined as the proportion of steam mass to the total mass of the steam-water mixture. Due to the temperature and pressure limitations, the steam quality of pressurized water reactors is lower than the steam produced by a combustion of hydrocarbon fuels. Therefore, the steam may be further heated in a fuel fired heat exchanger to improve the efficiency of the heat engine.

4.5.6.2 Boiling Water Reactor (BWR)

The second largest number of nuclear reactors used in nuclear power plants is BWR. In this type of reactor, light water serves as moderator, coolant, neutron reflector, and working fluid for the heat engine that turns the electric generator. Such reactors are fueled with enriched uranium oxide that contains 1.9–2.6% U-235 isotope. The reactor core has approximately 800 fuel assemblies, each consisting of a matrix of fuel rods with empty spaces left for control rods.

The cooling water is heated to nearly 300 °C at 70 bar pressure, which are lower values compared to pressurized water reactors. Unlike the PWR, cooling water directly circulates through the turbine loop. A schematic view of the BWR is shown in Figure 4.9. Since the

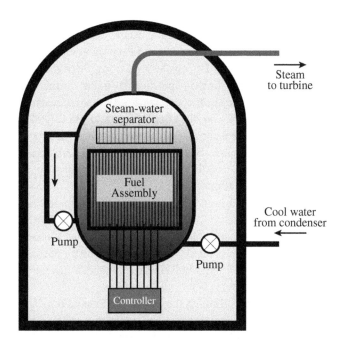

Figure 4.9 Boiling water reactor.

water circulating through the steam turbine is boiled directly in the reactor, this type of power plant does not need a secondary steam loop, therefore their structure is simpler. However, the steam bubbles in the boiling water affect the reactor power and water particles must be separated from steam. A steam-water separator and drying equipment is placed in the space above the core. Thus, a controller must be placed below the pressure vessel and control rods must enter from the bottom. In the case of emergency, the rods cannot be dropped in the fuel assembly by gravity. Instead, a hydraulic mechanism moves the rods upward. Dry steam separated from boiling water is sent to the turbine while saturated water flows down and mixes with the water coming from the condenser of the turbine loop. Cooling water circulates either naturally by density difference and gravity or forced by a pump. BWRs are equipped with a water demineralizer.

In a BWR, steam pressure and temperature are significantly lower compared to pressurized water reactors. A reheater may be needed to increase the steam temperature and consequently the efficiency of the steam turbine. The reheater is usually electric powered consuming some small part of the generated electricity. In a BWR, the amount of steam bubbles controls the reactor power since the water coolant also moderates the reaction. Increasing steam pressure in the core displaces water around the fuel elements and the reactor power increase. Control rods are then inserted to regulate the power. A small secondary steam generator may be used to help the regulation. Recent designs of single steam cycle BWR reactor power is controlled by changing the water circulation rate by a series of speed-controlled pumps.

4.5.6.3 Pressurized Heavy-Water Reactor (PHWR)

The pressurized heavy water reactor design developed in 1950s in Canada is known as CANDU (Canadian Deuterium Uranium). Heavy water (deuterium oxide D_2O) acts as moderator and neutron reflector. The coolant is also heavy water circulating through a separate loop. Heavy water is costlier but absorbs less neutrons, these reactors do not need enriched fuel and can operate on reactor grade purified natural uranium. The cost of heavy water is then compensated by the lower fuel cost. In addition, these reactors have a breeding ratio higher than unity, hence they produce fissile material from U-238.

CANDU reactors have a large horizontal steel tank of about 6 m diameter and 8 m length, known as a calandria that houses the reactor core. A calandria is not pressurized and remains at a lower temperature, which makes fabrication easier and cheaper. Heavy water filled in the calandria serves only as moderator.

The fuel bundle of a typical CANDU reactor is around 50 cm long and contains some 20 kg of low-enriched uranium. Fuel bundles are placed in pressure tubes with a 10 cm diameter, made of Zircaloy. Each pressure tube is enclosed in larger tubes made of the calandria material and the space around the pressure tubes is filled with carbon dioxide for thermal insulation. CANDU reactors can be refueled by adding small amounts of reactor fuel while the reactor is operating at full power (Rouben 1997).

The pressurized heavy water coolant circulates through the pressure tubes and transfers the heat produced by nuclear reaction to a heat exchanger in the steam generator. Steam at 40 bar pressure and 250 °C temperature circulates through the turbine loop producing mechanical power to drive an electric generator. Lower temperature moderator heavy water is separately circulated through a heat exchanger to recover the heat leaked from the pressure tubes (Gupta 2012).

The CANDU reactor design has been modified after the 1980s to improve safety, efficiency, initial cost, and performance. The major improvement is transition from heavy water to light water as coolant. The Advanced Canadian Deuterium Uranium Reactor (ACR) is fueled with sintered pellets of slightly enriched uranium oxide that contains 2.4% of U-235 (AECL 2010).

4.5.6.4 Gas Cooled Reactor (GCR)

Reactors cooled with carbon dioxide or helium use graphite as a moderator. Carbon dioxide is a cheaper coolant, but helium is a chemically inert gas. Because the melting point of graphite is high, the temperature of gas exiting the reactor can be as high as 800–900 °C. The coolant can be supplied to a gas turbine and the high temperature increases the overall efficiency of the electric generation. At the end of 2017, 14 of the operational commercial reactors were of this type.

Gas cooled reactors (GCRs) can operate either as a thermal or FBR. Gas cooled FBRs use helium as a coolant because of its lower neutron absorption property. Such reactors can breed fissile material from a fuel composed of U-233 and Th-232.

4.5.6.5 Light Water-Cooled Graphite Reactor (LWGR)

LWGR reactors are similar to BWRs except that graphite is used as a moderator. Since water is not the moderator, fuel elements are placed in vertical pressure tubes in which light water coolant circulates. Temperature of the water exiting the pressure tubes is around 300 °C. Like BWR, the heat exchanger in the steam generator produces steam, which is circulated in the turbine loop. Like GCR, this type of reactor is one of the least common types among the operating nuclear reactors. At the end of 2017, only 13 of the 450 reactors worldwide were light water-cooled graphite moderated reactors.

4.5.6.6 Sodium Cooled Fast Breeder Reactor (FBR)

Sodium is a convenient metal coolant for nuclear reactors because it remains in liquid state roughly between 97 and 1156 °C. Its low mass number (11) reduces the probability of capturing free neutrons. In addition, because of its high boiling point, there is no need to pressurize the core. However, sodium cooled reactors require additional loops for heat transfer and steam production.

The first nuclear reactor that produced electricity was a sodium cooled FBR named *Experimental Breeder Reactor I* (EBR-I), built in Idaho between 1948 and 1951 (see Figure 4.14). Today, only three commercial FBRs are operational worldwide.

General properties of reactor types and the number of each type in operation as of December 2017 are summarized in Table 4.5. Pressurized water and BWRs are the most common, and together they constitute 82% of all reactors operating in the world (IAEA 2019).

Figure 4.10 shows the share of reactor types in the electric generation of all operational nuclear power plants in the world. In 2017, pressurized and boiling light water reactors provided 90% of the total electric generation of nuclear power plants. The share of PHWR was 6% and the total contribution of GCR, LWGR, and FBR to nuclear electric generation remained around 5%.

Table 4.5 Summary of reactor types.

Type	Coolant Material	Pressure (bar)	Temp. (°C)	Moderator	Fuel	Number	Worldwide share[a] (%)
PWR	H_2O	155	315	H_2O	Enriched UO_2	298	66.2
BWR	H_2O	70	285	H_2O	Enriched UO_2	73	16.2
PHWR	D_2O	40	250	D_2O	UO_2	49	10.8
GCR	CO_2 or He	—	800–900	Graphite	U-233 Th-232	14	3.1
LWGR	H_2O	150	300	Graphite	Enriched UO_2	13	2.8
FBR	Liquid Sodium	—	800–900	Graphite	UO_2	3	0.006

a) Share in total number of operational nuclear reactors worldwide as of December 2017.
Source: IAEA (2019).

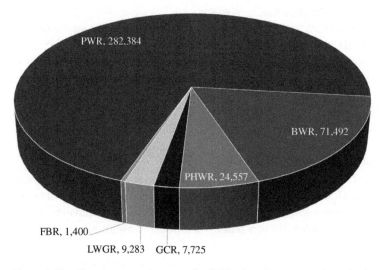

Figure 4.10 Electric generation capacity (MW) of nuclear power plants in the world by reactors types in 2017. Source: IAEA (2019).

4.6 Safety of Nuclear Power Plants

Since the installation of the first commercial nuclear power reactors in the mid-1950s, safety has always been the primary concern. Three major nuclear power plant accidents have shown that a nuclear accident is not just a local incident. In addition to deaths and immediate damage on directly involved individuals and the surrounding environment, a nuclear accident has long-term impacts on the global ecosystem, economy, and development of the energy system. Although nuclear power plants are among the safest and most secure facilities, accidents may happen. Any nuclear reactor failure creates widespread concerns that impact the public acceptance.

4.6.1 Nuclear Safety Concepts

The primary goal of nuclear reactor safety is to minimize the risk of accidents and, in the worst case of an emerging abnormal incident, to mitigate the harmful effects of possible radiation release on the population living nearby and the environment. The International Atomic Energy Agency (IAEA) publishes safety standards, guidelines, and safeguards to strengthen the safety of nuclear power plants.

During the planning phase, the site selection is crucial for safety of a nuclear power plant throughout its operation life. Possibility of earthquakes, volcanic activities, flooding, tsunami, tornedo, hurricanes, and meteorological hazards are among the major factors in risk assessment. Transportation and security are factors indirectly related to safety. Accessibility of the plant is important for transportation of nuclear fuel and radioactive waste material. On the other hand, power plants operating near highly populated areas may present evacuation issues in the worst case of a nuclear accident. Evacuation scenarios and emergency plans are also considered in site selection. Security of the nuclear plant is also indirectly related to safety. Military attacks or terrorist activities can compromise the safety of a reactor by disabling the reactor protection system. In addition to external attacks, insider intervention and cyber-attacks may risk the safe operation of nuclear power plants.

Modern nuclear reactors are designed with multiple independent layers of protection. Diversity, redundancy, and independence of protection equipment are key in reactor design. Regulations and standards mandate comprehensive precautions to ensure the safety of workers, the public, and the environment during construction, equipment installation, testing, and commissioning phases.

During the operation of a plant, nuclear fuel needs to be handled at a high level of security through the entire cycle. Fuel storage, refueling, discharge of used fuel, reprocessing, and storage of radioactive waste must be coordinated to prevent possible radioactive contamination. Nuclear waste management includes collection and sorting, concentrating liquid waste, and immobilizing and packaging nuclear waste before storage and disposal.

Human and organizational factors play a crucial role in safe operation of nuclear facilities. All personnel involved in the operation of a nuclear power plant need to be continuously trained on nuclear safety to establish a collective safety culture. National and international agencies require power plant managements to develop a safety plan and effectively enforce its implementation.

The September 11, 2001 terrorist attacks increased the concerns on nuclear reactor security. While the thick walls of the confinement shell provide sufficient protection against external explosions, nuclear power plants are still vulnerable targets to terrorist attacks. A strategic damage to the auxiliary equipment outside of the reactor building can disable the cooling water supply and lead to an explosion or meltdown. Pumps, valves, piping, and emergency generators must be secured. As well as an external attack, insider actions and cyber-attacks are also possible. Therefore, safety of nuclear power plants is intricately linked to reactor security.

4.6.2 Reactor Protection Systems

Safe operation of a nuclear power plant relies on the stability of the chain reaction and proper cooling of the reactor core. Modern reactors are designed to maintain the chain reaction stability by moderating the fast neutrons. When the reaction rate increases, control rods are inserted between fuel rods to bring the reaction back to the desired rate. In the case of an unexpected rapid increase of the chain reaction, the control rods are completely inserted to stop the reaction. However, as we will discuss in the next section, the Chernobyl accident occurred because the chain reaction became out of control, or in other words, the reactor became supercritical. If the reaction rate continues to increase, the core temperature reaches extremely high levels causing reactor meltdown. A reactor explosion can release radioactive material in the air turning the accident into a disaster with long-term threats on human life.

Loss of cooling is another major cause in nuclear accidents. If the cooling fluid is not supplied properly, the core temperature may increase leading to a partial or complete reactor meltdown. Cooling fluid is circulated through the fuel rods by pumps. A failure of the pumping system or loss of power causes lower circulation rate and the coolant temperature increases. In most reactors, cooling water also serves as moderator. Steam bubbles in the reactor reduce the moderator effect of water. The control rods may compensate this change, but sudden increase of the reaction rate can be extremely dangerous. In Three Mile Island, cooling water circulation was turned off due to an instrumentation failure. In the Fukushima accident, cooling water could not be supplied because the circulation pumps lost electric power.

The goal of the reactor protection system is to automatically maintain the steady operation of the reactor and, if an emergency arises, shut down the reactor to prevent catastrophic meltdown and release of radioactive particles into the environment. The US Nuclear Regulatory Commission (NRC) defines the concept of defense-in-depth as creating multiple independent and redundant layers of protection responding to failures, accidents, or fires in power plants.

The protection system consists of sensors, data acquisition computers, relays, valves, and various types of actuators. In modern reactor design, the protection system is diversified by using different techniques to check the same parameters. Diversity increases the reliability of the instrumentation because it eliminates the risk of hardware or software failure that might affect data collection and system monitoring.

In reactor protection systems more than one device is used for the same purpose. Redundancy of equipment increases the reliability because if one device fails to respond in an emergency condition, another one would perform the required task.

Protection devices are powered from separate uninterruptible power supply systems with adequate energy storage to avoid loss of power during an accident or natural disaster. In addition, the operation of subsystems must be independent from each other such that the failure of a group of elements does not affect the operation of other protection equipment.

4.6.3 Major Nuclear Power Plant Accidents

Although major accidents have generated frightening memories and public reactions to nuclear reactors, only around 100 accidents were recorded over six decades, exceeding 18,000 cumulative reactor-years of operation worldwide (IAEA 2019).

Three major reactor accidents occurred in the history of the civil nuclear power industry: Three Mile Island, Chernobyl, and Fukushima. Other than the Chernobyl disaster, there was no immediate death during any of the nuclear accidents. Whereas no direct correlation has been established so far between nuclear power plant accidents and the increase of radiation-related diseases or genetic damage (UNSCEAR, 2016), there are controversies about the long-term consequences of released radiation. Evacuation and recovery process, however, usually cause panic, accidents, indirect deaths, sicknesses, and traumatic experiences.

4.6.3.1 Three Mile Island Accident

The Three Mile Island (TMI) nuclear power plant accident occurred on March 28, 1979 in the United States, near Harrisburg, the capital city of Pennsylvania. The power plant has two pressurized light water reactors (PWR). The first unit (TMI-I) was commissioned on September 2, 1974 and the second unit (TMI-II) on December 30, 1978. Four months after operation started, on March 28, 1979, an electromechanical failure disabled the pumps feeding water to the steam generators, which extract heat from the reactor core. This caused an automatic shutdown of the turbine-generator system and consequently the reactor. The pressure in the primary loop immediately began to increase due to high core temperature. To control the pressure, the pilot-operated relief valve on the top of the pressurized tank opened. This valve was supposed to close when the pressure fell to normal level, but it stayed stuck open. However, the instruments in the control room incorrectly indicated that the valve had closed. Therefore, the operators in the control room did not realize that the plant was experiencing a loss-of-coolant accident (LOCA) because the cooling water mixed with steam was discharged from the stuck-open pressure valve.

Under normal conditions the pressure vessel that holds the reactor core should be filled with water, but there was no water level indicator in the vessel. In addition, the available instruments displayed misleading readings. Based on the available information, the operators thought that the core was properly covered with water, which was not true. Reactor staff took a series of actions that mistakenly lowered the cooling water level in the core. Without adequate coolant, the reactor started to overheat. The Three Mile Island accident ended with a partial meltdown of the reactor, but radioactive contamination remained in the reactor, except for a small leakage into the environment. After the accident, TMI-II reactor was permanently shut down and 99% of its fuel was removed. The reactor coolant system was completely drained, and decontaminated water was evaporated. The reactor fuel and radioactive debris was shipped to a disposal area off-site (NRC 2018).

The first unit continued operation until 2019. Three Mile Island power plant was permanently closed on September 20, 2019. The decommissioning process is estimated to take about 60 years.

Design deficiencies, component failures, and wrong decisions based on misleading information received from instruments combined with communication and coordination difficulties escalated the accident to a near-disaster level. The accident emerging through several days presented the risk of an uncontrolled chain reaction that could lead to a nuclear explosion, threatening accumulation of explosive hydrogen gas in the dome of the reactor, and contamination risk of the adjacent Susquehanna River. The highly populated area around the power plant was evacuated. Although the Three Mile Island accident did not cause any immediate death or long-term environmental damage, the cost of cleanup was enormous, reaching approximately one billion dollars. The accident became a turning point in nuclear power development in the United States. The nuclear power industry has significantly improved the safety of reactors in new designs. The US Nuclear Regulatory Commission (NRC) has increased safety requirements in nuclear power plants and has been extremely strict in issuing permits for new power plants and capacity increases of existing plants.

4.6.3.2 Chernobyl Nuclear Accident

Seven years after the TMI accident, on April 26, 1986, the Chernobyl nuclear power plant near a major city, Kiev, Ukraine, in the former Soviet Union, became unstable and exploded because of an unusual test. The unit 4 of the power plant was shut down for regular maintenance. Engineers decided to test whether the plant equipment could provide sufficient power to run the reactor cooling system and emergency equipment during the transition from loss of main power and startup of the diesel engines and generators. The reactor used in Chernobyl, named RMBK, has an inherent design weakness that causes instability at low load conditions. During the routine maintenance the engineers performed an experiment on the electrical control system while the reactor was shut down to test the reactor's possible response to an electric power failure. The operators, in violation of safety regulations, intentionally turned off the control systems and allowed the reactor to reach unstable low power conditions. A sudden power surge caused a steam explosion, which ruptured the reactor vessel. Fuel-steam interactions destroyed the reactor core and damaged the reactor building. An intense graphite fire followed the explosion burning the power plant for about 10 days and releasing the radioactive fission products into open air.

Highly populated areas of Belarus, Russian Federation, and Ukraine were heavily contaminated with radioactive materials. Northwest wind at the time of the explosion carried radioactive dust to the neighboring countries. In subsequent days, wind came from all directions and the contamination eventually spread up to northern Europe and Black Sea region of Turkey. Radionuclides from the Chernobyl release were measurable in all countries of the northern hemisphere (UNSCEAR 2008).

Emergency response to the accident included pouring sand and boron on the reactor debris from helicopters. The purpose of sand was to stop the fire and prevent additional release of radioactive materials. The boron was dumped to prevent continuation of nuclear reactions. A few weeks after the accident, the damaged unit was temporarily enclosed in a concrete structure. The area within 30 km (18 mi) of the plant was officially closed, and 115,000 people were evacuated in 1986 from the most heavily contaminated areas. In subsequent years, the government sequentially evacuated 220,000 persons (UNSCEAR 2008).

Two immediate deaths were reported due to the explosion and lethal doses of radioactivity, and 134 employees were hospitalized and died within days to months due to radioactivity. The actual number of deaths and genetic damage caused by the accident remains unclear since some estimates are either biased or controversial. The Chernobyl accident is considered the worst disaster in the history of nuclear power. Beside deaths and defective births, the disaster caused social and economic damage. Pripyat in northern Ukraine, where the Chernobyl power plant is located, became a ghost town. The disaster created strong public reactions against nuclear power development. Significant quantities of agricultural products in many countries became unusable because of the radioactive contamination.

4.6.3.3 Fukushima Daiichi Nuclear Accident

Another nuclear accident was primarily initiated by a natural disaster in Fukushima Daiichi nuclear power plant in Japan. On March 11, 2011, the devastating Tohoku earthquake and the resulting tsunami, which killed nearly 16,000 people, flooded the area where the nuclear power plant was located. Protection equipment of the power plant automatically shut down the nuclear reactors. However, the disabled emergency generators failed to power the pumps that should have continued to circulate the cooling water. The insufficient cooling caused the meltdown of three reactors, explosions of hydrogen gas mixed with air, and release of radioactive material. While no immediate death was reported, some controversial estimates claim 1600 deaths due to evacuation and radiation related diseases following the accident. Based on the International Nuclear Event Scale the Fukushima disaster is rated Level 7, as is Chernobyl. Cleanup and decommissioning of the power plant is estimated to take about 30–40 years. The Fukushima incident showed that despite high-end technology and proper operation of protective equipment, unforeseen factors can still lead to nuclear disasters.

4.6.4 Consequences of Nuclear Accidents

Nuclear accidents have always been the focus of media and social groups opposing to nuclear power. Every nuclear disaster reduced the public confidence and triggered public reactions. At the same time, lessons learned helped reactor designers, operators, and regulatory agencies to improve safety and performance of nuclear power plants. Countries revised their nuclear safety regulations after all major accidents.

The Three Mile Island Unit 2 (TMI-II) nuclear power plant accident had the greatest impact on nuclear regulation in the US. Since the TMI accident, the US Nuclear Regulatory Commission expanded emergency preparedness and placed NRC resident inspectors at each plant site in the US. In addition, new regulations mandated all nuclear power plants to notify NRC immediately of critical events. An NRC Operations Center was formed for identification, analysis, and publication of plant performance information. Human performance has been recognized as a critical component of plant safety and training of personnel involved in the operation of nuclear reactors has been enhanced.

Design improvements were focused on reducing the risk of reactor damage due to a LOCA. Consequently, the average number of reactor events has significantly dropped in the US following the TMI accident. The NRC reported that radiation exposure levels to plant workers has steadily decreased to about one-sixth of the 1985 exposure levels, and are well below federal limits.

The Chernobyl disaster showed the cross-border impacts of nuclear accidents. Consequences of the Chernobyl disaster have been discussed in many international meetings. The accident impacted heavily the life, health, and the environment in Ukraine, Belarus, and Russia and caused anxiety around the world. The long-term health effects of radiation, economic losses, and social impacts of dislocation were severe in the region.

The Chernobyl accident, which took the lives of 30 workers during and after the explosion, also caused the hospitalization of hundreds of other emergency workers and exposed five million people to radionuclides. This has led to a sharp increase in thyroid cancer incidences, especially among children living in the contaminated areas. Apart from radiation-induced effects, many people have also suffered psycho-social stress and significant disruption in their lives, affecting their physical and mental health and well-being.

At the time of the accident, there was a great deal of fear and mistrust because of lack of information. Since then, massive efforts have been made by the governments and health authorities of Belarus, the Russian Federation, and Ukraine to mitigate the effects of the accident.

The Fukushima Daiichi disaster showed that advanced technology and proper implementation of safety procedures are not sufficient to prevent the worst-case scenario. Although no life was lost during the accident, evacuation process caused indirect deaths and health issues. In total 18 reactors have been permanently shut down in Japan after the Fukushima disaster.

In addition to the three disasters at Three Mile Island, Chernobyl, and Fukushima, there has been at least one nuclear incident every year causing interruption of power plant operation and substantial recovery cost (IAEA 2019). In 1990, the International Atomic Energy Agency (IAEA) introduced *International Nuclear and Radiological Event Scale* (INES) to assess the severity of nuclear incidents. INES classifies nuclear events in seven levels ranging from Level 1 (anomaly) to Level 7 (major accident). The levels on the INES scale are determined in terms of the impact of the events on people and the environment, radiological barriers and control, and the impact on defense-in-depth. Events without safety significance are rated as Below Scale (Level 0). Events that have no safety relevance with respect to radiation or nuclear safety are not rated on the scale. The rating scale is logarithmic, which means that the severity of an event increases 10 times at each level. INES levels are listed in Box 4.1 as described in (IAEA 2008).

Box 4.1 International Nuclear and Radiological Event Scale (INES)

- *Level 7 – Major accident.* Major release of radioactive material with widespread health and environmental effects requiring implementation of planned and extended countermeasures.
- *Level 6 – Serious accident.* Significant release of radioactive material likely to require implementation of planned countermeasures.
- *Level 5 – Accident with wider consequences.* Limited release of radioactive material likely to require implementation of some planned countermeasures. Severe damage to reactor core. Release of large quantities of radioactive material within an installation with a high probability of significant public exposure. This could arise from a major criticality accident or fire. Several deaths from radiation.

(Continued)

Box 4.1 (Continued)

- *Level 4 – Accident with local consequences.* Minor release of radioactive material unlikely to result in implementation of planned countermeasures other than local food controls. At least one death from radiation. Fuel melt or damage to fuel resulting in more than 0.1% release of core inventory. Release of significant quantities of radioactive material within an installation with a high probability of significant public exposure.
- *Level 3 – Serious incident.* Exposure in excess of 10 times the statutory annual limit for workers. Non-lethal deterministic health effect (e.g. burns) from radiation. Exposure rates of more than 1 Sv/h in an operating area. Severe contamination in an area not expected by design, with a low probability of significant public exposure. Near accident at a nuclear power plant with no safety provisions remaining. Lost or stolen highly radioactive sealed source. Misdelivered highly radioactive sealed source without adequate radiation procedures in place to handle it.
- *Level 2 – Incident.* Exposure of a member of the public in excess of 10 mSv. Exposure of a worker in excess of the statutory annual limits. Radiation levels in an operating area of more than 50 mSv/h. Significant contamination within the facility into an area not expected by design. Significant failures in safety provisions but with no actual consequences. Found highly radioactive sealed orphan source, device, or transport package with safety provisions intact. Inadequate packaging of a highly radioactive sealed source.
- *Level 1 – Anomaly.* Overexposure of a member of the public in excess of statutory limits. Minor problems with safety components with significant defense in depth remaining. Low activity lost or stolen radioactive source, device, or transport package.

The Chernobyl and Fukushima Daiichi accidents are rated at Level 7, and the Three Mile Island accident is rated at Level 5, based on INES. In addition to the three major accidents discussed in this chapter many other nuclear and radiological events have occurred in civilian nuclear facilities.

While nuclear safety has continuously improved, strong public reaction still exists. The chart in Figure 4.11 shows the trend of nuclear power development in the world. The number of new power plant grid connections decreased after every major nuclear accident. Decreases are partially due to the public reactions and partially the cost of new safety procedures to comply with the strengthened safety standards and regulations.

4.7 Status of Commercial Nuclear Power

According to the International Atomic Energy Agency (IAEA) PRIS database, 450 nuclear reactors were operating in 30 countries in 2019, and in total 55 reactors were under construction in 18 countries (IAEA 2019). Grid connection years of the operating 450 nuclear power plants are shown in Figure 4.11.

In 2017, nuclear power plants generated 2636 TWh electric energy, which is about 10% of total electric generation worldwide. The share of electric generation by nuclear power

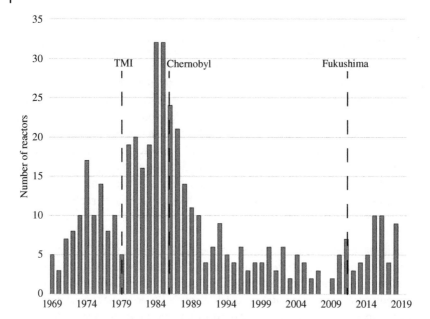

Figure 4.11 Grid connection years of 450 nuclear power plants operational as of December 2018. Source: IAEA (2019).

plants was 17.7% in OECD countries, and 4.7% in non-OECD countries. Over the last decade, net nuclear generation has decreased 4% worldwide. Permanent reactor shutdowns resulted in a 15% drop in the total net generation in OECD countries while non-OECD countries boosted their nuclear generation by about 54% (IAEA 2019).

Figure 4.12 shows the trend of nuclear electric generation by geographic regions (BP 2018). While many reactors permanently shut down in North America and western Europe, Asian countries, especially China and India have increased their nuclear capacity significantly.

Over the period of 2007–2017, net nuclear generation in North America remained nearly constant. Permanent reactor shutdowns in European countries caused 11% decrease in their net nuclear generation (BP 2018). Over this period, nuclear generation increased significantly in Central and South America (12%) and former Soviet Union countries in Central Asia (CIS), including Russian Federation (14%).

The trend of nuclear power in Asia Pacific countries has been diverse over the same decade. Japan permanently shut down a large number of reactors after the Fukushima disaster, causing a decrease of 90% in electric generation from 279 TWh in 2007 to 29 TWh in 2017. Taiwan decreased its nuclear generation from 40 to 22 TWh (45%) over this period. On the other hand, nuclear generation substantially increased in China, India, and Pakistan. In China, nuclear generation quadrupled from 62 to 248 TWh. Pakistan tripled and India doubled electric generation in nuclear power plants.

Charts in Figure 4.13 show the worldwide status of nuclear capacity in 2019. The top chart shows the number of operational and permanently shut down reactors as well as new reactor constructions. The bottom chart displays the electric generation capacity of these reactors (IAEA 2019).

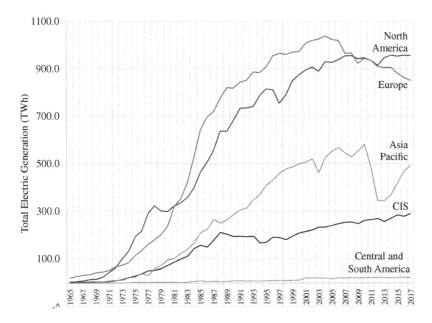

Figure 4.12 Nuclear electric generation trend worldwide. Note: *Commonwealth of Independent States* (CIS) includes Armenia, Azerbaijan, Belarus, Kazakhstan, Kyrgyzstan, Moldova, Russian Federation, Tajikistan, Turkmenistan, Ukraine, and Uzbekistan. Source: BP (2018).

Table 4.6 shows the status of commercial nuclear power in the world (IAEA 2019). The US has the largest fleet of commercial nuclear reactors in the world, accounting for one-fifth of all nuclear reactors worldwide. Many of the reactors currently operating in the US were designed and built after 1970. As a consequence of the research and development diversity and competitive market conditions, diverse nuclear power plant technologies were implemented in the US (IAEA 2018a). Diversity, on one hand, enhances reliability of the electric generation because if a design flaw is discovered in one of these technologies, stopping the reactor until the issue is resolved would not have significant effect on overall electric generation. On the other hand, diversity increases the manufacturing, maintenance, and personnel training costs.

In the US, out of 99 nuclear reactors, 65 are pressurized water reactors (PWR) and 34 are BWRs. The economic circumstances in mid-1970s, higher cost of nuclear power plants, combined with safety concerns increased by the Three Mile accident significantly slowed down nuclear power plant development in the United States. Currently nuclear power plants generate around 20% of the nation's electricity.

France has the second largest number of nuclear reactors after the US and generates about 70% of electricity from nuclear power (IAEA 2018a). Several western European countries including Germany are reducing their dependence on nuclear power and shifting electric generation to renewables.

Safety concerns, performance characteristics, and nuclear waste disposal are the major factors that have driven the evolution of commercial nuclear power plants around the world. Every nation using nuclear energy developed a reactor fleet according to the particular characteristics of its national energy policy and public acceptance. In addition,

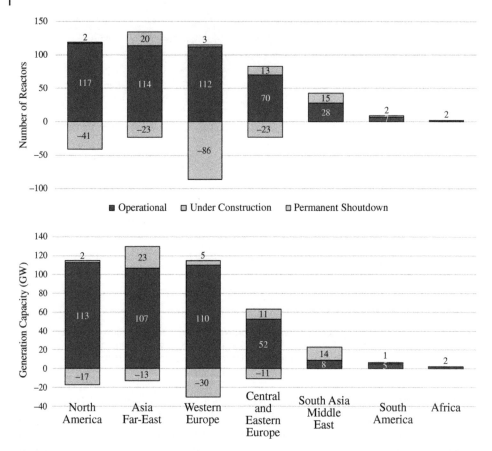

Figure 4.13 Status of commercial nuclear power plants in the world as of 2019. Source: IAEA (2019).

transparency of decision-making processes differs among the nations. In liberal societies, public opinion has more weight on plans regarding nuclear power plant development, while in totalitarian countries decisions are made by official committees.

Western Europe started to use nuclear power around 1960, several years after the first reactor prototypes were developed in the US and Soviet Union. European reactors initially benefited from the experience gained from the development of the first reactor prototypes. The 1974 energy crisis motivated Western countries to increase the share of nuclear energy in electric generation. In continental Europe, pressurized water reactors (PWR) dominate the commercial nuclear industry (IAEA 2018b).

In European countries nuclear power plants are more uniform compared to the US (IAEA 2018a). Standardization reduces the initial cost and simplifies the maintenance and personnel training. Since expertise is more focused on whether a certain technology and essential equipment are compatible, coordination of technical services is more efficient. However, in the case a design or material problem arises, several reactors might need to be shut down during the investigation, troubleshooting, and repair works. Considering that in most

Table 4.6 Status of nuclear power generation in the world.

Country	Operating[a] reactors Number	Net electric capacity (MW)	Under construction[a] Number	Net electric capacity (MW)	Nuclear generation (MWh)	Share in total electric generation (%)
USA	98	99,061	2	2 234	805,327.20	19.7
France	58	63,130	1	1 630	386,452.90	72.3
China	46	42,800	11	10 982	228,290.00	17.3
Japan	39	36,974	2	2 653	17,537.10	2.2
Russia	35	27,241	6	4 573	184,054.10	17.1
Rep. of Korea	24	22,444	5	6 700	154,306.70	30.3
India	22	6255	7	4 824	35,006.80	3.4
Canada	19	13,554			95,650.20	15.6
Ukraine	15	13,107	2	2 070	76,077.80	52.3
UK	15	8923	1	1 630	65,149.00	20.4
Sweden	8	8612			60,647.40	40
Belgium	7	5918			41,430.50	51.7
Germany	7	9515			80,069.60	13.1
Spain	7	7121			56,102.40	21.4
Czech Rep.	6	3932			22,729.90	29.4
Pakistan	5	1318	2	2 028	5438.90	4.4
Switzerland	5	3333			20,303.10	34.4
Finland	4	2784	1	1 600	22,280.10	33.7
Hungary	4	1902			15,183.00	51.3
Slovakia	4	1814	2	880	13,733.40	54.1
Argentina	3	1633	1	25	7677.40	5.6
Brazil	2	1884	1	1340	14,970.50	2.9
Bulgaria	2	1966			15,083.50	35
Mexico	2	1552			10,272.30	6.2
Romania	2	1300			10,388.20	17.1
South Africa	2	1860			15,209.50	6.6
Armenia	1	375			2194.90	31.4
Iran	1	915			5924.00	2.1
Netherlands	1	482			3749.80	3.4
Slovenia	1	688			5431.30	35.2
UAE			4	5380		
Bangladesh			2	2160		
Belarus			2	2220		
Turkey			1	1114		
Total	**450**	**396,841**	**55**	**56,643**	**2,476,671.20**	

a) As of December 2017.
b) As of December 2016.
Source: IAEA (2019).

countries a large fraction of electric power is delivered by a particular type of nuclear power plant, such standardization increases the risk of electric power shortages.

4.8   Outlook for Commercial Reactors

The first commercial reactors built in the 1950s in the United States were influenced by the PWR and BWR type reactor models developed for the US nuclear navy program. Early prototypes developed in the US and UK are known as Generation I reactors. Most of the first-generation reactors built before the early 1970s are now permanently shut down.

Higher capacity Generation II reactors emerged in the 1970s, mostly with the expectations of inexpensive abundant electricity independent from foreign oil. Reactors developed during this period were mostly LWR, PWR, and BWR types using enriched fuel, except the Canadian Deuterium Uranium (CANDU) heavy-water reactors that use low enriched or natural uranium. Most of the reactors operating worldwide are second-generation reactors.

Advanced light-water (ALWR) and advanced boiling water (ABWR) reactors deployed in the 1990s are known as the Generation III reactors. Such reactors improve the safety by implementing passive cooling systems that reduce the "loss of coolant accident (LOCA)" risk.

Evolving nuclear reactor designs that demonstrate commercial viability and potential for the near future are considered as Generation IV. The goals for development of the future reactors set forth by the international organization named *Generation IV International Forum* (GIF) include sustainability, safety, reliability, economic competitiveness, proliferation resistance, and physical protection (Abram 2002). Six potential reactor technologies currently considered as Generation IV are:

- Gas-cooled Fast Reactor (GFC)
- Lead-cooled Fast Reactor (LFR)
- Molten-salt Reactor (MSR)
- Sodium-cooled Fast Reactor (SFR)
- Supercritical-water-cooled Reactor (SCWR)
- Very-high Temperature Reactor (VHTR).

The Fukushima Daiichi nuclear disaster demonstrated the need for the highest level of reliability in case of accidents and malfunctions. Safety concerns have led to several design considerations including improved residual heat removal, the use of non-water coolants, and higher operational temperatures. Higher reactor power density is a desired characteristic of Generation IV reactors. In addition, the nuclear fuel processing and recycling facilities would be either integrated into or located near the nuclear power plants. Engineering-scale prototypes and performance tests on several new reactor designs are expected to be completed by 2025 and the first commercial models should be operational by 2030 (GIF 2014).

One of the prospective designs is Very-High Temperature Reactors (VHTR), which use helium as a coolant and graphite as a moderator. Helium is an inert gas and does not cause corrosion problems. The pressure is relatively lower compared to water-cooled reactors, and the gas temperature around 1000 °C allows the use of Bryton cycle gas turbines discussed in

Chapter 7. Highly enriched uranium oxide fuel is enclosed in spherical shells called "pebble bed" made from graphite and silicon carbide. Such reactors are developed for dual purposes to generate both electricity and hydrogen as energy carrier. The VHTR development program is sponsored by Idaho National Engineering and Environmental Laboratory, where the first Experimental Breeding Reactor EBR I was developed.

4.9 Benefits and Challenges of Nuclear Power Plants

Civilian use of nuclear power for electric generation has been the focus of various public reactions and debates since the early 1950s. On one hand, the terrifying experience of the two atomic bombs was fresh in people's memories. On the other, nuclear energy offered new possibilities to meet the growing energy need, especially of developed countries.

Starting from the oil crisis of 1974, nuclear power was a way to ensure energy security in the Western world dependent on petroleum. The Three Mile Island accident on March 28, 1979 was a terrifying experience in the densely populated eastern coast of the US, and justified the concerns of anti-nuclear social groups. The Chernobyl accident in 1986 and the Fukushima disaster caused primarily by a tsunami increased the concerns about nuclear power.

Nuclear power plants have been inevitable despite continuous public reactions because of their benefits, especially as the air pollution and climate change issues introduced significant constraints on fossil fuel burning power plants.

The principal benefits of nuclear power generation are summarized below.

- *Abundant source of energy.* Uranium reserves are available in different parts of the world. Heat released by a fission chain reaction of 1 kg uranium is approximately equivalent to the heat released by burning 4500 t of coal. Therefore, fuel consumption is extremely small compared to fossil fuel burning power plants.
- *New fuel production and recycling.* Some nuclear reactors produce new fissile material during the chain reaction. Used nuclear fuel materials can be reprocessed to be reused in other power plants or other nuclear applications.
- *No GHG emission.* Nuclear power plants do not release greenhouse gases or particulate matter to the atmosphere like fossil fuel fired power plants. They are virtually carbon free.
- *Small footprint.* Nuclear power plants are more compact compared to other power generation units with similar power output. In addition, they do not require large amount of continuous fuel supply and storage.
- *Cost of the energy production.* The cost of generation in nuclear power plants has decreased with the development of new technologies. The current cost per unit energy generated is comparable to other power generation plants, and even less than most renewable energy sources.
- *Economic stability.* Cost of energy generated by nuclear power plants is less sensitive to market fluctuations compared to petroleum, natural gas, and coal.
- *Energy independence.* Countries using nuclear power are less dependent on international political and economic tensions in providing continuous energy supply to the end users.

Despite the advantages listed above, nuclear power plants face many challenges too.

- *Risk of undesired nuclear activity.* While nuclear power plants are designed and operated with high-level reliability, the magnitude of damage due to a nuclear malfunction can turn to a disaster with long-term effects. Whereas the life losses and injuries in nuclear power stations are much smaller than in coal mines and oil fields, long-term effects of nuclear accidents increased the risk of cancer for persons exposed to the released radioactivity.
- *Location.* Site selection for nuclear power plants may cause intense public reactions and concerns. In earlier years of nuclear power, many generation plants were constructed in populated areas of developed countries. For the safety of residents, the newer nuclear power plants are built in less populated and remote areas. However, the site must have access to water, roads, and electric grid.
- *Water use.* Cooling water is necessary for the operation of steam turbines. The steam quality produced by nuclear reactors is not as high as coal and natural gas fired plants. Nuclear power plants require more cooling water compared to other similar size power plants.
- *Output power control.* Power of a nuclear reactor cannot be controlled as easily as other types of generation units. The generated electric power must always be equal to the consumption. The electric energy demand of users changes every instant. Nuclear power plants can provide base power since their power output cannot follow the fluctuations of demand.
- *Nuclear waste disposal.* Nuclear power plants produce two types of radioactive materials. High-level nuclear waste is produced during recycling of used nuclear fuel. Low-level nuclear waste is the tools and parts used during the operation and maintenance of nuclear power plants. In one year, a typical nuclear power plant generates 20 metric tons of used nuclear fuel. The nuclear industry generates a total of about 2000–2300 metric tons of used fuel per year (NEI 2018). Radioactive waste is temporarily stored in pools on-site but eventually it is transported to a permanent disposal facility. Radioactive waste disposal has always been one of the major concerns about nuclear industry.

Box 4.2 Experimental Breeding Reactor EBR-I

In 1942 Enrico Fermi and Leo Szilard demonstrated that a fission chain reaction could be controlled in a reactor. Enrico Fermi and his colleague Walter Zinn pursued the idea of a "breeding reactor" to maximize the useful energy that could be obtained from natural uranium composed of 99% of U-238 and only 1% of U-235, which is the fuel of most nuclear reactors today. Zinn, who was the director of the Argonne National Laboratory in Chicago, designed the first experimental breeding reactor EBR-I to prove the concept. EBR-I was built in Idaho National Laboratory between 1948 and 1951. On December 20, 1951 EBR-I generated the first electricity to power four 200 W light bulbs. The next day electric generation rose to 100 kW, sufficient to supply the power need of the facility (Goff 2019). Figure 4.14 shows selected photos taken in the facility. The top left corner is the turbine-generator area (a). Photo (b) shows

---

(Continued)

Box 4.2 (Continued)

the "breeding blanket" where fissile plutonium isotope is produced. The explanation note displayed under the breeding blanket reads: "*The EBR-I core had both an inner and outer blanket of fertile Uranium-238 in which fuel breeding took place. The inner one surrounded the fuel inside the reactor vessel. The outer one, shown here, formed a movable cup around the lower end of the vessel which served the dual purpose of capturing some neutrons for breeding and reflecting others back into the core to feed the nuclear chain reaction. Uranium-238 atoms could not fuel the chain reaction but could capture excess neutrons from the process to breed plutonium-239, an element that could be used as a reactor fuel.*"

Figure 4.14 Experimental Breeding Reactor: EBR I. Source: © Soysal.

Photo (c) shows the logbook by Walter Zinn. The open page notes the first plutonium-239 breeding observation. Photo (d) shows the four light bulbs lit for the first time when EBR-I started to generate the electricity.

In 1953, measurements confirmed that EBR-I had proven the basic principle of breeding and on November 27, 1962, the EBR-I produced electricity using a plutonium core. The EBR-I operated for 12 years, until December 30, 1963, then it was officially shut down. On August 26, 1966, US President Lyndon B. Johnson designated the retired reactor a Registered Historical National Monument (for more information, visit http://www4vip.inl.gov/ebr).

4.10 Chapter Review

This chapter discusses nuclear power options developed for energy systems. Currently the primary fuel for all commercial nuclear reactors is uranium. Thorium is considered as an alternative nuclear fuel but currently no nuclear power plant operates only on thorium as fuel.

Nuclear energy is a non-renewable source since the uranium and thorium resources are limited in the Earth. Unlike fossil fuel fired power plants, nuclear reactors do not emit greenhouse gases, and do not cause air and water pollution during normal operation.

The atoms of a chemical element have nuclei that contain a certain number of protons and neutrons jointly called nucleons. Chemical properties of an element depend on the number of protons. Atoms with the same number of protons but different number of neutrons are called the isotopes of that same element.

Binding energy holds the nucleons together against the electrostatic force between positively charged protons. Binding energy is larger in stable elements with medium atomic mass such as iron, nickel, copper, etc. Lightweight atoms with small atomic numbers (hydrogen, helium, lithium) and heavier atoms with larger atomic numbers (uranium, plutonium, thorium) have less binding energy than medium range atoms. Due to the binding energy the actual mass of an atom is less than the total mass of the subatomic particles. This difference is called *mass deficiency*.

Nuclear reactions change the identity of the atoms involved by either combining them in a larger atom (fusion) or splitting into two smaller atoms (fission). Whereas the total number of protons and neutrons remain unchanged, energy is released due to the change of the mass deficiency in the resulting new atoms.

Three types of nuclear reactions are *radioactive decay*, *fusion*, and *fission*. All forms of nuclear reactions release energy.

Radioactive decay is the transformation of unstable isotopes until they become stable by spontaneous emission of particles. Three types of particles associated with radiation are named *alpha*, *beta*, and *gamma* particles. Radiation of alpha particles can be stopped by a thin layer of any substance, whereas a thick layer of concrete, lead, or water can stop gamma particles. Radioactive decay is described by an exponential function usually characterized by *half-life*, which is the time interval needed to reduce the radiation to half of its initial value. Ionizing radiation may have severe health effects if absorbed in the tissues of a living body.

Fusion reaction combines two lightweight atoms such as hydrogen to form a heavier atom. Fusion reactions require extremely high temperatures and pressure, which are naturally available in the Sun. A theory to produce fusion reaction on Earth has been developed but it can be used only in hydrogen bombs that can be triggered by the fission reaction of an atomic bomb. Currently, fusion reaction cannot be controlled to be used in a nuclear power reactor.

Fission reactions are based on splitting a heavier atom by neutron bombardment into two medium weight atoms. Fission of certain isotopes such as uranium 235 and plutonium 239 release free neutrons. Isotopes that can be split by neutron bombardment are called *fissile*. A free neutron absorbed in the nucleus of a fissile isotope can cause new fission if it has sufficient energy. If the number of free neutrons decrease, the fission reaction cannot sustain; such a reaction is called subcritical. Increasing the number of free neutrons that cause more fissions than needed causes the reaction to become supercritical and destructive. A critical number of new fissions can sustain the reaction at a certain rate. Such chain reactions are called critical state. A chain reaction can be controlled by modifying the velocity, and consequently the kinetic energy of free neutrons.

Uranium oxides found in mineral deposits are the primary fuel sources for nuclear reactors. Natural uranium contains more than 99.27% U-238, 0.72% of U-235, and a trace of U-234. U-238 is not a fissile element, but by absorbing a neutron it can transform into a fissile isotope of plutonium Pu-239. An isotope that can produce a fissile isotope is called fertile. For efficient operation, most nuclear reactors require enriched uranium in which the amount of U-235 is increased by an enrichment technique. Practical enrichment techniques include EMIS, centrifugal separation, and gaseous diffusion methods.

Main elements of a nuclear power reactor are fuel assembly, moderator, control rods, and coolant. Fuel assembly contains the nuclear fuel, usually uranium oxide pellets. A moderator is a substance where fast neutrons are slowed to a velocity range suitable for a controlled chain reaction. Control rods are inserted in the holes of the fuel assembly to control the reaction rate. Coolant is a liquid or gas used to transfer energy released from nuclear reaction to external heat exchangers. A reactor core is enclosed in a pressure vessel where the coolant is circulated. A neutron reflector returns the scattered free neutrons back to the reaction and radiation shell serves as shields to the radioactive particles.

In homogeneous reactors the fuel is mixed with a moderator. In a heterogeneous reactor, the fuel assembly is a matrix of fuel rods with openings left for control rods.

Light water reactors in which regular water is used both as moderator and coolant dominate the commercial nuclear power industry worldwide. In pressurized light water reactors (PWR) high pressure maintains water in liquid state at higher temperature levels, and a steam generator produces steam for turbines. BWRs produce steam supplied to turbines directly in the core. PWR and BWR reactors use enriched nuclear fuel.

Heavy water (deuterium oxide) reactors also known as CANDU can use slightly enriched natural uranium. GCRs use helium, air, or carbon dioxide as a coolant. These reactors are moderated typically by graphite.

Three major nuclear accidents created intensive public reactions and safety concerns. However, third and fourth generation reactors increased the safety, improved the performance, and reduced the nuclear waste generation by on-site fuel recycling. Security and nuclear proliferation are additional concerns regarding commercial nuclear power plants.

Nuclear power was initially seen as solution to sustainable, clean, and inexpensive energy sources, and developed countries developed their nuclear reactor fleets to enhance energy independence from foreign oil. Public reactions and safety concerns flared by three major nuclear power plant disasters, economic circumstances, and the development of unconventional oil and gas recovery technologies slowed down use of nuclear power in North America and Western Europe. In the US and European Union nuclear electric generation has been decreasing. Former Soviet Union countries grouped as the Commonwealth of Independent States are steadily increasing their nuclear electric generation capacity. In the Asia Pacific region, while Japan has permanently shut down many reactors after the Fukushima disaster, China and India are significantly increasing their commercial nuclear reactor capacity.

Review Quiz

1 Which statement below is true for the nucleus of $_6C^{15}$?
  a. The number of nucleons is 6, the mass of all nucleons is 15
  b. The number of protons is 6, the total number of protons and neutrons is 15
  c. The number of neutrons is 6, the mass number is 15
  d. The number of protons is 6, the number of neutrons is 15

2 Binding energy is
  a. energy that holds nucleus's together.
  b. energy that holds molecules together.
  c. energy that keeps electrons on their orbit.
  d. energy that forms crystal structures.

3 Bonding energy is bigger in
  a. radioactive isotopes.
  b. fissile isotopes.
  c. unstable nuclei.
  d. more stable nuclei.

4 Nuclei with which atomic mass number (A) are most likely to be involved in a fusion reaction?
  a. $A < 25$
  b. $25 < A < 50$
  c. $100 < A < 150$
  d. $A > 150$

5 In a fission reaction
  a. the total mass of fission products increases.
  b. the total mass of fission products decreases.
  c. the total mass of fission products remains unchanged.
  d. some of the atoms are destroyed.

6 The source of energy released during a fission reaction is
  a. the mass of the destroyed neutrons.
  b. the mass of the destroyed protons.
  c. an increase of the mass deficiency.
  d. a decrease of the mass deficiency.

7 Which uranium isotope below is fertile?
  a. U-233
  b. U-234
  c. U-235
  d. U-238

8 Which uranium isotope below is fissile?
   a. U-235
   b. U-234
   c. U-236
   d. U-238

9 Emission of which particle below is most difficult to stop?
   a. Alpha
   b. Beta
   c. Gamma
   d. Photon

10 In a radioactive decay
   a. a stable isotope becomes unstable.
   b. an unstable isotope is transformed until it becomes a stable isotope.
   c. a fertile nucleus is split into smaller subatomic particles.
   d. a stable element transforms into a more stable element.

Answers: 1-b, 2-a, 3-d, 4-a, 5-b, 6-c, 7-d, 8-a, 9-c, 10-b.

Research Topics and Problems

Research and Discussion Topics

1 Discuss the challenges of using controlled fusion for electric generation.

2 Discuss in a research essay the possible factors impacting the evolution of the nuclear power plant technologies.

3 Why do the pressurized light water reactors (PWR) and boiling water reactors (BWR) dominate the nuclear power plant industry worldwide despite their challenges compared to other reactor types?

4 Why are there so few fast breeder reactors operating in commercial nuclear power plants in the world although they can produce new fissile material from natural uranium?

5 Research the recent developments in the nuclear reactor technologies and draft a technical report on the future of nuclear power.

6 What are the benefits and challenges of electric generation from nuclear power?

7 What were the impacts of Three Mile Island, Chernobyl, and Fukushima on the evolution of the nuclear power plant industry?

8 What is nuclear proliferation?

9 Can nuclear power plants produce weapon-grade nuclear fuel?

10 What are the environmental benefits and drawbacks of generating electricity using nuclear power?

Problems

1 What is the half-life of an isotope if its mass decreases 0.1% every year due to radioactive emission?

2 What is the total binding energy in the nucleus of $_1H^2$?

3 What is the total binding energy in the nucleus of $_2He^4$?

4 Find the energy that would be required to remove a neutron from the nucleus $_{31}Ga^{60}$ given that the atom has a mass of 68.925581 a.m.u.

5 What is the average binding energy per nucleon in the nucleus $_{94}Pu^{238}$ given the atomic mass is 238.0496 a.m.u.?

6 Natural uranium contains 99.27% of U-238 and 0.72% of U-235. Using the half-life of these isotopes given in Table 4.3, estimate how long ago these two isotopes were in equal quantities in natural uranium deposits.

7 Calculate the energy released from the complete fission of all the atoms in 1 g of U-233.

8 How many barrels of crude oil with heating values of 129,670 Btu/gal produces heat equivalent to 1 ton of 3% enriched uranium?

9 Given the average heat content of bituminous coal used in most power plants is 22.5 million Btu/ton, how much coal produces the heat that would be produced by a complete fission reaction of 1 kg of natural uranium that contains 0.72% of U-235?

10 Given the lower heating value (LHV) of natural gas 983 Btu/ft^3, estimate the amount of natural gas to produce heat equivalent to 1 kg of nuclear fuel containing 2.5% of U-235 isotope?

11 Calculate the energy released by splitting U-235 isotope into Sr-96 and Xe-140.

12 Calculate the change of mass defect when U-235 is split into Ba-137 and Kr-97.

13 A nuclear power plant with 600 MW electric power generation capacity uses 2.5% enriched uranium and operates at an average capacity factor of 90%. Assuming that the overall efficiency of the heat engine and generator is 35%, estimate the amount of nuclear fuel consumed in this facility over one year.

14 In 2017, nuclear power plants generated 2635.6-TWh electric energy worldwide. Assuming that the average plant efficiency is 35%, estimate the amount in average 3% enriched uranium consumed in the world.

Recommended Web Sites

- International Atomic Energy Agency (IAEA): https://www.iaea.org
- IAEA Power Reactor Information System (PRIS): https://pris.iaea.org/PRIS
- Atomic Energy of Canada Limited (AECL): https://www.aecl.ca
- US Nuclear Energy Regulatory Commission (NRC): https://www.nrc.gov
- Nuclear Energy Institute (NEI): https://www.nei.org/home
- United Nations Scientific Committee on the Effects of Atomic Radiation (UNSCEAR): http://www.unscear.org/unscear
- Idaho National Laboratory Experimental Breeding Reactor web site: https://inl.gov/experimental-breeder-reactor-i

References

Abram, T. (2002). *Technology Roadmap Update for Generation IV Nuclear Energy Systems*. s.l.: USDOE/GIF-002.

AECL (2010). *ACR-1000 Technical Description Summary*. Mississauga, Ontario, Canada: Atomic Energy of Canada Limited (AECL).

Andrews, J. and Jelley, N. (2007). *Energy Science*. Oxford, UK: Oxford University Press.

Audi, G. and Wapstra, A.H. (1995). The 1995 update to the atomic mass evaluation. *Nuclear Physics A* 595 (4): 409–480.

Audi, G., Wapstra, A.H., and Thibault, C. (2003). The AME 2003 atomic mass evaluation: (II). Tables, graphs and references. *Nuclear Physics A* 29 (1): 337–676.

BGR, (2017). *BGR Energy Study 2017 - Data and Developments of German and Global Energy Supplies*. Hannover: s.n.

BP (2018). *BP Statistical Review of World Energy 2018*. London, UK: BP.

Close, F.E. (2015). *Nuclear Physics: A Very Short Introduction*. Oxford, U.K.: Oxford University Press.

Culp, A.W. Jr., (1979). *Principles of Energy Conversion*. s.l.: McGraw-Hill.

Ferguson, C.D. (2011). *Nuclear Energy: What Everyone Needs to Know*. Oxford, UK: Oxford University Press.

GIF, (2014). *Technology Roadmap Update for Generation IV Nuclear Energy Systems*. s.l.: OECD Nuclear Energy Agency for the Generation IV International Forum.

Goff, M., (2019). *EBR-I lights up the history of nuclear energy development*. [Online] Available at: https://inl.gov/article/ebr-i-lights-up-the-history-of-nuclear-energy-development/ [Accessed 12 2019].

Gupta, M.K. (2012). *Power Plant Engineering*. New Delhi: PHI Learning.

IAEA (2008). *The International Nuclear and Radiological Event Scale – User's Source*. s.l.: International Atomic Energy Agency.

IAEA (2009). *World Distribution of Uranium Deposits (UDEPO) with Uranium Deposit Classification*. Vienna: IAEA.

IAEA (2018a). *Country Nuclear Power Profiles*. Vienna: International Atomic Energy Agency.

IAEA (2018b). *Nuclear Power Reactors in the World*. Vienna: International Atomic Energy Agency.

IAEA, (2019). *Power Reactor Information System (PRIS)*. [Online] Available at: https://www.iaea.org/PRIS [Accessed 8 April 2019].

IEA (2017). *Key World Energy Statistics*. Paris: International Energy Agency.

IUPAC, (1997). *Standard average atomic weights*. [Online] Available at: https://iupac.org/standard-atomic-weight-of-ytterbium-revised [Accessed 31 March 2019].

Jelly, N.A. (1990). *Fundamentals of Nuclear Physics*. Cambridge: Cambridge University Press.

NEI, (2018). *Knowledge Center: On site storage of nuclear waste*. [Online] Available at: https://www.nei.org/Knowledge-Center/Nuclear-Statistics/On-Site-Storage-of-Nuclear-Waste [Accessed 20 January 2018].

NRC, U.N.R.C., (2018). *Backgrounder on the Three Mile Island Accident*. [Online] Available at: https://www.nrc.gov/reading-rm/doc-collections/fact-sheets/3mile-isle.html#tmiview [Accessed 24 April 2019].

Rouben, R. (1997). *Fuel Management in CANDU*. Bangkok, Thailand: Chulalongkorn University.

UNSCEAR (2008). *Sources and Effects of Ionizing Radiation*. New York, NY: United Nations Scientific Committee on the Effects of Atomic Radiation.

UNSCEAR (2008). *Report to the General Assembly with Scientific Annexes*. New York, NY: United Nations Scientific Committee on the Effects of Atomic Radiation.

UNSCEAR, United Nations Scientific Committee on the Effects of Atomic Radiation (2016). *Developments Since the 2013 Unscear Report on the Levels and Effects of Radiation Exposuredue to the Nuclear Accident Following the Great east-Japan Earthquake and Tsunami*. New York, NY: United Nations Scientific Committee on the Effects of Atomic Radiation.

5

Renewable Energy Sources

The Dalles Hydroelectric Power Plant on the Columbia River, between Washington and Oregon in the US. The Dalles plant has 2160-MW nameplate capacity and 2058-MW summer capacity. It has a lock to move cargo vessels, and fish ladders to allow various species of fish to migrate up- and downstream. Overhead wires above the river protect migrating fish from avian predators.

The Mountain Ridge Wind Farm in the background has 50-MW nameplate capacity. The facility operated by the Los Angeles Department of Water and Power consists of 25 wind turbines of 2-MW rated power each. Wind turbines are installed on 125-m towers.

Energy for Sustainable Society: From Resources to Users, First Edition. Oguz A. Soysal and Hilkat S. Soysal.
© 2020 John Wiley & Sons Ltd. Published 2020 by John Wiley & Sons Ltd.

5.1 Introduction

Renewable energy sources were sufficient for all human functions in early civilizations. Sunlight, human power, animal work, blowing wind, and flowing water satisfied all energy needs for lighting, heating, hunting, agriculture, production, and transportation, as well as creating marvelous art, cities, and landmarks.

Humanity has, however, always challenged nature. As societies evolved, the need for energy has continuously increased to produce artificial light and heat when sunlight was insufficient or unavailable, produce more goods with less labor, and reach farther places in a shorter time. The industrial revolution in the mid-nineteenth century answered the quest of humanity for more power. Since then, fossil fuels have become the essential sources of energy, while renewables turned out to be their alternatives. Fossil fuels provided great amounts of energy whenever needed and in the amounts needed, whereas energy flow from renewable resources was often unpredictable, variable in time, and insufficient for industrial development.

Currently about 80% of the world's energy is supplied by burning non-renewable natural resources. Over the last several decades of the twentieth century, however, the pursuit of sustainable and cleaner renewable energy options has accelerated because depletion of fossil fuel reserves has become more evident, and public awareness of environmental pollution and climate change increased. In addition, energy security concerns raised by dependency on foreign fossil fuels, especially petroleum, urged industrialized countries to develop energy policies requiring an increased use of renewable energy.

Natural processes recreate continuously enormous amounts of energy that can be harnessed in many forms. Energy from sunlight, wind, water, hot springs, and combustible materials obtained from plants and animals are the most common resources that can be converted into practical secondary sources to supply energy need for various applications.

Renewable energy sources are always replenished by natural processes at a rate that exceeds their rate of use; thus, they are not depleted by consumption. Potential energy of water, kinetic energy of wind, and radiant energy of sunlight are examples of renewable sources. Conventional hydroelectric power is the leading clean energy source among all renewables. The share of wind and solar power is growing in all regions of the world. Biofuels are consequences of solar energy and the natural water cycle. Plants fed by groundwater produce hydrocarbons by photosynthesis. Firewood, sugarcane, maize, soybean, palm, and many other plants are renewable biomass reproduced by natural water and carbon dioxide cycles. Such carbohydrate-rich plants are used to produce ethanol, methanol, and other biofuels. Biodegradable waste can be transformed into biodiesel, which can be directly used in transportation vehicles. Ethanol and biodiesel are potential alternatives to petroleum products in the transportation sector. Use of combustible waste is increasing in district heating, electric generation, and synthetic fuel production.

The technical potential of renewable energy is forecasted to exceed the projected energy demands of the world until the year 2100 (UNDP 2000). The major challenges of all renewables are efficient and effective management of resources and their integration in the existing energy supply chain.

In this chapter, we will first review the common features of all renewables, then discuss the specific characters, limitations, challenges, and potential of each particular renewable resource in separate sections. Energy conversion techniques used to transform each one of the renewable sources into usable forms of energy will be discussed further in later chapters.

5.2 Common Features of Renewables

Renewable energy sources can be separated into two categories; combustible or non-combustible. Noncombustible sources are sunlight, wind, flowing water, and geothermal energy. Combustible sources are hydrocarbons obtained from plants, animals, or microorganisms used as alternatives to fossil fuels. General properties of renewable sources are summarized below.

- *Quantity.* All forms of renewable energy sources are reproduced naturally, and their quantity is practically independent of their consumption, but the quantity available for conversion is limited. The amount of solar and wind energy that can be converted into a usable form depends on the size of the land where they are captured. The total potential energy of a river is limited to the elevation of its source from sea level. Energy of water in a reservoir or lake is restricted by the surface area, depth, and height above the conversion system. Evaporation and leakages reduce the available energy from water. Energy produced from plants depends on the land area where they grow and their growth rate. Geothermal heat has thermodynamic limitations.
- *Availability.* Renewable energy is available at every location in the world in various amounts depending on the climate, elevation, and topography. Resource and reserve classifications developed for fossil fuels and nuclear energy do not apply to renewable energy sources since they reproduce by natural processes. Rather than their available amount, the flow rate available for conversion is more significant in estimating their potential.
- *Intermittency.* The amount of wind, solar, and hydro energy available for conversion into a usable form changes randomly over time. These sources significantly depend on local climate, atmospheric conditions, weather changes, and precipitation. Whereas yearly total amount, average, and frequency distribution of such resources can be predicted at a certain level of confidence, forecast of short-term variations and predictions of changes in longer timescale of several years or decades are based on assumptions and complex models.
- *Conversion.* Most renewable sources are delivered to consumers in the form of electricity or heat. Whereas windmills and watermills produce directly mechanical work, such applications are negligible in modern energy systems.
- *Energy transfer.* Biofuels are transported by conventional methods like fossil fuels. However, other renewable sources are transferred through the energy system and delivered to end users in the form of either electricity or heat. Whereas wind and hydro energy can be used directly to produce mechanical work, in modern energy systems these applications are negligible. Worldwide, about three-quarters of renewables are delivered to energy users in the form of electric power and one-quarter as heat (REN21, 2019). Electricity is a convenient energy carrier to transmit renewable energy to consumers for a broad range of end uses. Electric generation, transmission, and distribution systems transform uncontrollable and sporadic renewables into predictable, continuous, reliable energy supplies

for end users. In the form of electric energy any type of renewable source can supply lighting, appliances, computers, communication devices, and many other devices.

- *Energy storage.* Among renewable energy sources, only biofuels can be stored directly in fuel stacks or tanks similar to fossil fuel storage. Hydro and tidal power is stored in reservoirs or lakes. Storage of wind and solar energy requires specialized advanced conversion techniques. Both can be stored in the form of chemical energy in battery banks after being converted into electricity. Batteries are convenient for either stationary or portable storage. While the energy density and unit cost of modern battery types (such as lithium ion, metal hydride, etc.) is continuously improving, storage of wind and solar energy still present practical and economic challenges. Concentrated solar power (CSP) plants and solar thermal systems can store solar energy in the form of heat. Wind power can be stored in the form of potential energy by pumping water in a tank elevated from the ground or in the form of kinetic energy using flywheels. Hydrogen production and storage using wind or solar energy is an emerging technology. However, such advanced storage techniques are still at the experimental phase.
- *Integration.* The intermittent nature of wind and solar energy presents challenges in their integration into the conventional electric grid. Because the storage of electric energy is expensive, generation must always be equal to the consumption in an interconnected electric grid. Penetration of large amounts of electric power generated from uncontrollable, time variable, and discontinuous sources may cause stability issues. Electric utility companies use advanced forecasting and energy management techniques to balance generation and demand at all times. Hydroelectric generation is the only renewable that has been part of the conventional electric power system since the early years. Biofuels are also controllable energy sources and can be used to power steam or gas turbines similar to the ones used in fossil fuel fired electric generation.
- *Environmental impacts.* In common culture, renewable energy is often considered as a clean, or green source of energy. This is true for wind, solar, and hydro energy, which are completely carbon-free and do not pollute air and water with toxic chemicals, particles, and solid waste. Biofuels are cleaner than fossil fuels, but burning them still produces carbon dioxide. Geothermal electric generation plants emit toxic gases, including considerable amounts of sulfur dioxide. Even non-combustible renewables impact the environment like any large-scale energy conversion unit. Wind and solar farms impact the ecology, vegetation, bird population, and animal migrations. Hydroelectric and tidal dams present obstacles to fish migration. Large hydroelectric projects require relocation of a population and historic landmarks. However, dams constructed on rivers regulate the water flow, prevent flooding, provide irrigation water for agriculture, and create recreational parks.
- *Land use.* Renewable sources require greater land area for the same amount of energy produced by a fossil fuel fired or nuclear power plant. Turbines on a wind farm can be arranged to allow use of land for roads and agriculture. Commercial scale solar generation units, hydroelectric reservoirs, and fields used to grow plants for bioenergy occupy huge land area that cannot be used for other purposes.
- *Public acceptance.* Social reactions to renewable energy development are often controversial. Groups that support renewable energy development at large expect reduction of

carbon dioxide and toxic gas emissions as well as air and water pollution. Populations affected by wind farm development often raise concerns about visual impact, noise, impacts on bird population, and change of vegetation. Reactions to solar generation units include land use, tree removals, and impacts on ecology. Development of large hydroelectric power plants generates public reactions because of relocation of local population, flooding of cultural treasures, fish migration, and use of water for energy rather than irrigation of farmlands.

5.3 Energy Supply from Renewable Sources

Renewable energy sources have immense potential to supply all the energy need of humanity, even considering future increases. Such resources, however, are scattered in a wide area and often available at remote places. Potential, kinetic, thermal, or radiant energy available in renewable resources is not suitable for transmission to long distances or direct use for the broad range of practical applications. Diverse conversion systems transform nature's raw energy into high quality energy carriers in the form of heat, electricity, and fuels.

In energy vocabulary, the term *capacity* means the maximum energy that a conversion system can produce under its design conditions. It is also stated as *installed capacity* or *nameplate capacity*. Generator capacity is usually expressed in megawatts (MW). Capacity of thermal converters can be expressed in British thermal units (Btus) or a metric multiple of Joule [megajoule (MJ), gigajoule (GJ), exajoule (EJ)].

Installed capacity does not necessarily imply that the conversion system will always produce that power. In fact, because renewable power available for conversion continuously changes, the output power also changes. Mathematically, delivered energy is the integral of power converted over a time interval. In the first expression in Eq. (5.1) the initial time is 0 and the final time is T, $P(t)$ represents the value of power at the instant t.

$$W_{out} = \int_0^T P(t)dt = \sum_{i=1}^{8766} P_i \cdot \Delta t_i \qquad (5.1)$$

In energy calculations, yearly energy output is more relevant since natural processes can be assumed to repeat every year if climate variations from one year to another are neglected. In practice, energy output of a conversion unit is calculated by adding up the products average power over a certain time interval and the length of the interval. In most practical calculations, one-hour intervals give sufficiently accurate results. In yearly energy calculations, the total length of time is $24 \times 365 + 6 = 8766$ hours.

5.3.1 Installed Renewable Power Capacity

According to the International Energy Agency (IEA), the share of renewable energy in the world's total primary energy supply (TPES) was about 14% in 2016, including solid biofuels extensively used in developing countries for conventional residential heating and cooking (IEA 2018b). In modern energy systems, renewable energy sources are used at a commercial scale either for electric generation or industrial heat production. If residential use of conventional solid biofuels (firewood, charcoal, manure, etc.) is excluded, modern renewables

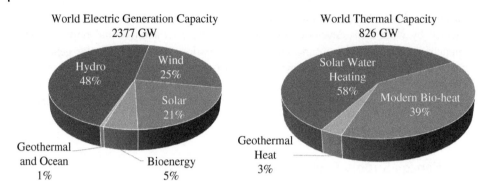

Figure 5.1 Total electric generation and direct heating capacity of renewable energy sources in the world as of 2018. Source: Global Status Report 2019 (REN21 2019).

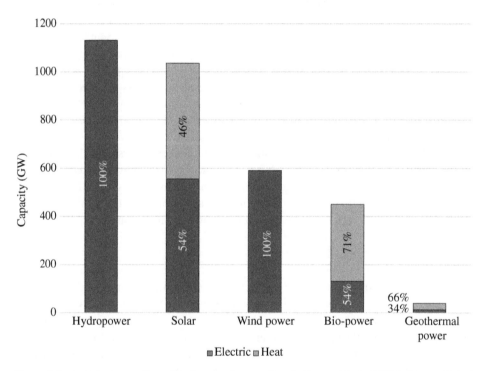

Figure 5.2 Installed capacity and end-uses of renewables in the world as of 2018. Source: Global Status Report 2019 (REN21 2019, p. 186, Table R1).

account for 10.6% in the world's TPES (REN21 2019, p. 31, fig. 1). The share of renewable sources in electric generation and heat production are shown in Figure 5.1 (REN21 2019). Figure 5.2 shows the installed renewable energy capacity of electric and thermal conversion units in 2018 (REN21 2019, p. 186, Table R1).

Hydropower traditionally has been the biggest source of renewable energy. In modern energy systems hydropower is delivered to end users solely as electric energy.

Use of solar energy has significantly grown since the beginning of the twenty-first century and almost reached hydropower. In 2018, the world's solar energy conversion capacity was composed of 54% photovoltaic (PV) electric generation facilities and 46% solar thermal units. The largest amount of solar thermal energy is used for water heating by residential scale solar collectors. Commercial scale CSP facilities produce heat for industrial processes or electric generation.

Wind power is the third renewable source but second in electric generation because, like hydropower, wind power is also entirely used for electric generation. Direct conversion of wind force into mechanical work is negligible since sailboats are no longer used for transportation, and mechanical windmills have become antique.

The largest amount of bio-power is converted into heat. More than two-thirds of biofuels are used for heat production; including liquid biofuels used in transportation. The rest of bio-power capacity is used for electric generation.

Although the Earth's crust contains immense geothermal power, only a small portion of it is currently converted to usable forms of energy. About two-thirds of geothermal capacity produces heat for end-users. Direct end-use of geothermal power is either through thermodynamic heat pumps for space heating or hot springs for recreational purposes. At a commercial scale, geothermal energy is delivered to end users in the form of electricity. Geothermal electric generation systems use thermal energy from the Earth to produce steam, and then generate electricity by turbine-generator units similar to conventional thermoelectric plants.

Among all renewables, bio-power and industrial scale geothermal power plants emit greenhouse gases. Small-scale geothermal heat pumps (GHPs) used for space heating have negligible carbon footprint and other gas emissions. Other renewables are carbon-free and do not emit greenhouse gases during their operation.

Figure 5.3 shows the use of renewable energy sources by geographic regions in 2016 (IRENA 2019). In all regions, hydropower is the leading renewable source for electric generation. Asian countries, in particular China and India, exploit the highest renewable energy potential. Hydroelectric generation in Asia is more than double that of any other region, and solar electric generation is also highest in Asia. Western European countries use wind power more than all other regions. Electric generation from biofuels is also greatest in western Europe. Use of wind, solar, and geothermal energy is insignificant in Eurasia, which includes the Russian Union and former Soviet Union countries in Europe. In the chart, "others" includes Oceania, Central America, and the Caribbean Islands, where the combined renewable electric generation is less than all other regions.

The share of renewable energy in electric generation has considerably increased since the 1980s, and reached 25% in 2017 (BP 2018). The trend of yearly electric generation from major renewable energy sources in the world is shown in Figure 5.4. At the end of the twentieth century, hydropower was generating more than 90% of the renewable electricity and the contribution of other renewable sources was less than 10%. Between 2000 and 2016, wind, sun, and biomass contributed 30% to electricity generated worldwide from all renewable sources.

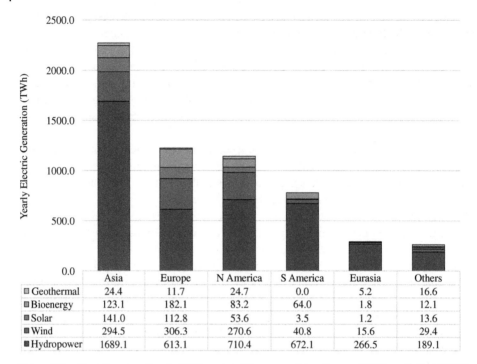

	Asia	Europe	N America	S America	Eurasia	Others
Geothermal	24.4	11.7	24.7	0.0	5.2	16.6
Bioenergy	123.1	182.1	83.2	64.0	1.8	12.1
Solar	141.0	112.8	53.6	3.5	1.2	13.6
Wind	294.5	306.3	270.6	40.8	15.6	29.4
Hydropower	1689.1	613.1	710.4	672.1	266.5	189.1

Figure 5.3 Electric generation in 2016 powered by renewables. Source: IRENA statistics (IRENA 2019).

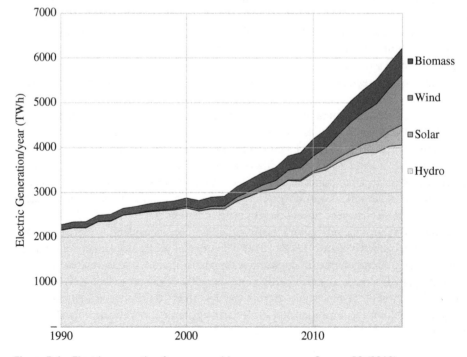

Figure 5.4 Electric generation from renewable energy sources. Source: BP (2018).

5.3.2 Capacity Factor

Since renewable sources are not regularly available at all times, the energy output of a renewable conversion unit changes continuously. Such units, operating at variable power, cannot deliver the maximum energy that corresponds to their continuous operation at full capacity. For example, consider a generation unit that has 1-MW generation capacity that operates only four hours a day and does not produce any energy for the rest of the day; it would produce $4 \times 365 = 1460$ MWh of energy. On the other hand, if this unit operated at full capacity at all times in a year, it would deliver 8766 MWh of energy. Thus, such a unit is equivalent to a smaller unit with a capacity of $1460/8766 = 0.166$ MW capacity.

The capacity factor of an energy conversion unit is the ratio of the electrical energy produced over a certain time interval to the electrical energy that the unit could have produced if it operated continuously at full power during the same period. Suppose that an electric generation facility with P_c kW installed capacity generates W_{out} kWh output energy over one year. The yearly capacity factor is obtained using the Eq. (5.2), where the factor 8766 is the number of hours in a calendar year, P_c is the rated capacity, and W_{out} represents the total actual energy output.

$$F_c = \frac{W_{out}}{W_{max}} = \frac{W_{out}}{P_c \cdot 8766} \tag{5.2}$$

Energy output of generation units powered by fossil fuels, and nuclear reactions are more predictable and easier to control. Biofuel fired plants, conventional hydroelectric power plants, and pumped storage hydroelectric units also deliver steady output energy since their input is controllable. The capacity factor of such generation plants depends on the control strategies applied by system operators to balance electric generation and demand on the interconnected network.

On the other hand, wind and solar powered units, as well as small hydropower plants without a reservoir produce variable output energy depending on the uncontrollable input they receive. The capacity factor of such conversion units depends on the frequency distribution of the renewable energy source. Therefore, conversion facilities with the same capacity at separate locations may produce different yearly total energy.

Table 5.1 shows the regional capacity factors of electric generation plants powered by various types of renewable energy sources. Geothermal, biogas, and large hydroelectric power plants are typically used at a higher capacity factor. Solar and wind powered generation units, however, have a lower capacity factor due to the occurrence cycles and frequency distribution of natural phenomena that produce them. Generation stations powered by waste and bioenergy usually have a smaller capacity compared to fossil fuel, nuclear, and conventional hydroelectric plants. They are mostly used to balance the electric generation and demand. This is reflected by the lower capacity factor of such stations.

5.4 Renewable Resource Potential

Resource and reserve definitions used for fossil and nuclear fuels do not apply to renewable energy. The amount of energy in renewable resources is enormous but its conversion into usable forms depends on many factors including existing technologies, available land, and

Table 5.1 Regional yearly capacity factors of electric generation plants in 2016.

	Hydro power	Wind	Solar	Bioenergy	Geothermal
World	0.38	0.23	0.13	0.51	0.77
Africa	0.44	0.31	0.18	0.23	0.81
Asia	0.37	0.18	0.12	0.48	0.68
Central America and Caribbean	0.37	0.18	0.12	0.48	0.68
Eurasia	0.37	0.31	0.14	0.12	0.66
Europe	0.44	0.31	0.18	0.23	0.81
North America	0.41	0.32	0.16	0.58	0.82
Oceania	0.33	0.33	0.15	0.50	0.87
South America	0.48	0.36	0.25	0.44	—

Source: Renewable Capacity Statistics (IRENA 2019).

population density. End-users are more interested in energy services rather than the energy itself. In assessment of a renewable resource potential, rather than the actual total amount that the resource contains, energy that can be conveyed to the consumers is more relevant. Availability, accessibility, and time variations are among the primary considerations in assessment of a renewable energy resource.

5.4.1 Assessment of Non-combustible Resources

Wind and hydro energy are mainly delivered to end users in the form of electricity. Currently, more than two-thirds of the harnessed solar energy is converted into electric power and one-third is used for water heating. Yearly total energy potential of a wind, hydro, or solar energy resource can be roughly estimated using the yearly average or mean values. Since wind speed, solar irradiance, and water flow of rivers continuously change in time, a more realistic evaluation should be done based on frequency distributions rather than the long-term averages. Depending on the type of the renewable source, the maximum, minimum, average, and standard deviation of the variables used in energy calculations are measured and recorded over a monthly or yearly cycle. In practice, 10-minute or 1-hour data sampling intervals are adequate for a realistic evaluation.

5.4.2 Assessment of Biomass Resources

Biofuels are evaluated based on their heating value, similar to fossil fuels. The difference between bioenergy sources from fossil fuels is that they are cultivated instead of being extracted. The capacity of a field to grow biofuel resources is estimated similarly to other agricultural products. Land area, water availability, and transportation are the main considerations. Existing farmlands are sometimes used to grow sugar cane, soybean, corn, or other carbohydrate-rich plants for energy production. Seasonal and yearly production of bioenergy can be estimated at high confidence levels. Forecasting the harvest variations due to climatic changes over several years and even decades is more complicated and depends on projections and assumptions.

Energy production from waste relies on the amount, disposal rate, and composition of the collected waste. Industrial and municipal wastes have different heating values. For example, timber and furniture industries produce considerable amounts of sawdust, wood chips, and pellets, which can be directly used for heating. Municipal waste contains a mixture of combustible and non-combustible substances. The direct heating value of municipal waste is not suitable for direct energy conversion. Municipal waste must be screened and processed to obtain a liquid or gas fuel. Energy produced from waste depends on the fraction of combustible components and process technologies used to convert waste into synthetic fuel.

The potential of waste to energy in a region is assessed by the type of waste and the rate it is produced. In densely populated areas like European cities, municipal waste has greater potential for district heating. In sparsely populated areas, it is more convenient to produce either biogas or generate electric energy from waste.

5.5 Benefits and Challenges of Renewable Energy

Non-combustible renewables such as wind, solar, hydro, and marine energy do not produce carbon dioxide and toxic gas emissions, water pollution, solid waste, or particulate matter, nor do they contribute to climate change by greenhouse gas emissions. Their inevitable impacts on the natural balance are less severe compared to fossil and nuclear fuel cycles.

Combustible renewables such as biofuels and waste are cleaner than coal or petroleum, but they are not carbon-free since burning any hydrocarbon produces carbon dioxide. Any source of energy provided at the scale required to supply modern societies impacts the balance of the nature. Hydroelectric dams change the initial ecosystem in a large area.

Wind and solar energy are available everywhere in the world from deserts to Polar Regions. In rural and suburban areas where energy services are unavailable, they provide electric power independent from energy providers like electric utilities and gas stations. Small PV units and wind turbines can serve as portable power supply for remote places, small islands, camping areas, or sailing boats. Use of renewable resources locally available in a nation's territories reduces dependence on foreign oil and gas, and thus enhances national energy security.

Decentralized generation systems developed by the residential, commercial, and industrial users can harness renewable energy on private properties. Electric generation by small-scale units spread on a wide area, rather than centralized conventional power stations, is known as distributed generation. Generation at the point of use avoids the losses on transmission lines and increases the reliability of the distribution network. Consumers benefit from their individual energy generation from renewables by avoiding service surcharges, energy related taxes, and increasing energy costs. Local generation backed up with energy storage provides energy security to consumers. In many countries, governments are developing policies to encourage distributed generation to take advantage of small-scale renewable energy development.

The cost of energy obtained from renewable resources is independent of oil price fluctuations, economic circumstances, or political tensions. Once the conversion facility has been developed, the fuel is free over the economic lifetime of the project.

In spite of many benefits, renewable energy has also limitations and challenges. Renewable energy resources are not always available for conversion. Sunlight is unavailable at night, the amount of solar energy per unit area changes during the day. Daily total solar energy that can be harvested depends on the daylight duration, hence changes every day of the year. In addition, solar energy received on the surface depends on the air clarity and cloudiness of the atmosphere. Air pollution, smog, and humidity affect the power of the solar radiation per unit area called irradiance. Clarity and cloudiness are difficult to predict, therefore solar irradiance at a location changes randomly within a range. The variation of wind speed is hard to predict as well. Availability of both solar and wind power are probabilistic and uncontrollable variables.

The time-rate of energy that can be extracted from a renewable resource is limited in time and space. Whereas biofuels and combustible waste can deliver controllable and consistent energy flow like fossil fuels, energy yield of non-combustible renewables is uncontrollable and randomly changes in time unless it is regulated by some kind of storage technique.

At any location, the amount of solar energy received on unit area changes continuously during the day, reaching its maximum when the sun is at the highest point in the sky. Therefore, solar energy that can be harnessed depends on the surface area, time of the day, and day of the year. Electric power that a wind turbine can generate is a function of the wind speed and the swept area of its blades. Although wind is available at any time of the day, wind speed may change randomly within a few minutes to hours. Like solar power, wind power also varies in daily or seasonal cycles. The total amount of hydraulic energy that can be produced on a river depends on its flow rate as well as the elevation of its source from the sea level. The river flow changes daily or seasonally with precipitation and melting of the snow in the mountains.

Availability of all forms of renewable energy depends on location and time. The amount of electric power generated by wind and solar units change instantly in a time scale as short as a few minutes and vary significantly on daily or seasonal cycles. Hydroelectric resource potential depends on seasonal precipitation and water level in reservoirs. Drought conditions have considerable effect on the energy production of hydroelectric power plants.

The amount of precipitation and other weather conditions change every seven or eight years by natural cycles known as El Niño and La Niña. Yearly rainfall affects the flow of rivers and consequently the amount of water stored in reservoirs. Seasonal changes also affect renewable energy resources; for example, more hydropower is available in spring, when melting mountain snow and rainwater supply the rivers. During the summer, however, the limited water flowing in rivers or stored in dams is needed more for agricultural irrigation. Solar energy incident on Earth's surface depends on its angle entering the atmosphere, and therefore changes by the time of day and day of the year. Wind power also changes with local temperature changes during a day or from seasonal atmospheric changes. Biofuel production may change with yearly harvest variations in agricultural fields where hydrocarbon rich plants are cultivated.

Large amounts of renewable energy are generally harnessed at remote locations such as mountain ridges for wind, deserts for solar power, and river basins for hydropower. The energy produced from such resources can be transferred to end users only in the form of electricity. The intermittent character of wind, solar, and small hydropower presents challenges for interconnected electric networks. Large amounts of energy generated from irregular renewables must be stored in various forms to stabilize the electric generation.

Renewable energy is not necessarily sustainable. For instance, biofuels are sustainable only if they are consumed at a rate equal or less than they are reproduced. Firewood is an example of a renewable source but not necessarily a sustainable energy source; it takes many years for a redwood, oak, pine, or any other tree to grow but they can be consumed in a short time as a heat source. Before petroleum, whale oil was used as fuel to light lamps, and there was a time in the past when whales were near extinction because of lamp oil production. As a consequence, whaling has been banned worldwide to prevent extinction of whales.

Integration of renewable energy into interconnected electric power systems presents technical challenges. Controllable and predictable energy sources like hydropower, geothermal energy, and bioenergy are easier to integrate into electric transmission networks. Uncontrollable and probabilistic variation of solar and wind energy present significant challenges for electric power systems supplied by various conventional centralized generation plants.

Renewable resources are delivered to consumers in the form of electric power, direct heat, or liquid fuels used in the transportation sector as alternatives to petroleum. Electricity is the most efficient, clean, and convenient energy carrier to deliver most renewable resources. In an electric power system, generation and consumption must be equal at all times. The interconnected transmission network is supplied by a large number of generators operating synchronously to maintain the system frequency constant. If the generation exceeds the consumption at any time, the frequency tends to increase and some of the generators temporarily experience electromechanical oscillations until all the system frequencies settle to a new value. When the generation suddenly decreases, the prime movers and generators experience similar transient oscillations.

River dams block the migration path of fish that need to swim upstream to lay their eggs, and for the newborn fish to swim downstream to reach habitat to grow. Hydroelectric projects also affect river transportation and regional trade activities. The Dalles Dam and hydroelectric power plant shown in the photo on the title page of this chapter has a lock for river transportation vessels and fish ladders to allow various fish species migrate upstream or downstream. Figure 5.5 shows a fish separation facility installed at McNary Dam near Umatilla, Oregon. Many dams on the Columbia River in the northwest United States are designed similarly. Reservoirs of large hydroelectric power plants flood a vast area, causing thousands, and sometimes millions, of persons living in the area to be relocated. In addition, cultural heritage, archeologic, and historic sites must be protected or transported to other places.

Energy conversion systems supplied by renewable sources have a larger footprint compared to fossil fuel burning and nuclear power plants. Based on data generated from the US Energy Mapping System (EIA 2019b), yearly electric generation per unit land area of a typical coal burning power plant is about 8000 times the generation per unit area of a solar PV facility in the Midwest region of the US. A 103-MW capacity wind farm in West Virginia

Figure 5.5 Fish separation facility at McNary Dam, Oregon. Source: H. & O. Soysal.

that generates an average 300,000-MWh per year has 45 wind turbines, each with 2.1-MW rated power, installed on a 5-mi corridor.

Increased use of renewables is directly reflected on electric generation. Until the 1980s, the main renewable energy supply was from hydroelectric generation. Large amounts of water stored in a reservoir allow conventional hydroelectric power plants to operate in steady conditions. Electric power output of hydroelectric generators can be controlled easily to match the electric demand.

Use of solar, wind, and biomass energy has increased sharply in the twenty-first century. Vigilance of climate change issues played the primary role in increasing use of renewable energy systems. At the same time, advanced conversion systems, especially regarding solar and wind energy, made renewable energy more affordable and economically feasible.

Geothermal energy is also a potential alternative to fossil fuels and nuclear power for electric generation. Heat produced in deep layers of the ground by the decay of radioactive minerals is not renewable in principle, but using geothermal heat does not accelerate the consumption of such minerals. On the other hand, the temperature difference between the above and below ground water converted into heat by geothermal heat pumps (GHP) for direct heating or cooling is a renewable resource with significant potential.

The initial investments needed to develop renewable energy projects are generally higher than developing comparable fossil fuels and nuclear power plants. Development of such plants needs financial incentives. Historic trends of wind and solar powered generation systems reflect the political decisions of governments.

5.6 Solar Energy

The Earth receives an immense amount of energy from the Sun; in fact, life on Earth is powered by the Sun. Solar energy is sustainable and available at all locations. Sunlight is the source of all types of energy resources, except nuclear and geothermal energy. Energy of the Sun reaches the Earth in the form of electromagnetic radiation. *Irradiance* is the incident flux of radiant energy per unit area. Solar irradiance at the top of the atmosphere is called solar constant and often represented by S. While this value changes day by day as the distance of the Earth to the Sun changes, 1367 W/m^2 is a commonly accepted value for solar constant.

The total energy flux of solar radiation incoming to the Earth is obtained by multiplying the solar constant S and the area of a circle with the radius of the Earth, that is πR^2. The average flux on the external surface of the atmosphere can be calculated by dividing this number by the total surface area of the globe, Eq. (5.3).

$$S_{av} = S\frac{\pi R^2}{4\pi R^2} = \frac{1367}{4} = 342 \text{ W/m}^2 \tag{5.3}$$

This is the total energy per second received on the Earth. Note that since the total surface is considered, this average power density is constant at any time, independent of the Earth's rotation. Out of the 342 W/m^2, 105 W/m^2 is reflected back into space and the remaining 237 W/m^2 is absorbed to heat the atmosphere, oceans, and land, and powers the photosynthesis in green vegetation (Markvart 2000).

During the daytime, the extraterrestrial irradiance above any point on Earth is the same. Solar energy is attenuated as sunlight passes through the atmosphere due to the absorption and scattering caused by gases, water vapor, and solid particles. Average power density received at sea level around noon on a surface perpendicular to the sunbeam is approximately 1 kW per square meter.

Solar energy available on Earth is well above the total primary energy consumption worldwide. Obviously, the actual amount of solar energy that can be harnessed is limited due to several reasons. First, solar energy is unavailable at night, and when available, it changes during the day and throughout the year. Second, it changes with latitude and altitude. Finally, clarity of the sky due to cloudiness, fog, haze, and pollution affects the radiant energy received on the Earth's surface.

5.6.1 Solar Resource Potential

Maximum irradiance at a location mainly depends on the latitude, elevation from sea level, and clarity of the air. Direct normal irradiance (DNI) on the Earth's surface at a certain location is the power per unit area orthogonal to the solar beam, measured in W/m^2. DNI is equal to the solar radiation that reaches the top of the atmosphere minus atmospheric losses. The length sunlight travels in the atmosphere changes during the day by the solar elevation called zenith angle. In addition, clarity of the sky changes with atmospheric conditions, motion of the clouds, pollution, and precipitation. Consequently, DNI significantly changes in time.

Diffuse horizontal irradiance (DHI) is the radiation received on the Earth's surface from sunlight scattered in the atmosphere by clouds, fog, or haze. DHI is also measured on a horizontal surface in kW/m^2 and its value depends on the opacity of the air and reflection of the surrounding ground. The radiation emitted from the sun disk (circumsolar radiation) is subtracted from the radiation diffused in the atmosphere. The maps in Figure 5.6 show long-term average of daily and yearly sums of DNI and global horizontal irradiance (GHI) (Solargis 2017).

GHI is the total radiant power received from the sun on a horizontal surface on Earth. It is the sum of DHI and the component of the DNI perpendicular to the horizontal surface. At an instant when the incident angle of the Sun's ray with the perpendicular to the surface (zenith angle) is z, GHI can be calculated using Eq. (5.4).

$$GHI = DHI + DNI \times \cos(z) \quad [W/m^2] \tag{5.4}$$

Solar energy at a certain location is defined as *solar irradiation* or simply *insolation* and measured in kWh/m^2. Insolation changes throughout the year with the length of daylight time. Yearly total insolation depends on the latitude and clarity of the sky at a particular location. The average of daily irradiations over a year allows a rough estimation of the yearly total solar energy received at a location. Solar irradiation received at a location is obtained by integrating GHI over a certain time interval, typically one day. If DHI and DNI are recorded with a sampling interval of Δt and GHI is calculated for each record, then the integral operation can be simplified with a summation as in the second part of Eq. (5.5).

$$S = \int_0^T GHI(t)dt = \sum_0^k GHI_k \Delta t \quad [Wh/m^2] \tag{5.5}$$

The available solar potential on Earth is estimated to exceed all global energy needs, even considering the projected energy use until 2100 (UNDP 2000). Development of utility scale solar generation facilities depends on the availability of suitable land and the population density of the region. In addition, air quality and shading considerably affect the energy output. Figure 5.7 shows the range of technically available solar energy potential by geographic regions based on UNDP (2000, Table 5.19). According to the *Special Report of the Intergovernmental Panel on Climate Change* (IPCC 2011), the total global solar potential is estimated between 1,575 and 49,837 exajoules (EJ) which is well above the current primary energy supply of 576 EJ worldwide reported by International Energy Agency (IEA 2018a).

The estimated amounts reflect the availability of solar energy as a primary resource; thus, they do not consider the limitation and efficiency of solar conversion technologies, nor technical issues regarding the integration of solar electric generation to the interconnected grid. The minimum and maximum limits are based on the assumptions of the land available for solar generation.

5.6.2 End-use of Solar Energy

End-uses of solar energy include heating, lighting, and electric generation. Passive solar technologies use natural sunlight without any energy conversion element or system. Active

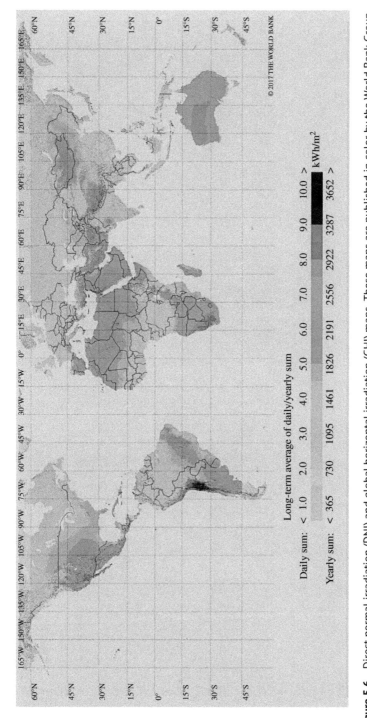

Figure 5.6 Direct normal irradiation (DNI) and global horizontal irradiation (GHI) maps. These maps are published in color by the World Bank Group, funded by ESMAP, and prepared by Solargis. For more information and terms of use please visit http//globalsolaratlas.info. Source: © 2017 The World Bank. Solar resource data: Solargis; Black and white rendering; Soysal.

Long-term average of daily/yearly sum

Daily sum:	< 1.0	2.0	3.0	4.0	5.0	6.0	7.0	8.0	9.0	10.0 >	kWh/m²
Yearly sum:	< 365	730	1095	1461	1826	2191	2556	2922	3287	3652 >	

© 2017 THE WORLD BANK

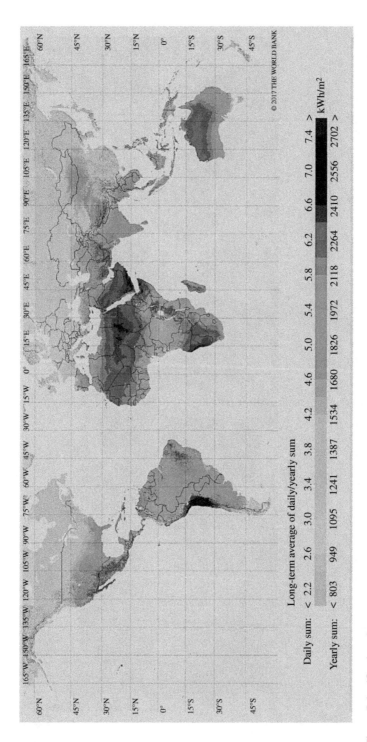

Figure 5.6 *(Continued)*

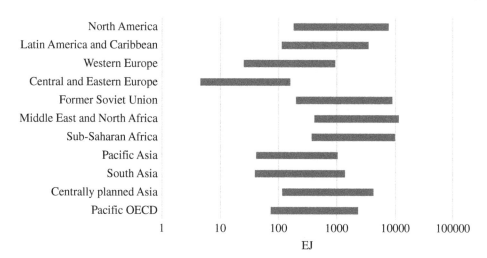

Figure 5.7 Estimated annual solar potential. Source: UNDP (2000, p. 174, Table 5.19).

solar technologies convert solar irradiance into heat or electricity for transfer to a point of use.

5.6.2.1 Passive Solar Buildings

All buildings benefit of the sun for natural interior lighting. Sunlight reduces the cost of artificial heating in cold climates and winter. In contrast, in the summer excessive sunlight entering the buildings is less desirable since it increases the indoor temperature. If the building has artificial ventilation and air conditioning, contribution of the sun to the interior temperature increases the energy consumption. Buildings designed with properly oriented large, high emissivity glass windows and efficiently insulated envelop are referred to as passive solar buildings. Such buildings are primarily heated by solar energy in winter, but they are designed to minimize the sunlight in summer. Solar energy potential for passive solar buildings depends on the climate of the location.

5.6.2.2 Heat Production

Solar thermal systems convert solar irradiance directly into heat using a solar thermal collector for space or water heating. Solar collectors may be stationary or moving to track the position of the sun in the sky. Produced heat is transferred to the point of use or to thermal storage by circulating a working fluid.

Contemporary solar thermal conversion systems are mainly used for space and water heating, desalination of seawater, and electric generation. Industrial solar thermal systems are used for process heating or metal melting. Smaller scale commercial and residential solar thermal applications include desalination, greenhouse heating, drying, pool heating, water heating, and space heating. Solar thermal conversion has the highest efficiency among all solar conversion technologies.

5.6.2.3 Solar Electric Generation

Solar energy can be converted into electricity either directly by using PV arrays or indirectly by using solar thermal collectors to produce steam and convert the thermal energy into electricity through turbine-generator units similar to conventional electrothermal power plants.

PV arrays consist of modules mounted on the ground, a pole, or building roofs. PV modules are a series and parallel combination of semiconductor cells that convert solar irradiance directly into electric power. CSP plants generate electric energy indirectly from the solar heat. In a CSP plant sunlight is reflected by mirrors on a tower where a fluid is heated. The heat transfer fluid circulated through heat exchangers boil water, and steam turbines drive electric generators. Such power plants can store solar energy as heat and continue electric generation at night.

Direct PV conversion does not suffer from thermodynamic efficiency of the heat engines, but the efficiency of practical PV cells is in the range between 15% and 25%. CSP generation has also low efficiency due to the thermodynamic cycle of the heat engines and electromechanical losses in the generators. Both types of generation units have lower running costs compared to conventional power plants since there is no fuel cost or operational cost due to fuel transportation and waste disposal. The maintenance and service costs of PV generation units are significantly lower, hence, the efficiency impacts mainly the initial investment for material and procurement of the installation area.

Commercial-scale solar generation units are generally mounted on the ground. In densely populated areas, the land dedicated for PV array installation is a considerable part of the initial cost. Uninhabited lands with adequate solar exposure are more suitable for large-scale solar power plants. However, if the location is not close to the interconnected grid, the cost of additional high voltage power transmission systems increases the initial investment.

PV generation has the benefit of being modular and can be easily installed on the building roofs and unused land. Grid-tie residential and small commercial PV systems supply the excess power they generate at times of low consumption back to the electric grid. Distributed generation by small-scale units spread in a wide area reduces the transmission losses of the electric grid.

5.6.3 Strengths and Challenges of Solar Energy

Sunlight is the most environmentally friendly energy source. Solar conversion systems do not release carbon dioxide, any toxic gas, any form of pollutant, or water vapor during their operation. Unlike electromechanical energy conversion systems, PV generation units do not have any moving parts (other than a solar tracker if installed). Therefore, they create no noise and do not need water for cooling. CSP generation systems that use steam turbines are similar to other thermal energy conversion systems in water use, but they are cleaner since no combustion is involved.

Solar thermal and PV systems present minimal risk of accident, and if even an accident occurs, it remains local; unlike nuclear accidents and oil spills, which present life-threatening and long-lasting contamination of a large area. Solar energy does not cause any land or water pollution like fossil fuels, or radioactive contamination like nuclear power plants.

Environmental impacts of large-scale solar energy systems are mainly related to land use. A commercial size PV array changes the vegetation and animal habitat on the land it covers, thus animal migration and biodiversity may be affected of the fragmentation of the land and change of vegetation.

The efficiency of commercially available PV modules ranges between 10% and 20%, which seems lower compared to other energy conversion systems. The significance of efficiency is, however, different for renewables from fossil fuels. A 30–40% efficient fossil fuel powered generation unit burns approximately three units of primary-source energy to deliver one unit of energy. Solar power plants, on the contrary, do not diminish the available solar energy on site. In addition, on-site PV generation has no cost to transfer of the energy source to the point of use. Since the fuel is freely available independent of the production rate, efficiency affects the initial cost and the area dedicated to the conversion unit. Lower efficiency PV modules are generally less expensive, therefore, rather than the efficiency of a particular module type, the cost per unit power is more relevant when comparing different technologies. Higher efficiency is particularly desirable at locations where the area is limited, like the roof of a building or the land available for larger commercial solar farms. In some applications, solar trackers may be used to maximize the energy conversion by continuous position control to maintain the collector or module orthogonal to the incident sunlight. The additional cost of a solar tracker versus using fixed solar array is a decision to be made based on a comprehensive economic cost analysis.

Unavailability of sunlight at night and variable irradiance during the day is the biggest challenge of PV generation. The magnitude of daily, monthly, and yearly insolation is predictable at any location. However, the short-term forecast of fluctuations of irradiance is more complicated and always includes uncertainties.

Integration of PV generation to the interconnected grid supplied by conventional generation systems requires comprehensive analysis and monitoring techniques. In an interconnected network supplied by diverse power plants, sudden changes of the supply or demand can cause electromechanical oscillations that may lead to serious instability problems. Random fluctuations of electric power injected into the distribution system is called penetration. To prevent the risk of instability, utility companies restrict the penetration ratio according to the power variations that can be tolerated by other types of conversion systems. Electric supply cannot depend on solar PV generation unless some form of energy storage provides backup energy. CSP generation systems regulate the short-term fluctuations by storing thermal energy in the working fluid. Small-scale off-grid PV systems are generally supported by a battery backup.

5.7 Wind Energy

Wind power has been an essential driving force of economic development since the early civilizations. Sailboats carried voyagers and merchandise on rivers, lakes, seas, and oceans for thousands of years. Throughout centuries, windmills powered water pumps, grain grinding mills, and various mechanical equipment.

Windmills, like waterwheels have supported the economy for many centuries. In mid eighteenth century, however, they could not compete with the power of steam during the industrial revolution as fossil fuels became preferred sources of energy. Whereas the idea of using windmills to generate electricity stems back to late nineteenth century, wind power appeared to be a feasible option in 1970s during the energy crisis. As Figure 5.4 shows, wind has become the fastest growing primary energy source in renewable electric generation since 2000. This section will focus on the electric generation potential of wind power, its benefits, strengths, and challenges.

5.7.1 Electric Generation Potential of Wind Resource

Wind energy is basically the kinetic energy of a moving air mass, which can be either used directly as mechanical work or converted into electric power. For example, linear mechanical work caused by the force of the wind moves a sailboat. Windmills transform wind power directly into mechanical power to turn a grinder, pump, or a mechanical apparatus. Wind turbines drive an electric generator to convert the kinetic energy of the wind into electrical energy.

Although in earlier applications wind power was used to directly produce mechanical work, modern conversion systems are mostly based on electric generation. For cost-effective generation, wind turbines must be exposed to consistent, steady, and sustained wind with a suitable average velocity. Energy output of a turbine depends on the wind regime at the site and characteristics of the selected turbine type. Fast changes of wind velocity, gusts, and turbulence are not desirable for wind powered electric generation.

Electric generation potential of wind resource in a region is generally estimated by using the yearly average wind power density and the land area available for wind farms. Wind power density is the power delivered by a wind turbine per unit area swept by the blades. Realistic assessment of the wind potential, therefore, highly depends on the mean wind speed and frequency distribution at the height at which the wind energy is harnessed, as well as the available energy conversion technologies. Figure 5.8 shows the estimated wind power density at 100-m above the ground (DTU 2018). The map is obtained by downscaling meteorological data and interpolating wind speed, considering terrain topography and surface roughness.

The maximum electric generation potential of global wind resource is estimated about 6000 EJ, assuming that development of a wind generation facility is feasible at regions where the average wind power density is at least 250–300 W/m² at 50-m height, and all available area can be used for wind farms (Grubb and Meyer 1993). This estimation considers an overall efficiency of 30% representing wind turbine and transmission line losses.

Wind farms consist of arrays of properly spaced wind turbines. Distance between wind turbines affect the overall efficiency because of the turbulence and wake created by rotating blades. Land area required for commercial wind farms depends on the number of turbines, selected turbine types, rotor diameter, and tower height. For a reasonable generation efficiency, the minimum distance between turbines must be about seven times the rotor diameter, which can be as large as 110–130 m for a 2- to 2.5-MW turbine. A 50-MW wind power

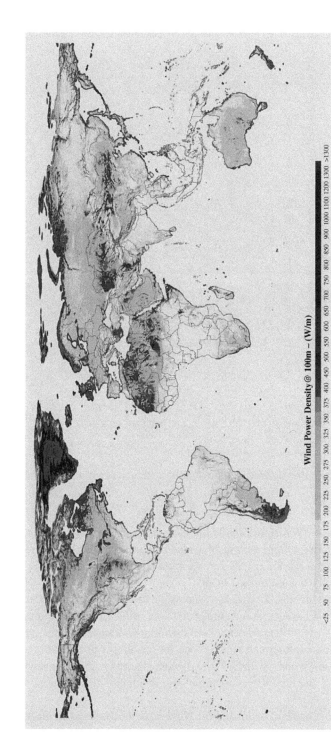

Wind Power Density @ 100m – (W/m)

<25 50 75 100 125 150 175 200 225 250 275 300 325 350 375 400 450 500 550 600 650 700 750 800 850 900 1000 1100 1200 1300 >1300

Figure 5.8 Wind power density in W/m^2 estimated for wind speed at 100-m. Map obtained from the *Global Wind Atlas 2.0*, a free, web-based application developed, owned, and operated by the Technical University of Denmark (DTU) in partnership with the World Bank Group, utilizing data provided by Vortex, with funding provided by the Energy Sector Management Assistance Program (ESMAP). Source: © Technical University of Denmark ("DTU"), a university registered in Denmark CC BY 4.0 (DTU 2018). B&W rendering: Soysal.

Figure 5.9 Top: Wind farm in the midwest, US. Bottom: Transportation of a wind turbine blade. Source: © Soysal.

facility utilizing 25 of the 2-MW turbines would therefore require more than ten square kilometers of land. During construction, each turbine blade is carried by one truck and towers are shipped in separate sections. Figure 5.9 shows a wind farm in the midwestern plains of the US and a truck carrying a wind turbine blade.

A commercial wind turbine site must be easily accessible from highways and located near high-voltage power lines for connection to the grid. Cities, difficult terrains, forests, and inaccessible mountain areas limit commercial-scale wind power development. Although lands used to install wind turbines can still be used for farming, a certain area needs to be secured for tower foundations switchgear and other equipment. Public acceptance, environmental concerns, and land use constraints impose additional restrictions to wind power development.

World Energy Council specialists estimated the wind resource available for electric generation by geographic regions as shown Table 5.2. The estimation assumes that 4% of the

Table 5.2 Estimated wind resource potentials for electric generation.

Region	Available land with wind class 3–7 (× 1000 km²)	Electric generation potential of wind resource if 4% land is used (TWh)	Primary Energy Equivalent (EJ)
North America	7876	5000	60
South America and Caribbean	3310	2100	25
Western Europe	1968	1300	16
Eastern Europe and former Soviet Union	6783	4300	52
Middle East and North Africa	2566	1600	19
Sub-Saharan Africa	2209	1400	17
Pacific Asia	4188	2700	32
China	1056	700	8
Central and South Asia	243	200	2
Total	30,200	18,700	231

Source: UNDP (2000, p. 164, Table 5.21).

area with sufficient wind resource can be used for commercial wind farm development. The primary energy equivalent is calculated by dividing the electricity generation potential by a factor of 0.3, which is a typical efficiency of wind turbines, including transmission losses (UNDP 2000, p. 164, Table 5.21).

Kinetic energy available from wind at a certain location depends on many factors including air density, elevation, topography of the land, vegetation, buildings, and large structures.

Utility-scale wind farms are feasible in areas where the wind speed is consistently in the operational range of available wind turbines. Average wind speed may be used for a rough estimate of the electric generation potential of the wind resource in a region. For a more realistic estimation, the frequency distribution of the wind resource and turbine characteristics should be considered. Methods for assessment of prospective wind turbine sites are described in Chapter 9.

5.7.2 Strengths and Challenges of Wind Energy

Wind is a clean, sustainable, and renewable source of energy. Wind powered generation, like solar and hydroelectric, is free from carbon dioxide or toxic emissions, particulate matter, solid waste, and land or water pollution. Unlike fossil fuel and nuclear power plants, wind powered generation units do not use any water.

Commercial wind farms have a larger footprint for the energy they generate. However, since the turbines are located at a sufficient distance to avoid wake and turbulence, the land under spaced turbine towers can be used for other purposes, including agricultural farms, roads, storage, and recreational or industrial parks.

Unlike solar irradiance, wind may be available at any time of the day. However, wind speed and direction are less predictable than solar irradiance. Because wind is a consequence of uneven solar heating, in some regions wind may complement solar energy in electric generation. Hence, wind-solar hybrid generation may produce steadier power output during a day or year around.

Wind turbine blades drive an electric generator which directly delivers three-phase AC power to the external circuit. While advanced electronic controls are still used to regulate voltage and frequency using an electronic converter, utility-scale units are directly synchronized with the power grid. Wind powered energy conversion does not suffer from low thermodynamic efficiency of fossil fuel fired generation. Depending on the design, the efficiency of a wind turbine can reach as high as the theoretical maximum efficiency of 59%, known as the Betz limit.

Effective use of available wind resources relies on technologic advancements. Composite materials used to make larger, lightweight, and stronger blades combined with advanced airfoil design improve the performance of wind turbines. Strong permanent magnets made from advanced magnetic materials led to the improved design of permanent magnet synchronous generators used in small wind turbines. Since such turbines have become commercially available at affordable prices, residential and small commercial wind generation propagated from rural areas to suburbs and even cities increasing the share of distributed generation in electric supply.

Generators used in large units have evolved from constant-speed induction machines with soft start to more controllable variable-speed machines. Advanced technologies to allow control of ramp-rate, output voltage, and low-voltage-ride-through (LVRT) ability are emerging (Smith 2005). Technologic evolution progressively reduces the adverse effects of the variable wind power generation on the interconnections and increases its reliability.

In addition to turbine and generator technologies, structural improvements allow the use of higher turbine towers. Increased tower height has several benefits. First, longer turbine blades can be used, which result in quadratically increased rotor-swept area, and consequently quadratic increase of the turbine power output. Larger single units increase the efficiency and unit cost of generated electricity. Turbines installed on higher towers are exposed to more uniform wind at higher velocity, hence generate electric power at a higher capacity factor.

5.7.3 Environmental Impacts of Wind Powered Generation

Although wind power generation has significantly smaller impact on the environment, development of wind farms near populated areas has created considerable public reaction. Public concerns to wind turbines can be combined in three groups; visual impact, interactions with wildlife, and audible noise.

5.7.3.1 Visual Impact

Wind farms and small individual wind turbines are more visible in a larger area compared to conventional power plants. Because most wind turbines are mounted on white monopole towers, they change the natural landscape. Rotation of the blades creates a flickering effect, especially early in the morning or late in the afternoon when the Sun is lower behind the turbines. At night, the blinking red lights to mark the tower tips for the aeronautical safety may be disturbing for the population living in rural areas.

5.7.3.2 Impacts on Wildlife
Since the early years of wind power development, impacts of commercial-size wind turbines on migrating birds and bats have been the focus of public debates. There are long-standing evidences of bird and bat deaths by collision with turbine blades. Environmentalists are particularly concerned about bat fatalities caused by wind turbines because bats can live up to 30 years of age or more. Their slow reproductive rates increase the risk of endangerment or extinction if large numbers of migratory bats are killed by turbine blades.

Scientists have recently speculated that some bat fatalities may occur for reasons other than hitting a turbine blade. An effect called barotrauma, which is caused by rapid and excessive pressure change, is another potential reason for bat fatalities. An atmospheric pressure drop around rotating turbine blades is undetectable for bats, but their respiratory system cannot handle such pressure change (Baerwald and D'Amours 2008).

Since wind turbines are necessarily distant from each other, wind farms do not present land fragmentation issues for migration of ground mammals. The large tower foundations and access road between turbines may, however, impact at some degree the diversity of the ecosystem.

5.7.3.3 Audible Noise
Wind turbine blades produce low frequency wobbling sound as they turn. Wind turbine noise has been one of the major sources of reaction to wind farm development, although its level is much lower compared to the ambient noise in cities, including train, roadway, aircraft, and construction equipment noise. Most household appliances, and even sometimes the background noise of the rustling leaves and falling water, is noticeable. Especially because the sound is periodic and continues day and night it is annoying for many people who live in quiet countryside.

Noise level is expressed as ratio of the sound pressure to a reference level on logarithmic scale denoted in dB (decibel). Since sound propagates in all directions from the point source, its pressure attenuates by the inverse square of the distance. Many countries and local governments around the world published standards, codes, and guidelines to restrict the distance of commercial wind farms from buildings based on the outdoor noise level created by wind turbines. Wind farm developers, therefore, estimate the geographic distribution of the noise level at the planning stage using comprehensive acoustic simulation software. Noise level limitations rightfully restrict the maximum exploitation of the wind resource at populated areas.

5.8 Hydraulic Energy

Hydraulic energy is a direct consequence of solar energy; the water cycle originates from natural evaporation of water by the Sun. Water evaporation rate per unit surface area is higher for oceans than land. In addition, oceans cover about 70% of the Earth's surface. Water mainly evaporated from the oceans and carried by wind over the land is the principal cause of the continuous water supply through precipitation. Water precipitated on the land eventually returns to the oceans by river outflow. The total amount of water on the Earth is finite, but because of the solar-powered cycle, the energy from water is always available.

Like solar and wind energy, the power of water has been used since ancient civilizations. Archimedes is credited to discovering the lifting force of water, water pressure, helical pumps, and many other applications of hydropower. Waterwheels have been used for centuries to drive sawmills, grain grinding mills, and many other machines. Ancient Greeks used waterwheels for grinding wheat into flour more than 2000 years ago. Before the industrial revolution, hydropower was converted into mechanical work for milling and pumping. In 1870, a small hydroelectric unit was installed in a house named Cragside, in Rothbury, Northumberland, England (IPCC 2011, p. 443). The first hydroelectric power station with 12.5-kW capacity started in 1882 in Appleton, Wisconsin (US Bureau of Reclamation 2016). Use of hydropower to generate electricity became popular because the first hydroelectric stations proved to be more efficient than fossil fuel burning power plants. Today hydroelectric power generation spans a broad range from a few watts to several gigawatts. The world's largest hydroelectric power station at Three Gorges Dam in China with 22.5-GW capacity has been generating approximately 100 TWh electric energy every year since 2012.

5.8.1 Hydroelectric Potential

Hydropower is not evenly accessible at every region of the world and most of the substantial resources are remotely located. Electric power is the only energy carrier to transmit these resources to populated areas. Moreover, international and intercontinental electric grid connections distribute hydropower resources worldwide.

Hydraulic resource potential is estimated by evaluating the world's annual water balance. About 577 billion (10^9) cubic meters of water evaporates every year from ocean and land surfaces. Out of this amount, 119 billion cubic meters precipitate on land. About two-thirds of the precipitation is absorbed by vegetation and soil but most of this water evaporates again. The estimated amount of runoff water, which is theoretically available for energy purposes is about 47 billion cubic meters (UNDP 2000, p. 153). The amount of runoff water by continent can be estimated considering the yearly amount of inland precipitation.

Assuming that the runoff water is evenly distributed across a region, the available potential energy of water should be proportional to the product of runoff volumes and average altitude ($W_p = mgh$; $g \simeq 9.81$). In reality, however, the runoff water is not evenly distributed. Moreover, seasonal variations of precipitation also influence the potential energy.

Technical potential depends on available technologies and feasibility of resources. Development of new technologies to exploit smaller hydroelectric sources increases the technical potential. Estimation of the technical potential varies by countries and is sometimes underestimated or inflated due to incomplete assessments or political reasons. Estimated regional theoretical and technical hydroelectric potentials are shown in Figure 5.10.

Conventional hydroelectric plants are typically designed, engineered, and developed by international consortiums. Less developed countries may not be able to afford the large initial cost of hydroelectric plants because of financial constraints. Moreover, socioeconomic impacts may increase the initial cost and delay construction of large hydroelectric plants.

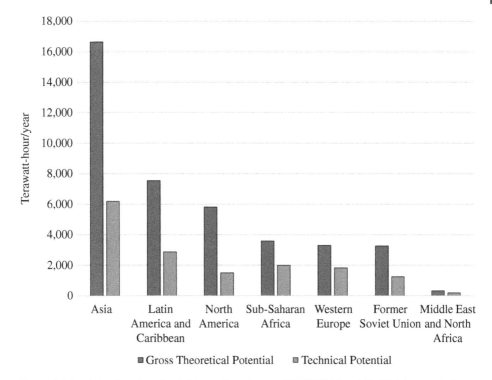

Figure 5.10 Hydroelectric potential by regions. Source: UNDP 2000 and World Atlas and Industry Guide, 1998 (https://www.hydropower-dams.com/world-atlas).

5.8.2 Strengths and Challenges of Hydroelectric Generation

Hydroelectric power plants are clean and reliable energy sources. Generating electricity by hydropower instead of fossil fuels avoids significant amounts of greenhouse gas emission, air pollution, and release of toxic elements to the environment.

Reservoir backup and pumped storage hydroelectric units add massive energy storage capability to the electric power system. Since the power output can be adjusted in a relatively short time, hydroelectric power plants conveniently adjust the electric generation to the variable demand in the interconnected grid.

Development of a hydroelectric plant is a multipurpose project. Dams regulate water flow in rivers and help flood control. Reservoirs improve the regional climate and vegetation. The area around dams is often used for recreational activities and increases the value of the region.

On the other hand, like any large project, hydroelectric power generation also has adverse environmental and social impacts. Construction of a dam impacts the biodiversity by blocking the fish migration path and changing the food chain. Certain fish species, like salmon, migrate downstream as they grow up and come back to the area where they were born to lay their eggs. Blocking the natural lifecycle of such species may cause a significant decrease in population, and even lead to their extinction.

Construction of a dam causes flooding to a vast area, which generates a wide range of public reactions depending on the population density, education and economic level of the population, existence of historic landmarks, and cultural heritage value of the land. Running river hydroelectric power plants control the river flow for electric generation, which may interfere with the benefits of farmers who need the water for irrigation. Dams constructed on larger rivers also affect transportation and trade along the river. Many dams constructed in the northwest United States were designed with fish ladders and locks to resolve the environmental and socio-economic issues.

Case Study 5.1

Hydropower Development in the Columbia River Basin in the northwestern US

About half of the electricity used in the northwestern region of the United States comes from hydropower harnessed in the Columbia River Basin. Today, 56 dams generate about 40% of all hydroelectric power in the US. In addition to electric generation, these dams regulate the water flow of the Columbia River and its contributories, prevent flooding, and support irrigation of farmlands without blocking the river transportation and fish passage.

The Columbia River, running 1243 mi from Canadian Rockies in British Columbia to the Pacific Ocean drains some 258,000 mi^2 (668,000 km^2). About 85% of the river's watershed is in the northwestern United States. The elevation of its source in British Columbia is 2700 ft (820 m). Water flow rate changes in a wide range from its peak in late spring and early summer when snow melts in the mountains, to the lowest level in autumn and winter, causing significant seasonal variations in electric generation (Marts 2019).

For thousands of years, the Columbia River and its tributaries have provided abundant food supply and transportation ways for indigenous tribes. The river is habitat for salmon, steelhead trout, and diverse wildlife including beaver, elk, deer, bear, bighorn sheep, osprey, hawk, falcon, eagle, and many other species.

The Columbia River and its tributaries have been a vital part of the regional economy providing irrigation water for farmlands and a route for commercial river transportation. Tidewater from the Pacific Ocean flows upriver for 140 mi (225 km). A series of locks and channels allow maritime navigation from the mouth in Astoria, Oregon upstream to Portland, Oregon and Vancouver, Washington (Marts 2019).

Beginning in the late nineteenth century, population growth, urban development, and industrial progress changed the needs and expectations of society. The first hydroelectric facility in the lower Columbia River Basin, T. W. Sullivan Dam, was built in 1888 at Willamette Falls in Oregon City. In 1890, just 15 years after the city of Spokane was founded, the Monroe Street Dam was completed on the Spokane River in downtown Spokane, following a fire that destroyed a big part of downtown business district.

The city of Idaho Falls built the first hydroelectric dam in 1904 on the Snake River, a major contributory of the Columbia River. Development of hydroelectric power facilities

(Continued)

Case Study 5.1 (Continued)

on the mainstream of the Columbia River began in early 1930s. The first was Rock Island Dam built near Chelan, Washington between 1929 and 1933. Bonneville Dam followed in 1938, and construction of dams on the Columbia and Snake Rivers continued into 1970s.

Bonneville Dam is the last facility downstream before the Columbia River flows into the Pacific Ocean. It is located 40 mi east of Portland, Oregon at the 145th mile from the mouth of Columbia River. The run-of-river type dam and reservoir provide water for two powerhouses, both constructed and operated by the US Army Corps of Engineers. Figure 5.11(a) is an aerial view of the facility.

(a)　(b)

(c)　(d)

Figure 5.11 Bonneville hydroelectric power station on Columbia River between Washington and Oregon States, US. (a) Aerial view. (b) Kaplan turbine rotor formerly used on one of the turbines. (c) Generators in the second powerhouse. (d) Fish ladders. Source: Courtesy of US Army Corps of Engineers, public domain. Original image available at: https://commons.wikimedia.org/wiki/File:Corps-engineers-archives_bonneville_dam_looking_east.jpg; Photos b, c, and d: © Soysal.

Currently 77 multiple purpose projects, 56 of which are hydroelectric generation facilities, are currently operating in the Columbia River Basin. Main features of the major facilities are summarized in Table 5.3.

(Continued)

Case Study 5.1 (Continued)

Table 5.3 Characteristics of major dams on mainstream Columbia River.

Project	Distance from Pacific (km)	Year	Capacity (MW)	Annual generation (GWh)	Head (m)	Av. flow rate (m³/s)
Bonneville	134.1	1938	1227	5743	18.2	184,900
Dalles	308.1	1957	2100	8339	26.8	179,500
John Day	346.9	1968	2200	11,482	32.0	174,000
McNary	469.8	1953	980	5791	22.8	171,600
Priest Rapids	638.9	1959	995	3885	23.8	120,200
Wanapum	669.0	1963	1038	4611	24.4	120,000
Rock Island	729.5	1933	623.7	2600	12.5	120,000
Rocky Reach	761.1	1961	1300	5806	27.7	116,400
Wells	828.8	1967	840	4108	21.0	114,200
Chief Joseph	875.5	1955	2620	12,263	53.6	109,800
Grand Coulee	959.9	1941	6809	22,014	100.6	110,000

Source: USACE.

Construction of dams on the Columbia River mainstream created two major socioeconomic concerns; one was associated with river transportation, the other fish migration. The dams upstream were designed with locks to allow passage of barges. River transportation is now possible through locks more than 460 mi (740 km) from the river mouth to Lewiston, Idaho.

The concerns about salmon migration came from Native American groups that economically depend on salmon fishing. Salmon and steelhead trout hatch in the tributaries of the Columbia River, migrate downstream to the ocean as they grow, and come back as adults to the same place they hatched to spawn and die. Blocking their migration path would decrease their population, possibly to extinction. Dams were designed with fish bypass systems to allow the juvenile passage downstream. Screens prevent juvenile fish from entering the turbines and a bypass channel with flow deflectors direct fish downstream. Fish ladders allow migration of adult fish upstream to return to their spawning places. All dams downstream of Chief Joseph Dam on the Columbia and the Hells Canyon Dam on the Snake River have fish passage facilities.

The first powerhouse of the Bonneville Dam was completed in 1938 and the second one was added in 1982. The first powerhouse accommodates 10 and the second 8 turbine-generator units. One of the turbine generators is still operating at 4000 kW capacity. Two units were upgraded later to 54,000-kW, and 8 units to 80 000-kW. In the second powerhouse [Figure 5.11(c)], two units have 13,500-kW and 8 units have 76,000-kW nameplate capacity. The total generating capacity of both facilities is 1227 MW. All generators are driven by Kaplan turbines with propeller type rotors as shown in Figure 5.11(b). The rotation speed is 70-rpm, stator frame diameter of generators is 528 in. (13.4 m) and they are 32 ft (9.7 m) high. Each powerhouse has

(Continued)

Case Study 5.1 (Continued)

one fish passage facility [Figure 5.11(d)]. Both fish passage facilities have counting stations. (Source: US Army Corps of Engineers documents.)

Plans to construct Grand Coulee Dam began in 1920s, but strong debates between two groups delayed the start of construction until 1933. Farmers were supporting construction of a concrete canal to irrigate farmlands. Groups that supported electric generation for the industry and cities were proposing construction of a high dam with pumped storage. Although the dam supporters won the debate, the initial project was approved for a 290-ft (88-m) high dam, which would only generate electricity without supporting irrigation. One year after the construction started, President Franklin Delano Roosevelt endorsed the proposal to build a high dam during his visit to the site. The 550-ft (168-m) tall dam would provide enough electric power to pump water for irrigation of the farmlands. The dam has been serving since 1941 for flood control, electric generation, and irrigation. Grand Coulee power plant is the largest electric generation facility in the US with total nameplate capacity of 6809 MW (EIA 2019a, p. EIA-860 M). The conventional pumped-storage power plant operates 33 turbines, 27 of which are Francis turbines, and 6 pump-generators to pump water to a higher elevation at Banks Lake between Coulee City and the dam. The reservoir created by the Grand Coulee Dam is named after President Franklin Delano Roosevelt.

(Source: US Army Corps of Engineers (USACE) Portland District documents.)

5.9 Geothermal Energy

Geothermal energy is the heat stored in the Earth. The total thermal energy in the Earth's solid core is enormous. However, only a small fraction of this energy is natural heat that can be technically extracted from the Earth's crust up to a maximum depth of 10 km. In this region, the temperature increases by depth about 30 °C every kilometer on average. The vertical temperature gradient is not uniform and can change in range from half of the average value to as much as ten times more. For example, at 5-km depth, in one zone the temperature may be only 70 or 80 °C while in another it may exceed 500 °C (Palmerini 1993).

Heat coming from the ground has been known since ancient civilizations and used for bathing, therapy, or rituals. Exploiting geothermal steam started about 1827 to support boric acid extraction from volcanic mud in Italy. In the following years, geothermal heat was used for industrial drying processes and space heating. In early years of the twentieth century, geothermal energy provided mechanical power and electricity. The first geothermal electric generation powered four light bulbs in 1904 at the Larderello dry steam field in Italy. The first industrial power plant, with 250-kW generation capacity, started operation in 1913 in the same place (Tiwari and Ghosal 2005). Italy continued to be pioneer in geothermal energy use and increased its geothermal electric capacity to 127-MW by 1944.

In the US, the first geothermal power plant with a 250-kW generator was installed in Geysers, California in 1925. Currently, 24 countries use geothermal electric generation to supply base load. In 2008, the estimated worldwide yearly geothermal generation was 67.2

TWh. Direct heat produced from geothermal resources in 78 countries is estimated about 0.4 EJ/year equivalent to 121 TWh/year (UNDP 2000).

5.9.1 Sources of Geothermal Energy

Thermal energy in the Earth comes from two sources. One of them is extremely hot magma that contains molten or semi-molten rocks, the other is the radioactive decay of unstable elements. Magma reaching the surface directly through tectonic faults can result in volcanic eruptions, which contain huge uncontrollable thermal energy. The aquifers or ground water heated by magma or radioactive decay can rise to the surface either mixed with steam or gases, and erupt as geysers or simply form hot water springs. Natural discharges of geothermal energy are generally uncontrollable and intermittent. Thermal energy available in the earth can, nevertheless, be extracted in a controllable way by drilling until reaching a feasible heat source, depending on the intended commercial application.

Underground heat sources are transferred to the ground near the surface by either conduction or convection. The temperature gradient in the ground is non-uniform and highly variable depending on the geological structure of the region. Geothermal energy resources can be grouped in four types (UNDP 2000):

- *Hydrothermal resources.* Hot water or steam found at moderate depths, typically less than 10 km.
- *Geo-pressured resources.* Deep hot-water aquifers containing dissolved methane under high pressure.
- *Hot dry rocks.* Extraordinarily hot geologic formations with little or no water.
- *Magma.* Molten or semi-molten rocks at much higher temperatures than the average.

Exploration of geothermal sources is similar to oil and gas exploration. Because of the high cost of complete exploration, more promising zones are determined by regional geological and geochemical studies. The structure of the potential area is first thoroughly surveyed by detailed geological, geochemical, and geophysical studies. If commercially valuable resources are located within a reasonable technical certainty, then several test wells are drilled to obtain a more realistic geothermal model of the prospective site. The depth of the exploratory drilling depends on location of the reservoir and the intended use of the resource. Smaller diameter exploratory bores called "slim holes" reduce the initial investment, but their flow rate is significantly lower than standard wells. The diameter of production wells is between 9 and 13 in. (23–33 cm) to provide a fluid flow in the range from tens to hundreds of tons per hour. Test wells normally have the same profile as the production wells. Some of the test wells, therefore, may be large enough to start a small-scale production that can offset part of the investment for exploration studies. The goal of the test wells is to obtain the physical and chemical characteristics of the reservoir. The temperature and steam/water ratio of the geothermal fluid determines the type of application. Pre-production wells are operated at different production conditions to assess the probable production for the following 12–20 years. In addition, productive-size exploratory drilling can produce a quantity of fluid needed for the first operation of the commercial-size power plant, reducing the overall capital cost of the project. During the exploration phase, the prefeasibility studies are conducted to estimate the cost/benefit ratio of the site. Based on the findings, the project may be terminated or move on to the commercial-scale development.

Current technologies allow exploitation of hydrothermal resources, which are a small fraction of the geothermal resources available in the Earth. Enhanced geothermal system (EGS) technologies are emerging to reach less accessible resources, lower permeability formations, hot dry rocks, and magma by hydraulic stimulation and fracking (IPCC 2011).

5.9.2 Geothermal Energy Potential

The theoretical global potential of geothermal energy is about 140,000,000 EJ, but 600,000 EJ is classified as useful accessible resource base. The portion of the accessible resource base expected to become economical within 40–50 years is estimated to be about 5000 EJ. Reserves accessible with current and emerging new technologies in a relatively shorter timeframe are only 500 EJ (UNDP 2000). Still, geothermal energy has enormous potential; even the technically and economically accessible reserves (about 434 EJ), exceeds the total annual global primary energy consumption. The regional distribution of geothermal energy potential is shown in Figure 5.7 and Table 5.4.

Like all other renewable resources, availability of geothermal energy also varies by location and the only way to transmit this energy to remote consumers is conversion to electricity. Geothermal power generation has many limitations. First, it suffers from low thermodynamic conversion efficiency, therefore large amounts of geothermal fluids at relatively higher temperatures must be extracted. In addition, use of geothermal energy results in significant air and water pollution. Therefore, rather than the quantity of available energy, conversion technologies and mitigation of environmental impacts determine the exploitation of geothermal energy in power generation.

5.9.3 End-uses of Geothermal Energy

Geothermal energy is used for direct heat production, electric generation, or combined heat and power (CHP) cogeneration. Direct heat production is more efficient but transmission

Table 5.4 Regional geothermal energy potential.

Region	Energy (EJ)	Share (%)
North America	26,000,000	18.6
Latin America and Caribbean	26,000,000	18.6
Western Europe	7,000,000	5.0
Eastern Europe and former Soviet Union	23,000,000	16.4
Middle East and North Africa	6,000,000	4.3
Sub-Saharan Africa	17,000,000	12.1
Pacific Asia (excl. China)	11,000,000	7.9
China	11,000,000	7.9
Central and South Asia	13,000,000	9.3
Total	140,000,000	100.0

Source: UNDP (2000, p. 165, Table 5.23).

of heat by steam is limited to short distances from the point of use. Electric generation is more valuable, flexible, and controllable, but less efficient due to the lower thermodynamic efficiency of the conversion of heat into mechanical work. Heat and power cogeneration plants are the most efficient conversion systems since the heat rejected as the part of thermodynamic cycle is used for other purposes.

5.9.3.1 Geothermal Heating

Heating and cooling applications using a geothermal heat pump (GHP) should not be confused with conversion of geothermal energy sources into heat and power. Also known as "earth-coupled," and "ground-source," such systems operate based on the temperature difference between the air above the surface and ground or water below the surface. Depending on latitude, the temperature below the ground or water surface stays constant year-around in the range between 45 °F (7 °C) and 75 °F (21 °C). In winter, when the air temperature is lower, a thermodynamic heat pump transfers energy from higher temperatures below the surface to heat the air in a building space. In summer the energy flow is reversed from the higher temperature in a building to cooler ground or water.

Geothermal energy is used in a broad range of direct heating applications depending on the temperature of the geothermal fluid either discharged naturally or extracted by drilling a well. Natural hot springs are among touristic attractions in touristic, therapeutic, and recreational resorts around the world. Space heating in residential districts, commercial buildings, and workshops is common in regions where hot springs are available. Industrial and commercial applications can be grouped as drying, evaporation, distillation, refrigeration, process heating, industrial-space air conditioning, and other processes such as extraction, washing, baking, etc.

Geothermal heating is achieved by simply direct contact with the pipes that carry geothermal fluid or, in more advanced systems, using a heat exchanger to transfer geothermal steam or water to a fluid that circulates in a secondary loop. Drying, evaporation, and distillation are based on this technique.

Timber and wood preparation, pulp and paper processing, textile industry, leather and fur treatment, synthetic fuel production and enhancement, chemical processes, and mineral production are among common applications of geothermal heating. Agricultural applications include greenhouse and soil heating, mushroom cultivation, fish hatching, food processing, food drying, and canning.

An evaporation process is used to concentrate solutions. One of the common applications is used salt production. Distillation is based on evaporation and controlled condensation of a liquid. Seawater desalination, water purification, heavy water (deuterium) production, and liquor production are traditional distillation applications. Fractional distillation is used to separate components of a liquid mixture that have different boiling temperatures. The hydrocarbon industry uses fractional distillation to separate liquid petroleum gases (LPG) from crude oil and raw natural gas. Geothermal energy can be also used for cooling and refrigeration by using an appropriate thermodynamic system. Figure 5.12 shows heating fluid temperatures needed for various industrial and agricultural operations where geothermal energy can be directly applied.

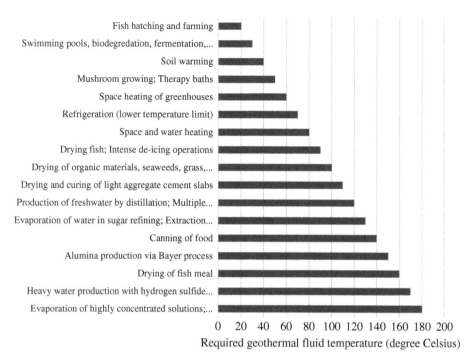

Fish hatching and farming
Swimming pools, biodegredation, fermentation,...
Soil warming
Mushroom growing; Therapy baths
Space heating of greenhouses
Refrigeration (lower temperature limit)
Space and water heating
Drying fish; Intense de-icing operations
Drying of organic materials, seaweeds, grass,...
Drying and curing of light aggregate cement slabs
Production of freshwater by distillation; Multiple...
Evaporation of water in sugar refining; Extraction...
Canning of food
Alumina production via Bayer process
Drying of fish meal
Heavy water production with hydrogen sulfide...
Evaporation of highly concentrated solutions;...

0   20   40   60   80   100 120 140 160 180 200
Required geothermal fluid temperature (degree Celsius)

Figure 5.12   Average geothermal fluid temperatures required for various applications. Source: Palmerini (1993) and Lindal (1973).

5.9.3.2   Geothermal Power Generation

Industrial scale geothermal power generation needs higher temperatures and flow rates compared to direct-use for heat. In practice, two types of reservoirs grouped as steam-dominated and water-dominated are exploited for electric generation or electric-heat cogeneration. Figure 5.13 illustrates the operating principle of four basic systems used for geothermal power generation (Palmerini 1993).

Steam-dominated reservoirs contain a superheated fluid at high temperature and pressure such that dry and saturated steam extracted at the wellhead can be directly piped to the generation system. In the simplest configuration, dry steam is fed directly to the turbine and exhausted to the atmosphere as shown in Figure 5.14(a). Despite its lower initial cost, such systems have several drawbacks. The steam coming out of the ground generally contains a mixture of gases including carbon dioxide, hydrogen sulfide, and numerous toxic gases. Part of these gases can be condensed and processed to obtain liquid carbon dioxide, sulfur, and other industrial products. The system shown in Figure 5.14(b) uses a gas extractor to separate condensate and non-condensable gases. More advanced systems use one or more upstream surface heat exchangers/evaporators to separate practically all gas content from steam.

Water-dominated reservoirs that contain hot water or a mixture of water and steam are more common. In the reservoir the fluid is under high pressure, as it rises in the well, the

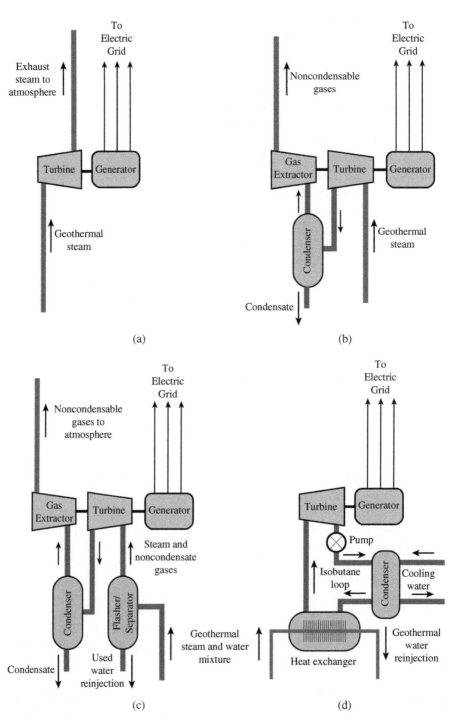

Figure 5.13 Operating principles of four different types of geothermal electric generation systems.

pressure drops and a mixture of liquid and steam reaches the wellhead. The fluid can be maintained in liquid phase by increasing the pressure by pumps in the well or extracted in a mixed phase. Depending on the reservoir temperature, either pressurized or mixed phase fluid can be used for power generation.

Pressurized fluid evaporated in a flasher/separator the geothermal emits a mixture of steam and gases. In a flashed system shown in Figure 5.14(c) steam and non-condensate gases extracted from the geothermal fluid enters the turbine and residual water is reinjected into the ground. The reinjected fluid can be as high as 80% of the extracted geothermal fluid. The mixture of steam and gases exiting the turbine is processed in a condenser and gas extractor to separate the condensate and non-condensate components.

Figure 5.14(d) illustrates the operation of a "binary system" that converts energy from mixed-phase fluid. A heat exchanger transfers heat to a working fluid, which circulates in a secondary loop through the turbine for thermodynamic conversion. The cooled geothermal fluid exiting the heat exchanger is reinjected into the ground. Such systems utilize a special type of turbine that operates at a lower temperature range than the turbines fed by steam generated in fossil-fuel fired boilers. The thermodynamic cycle implemented in such systems is named "organic Rankine cycle" (ORC) because it uses an organic compound, such as isobutane, with a lower boiling temperature than water at atmospheric pressure. ORC allows operation of the turbine at the temperature range available from geothermal sources.

The efficiency, which is the ratio of the useful energy output to the thermal energy available from the source strongly depends on the heat-sink temperature limited by the ambient temperature. Geothermal sources used for power generation can increase the working fluid temperature up to a few hundred degrees Celsius. In the case of low-temperature sources, especially below 100 °C, the efficiency can be as low as 10–20%. Power-heat cogeneration systems increase the overall efficiency by recovering the excess heat for direct heating applications.

Depending on the physical characteristics of the heat source, practical geothermal conversion systems may combine multiple stage upstream reboilers, pre-flash or double flash processes or combinations of flashed and binary cycles (Palmerini 1993).

Geothermal fluid requirement of various power plant types depends on the reservoir temperature, as shown in Table 5.5. The lowest cost plant type is exhausting into the atmosphere, but it requires the greatest amount of geothermal fluid extraction. Single flash power plants are the lower cost, but they require greater amounts of fluid extraction. They are not feasible for reservoir temperature less than 150 °C. Multiple flash and binary cycle facilities need higher initial investment, but they can operate with less geothermal fluid. Direct steam dominant power plants are preferred for reservoir temperature is above 200 °C

From efficiency and fluid consumption standpoint, direct use of geothermal heat is clearly more advantageous than electric generation. Direct heating applications are, however, limited by difficulty of transfer to end-users. However, electric power is more flexible and controllable.

Table 5.5 Geothermal fluid requirements of various power plants.

Cycle	Required GT fluid supply (kg/kWh)			
Reservoir temperature (°C)	120	150	200	250
Single flash				
Free exhausting (Figure 5.13a)	—	650	150	80
Condensing (Figure 5.13b)	—	150	80	50
Binary cycle (Figure 5.13d)	400	140	70	—

Source: Palmerini (1993).

5.9.4 Strengths and Challenges of Geothermal Energy

Geothermal energy resources are considered renewable since the heat tapped from an active reservoir is continuously restored by natural heat production, conduction, and convection from surrounding hotter regions. The extracted geothermal fluids are naturally replenished or returned to the reservoir after heat has been transferred to the conversion process.

Industrial power plants, on the other hand, may cause local declines in pressure and temperature within the reservoir over their economic lifespan. The cooled zones with reduced pressure are eventually reformed from surrounding regions after extraction ends.

Initial and operating cost of geothermal power plants are typically higher than other types of generation facilities because of the special material used to prevent corrosion, as well as periodic maintenance requirements of pipes and equipment. Electric generation in such plants is, however, steady and controllable. In general, geothermal power plants serve as base load supply in interconnected electric systems.

Turbines used in geothermal electric generation operate on the Rankine cycle, which has thermodynamic efficiency limited by the temperature of the heat source above ambient temperature. Power generation units fed from lower temperature reservoirs have overall efficiency from single digits to about 20%. To increase the efficiency, steam leaving the turbine must be cooled either by water or air. Water-cooled systems evaporate considerable amount of fresh water.

Geothermal energy is not as clean as solar, wind, and hydropower. Environmental impacts of geothermal energy can be divided in two groups. Temporary issues are related to pollution during the exploration and drilling phase. Permanent environmental impacts are associated with maintenance, make-up drilling, and pollution that result from power plant operation. Geothermal fluids contain sulfur, ammonia, and heavy metals such as mercury. Although geothermal energy conversion processes release carbon dioxide (a mixture of polluting gases including methane, nitrogen oxides, and hydrogen sulfide), they are less than greenhouse gases released by fossil fuel combustion (Goldstein, Hiriart, Bertani et al. 2011) Whereas toxic elements such as mercury and arsenic are also released, they are in trace quantities and generally do not create a significant environmental concern. The operational pollution created by geothermal power plants, however, is still less than fossil fuel fired energy production. On the other hand, numerous geothermal byproducts that have commercial value can be retrieved from geothermal fluid. Some examples are boric acid, sulfur, carbon dioxide, potassium salts, and silica. Processing of exhaust gases has the double benefit of production of valuable chemicals while reducing the toxic emissions.

Residual water from geothermal power plant operation is reinjected into the same reservoir to be reheated by the natural process. The production wells and reinjection bores must be effectively isolated by casing to prevent pollution of fresh ground water. In addition, extraction and reinjection of the thermal fluid may cause seismic activity and vibrations that can trigger earthquakes.

Because geothermal energy is a domestic resource, its use to replace foreign oil and natural gas enhances energy security of nations. Like other renewable resources, geothermal energy is also independent from volatility of energy prices due to economic and political conditions.

5.10 Biomass Energy

Substances derived from plants, animals, and microorganisms are commonly known as biomass. Since such materials are replenished by natural processes, energy extracted from any kind of biomass is considered renewable. Biomass is essentially a hydrocarbon that produces heat by combustion like fossil fuels. In fact, all fossil fuels are ancient biomass converted over millions of years by natural processes. Bioenergy refers to a broad range of fuels in solid, liquid, and gas form either used directly or after an industrial process.

Biomass has always been the main source of energy for humanity. Wood, animal dung, and plants have been burned for cooking and heating. Still in rural parts of the world and technologically less advanced regions, biomass is the primary energy source for heating. Modern biofuels are obtained from a wide range of biological feedstocks using advanced chemical processes.

In 2016, the share of biofuels and waste constituted 9.5% of the world's TPES, representing 62.4% of global supply of renewables (IEA 2018b). In developing countries, traditional solid biofuels (wood, charcoal, etc.) are extensively used for residential heating and cooking. In poor countries, use of unprocessed biomass can be as high as 90% of the energy supply (UNDP 2000). Bioenergy may be used directly as a heat source or converted to mechanical work by heat engines at the point of use. Firewood, wood pellets, sawdust, and charcoal are used for residential heating.

Some industrial facilities burn combustible waste produced on-site for process heating. Many thermoelectric power plants use biofuels, industrial waste and municipal waste are used in to generate electricity. Cogeneration plants, in addition to generating electricity, transmit residual thermal energy in the form of steam for industrial processes, space heating in residential districts or commercial building complexes. Electricity is a convenient and controllable energy carrier to transmit bioenergy to consumers for a broad range of end uses.

5.10.1 Biomass Sources

Bioenergy is produced by processing a wide variety of materials commonly known as biomass, obtained from living organisms such as plants, animals, algae, and bacteria. Such

Figure 5.14 Ethanol production in the midwest region of the US. Source: © Soysal.

materials are mainly hydrocarbons produced by either photosynthesis or biosynthesis. Biomass resources are abundant in most parts of the world.

Plant-based biomass includes woody or non-woody substances, processed waste, or processed fuel. Trees, shrubs, bushes, bamboo, palms are woody biomass. Non-woody biomass, mainly used for biofuel production, is from energy crops such as sugarcane, soybean, and corn. Other examples are residues, and clippings and roots of industrial plants such as cereal straw, cotton, tobacco, bananas, potatoes, etc. Processed wastes comprise industrial byproducts like sawmill chip and dust, wood bark, plant oil cake, black liquor from pulp and paper mills, and paper products in collected trash. Processed fuels are charcoal, methanol, ethanol, plant oils, biogas, and similar biomass ready to be used as fuel.

Residues from forest industries are valuable byproducts to produce biofuels. However, Biomass feedstock production and regeneration of the forests must be balanced by replanting and sustainable management to allow natural propagation. Agricultural residues can be also used for biofuel production, but some part of these residues must remain at the site for fertilization and soil conditioning.

Ethanol made from sugar cane, soybean, and corn is a potential alternative substitute to petroleum-based fuels. The fuel-ethanol industry is extremely water- and energy-intensive. Especially, the cultivation of corn for ethanol production requires vast agricultural land, substantial amounts of ground water, and energy for heavy agricultural equipment. Depending on the climate, production of one unit volume of corn ethanol requires between 3 and 160 units of fresh water from farming to processing (Wu and Chiu 2011). Interactions of biofuel production with ground water and food supply are discussed in more detail in Chapter 12. A cornfield and an ethanol production facility in the Midwest US are depicted in Figure 5.14. In the US, retail gasoline sold at most gas stations contains 10% corn-based

Figure 5.15 Spittelau Waste Incineration CHP plant in Vienna, on the Danube River. Source: ©
Soysal.

ethanol. Regular vehicle engines can burn gasoline that contains up to 15% (E-15) or can
be modified to use a mixture of 15% gasoline and 85% ethanol (E-85). The gas pump in
Figure 5.14 displays retail prices of various gasoline-ethanol mixtures in 2017.

Municipal solid waste is a renewable source indirectly associated with biomass since
it contains paper, cardboard, vegetable residues, discarded food, etc. Energy potential of
municipal solid waste depends on the composition of the refuse materials and recycling
policies. Industrialized countries generate 0.9–1.9 kg per capita of municipal solid waste
every day. Energy contents range from 4 to 13 MJ/kg and reach as high as 15.9 MJ/kg in
Canada and the United States (UNDP 2000).

In many developed countries special power plants burn municipal solid waste to produce
steam for electric generation and/or district heating. Municipal waste may also be processed
to produce methane and synthetic fuels.

Because disposal of municipal solid waste for landfill in densely populated urban areas is
increasingly constrained, energy conversion becomes more profitable. Figure 5.15 shows a
waste-burning CHP plant constructed on the Danube River in a highly populated district of
Vienna. The Spittelau waste incineration plant processes 250,000 tons of municipal waste
every year to generate 120,000 MW-hours of electricity and 500,000 MW-hours of district
heating.

Separating and recycling of non-combustible contents improves the energy output of
waste-to-energy power plants. Municipal solid waste incineration requires advanced pollution abatement equipment to prevent harmful emission of toxic and polluting gases.

Discarded animal parts from food industry, animal fat, and manure are animal-based
biomass resources. Dung, which is a traditional solid biofuel in rural parts of some poor

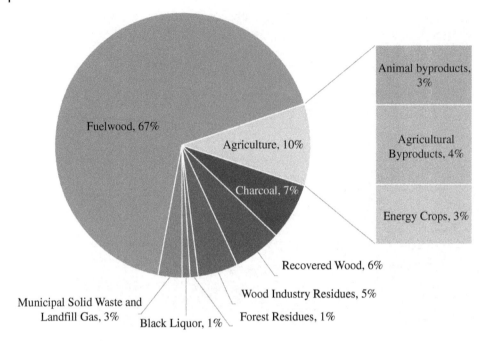

Figure 5.16 Share of various sources of biomass in the primary bioenergy mix. Source: IPCC (2011, p. 217, Chapter 2: Bioenergy).

countries, can be processed to obtain biogas, which is a valuable modern renewable. Whale oil used to be the main fuel for lighting before petroleum. All animal residues can be used to produce liquid or gas biofuels. Biogas extracted from sewage, part of the byproducts of the food industry, and most of municipal wastes are animal-based biomass. Figure 5.16 shows the share of biomass sources in the primary bioenergy mix (IEA 2009).

5.10.2 Energy Potential of Biomass Resources

Estimation of energy conversion potential of biomass resources is generally based on the size of land available for biomass plantation, the regional distribution of this land and distances to consumption centers, the productivity of the land for biomass production including the technical and economic performance of conversion technologies, and net energy balance.

The growth of biomass supply in industrialized countries is mainly based on the development of cost-effective biomass production, collection, and conversion systems to create biomass-derived fuels that can compete with fossil fuels in the energy supply mix with relatively less contribution to the climate change. Given the land area available for agriculture and advanced agricultural methods implemented in developed countries, the biomass production opportunities are generally considered substantial.

In developing countries, biomass production may conflict with land-use for food production. Moreover, low-tech use of traditional biomass is inefficient and may cause energy-related health problems. Exposure to particulate matters and carbon monoxide discharged from incomplete combustion of biomass in closed spaces can cause respiratory infections, problems in pregnancy, and even death. Large amounts of undocumented traditional biomass in poor areas are difficult to estimate.

Several researchers estimated biomass energy potential using integrated models. The projections of primary bioenergy supply in 2050 differ in a wide range from 40–50 to 500 EJ. The variation is mainly due to assumptions and difficulties in accounting traditional applications, which are usually not officially recorded. Regional geographic differences, support policies, and degree of agricultural development are also sources of uncertainties.

According to IEA energy statistics, TPES from bioenergy sources totaled between 48 and 65 EJ in 2008 (IEA 2010). The uncertainty stems from the insufficient data about the consumption of traditional biomass (wood, straw, dung, etc.) in less developed regions using low tech conversion methods. Between 37 and 43 EJ global primary energy from traditional biomass was converted with 10–20% efficiency yielding only 3–6 EJ/yr secondary energy.

Total primary energy of 11.3 EJ supplied from modern bioenergy (processed solid and liquid biomass, biogas) was converted to secondary sources with an average efficiency of 58%. The conversion efficiency was as high as 80% in heating applications, followed by 60% in the use of biofuels for transportation. Efficiency in electric generation is around 30% because of the lower thermodynamic efficiency of heat engines. Overall efficiency of traditional and modern biomass global consumption is from 21% to 27%.

Strapasson and others estimated total global bioenergy production including traditional biomass and residues as 60 EJ/yr in 2015 (Strapasson, Woods, Chum et al. 2017). Based on the integrated models, they project the global bioenergy production in 2050 to be between 70 and 360 EJ/yr, depending on different land-use prospects, food diet patterns, and climate change mitigation efforts. Data about the global potential of biomass energy is being continuously updated to reach a consensus on a reliable database (Rosillo-Calle 2016).

5.10.3 Bioenergy Conversion Technologies

Biomass resources are abundant in most parts of the world. The challenge is their sustainable management, efficient conversion, and delivery to the consumers in the form of modern and affordable energy services. Biomass resources can be converted to chemical fuels by several methods.

Feedstock for bioenergy production can be grouped as oil crops, sugar and starch crops, lignocellulosic biomass, biodegradable waste, and photosynthetic microorganisms. Throughout a series of conversion steps, raw biomass is transformed into a final secondary energy source such as heat, electricity, or biofuel. Because of the wide range of biomass feedstock and diversity of end-uses, many bioenergy process technologies have been developed.

The simplest way to use biomass is direct combustion to obtain heat. Forest wood, bushes, straw, leaves, dung, and other materials are being burned directly in many parts of the world. Advanced bioenergy routes necessitate some sort of screening, pretreatment, upgrading, and conversion steps. For example, firewood sold on the market is selected, cleaned, cut to a convenient size, perhaps treated, and packaged. Charcoal used for direct heating and cooking is basically wood processed to enhance the heating value.

Raw biomass has lower energy density compared to fossil fuels. Moreover, the moisture content may vary considerably. Transportation of biomass is, therefore, more expensive than fossil fuels that contain comparable amount of energy. Biomass must be preprocessed and treated to enhance the fuel quality.

In general, liquid and gas biofuels are cheaper to transport, store, and handle than solid biofuels. The goal of advanced processes is to convert raw biomass into high energy density synthetic fuels that can be substituted to petroleum products and natural gas. Modern bioenergy process chains can be grouped in three main categories identified as thermochemical, physicochemical, and biologic routes.

5.10.3.1 Thermochemical Conversion

Burning raw biomass is not energy-efficient and produces smoke that contains considerable amounts of particulate matter. In addition, its higher water content reduces the energy density, thus increasing transportation and handling costs. To be more marketable, solid biomass is dried, chopped into small pieces, and transformed into pellets by compression. Pelletized sawdust and wood shaving are byproducts of forest, lumber, and furniture industries. Wood pellets are common as heating fuel both in households and industry. Pellets are convenient for long distance transportation and even traded internationally. However, they tend to absorb moisture during transportation and storage.

In thermochemical conversion processes biomass undergoes chemical degradation under high temperature. Combustion, gasification, pyrolysis, and torrefaction are the main technologies used to upgrade the fuel properties of biomass. Essentially, they are all oxidation reactions occurring at different temperature ranges with different amounts of oxygen entering in the reaction.

Torrefaction transforms biomass into a dry product with high energy density that looks like coal. This upgraded product does not absorb water in transportation and storage. It can be also used to produce pellets, which further reduce transportation and handling costs.

Pyrolysis is controlled thermal decomposition of biomass at 500 °C in the absence of oxygen. Pyrolysis of biomass yields charcoal, liquid bio-oil, and a mixture of gas named syngas. Bio-oil can be further upgraded to be used as a transport fuel by hydrothermal upgrading using water and solvents at a temperature range between 300 and 400 °C.

5.10.3.2 Physicochemical Conversion

Physicochemical processes are mainly used for liquid biofuel production. Oil crops like rapeseed, soybean, and palm are first pressed to extract oil, then converted into liquid biofuel by a catalytic process known as transesterification. Fatty acid methyl/ethyl ester (FAME and FAEE) are commonly known as biodiesel; they have lower energy content than diesel fuel obtained from petroleum, and have blending limitations in some applications. Biodiesel can also be produced by hydrotreatment process of vegetable oils and animal fats. FAME and hydrotreated diesel are classified as first-generation biodiesel. Second-generation of biodiesel, also known as green diesel or synthetic diesel, is obtained by gasification of biomass (wood, straw, etc.) followed by Fischer Tropsch (FT) synthesis (IEA 2009).

5.10.3.3 Biological Conversion

Biological processes use living microorganisms such as enzymes or bacteria to degrade the feedstock into liquid or gaseous fuels. Sugar and starch-based crops (sugarcane, sugar beet, corn, soybean, etc.) are fermented to obtain ethanol or methanol. Lignocellulosic feedstock (grass, wood, bamboo, etc.) can be also fermented for alcohol production. In anaerobic digestion microorganisms break down a biodegradable material in the absence of oxygen. Anaerobic digestion is used to obtain biofuels mostly from wet biomass. Bio-photochemical processes are emerging for hydrogen production using algae involving the action of sunlight.

A detailed description of different conversion technologies particularly developed for biofuel production is available in the main report on bioenergy published by International Energy Agency (IEA 2009).

5.10.4 Strengths and Challenges of Bioenergy

Biomass is a renewable energy resource either naturally grown or produced as agricultural product. Biofuels are in general cleaner than fossil fuels, but more pollutant compared to other renewable energy sources.

Raw biomass needs additional processes to be converted into consumable bioenergy products that can compete in the energy market. Such techniques involve a complex chain of advanced processes. Despite methods developed to enhance the fuel quality, biofuels have lower energy density compared to fossil fuels. This makes transportation and handling more expensive than fossil fuels.

Unlike other renewables, biomass is not a free energy source. The cost of biomass production often represents from 50% to 90% of the total cost of final energy products delivered to consumers.

Biofuels can be easily stored, transported, and delivered to consumers like conventional fuels. In contrast to intermittent wind and solar energy, biofuels can produce controllable, continuous, reliable, and steady energy output, similar to fossil fuels. Liquid biofuels are potential alternatives to petroleum and can be used either alone or blended with oil products. However, energy density of biofuels by volume is lower than all fossil fuels.

Cultivation of biomass crops requires substantial dedicated farmland area. Land requirements depend on energy crop yields, water availability, and the efficiency of biomass conversion into usable fuels. Land use for bioenergy production has been a controversial issue. While unused agricultural land areas in many developed countries could become significant biomass production areas, long-term projections are alarming (UNDP 2000).

Biomass production requires substantial amounts of water for irrigation. In addition, liquid biofuel and biogas production processes also use water. Water requirements for biomass depend on the climate conditions. It is estimated that in the US, ethanol production consumes 40–140 units of water per unit volume of finished product. Evidently, irrigation water returns to ground even though some part evaporates, and the total water on earth remain the same. However, as biomass production increases, in some regions the limited freshwater resources will have to be shared between agriculture of food products and biomass farming. IPCC studies indicate that even without considering water requirements for biomass production, water shortages are possible for about half the world's population as early as 2025. Thus, the water constraint for extended biomass production will likely be of importance, especially in the long term (UNDP 2000).

Large-scale biomass production may have negative impacts on soil fertility, biodiversity, and landscape. While use of biofuels instead of fossil fuels eliminate some part of the air pollution, emissions caused by farming equipment and transportation vehicles offset part of this benefit. On the other hand, biomass cultivation can protect land from erosion.

Farming machines and transportation from farmlands to processing plants consume energy, mostly oil products. Energy balance in biofuel production is described by the ratio of energy return on energy invested (EROI). Solid biomass used directly for heat produces 10–30 times greater energy output than the energy used to produce them, hence EROI for solid biofuels ranges from 10 to 30. Energy balance is not that favorable for liquid biofuels; some estimates suggest EROI as low as 1.3 for corn ethanol produced in the US. Nevertheless, ethanol production from sugarcane has higher EROI because of the higher energy content of the initial biomass. Advancements of biofuel process technologies tend to reduce the energy used to produce liquid biofuels.

Biofuel production relies on harvesting methods and process technologies. Cost of energy used for cultivation and processing impact the competitiveness of modern biofuels. Hence, fluctuations of oil prices, market conditions, and political decisions are major factors that affect the resource potential of biofuels in long-term.

5.11   Future Trend of Renewable Energy Development

The share of renewable sources in the energy mix depends on technologic innovations, climate change, and scenarios regarding social, economic, and political factors. In general, renewable energy development needs substantial capital investment and use of advanced technologies. Although most renewable energy sources are free, because of the initial investment, the cost of energy delivered to the end users is higher compared to the energy converted by burning non-renewable sources. Hence, development of facilities using renewable sources often need support of governments in the form of policies, standards, and financial incentives.

Climate change is the main motivation for replacing coal-burning power plants with cleaner energy sources. The International Energy Agency (IEA) has developed a comprehensive model to forecast long-term changes in the global energy mix. Figure 5.17 shows the projections of the use of renewables in electric generation and heat production based on three scenarios considered in the IEA's World Energy Outlook 2016 (IEA 2016). The current policies scenario assumes no change in policies and measures governments have developed before 2016. The new policies scenario considers the possible impacts of the consensus reached by 196 countries in the 21st United Nations Framework on Climate Change (UNFCC) convention known as COP21. The third scenario labeled as "450 scenario" incorporates measures aiming to limit the concentration of greenhouse gases in the atmosphere to about 450 ppm of carbon dioxide equivalent. Based on the research surrounding climate change mitigation, the 450 Scenario is now expressed as realizing a 50% chance of limiting warming to a 2 °C temperature rise above pre-industrial level by 2100.

In all scenarios, the share of renewable sources in the global energy supply is expected to increase, but at different rates depending on the implementation of policies. With existing

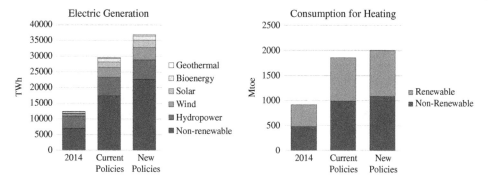

Figure 5.17 Projection of global total primary energy demand (TPED). Source: IEA World Energy Outlook 2016 (IEA 2016, p. 412, Table 10.1).

policies remaining in place, the share of renewables in the total primary energy demand (TPED) will increase from 8% to 13% of the TPED by 2040. Based on the "New Policies Scenario" of the IEA, the global share of renewable energy in the world's TPED will rise to 16% in the same period. To limit the global temperature rise and meet the air quality and universal access to modern energy goals set forth in UNFCC (Sustainable Development Scenario), the share of renewable energy must be at least 30% (IEA 2018c).

While energy consumption for heating will grow considerably, the proportions of conventional and renewable heat sources do not change much with the considered scenarios. Electric generation, however, grows in long-term projections with increasing shares of renewables. In addition to the energy-related climate policies, the cost of conversion from renewables to electricity is decreasing. While hydroelectric generation remains the largest source generation from all renewables grows, with solar PV and wind powered generation increasing most rapidly.

Commercial scale onshore and offshore wind power is the fastest growing type of renewable source. Continuously decreasing unit costs of PV modules stimulates the growth of residential scale electric generation. Advanced technologies that reduce the battery cost per unit energy stored help resolving the grid penetration issues related to intermittent renewable sources.

5.12 Chapter Review

Renewable energy is a primary energy source that replenishes forever by natural processes and is not depleted by consumption. Sunlight, wind, water, naturally heated liquids extracted from earth, photosynthesis, and biosynthesis are major forms of renewable energy that can be converted into heat, mechanical work, electricity, or fuels. All renewable sources, except geothermal energy are direct or indirect consequences of solar energy reaching the Earth.

Technical potential of renewable resources exceeds the current and projected TPES of the world. Renewable sources are available everywhere on the Earth but commercial scale conversion to usable forms depends on the quantity that can be extracted, time variations, energy flow rate, accessibility, and many socioeconomic factors. The main problem about renewables is their cost-effective conversion and integration in the existing energy system, rather than their available quantity.

Non-combustible sources include hydro, wind, solar, geothermal, and ocean energy. These sources are mostly delivered to end-users in the form of electric power. Solar and geothermal energy can be also converted directly into heat. CHP generation facilities can deliver excess heat in the form of steam for industrial processes or district heating. Combustible biomass can be used for direct heating or electric generation. Conversion to direct heat is more efficient than electric generation but transmission of heat to end users is limited. Electricity is a more convenient energy carrier and offers great practical flexibility to consumers.

Capacity factor is the ratio of the total electric energy generated by a facility over a time interval, and the total energy that the facility could generate if it were operated at full capacity over the considered interval, typically a year. Capacity factor of large conventional hydroelectric, geothermal, and biomass-fired power plants is higher than the intermittent wind, solar, and small hydropower facilities.

Solar power received per unit area is called irradiance and measured in W/m^2. Irradiance has two components named DNI and DHI. GHI is the total radiant power received on a unit horizontal surface. Solar irradiance at a location depends on the latitude, altitude, and clarity of the air. Solar irradiance changes rapidly with the passage of clouds. Solar energy captured over a certain interval on a unit's horizontal surface is defined as solar irradiation or insolation. Insolation changes throughout the year with the length of the daylight time.

Solar energy can be directly converted into electricity using PV modules or into heat using solar collectors. PV modules are rated for standard test conditions (STC), defined as 1-kW/m^2 irradiance and 25 °C ambient temperature. Although PV cell efficiency has significantly increased by advanced technologies, commercial modules have a maximum efficiency in the order of 20%. Integration of large-scale PV generation in an electric grid presents penetration issues due to the rapid changes of generated power throughout the day and unavailability at night. Solar collectors convert solar energy into heat, mostly used for space or water heating. The efficiency of direct thermal conversion is higher than electric conversion. CSP plants produce steam, which is used for electric generation.

Hydropower relies on the water cycle. Conventional hydroelectric power plants have the biggest share in renewable electric generation. Large dams allow huge energy storage and provide multiple uses including agricultural irrigation, river flow regulation, flood prevention, and development of recreational sites. Major socioeconomic impacts of large hydroelectric power plants are the relocation of local communities and flooding of cultural treasures. Run-of-river facilities can be constructed in a broad range of power capacities from kilowatts to gigawatts. They do not require a large reservoir and can be constructed in a shorter time with smaller capital investment, but their output power shows daily or seasonal variations. Run-of-river power plants may have adverse impacts on fish migration, river transportation, and amount of water needed for agricultural irrigation.

Wind results from uneven heating at different locations. Wind turbines convert the kinetic energy of moving air into electricity. The power output of a wind turbine is proportional to the cube of the wind speed, hence it changes rapidly with the wind speed variations. Integration of large-scale electric generation from wind presents similar penetration issues as PV generation because of the fast and often unpredictable variations of the generated power.

Geothermal energy is based on the ground fluids heated by magma or decay of radioactive minerals in the Earth. Geothermal energy can be used for direct heating applications or electric generation. Geothermal power plants convert heat from the extracted fluid into mechanical work through ORC for electric generation. Efficiency of geothermal electric generation is limited with the low thermodynamic efficiency of ORC. Although conversion of geothermal energy has many environmental impacts due to the carbon dioxide and toxic gas emissions, it is cleaner than fossil fuel burning power plants.

Biomass and hydrocarbon-based waste are combustible renewable sources that are used to generate electricity, produce heat, or both. Large amounts of biomass are directly burned in less-developed regions for conventional heating and cooking. Modern biomass is processed to produce solid or liquid fuels or biogas.

Long-term projections based on several scenarios show that fossil fuels will continue to dominate primary energy supply, and the share of renewables will significantly increase.

5.13 Review Quiz

1 Which one of the following statements is not true for renewable energy?
  a. Renewable energy sources are available in limited quantities.
  b. Renewable energy is available at every location in the world in various amounts.
  c. Most renewable sources are delivered to consumers in the form of electricity or heat.
  d. Intermittent nature of wind and solar energy presents challenges in their integration into conventional electric grid.

2 Which renewable source is dominant in electric generation?
  a. Hydropower
  b. Biofuels
  c. Solar power
  d. Wind power

3 Which renewable energy technology is dominant in direct heat production worldwide?
  a. Solar thermal collectors
  b. Photovoltaic cells
  c. Geothermal heat pumps
  d. Biomass combustion

4 Which renewable energy source below is more dominant in the transportation sector?
  a. Biofuels
  b. Biomass

    c. Waste

    d. Hydroelectric

5 Which renewable energy technology has the biggest share in the global thermal capacity?

    a. Modern bio-heat

    b. Solar water heating

    c. Geothermal heat

    d. Ocean thermal energy

6 Which region below generates the greatest amount of electric energy from hydropower?

    a. North America

    b. South America

    c. Asia

    d. Europe

7 Which region below generates the greatest amount of electric energy from wind power?

    a. North America

    b. South America

    c. Asia

    d. Europe

8 Renewable energy generation from which renewable source below has had the highest increase rate since year 2000?

    a. Biomass

    b. Wind

    c. Solar

    d. Hydro

9 A higher capacity factor indicates that

    a. maximum energy that the installed capacity can generate is larger.

    b. the generation facility has larger installed capacity.

    c. generated energy is closer to the maximum energy the installed capacity can generate in a certain time interval.

    d. the cost of the generated energy per installed capacity is smaller.

10 Which electric generation plants operate at the highest capacity factor worldwide?

    a. Hydroelectric

    b. Geothermal

    c. Wind farms

    d. Bioenergy

Answers: 1-a, 2-a, 3-d, 4-a, 5-b, 6-c, 7-d, 8-b, 9-c, 10-b.

Research Topics and Problems

Research and Discussion Topics

1 What are the social, economic, and environmental impacts of building a large dam?

2 What are the major uses of a hydraulic dam?

3 Why is the efficiency of a hydroelectric power plant higher than a fossil fuel fired power plant?

4 What are the major reasons for public reactions to the construction of wind farms?

5 Compare the benefits and drawbacks of commercial-size photovoltaic solar farms and concentrated solar power plants.

6 Why can't electric power generation in a country solely depend on renewable energy?

7 What are the major challenges of large-scale electric generation using wind and solar energy?

8 Discuss the environmental impacts of biomass production.

9 What are the technical limitations of using geothermal energy in large-scale electric generation?

10 What are the main factors that affect the long-term growth of renewable energy development?

11 The state of Hawaii consists of islands and imports most of its primary energy supply either from mainland US or other countries in the form of fossil fuels. Research the renewable energy options locally available in Hawaii and write a proposal to supply the primary energy need completely from renewables.

12 Research statistical databases to estimate the solar energy potential in Death Valley, Nevada. How large an area would be needed to supply all primary energy needs in the state from solar energy? What are the possible challenges of such a project?

13 What measures can be taken to overcome the issues associated with the integration of large wind and solar power generation plants?

14 What are the positive and negative impacts of residential size PV and wind generation on the electric grid?

Problems

1 The Bonneville hydroelectric power plant generated 4.86 TWh electric energy in 2018. Using data given in Table 5.3 estimate its capacity factor.

2 A wind farm with installed capacity of 50-MW generates 265 GWh energy in one year. What is the capacity factor?

3 Determine the land area needed to develop a wind farm of 50-MW capacity using 2-MW wind turbines with 60 m blade length. Assume that sufficient wind resource is available to operate the turbines at rated power and the turbines must be spaced by seven times the blade length to avoid wake and turbulence.

4 The first offshore wind farm of the United States was installed near Rock Island in the state of Rhode Island. The installed capacity of the wind farm is 30 MW. How much energy will this unit generate in one year if it operates at a capacity factor of 60%.

5 A house has a south-facing roof area of 60-m^2 with 30° angle in respect to the horizontal plane. The daily global horizontal irradiation (GHI) averaged over one year is 4500-Wh/m^2 at the location. Estimate the yearly solar energy potential for this house.

6 The owner of the house described in Problem 5 would like to install a solar PV array on the roof. The rated power of the selected PV module is 250 W fort 1 kW/m^2 irradiance, and the dimensions are 40 cm width and 64 cm height. How much electric energy would this array generate in one year?

7 The heating values of pure gasoline and pure ethanol are 32.0 and 21.1 MJ/l respectively. How much heat per liter would produce combustion of (a) pure gasoline, (b) a blend of 90% gasoline and 10% ethanol, and (c) blend of 15% gasoline and 85% ethanol?

8 Prices on various gasoline products seen in the pump picture in Figure 5.14 are listed below:
   - Premium, unleaded gasoline, without ethanol: 2.999 $/gallon
   - Unleaded gasoline and 10% ethanol: 2.549 $/gallon
   - E-15: 2.4999 $/gallon
   - E-85: 2.399 $/gallon

   Using the heating values of Problem 7, calculate the unit price in $/MJ for gasoline without ethanol and the blends of gasoline containing 10%, 15%, and 85% ethanol.

Recommended Web Sites

- American Solar Energy Society (ASES): https://www.ases.org
- American Wind Power Association (AWEA): https://www.hydropower.org
- International Geothermal Association (IGA): https://www.geothermal-energy.org
- International Hydropower Association (IHP): https://www.hydropower.org

- International Renewable Energy Agency (IRENA): https://www.irena.org
- IRENA Global Atlas for Renewable Energy: https://irena.masdar.ac.ae/gallery/#gallery
- National Renewable Energy Laboratory (NREL): https://www.nrel.gov
- Renewable Energy Policy Network for the 21st Century (REN21): http://www.ren21.net
- The Intergovernmental Panel on Climate Change (IPCC): https://www.ipcc.ch
- US Energy Mapping System: https://www.eia.gov/state/maps.php
- US Environmental Protection Agency (EPA): https://www.epa.gov/energy
- World Energy Council (WEC): https://www.worldenergy.org

References

Baerwald, E. F. and D'Amours, G. H., (2008). Why Wind Turbines Can Mean Death For Bats. *ScienceDaily*, 26 August.

BP (2018). *BP Statistical Review of World Energy 2018*. London, UK: BP.

DTU, (2018). *Global Wind Atlas*. [Online] Available at: https://globalwindatlas.info [Accessed 13 June 2019].

EIA, (2019a). *Preliminary Monthly Electric Generator Inventory*. [Online] Available at: https://www.eia.gov/electricity/data/eia860m [Accessed 7 July 2019].

EIA, (2019b). *US Energy Mapping System*. [Online] Available at: https://www.eia.gov/state/maps.php [Accessed 7 May 2019].

Goldstein, B., Hiriart, G., Bertani, R. et al. (2011). *Geothermal Energy. In IPCC Special Report on Renewable Energy Sources and Climate Change Mitigation*. Cambridge and New York: Cambridge University Press.

Grubb, M.J. and Meyer, N.I. (1993). Wind Energy: Resources, Systems and Regional Strategies. In: *Renewable Energy: Sources for Fuels and Electricity* (eds. T.B. Johansson, H. Kelly, A.K.N. Reddy and R.H. Williams), 157–212. Washington DC: Island Press.

IEA (2009). *Bioenergy – A Sustainable and Reliable Energy Source, Main Report*. Paris: International Energy Agency.

IEA (2010). *Sustainable Production of Second Generation Biofuels: Potential and Perspectives in Major Economies and Developing Countries*. Paris: International Energy Agency.

IEA (2016). *World Energy Outlook 2016*. Paris: International Energy Agency.

IEA (2018a). *Key World Energy Statistics*. Paris: International Energy Agency.

IEA (2018b). *Renewables Information: An Overview*. Paris: International Energy Agency.

IEA (2018c). *World Energy Outlook 2018*. Paris: International Energy Agency.

IPCC (2011). *Renewable Energy Sources and Climate Change Mitigation – Special Report of the Intergovernmental Panel on Climate Change (IPCC)*. New York: Cambridge University Press.

IRENA (2019). *Renewable Capacity Statistics 2019*. Abu Dhabi: International Renewable Energy Agency (IRENA).

Lindal, G. (1973). *Industrial and Other Applications of Geothermal Energy*. Paris: Unesco.

Marts, M. E., (2019). *Encyclopedia Britannica*. [Online] Available at: https://www.britannica.com/place/Columbia-River [Accessed 8 July 2019].

Palmerini, C.G. (1993). Geothermal energy. In: *Renewable Energy: Sources for Fuels and Electricity* (eds. T.B. Johansson, H. Kelly, A.K. Reddy, and and R.H. Williams), 551–589. Washington DC: Island Press.

REN21, (2019). *Global Status Report 2019*. Paris: Renewable Energy Policy Network for 21st Century.

Rosillo-Calle, F. (2016). A review of biomass energy - Shortcomings. *Journal of Chemical Technology Biotechnology, Issue* 91: 1933–1945.

Smith, C. J., (2005). Winds of change: Issues in utility wind integration. *IEEE Power and Energy Magazine*, November/December, pp. 20–25.

Solargis, (2017). *Solar resource maps of World*. [Online] Available at: https://solargis.com/maps-and-gis-data/download/world [Accessed 16 May 2019].

Strapasson, A., Woods, G., Chum, H., Kalas, N., Shah, N., and Rosillo-Calle, F., (2017). On the global limits of bioenergy and land use for climate change mitigation. *Global Change Biology - Bioenergy (Open Access),* Vol. 9 (12), pp. 1721–1735.

Tiwari, G.N. and Ghosal, M.K. (2005). *Renewable Energy Resources*. Harrow, Middlesex, UK: Alpha Science International Ltd.

US Bureau of Reclamation, (2016). *Hydropower Program*. [Online] Available at: https://www.usbr.gov/power/edu/history.html [Accessed 17 July 2019].

Markvart, T. (2000). *Solar Electricity*, 2e. Chichester (West Sussex), UK: John Wiley & Sons.

UNDP (2000). Chapter 5: energy resources. In: *World Energy Assessment:Energy and the Challenge of Sustainability*, 136–171. New York, NY: United Nations Development Programme (UNDP), Bureau for Development Policy.

6

Electric Energy Systems

Synchronous generators at Bonneville Dam and Power Plant on the Columbia River, 40 mi east of Portland, Oregon. Each generator in the picture is rated 60,000-kW generation capacity with output voltage of 13.8 kV. Total generation capacity of the hydroelectric power plant is 1227 MW. Generated electricity is transmitted to the grid via 115 and 230 kV overhead lines. The construction of Bonneville power plant was completed in 1938, and the first powerhouse was upgraded in 1942 and 1943. The second powerhouse was added between 1972 and 1982.
Source: US Army Corps of Engineers®; © H&O Soysal

Energy for Sustainable Society: From Resources to Users, First Edition. Oguz A. Soysal and Hilkat S. Soysal.
© 2020 John Wiley & Sons Ltd. Published 2020 by John Wiley & Sons Ltd.

6.1 Introduction

Electricity is a convenient energy carrier and the most flexible secondary source. As we indicated in Chapter 1, nearly one-fifth of the world's energy is supplied in the form of electricity. From the energy system perspective, electric systems transmit sources that cannot be delivered directly to consumers such as nuclear energy and most renewables. All forms of energy can be converted to electricity by using a proper conversion system and electricity can be converted to virtually all other forms of energy. Electric energy can be conveniently transmitted over long distances. At the user end, equipment driven by electric machines are cleaner, quieter, more efficient, and more controllable than their engine-driven counterparts. Moreover, electronic devices achieve many functions such as telecommunication, computation, data processing, which can be powered only by electric energy.

The major challenge of electric energy is that it has to be generated in real-time as much as it is needed. Direct storage of large quantities of electric energy in batteries and capacitors is expensive. Electricity is generally stored in other forms of energy using bidirectional conversion systems.

An electric energy system consists of three principal parts; generation units, transmission lines, and a distribution grid. Electric generation stations convert various primary energy sources into electric power. Worldwide, the biggest fraction of electric power is generated from thermal energy using turbine-generator units. Transmission lines interconnect generation stations and convey electric energy to distribution systems. A distribution grid delivers electric power to consumers.

The first section below summarizes the evolution of electric energy systems. Then we briefly review the fundamental principles necessary to understand the presented topics. The subsequent sections present the structure, operation principles, and characteristics of essential components of an electric energy system.

6.2 Evolution of Electric Power Systems

Around 600 BC Greek philosopher-mathematician Thales of Miletus observed that a piece of amber would attract dust and small particles when it was rubbed on fur or cloth. Ancient Greeks called this phenomenon *electron*, meaning *amber* in ancient Greek language. This was perhaps the first scientific experiment with static electricity, which was not completely understood for about 2,500 years.

In the eighteenth century, only a few scientists were interested in the study of electrical and magnetic phenomena. Electrostatic generators were developed to produce electric voltage by friction. Static electricity, essentially used for experiments, could be stored in Leyden jars. Charles-Augustine de Coulomb was a pioneer in the development of electrostatic theory. Among several scientists, Benjamin Franklin did experiments on high voltage using Leyden jars stacked to produce higher voltage, which he called a "battery" in analogy to the military canons. Franklin's famous experiment led him to the invention of a lightning rod in 1749. Although frictional generators could produce high voltage, the current was too small to supply power for practical applications. No substantial electric current would be produced until Luigi Galvani and Alessandro Volta developed the first battery cells (voltaic piles) in Italy around 1800.

Voltaic cells were mainly used for primitive telegraph communication. In 1807, Sir Humphrey Davy made the first arc lamp using 2000 voltaic cells and two charcoal electrodes spaced 100 mm from each other. Arc lamps produced abundant light, but their use was limited due to the considerable electric power needed.

As new applications to be powered by electricity were developed, battery cells with higher energy density than voltaic piles became necessary. A more compact battery cell known as the Daniell cell, named after its inventor British chemist John Frederic Daniell, powered telegraph networks and arc lamps. The first batteries were bulky and unsafe since they contained corrosive compounds and sulfuric acid in fragile containers. In 1866 Carl Gassner patented a dry cell in Germany, an improved variation of the wet Leclanche cell.

Effects of electric current fascinated many scientists and inventors. In the first half of the nineteenth century, French physicist Andre Marie Ampere and German mathematician Karl Gauss established the mathematical foundations of electrostatic and electromagnetic field theories. Michael Faraday in England and Joseph Henry in the US independently discovered in 1831 that an electromotive force (EMF) was induced on an electric wire that encircles a time-varying magnetic flux. Discovery of electromagnetic induction phenomenon was a breakthrough leading to the development of electromechanical devices including all kinds of generators, transformers, and electric motors. Michael Faraday demonstrated to the public the first electric generator known as the Faraday disk.

Faraday's discoveries and experimental demonstrations inspired several inventors to develop rotational devices to produce electricity using permanent magnets. The first generators, called dynamo, were designed to produce DC since the focus was on replacing electrochemical batteries by more convenient and robust electromechanical converters. A commutator was necessary to allow unidirectional current flow between rotating coils to a stationary external circuit. A commutator is a cylindrical piece that consists of segments insulated from each other turning with a generator shaft. Current flows through stationary brushes contacting the surface of the commutator segments rotating with the shaft. The commutator-brush assembly reverses the coil terminals every time the coil axis passes the neutral region between magnetic poles, allowing unidirectional current flow.

While the first dynamo designs were ingenious, they could not generate sufficient power to supply several arc lamps and they were inefficient. After several different designs were developed in the 30 years after Faraday's discovery, three inventors independently built the first generators that could serve in industrial applications. British engineer Samuel Alfred Varley patented his invention on December 24, 1866; Ernst Werner von Siemens and Sir Charles Wheatstone announced their discoveries on January 17, 1867 in Germany and England, respectively. The main improvement over the previous designs was the use of electromagnets to produce the magnetic field instead of permanent magnets. The new design could generate more power and create the necessary magnetic flux by self-excitation (Jarvis 1955). Note that in the quest for more efficient ways to produce electricity, dynamo and dry battery cells were invented at about the same time.

6.2.1 Early Electrification Systems

Electric power can be produced either by solid-state or electromechanical conversion systems. Solid-state converters include batteries, fuel cells, and photovoltaic (PV) cells. The early solid-state converters were batteries. Electromechanical systems consist of a prime mover and an electric generator. Initially, reciprocating steam engines powered by burning coal dominated electric generation, although hydraulic turbines were also used at the first hydroelectric facilities. Later, steam and gas turbines replaced reciprocating engines. Today hydraulic turbines and wind turbines are also common as prime movers. Internal combustion engines are only used in small backup generation units. A generator converts the mechanical power produced by the prime mover into electric power. The prime mover is direct-coupled with the generator. The speed versus torque characteristics of the prime mover determine the mechanical structure of the generator. The electrical part of a generator is designed based on the voltage requirements of the transmission system. First generators were producing direct current, modern generators deliver three-phase alternating current (AC). Figure 6.1 shows a typical electromechanical generation unit.

In first generation systems, DC power generated by a dynamo was directly supplying the electric loads mainly dominated by arc lamps. Thomas Edison's invention of the phonograph (1877); development of a more efficient DC generator, which can be operated as a motor by reverse action (1879); and development of the commercial incandescent lightbulb (1880) widely popularized electric power. Edison had already established the "Edison Electric Light Company" with funding provided by J.P. Morgan and the Vanderbilt family (Josephson and Conot 2019). Electric-powered lightbulb systems were first installed separately in isolated buildings. Incandescent lightbulbs could be made with sufficiently high resistance such that they could be connected in parallel, as opposed to the series connected arc lamps. Parallel connection increased the reliability since failure of one of the lamps would not affect the others, whereas failure of one of the series connected arc lamps would turn off the whole circuit. The first electric generation system in the United States began operation in 1882 in Pearl Street Station in New York City. The 30-kW system supplied a load of Edison's incandescent lightbulbs over a 220/110 V three-wire DC line (Gross 1979). The early electric companies were known as "illuminating companies" since their only service was lighting. The major problem with lighting load is that the generators would start at dark when lights were on and become idle or underutilized during the day. DC motors that were already developed before the centralized generation systems began to replace steam engines to power industrial equipment because they were more convenient, cleaner, and smaller than comparable powered steam engines.

Independently operated local facilities were supplying local consumers at the same voltage as the generator outputs; thus, the output voltage of generators was determined by the voltage of commercial lightbulbs. However, increasing power resulted in increased currents

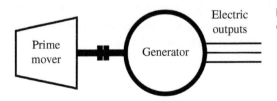

Figure 6.1 Schematic diagram of an electromechanical generation system.

which caused unacceptable voltage drops and fluctuations if the loads were located at a considerable distance from the generation facility. This issue forced the building of generation facilities closer to the loads, which became increasingly challenging because of the availability of convenient sites.

When electricity is conducted from one point to another, heat loss is proportional to the resistance of the wire and square of the current due to Joules heat effect $P = RI^2$. Obviously, as the transmission distances and the supplied power increased, losses also increased significantly. Since electric power is the product of voltage and current, increasing the voltage would clearly decrease the current at the same power. Transmission losses could be reduced by increasing the conductor diameter to decrease its resistance, but this would result in an enormous cost increase. The other option is to reduce the current by increasing the voltage level, but supplying high voltage directly to consumers is unpractical and unsafe. In addition, DC machines designed to operate at high voltage become too expensive because of the insulation cost. Development of AC generators and transformer technology in the late 1800s brought the idea of using AC for electric generation, transmission, and distribution to solve these practical and economic problems.

Serbian engineer Nikola Tesla had immigrated to the US and started to work for Thomas Edison's company in 1884. He was a strong supporter of AC generation and was working on developing induction machines. Edison was promoting DC power systems to protect the royalties from his patents. After working for about one year for Edison to improve the efficiency of DC machines, Tesla quit his job to continue his work independently.

Industrialist George Westinghouse already purchased the American patents of the AC transmission system developed by French inventors L. Guilard and G.D. Gibbs. In the winter of 1885–1886, an associate of Westinghouse installed an experimental AC distribution system which supplied 150 lamps in Great Barrington, Massachusetts. In 1890 the Westinghouse company installed a 13-mi (21-km) AC transmission line in the United States in Oregon, from Willamette Falls to Portland, to transmit electric power generated in a hydroelectric facility (Stevenson 1982). This transmission system consisted of a single-phase AC generator, a step-up transformer, a two-wire higher voltage transmission line, and a step-down transformer to supply low voltage lighting load and some single-phase motors.

On May 16, 1888, Tesla presented a paper describing two-phase induction and synchronous machines. In a polyphase AC system, power is transmitted over as many wires as the number of phases. All lines carry an AC voltage with the same magnitude and frequency, but each phase voltage and current are shifted by a time interval equal to the ratio of the period and the number of phases. One of the benefits of polyphase is that a return wire is not required since the sum of all sinusoidal currents is zero if the phases are equally loaded. In addition, a number of coils equal to the number of phases placed with equal angles around a circle supplied by a polyphase electric system produce a rotating magnetic field, which simplifies the structure of electric motors by removing the need for a commutator.

The advantages of polyphase AC systems were apparent and AC generation and transmission systems started to replace DC. Threatened by the rise of AC, Edison launched a campaign to discredit AC, which led to a bitter dispute known as "war of the currents." Edison tried to convince the public that AC was more dangerous than DC by electrocuting animals in public demonstrations and recommending New York State carry out death sentence execution by AC powered electric chair (Lantero 2014).

George Westinghouse bought several of Tesla's patents and they started a productive partnership. The Westinghouse company won the bid for electrification of the 1893 World's Columbian Exposition in Chicago. At the expo, Tesla demonstrated to the public a two-phase electric motor. The success at the exposition led Westinghouse to winning the bid at the end of that year to build an AC power station at Niagara Falls.

During this time, AC power generation and transmission gradually replaced DC systems; the war of the currents ended when Edison lost the control on his own company. Further contracts in the Niagara Falls hydroelectric project were awarded to General Electric to build AC transmission lines and transformers carrying electric power from Niagara Falls to industrial areas of Buffalo and highly populated New York City.

6.2.2 Development of Transmission Options for Growing Needs

Today, AC dominates the electric power generation, transmission, and distribution systems. A few exceptions are power generation to supply railroads and electrolysis-based electrochemical industries. Solar PV arrays produce DC, but for connection to the utility grid, their output is converted to AC through electronic inverters. At the user end, computers, LEDs, consumer electronics, and electric vehicles are supplied by DC power. With the advent of power electronics technology inverters, rectifiers, power supplies and other electronic devices have allowed cost-effective and efficient conversion between AC and DC at the point of use.

AC offers the advantage of using different voltage levels throughout the electric energy system for optimal generation, transmission, and distribution. Moreover, three-phase systems reduce the cost of generation facilities and transmission networks. In January 1894 there were five polyphase generation plants in the US, one of them two-phase and the four others were three-phase (Stevenson 1982). The first AC transmission lines were 10 kV. Starting from the beginning of the twentieth century, the power capacity and voltage of transmission lines have increased steadily. In the early 1930s, 500 kV overhead lines were constructed to carry electric power generated at the Hoover Dam from Nevada to southern California. In Canada, 735-kV lines were built in 1969 to connect remote hydroelectric facilities to Montreal. The first 765-kV line in the US was energized in 1970.

At the beginning of electrification, independently operated utilities selected arbitrary power frequencies between 25 and 140 Hz depending on the design and operational characteristics of their prime movers and generators. For example, the first generators at the Niagara Falls hydroelectric project built by the Westinghouse Company were 25 Hz to match the preselected turbine speed of 250 rpm. Power frequencies around 40 Hz were used in Europe, England, some parts of North America, and Australia. During the transition from DC to AC, electric utilities faced the practical challenges introduced by the selection of supply frequency.

Lower frequencies cause a disturbing flickering effect in lighting equipment, which were the essential electric loads in early electric systems. Increasing the supply frequency allows more uniform and comfortable lighting. In addition, AC motors can operate at higher speeds, which is desirable for many practical applications such as fans and water pumps. Another benefit of higher frequency is a lower manufacturing cost of transformers and AC motors. On the other hand, higher frequencies have adverse effects on the performance of longer AC transmission lines and cables due to the increased capacitive currents.

The Westinghouse Electric Company, using Tesla's patents to develop the first AC power systems in the US, standardized the power frequency at 60 Hz, considering the performance of the arc and incandescent lighting equipment used at the time and performance of induction motors and transformers. In Europe, the standard frequency of 50 Hz was selected. Standardization of the power frequency occurred in the first few decades of the twentieth century. The German Verband der Elektrotechnik (VDE), recommended 25 and 50 Hz as standard frequencies for electrical machines and transformers in 1902, but dropped the 25 Hz in 1914. The United Kingdom selected 50 Hz as early as 1904. In the United States, 60 Hz became standard frequency, except much of the southern California electric system operated at 50 Hz until 1948. In countries where electric equipment was imported from both Europe and the US, both 50 and 60 Hz were used in different electric grids.

Commercial generators used in electric power plants are typically designed to produce an output voltage between 11 and 25 kV; there is no standard generator voltage. A step-up transformer near the generators increases the voltage to a higher level, typically 33 kV. A substation collects power generated at several generators and increases the voltage to the 66–132 kV range. Depending on the length of transmission lines, the voltage may be further increased at substations.

After electric power is transferred from remote generation facilities to a primary distribution substation, the voltage is first reduced to a range of 34.5–138 kV. Secondary distribution substations further step down the voltage level to the 4–34.5 kV range, while 11–15 kV is more common in distribution systems. Some industrial and large commercial consumers are supplied at medium voltage levels depending on their needs. Such users adjust the voltage level according to their end uses and distribute by internal wiring. Individual commercial and residential buildings are supplied at a safe, standard low voltage level.

Although AC became the principal mode of electric power transmission at the beginning of twentieth century, the growing need to transmit larger amounts of electric power to longer distances introduced the option of DC transmission at high voltage. High voltage direct current (HVDC) transmission systems evolved after the 1950s. The first commercial HVDC line started operation in 1954 in Sweden. HVDC found wider applications when ratings of mercury arc converters approached 100-kV and 1000-A levels. Until 1972, mercury arc valves were used exclusively to perform AC–DC and DC–AC conversion. After high voltage and high current electronic switching elements became available, the total DC power transmission grew more rapidly. Another breakthrough that supported HVDC systems was the development of high voltage switchgear that use sulfur hexafluoride (SF_6) as insulating gas (Ellert et al. 1982).

In HVDC transmission, the generator voltage is first stepped up to the desired system voltage by a transformer. At the high voltage side of the transformer an electronic converter rectifies AC power to DC, which is transmitted a long distance by bipolar DC overhead lines (underwater DC transmission uses cables). At the receiving end, an electronic converter inverts DC back to AC, which is conveniently stepped down to lower voltage level for distribution. The converter stations consist of several units connected in series to operate at the selected line voltage.

HVDC systems can control the magnitude and direction of the power-flow by means of a sophisticated control system that varies the triggering sequence of switching elements of the converters. The choice of AC or DC power transmission is principally influenced by

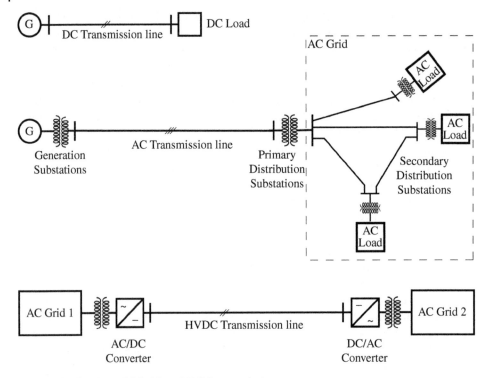

Figure 6.2 Outlines of DC, AC, and HVDC transmission systems.

the cost of the converters and other terminal equipment. The potential areas where HVDC transmission may be preferred over AC are:

- long distance overhead transmission;
- powering highly populated urban areas by underground cables;
- connection of grids that operate at different frequencies;
- underwater cable connection between islands and mainland.

Figure 6.2 shows simplified diagrams of DC, AC, and HVDC transmission systems. In early DC supply systems, the generator was connected by a two-wire line to the DC loads. Later, three-wire DC systems with two live and one ground wire were developed. Initial AC systems were single-phase with two-wire connections. In a three-phase system, generally three phase wires and one neutral is needed to complete the current path. If three phase lines are at the same voltage in respect to the ground (or each other) and carry the same current, the total current is zero at any time and the neutral is not needed. In long distance overhead lines, three wires are therefore sufficient, saving significant cost.

6.2.3 Interconnected Grid

Initially, utility companies were generating electricity in independently operating generation facilities and most of the load was lightbulbs. Capacity factors of the generation units was low because they were operating near their rated power only during evening hours,

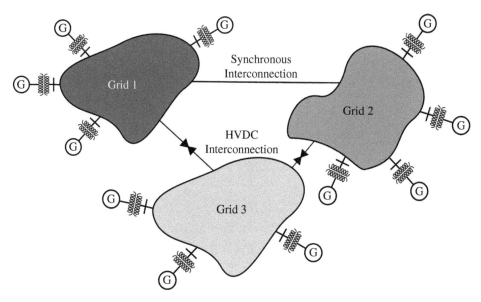

Figure 6.3 Schematic layout of an interconnected power system.

and at much lower capacity or not operating at all during the day. In addition, electric supply relied on a single generation facility and in the case of any electrical or mechanical failure, the whole area would be out of power until the fault was cleared. Reliability could be increased by building backup generation units, but this would add enormous capital and maintenance cost to the system. As the electric systems grew, separate networks were interconnected to increase reliability by supplying the consumers from several generation facilities such that if one of the units stops operating, the others could share the load. Interconnection of several power stations also increased the capacity factor of each power plant since the supply-demand balance could be managed in real time by coordinating the output power of separate generators. Figure 6.3 shows synchronous and HVDC interconnections between separate electric grids.

In AC interconnection, all generators supplying the grid must be precisely synchronized at a single frequency. Major electric failures (faults) may cause electromechanical oscillations throughout the interconnected system. Protective relays disconnect the area where the failure has occurred. In extreme cases, the phenomenon known as cascading, similar to the domino effect, may lead to a large area blackout. One of the techniques developed to resolve such issues is connecting separate AC grids through HVDC lines.

National interconnections spread to increasingly larger areas. Furthermore, countries interconnected their electric networks through cross-border and even intercontinental lines to form wider area synchronous grids. The continental US electric system consists of three interconnections, which are connected to each other, northern Mexico, and Canada at several substations. All western Europe countries are interconnected to each other and to northern Africa. Nordic countries of Europe and the former Soviet Union countries have separate interconnections linked to each other and western Europe. Several countries in southern Africa have combined their national grids. Figure 6.4 shows world's regions covered by interconnected electric grids.

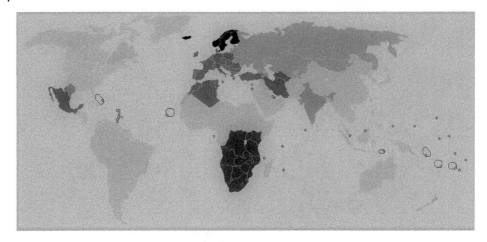

Figure 6.4 Wide area interconnected electric grids in the world (Source: Alinor at English Wikipedia available at https://commons.wikimedia.org/wiki/File:Wide_area_synchronous_grid_(Eurasia,_Mediterranean).png).

International electric grids take advantage of diverse resources available at geographically separate regions and allow international energy trade. Another advantage of wide area interconnection is the time differences, which naturally balance load distribution at peak hours. However, electric systems managed by separate entities using diverse power quality standards may cause technical issues.

6.3 Fundamental Concepts of Electric Circuit Analysis

As explained in the previous section, electric transmission networks started as DC, evolved to AC, and HVDC systems were developed for long distance transmission and asynchronous interconnection of electric grids. In current electric energy systems, distributed generation by PV arrays, fuel cells, and battery backup units resulted in large-scale integration of DC in AC-dominated electric grids. In this section, the terminology and basic concepts used in DC and AC circuit analysis are briefly reviewed.

6.3.1 Basic Definitions

An electric circuit comprises active and passive elements; active elements are voltage or current sources that deliver electric power to the circuit. An ideal voltage source has a known voltage across its terminals independent from its current. Similarly, an ideal current source delivers a known current independent from the voltage across its terminals. The current or voltage of a source may depend on a voltage or current of another element, in that case the term "dependent voltage source" or "dependent current source" is used to describe that element. Passive elements either consume or store electric energy.

Resistors, inductors, and capacitors are passive elements. Resistors convert electric energy into heat; hence, they are considered as energy consuming elements from an electric circuit perspective. Inductors store electromagnetic energy and capacitors store electrostatic energy. Although an inductor or capacitor can return the stored energy to the circuit, they are considered as passive elements since they do not produce electric power.

In electric circuits, a branch represents a single two-terminal element. A node refers to the connection point of two or more branches. A loop is a closed path in a circuit formed by starting at a node, passing through branches, and returning to the starting node without passing through any node more than once.

Two or more elements are said to be series connected if they carry the same current. Elements are in parallel if they are connected to the same pair of nodes and consequently have the same voltage across them.

6.3.2 Fundamental Laws

The conservation principles of physics are reflected to electric circuits by two fundamental laws named after German physicist Gustav Robert Kirchhoff (1824–1887). Kirchhoff's current law (KCL) states that *the algebraic sum of all currents entering a node, or a closed boundary, is zero*. KCL is based on the conservation of electric charge principle. Kirchhoff's voltage law (KVL) is based on the conservation of energy principle and states that *the algebraic sum of all voltages around a loop is zero*. Mathematical expression of the Kirchhoff's laws for a node joining n branches and a loop containing m branches are below.

$$\text{KCL}: \sum_{n=1}^{N} i_n = 0 \quad \text{KVL}: \sum_{m=1}^{M} v_m = 0 \tag{6.1}$$

Kirchhoff's laws apply to all types of circuits (linear or non-linear) and all variations of voltages or currents in time (DC, AC, or other waveforms) at any instant.

The electrical behavior of a circuit element is described by relationships between the current passing through it and the voltage across its terminals. Equations (6.2) are valid for any voltage and current waveforms applied to linear passive circuit elements.

$$v(t) = Ri(t)$$

$$v(t) = L\frac{di(t)}{dt}$$

$$i(t) = C\frac{dv(t)}{dt} \tag{6.2}$$

Note that the voltage across a linear resistor (R) is proportional to the current flowing through it. This property is known as Ohm's law. Terminal relationships of inductors (L) and capacitors (C) involve derivatives because of their energy storing property.

Electric power is the product of voltage and current at any instant.

$$p(t) = v(t)i(t) \tag{6.3}$$

Power absorbed by a resistor can be found by substituting either the voltage or current obtained from Ohm's law [Eq. (6.2)].

$$p_R(t) = [Ri(t)]i(t) = Ri(t)^2$$

$$p_R(t) = \left[\frac{v(t)}{R}\right]v(t) = \frac{v(t)^2}{R} \tag{6.4}$$

Energy processed by a circuit element is the integral of electric power over a time interval. For the general case where power is a function of time, energy is expressed by an integral.

$$W = \int_{t_1}^{t_2} p(t)dt \tag{6.5}$$

Energy stored in an inductor and capacitor is expressed as

$$W_L = \int_{t_1}^{t_2} \left[L\frac{di}{dt}i \right] dt = \int_{i_1}^{i_2} Li\ di = \frac{1}{2}L(i_2^2 - i_1^2)$$

$$W_C = \int_{t_1}^{t_2} \left[C\frac{dv}{dt}v \right] dt = \int_{v_1}^{v_2} Cv\ dv = \frac{1}{2}C(v_2^2 - v_1^2) \tag{6.6}$$

In electric energy systems, average power is more relevant than instantaneous power. If power changes in time at every instant, its average is determined by dividing the integral of the power function over the time interval to the length of the interval.

$$P_{av} = \frac{1}{t_2 - t_1} \int_{t_1}^{t_2} p(t)dt \tag{6.7}$$

6.3.3 DC Circuits

DC is a particular case for electric voltage and current. Ideal DC produced by a battery, PV module, or fuel cell is constant in time. In practice, the term DC is also used for a unidirectional current and voltage delivered by an electronic converter or DC generator, which change periodically in time. Using Fourier series, a periodic function can be represented as the sum of its average value and infinite number of sinusoidal components called harmonics. The DC component of a periodic voltage or current is expressed as

$$V_{av} = \frac{1}{T} \int_{\tau}^{\tau+T} v(t)dt \tag{6.8}$$

where T is the period. Harmonics are considered disturbances since they cause additional losses, reduce the system efficiency, and have adverse effects on system performances. DC generators are designed to minimize the harmonics on the output DC voltage. Electronic rectifiers incorporate filters to block AC components on the DC output. If the harmonics are properly eliminated or reduced to a negligible level, the supply can be considered pure DC equal to the average of the periodic voltage.

Non-ideal elements of an electric energy system such as sources, transmission lines, and loads are generally represented by equivalent circuits that contain resistors, capacitors, and inductors. In steady-state operation of a DC circuit, ideal inductors become a short circuit and capacitors do not conduct any current. Hence, the steady-state performance of a DC power system can be analyzed as a resistive DC circuit.

Since DC is constant in time, DC power is simply the product of DC current flowing through a circuit element and the voltage across its terminals.

$$P = VI \tag{6.9}$$

DC power is supplied from DC generators, batteries, fuel cells, and PV units.

6.3.4 AC Circuits

AC refers to an electric current that changes in time in the form of a sine or cosine function called a sinusoid. In an electric power network, all voltages and currents are in a sinusoidal form changing in time with the same frequency expressed in cycles per second or Hertz (Hz). In most parts of the world the standard grid frequency is 50 Hz, in North America and some Asian countries it is 60 Hz. Study of the power system performance under both normal and abnormal conditions requires thorough understanding of the AC circuit analysis methods.

6.3.4.1 Fundamental Concepts and Definitions

A sinusoidal waveform with amplitude X_m and frequency f is described by the following mathematical function, where x represents either a voltage or current.

$$x(t) = X_m \sin[(2\pi f)t + \theta] \tag{6.10}$$

The argument of the sine function in this expression is an angle in radians. Since periodic functions complete one cycle at 2π rad, angular frequency is expressed in radians per second (rad/s) and defined as

$$\omega = 2\pi f \quad [\text{rad/s}] \tag{6.11}$$

Figure 6.5(a) shows the time domain variation of a sinusoidal voltage and current. A sinusoid can be visualized as a vector rotating at the angular speed of ω as represented in Figure 6.5(b). The angle θ signifies the phase of the sinusoidal variation in time relative to a predefined reference origin. In this figure, the voltage is chosen as phase reference. The period of a sinusoid is the inverse of its frequency.

$$T = \frac{1}{f} = \frac{2\pi}{\omega} \quad [\text{s}] \tag{6.12}$$

In AC circuits, Kirchhoff's voltage and current laws apply at all instants. That is, at any time, the sum of all voltages around a closed loop is zero and the sum of all currents entering

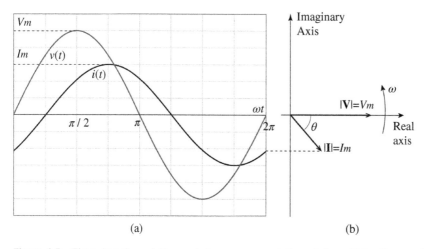

(a)   (b)

Figure 6.5 Time-domain variation and phasor representation of sinusoidal voltage and current.

a node is zero. Thus, all voltages and currents in an AC circuit are sinusoids with the same frequency.

In AC power systems the magnitudes of voltages and currents are always stated in effective values rather than amplitudes. Effective value, which is also denoted by the acronym "rms" (root-mean-square), is equal to the value of the DC voltage or current that would produce the same power on a resistor. For a periodic voltage changing in time with period T, the effective or root-mean-square (rms) value is the average of its values squared at all instants in one period.

$$V_{ef} = \sqrt{\frac{1}{T} \int_{\tau}^{\tau+T} v^2(t)dt} \qquad (6.13)$$

For a periodic current, I_{ef} is found by replacing the function $v(t)$ with $i(t)$ in the expression above. For the special case of pure sinusoidal voltages and currents, the effective value is obtained by dividing the amplitude to square root of two.

$$\text{rms}_{sinusoid} = \frac{\text{Amplitude}}{\sqrt{2}} \qquad (6.14)$$

In practice, test instruments are calibrated to display the "true rms" value for any waveform, nameplates of AC equipment state the rated values in rms voltages or currents. Given the rms value for a pure sinusoidal voltage or current, the amplitude is found simply by multiplying it with square root of two.

$$V_{max} = V_{ef}\sqrt{2}; \quad I_{max} = I_{ef}\sqrt{2} \qquad (6.15)$$

6.3.4.2 Phasor Quantities

In AC analysis, voltages and currents are represented by phasor quantities. A phasor is a vector of magnitude equal to the effective (rms) value, and oriented in the plane with an angle equal to the phase angle of the sinusoid as shown in Figure 6.6(a). Phasors are also expressed in complex numbers. A complex number corresponds to a vector on the complex

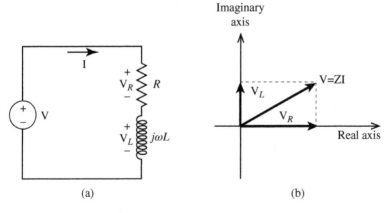

(a)   (b)

Figure 6.6 A simple AC supply circuit.

plane starting from the origin. Real and imaginary parts of a complex number equal the components of the corresponding vector on the real and imaginary axes respectively, as shown in Figure 6.6(b).

Because all voltages and currents in an AC system are measured or given as effective values, after this point we will drop the subscript and show all phasors with bold characters to distinguish from scalar values.

A sinusoidal quantity described by the time-domain function $x(t) = \sqrt{2}V \sin(\omega t + \theta)$ is expressed in the phasor-domain in either rectangular, polar, or exponential form shown below.

- Rectangular form: $\mathbf{X} = a + jb = |\mathbf{X}|(\cos\theta + j \sin\theta)$
- Polar form: $\mathbf{X} = |\mathbf{X}| \angle \theta$
- Exponential form: $\mathbf{X} = |\mathbf{X}|e^{\theta}$

where, $|\mathbf{X}| = \sqrt{a^2 + b^2}$ and $\theta = \tan^{-1}\frac{b}{a}$

The following expressions describe voltage-current relationships of circuit elements in terms of phasor quantities.

- Resistance: $\mathbf{V} = R \cdot \mathbf{I}$
- Inductance: $\mathbf{V} = (j\omega L)\mathbf{I} = \mathbf{X}_L \mathbf{I}$
- Capacitance: $V = \left(\frac{1}{j\omega C}\right)\mathbf{I} = \mathbf{X}_C \mathbf{I}$

The parameters $\mathbf{X}_L$ and $\mathbf{X}_C$ are called inductive and capacitive reactances. An RL load supplied from an ideal sinusoidal voltage source is shown in Figure 6.6(a).

$$\mathbf{V} = \mathbf{V}_R + \mathbf{V}_L = R\mathbf{I} + (j\omega L)\mathbf{I} \tag{6.16}$$

The voltage across the load equals the sum of two vectors. The resistive voltage drop is aligned with the real axis since the resistance does not change the angle of the current. The inductive reactance is an imaginary number; therefore, the voltage drop on the inductance is aligned with the imaginary axis.

Figure 6.6(b) shows the resistive and inductive parts of the total voltage phasor on the complex plane. The ratio of voltage and current phasors is defined as impedance. In the simple circuit of Figure 6.6(b) the impedance of the load consists of the resistance R and the inductive reactance $j\omega L$.

$$\mathbf{Z} = \frac{\mathbf{V}}{\mathbf{I}} = R + j\omega L \tag{6.17}$$

Impedance is a complex parameter that is generally obtained by series and parallel combination of resistances and reactances in an AC circuit. Ohm's law is generalized for AC circuits expressing the voltage phasor as the product of the complex impedance and the current phasor.

$$\mathbf{V} = \mathbf{Z} \cdot \mathbf{I} \tag{6.18}$$

AC loads are passive circuits generally represented by impedances. AC networks that contain one or more sources are active circuits. The modeling technique known as Thevenin equivalent is used to reduce a large linear active network into a simple circuit consisting of one voltage source and one equivalent series impedance. Figure 6.10 illustrates an electric power grid composed of several generation units, transformers, transmission lines, loads, and other electric equipment. To simplify analysis, such comprehensive power system is represented by its Thevenin equivalent as shown on the right side of the figure.

6.3.5 Three Phase Electric System

A three-phase electric system consists of a set of three sinusoids with the same magnitude and frequency, shifted in time by one-third of the period. Bulk electric generation and transmission is achieved all over the world by three-phase systems. A few exceptions include small-scale wind and solar powered units, small generation units used to supply a local single-phase load, and backup generators for emergency use.

Large industrial loads are generally supplied directly from a three-phase grid. Small motors, appliances, and electric lights are single-phase. In some distribution systems small consumers are also supplied by three-phase lines. If this is the case, internal wiring is designed to distribute the load equally between the phases. If small consumers are supplied in single phase or split-phase (such as 120-0-120 V), they are grouped in such a way that, as much as possible, equal power is supplied from each phase of the three-phase distribution system.

Figure 6.7 shows a three-phase supply system. Electric generation units produce sinusoidal voltages with equal effective values represented by phasors with equal magnitude and rotated by 120° from each other. The equivalent circuit of a three-phase supply consists of three ideal voltage sources corresponding to produced electromotive forces (EMFs), as indicated by circles on the diagram, and the internal equivalent impedance of the source. To illustrate the angles of the voltage and current phasors respective to each other, the Thevenin equivalent circuit of each phase of the source are rotated by 120° in the circuit

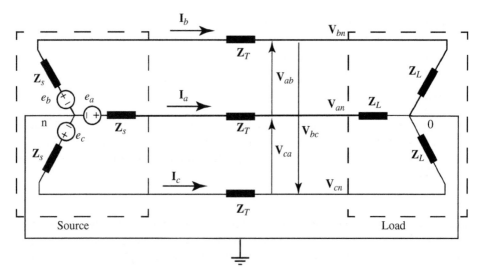

Figure 6.7 A three-phase supply system.

Figure 6.8 Star and delta configurations.

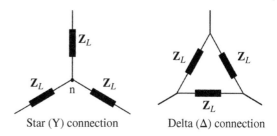

Star (Y) connection   Delta (Δ) connection

diagram. Obviously, the physical orientations of circuit elements do not affect their voltages and currents.

EMFs supplying each phase circuit are described by sinusoidal functions and phasors shown in Eq. (6.19).

$$e_a = E\sin(\omega t) \qquad\qquad \rightarrow \quad \mathbf{E}_a = E\angle 0°$$

$$e_b = E\sin\left(\omega t + \frac{2\pi}{3}\right) \quad \rightarrow \quad \mathbf{E}_b = E\angle 120°$$

$$e_c = E\sin\left(\omega t + \frac{4\pi}{3}\right) \quad \rightarrow \quad \mathbf{E}_c = E\angle 240° \qquad\qquad (6.19)$$

Generally, each phase of a three-phase supply system supplies equal load impedances. The configuration where one end of each phase circuit is connected at a common point as shown in the figure, is referred to as a star or wye (Y) connection, and the term "neutral" signifies the common connection point. Alternatively, the load impedances may be connected to form a closed loop, which is called a "delta (Δ) connection." Since there is no neutral in a delta connection, it is used when the supply and load sides are both balanced (Figure 6.8).

If the magnitudes of all source voltages and all impedances are equal, the three-phase system is said to be balanced. Since the source, transmission line, and load impedances in each phase circuit are identical, the angles between the voltage and current phasors are the same at each phase. Figure 6.9 shows the phasor diagram for the load side of a balanced three-phase system.

Voltages in respect to the neutral are referred to as "phase voltages." To simplify the expressions, we will represent the effective value of the phase voltages with V.

$$|\mathbf{V}_{an}| = |\mathbf{V}_{bn}| = |\mathbf{V}_{cn}| = V \qquad\qquad (6.20)$$

In a balanced three-phase system, the sum of the three sinusoids corresponding to phase voltages is zero at any instant. Consequently, the vector sum of the phase voltage phasors is also zero.

$$\mathbf{V}_{an} + \mathbf{V}_{bn} + \mathbf{V}_{cn} = V - 2V\cos 60° = 0 \qquad\qquad (6.21)$$

In electrical engineering, the potential difference between two lines connected to separate phases is defined as "line voltage" and usually denoted by U. In three-phase circuits it is generally cumbersome to denote so many variables by single subscript notation, which uses the polarity signs and current arrows. In power engineering, double-subscript notation is more convenient to indicate the directions of voltage and current phasors. When a voltage is written as $\mathbf{V}_{ab}$, the first subscript indicates where the voltage is positive with respect to

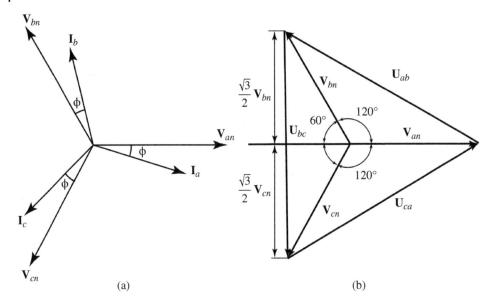

Figure 6.9 Phasor diagram balanced three-phase system.

the second. Changing the order of the subscripts produces a 180° phase shift in the variable. Thus, the phasor $\mathbf{V}_{ba}$ is in the opposite direction of $\mathbf{V}_{ab}$.

In a balanced three-phase system, the line voltage is obtained by multiplying the phase voltage with square root of three.

$$|\mathbf{V}_{ab}| = |\mathbf{V}_{ab}| = |\mathbf{V}_{ab}| = U = \sqrt{3}V \tag{6.22}$$

In the vector diagram of Figure 6.9(b) the line voltage phasors form a closed triangle, thus, the sum of all line voltages also adds up to zero.

$$\mathbf{U}_{ab} + \mathbf{U}_{bc} + \mathbf{U}_{ca} = 0 \tag{6.23}$$

Phasor relationships for voltages and currents are expressed in the vector relationships as shown in Eq. (6.24).

$$\mathbf{I}_a = \frac{|\mathbf{V}_{an}|\angle 0^o}{|\mathbf{Z}_L|\angle\phi} = |\mathbf{I}_a|\angle - \phi$$

$$\mathbf{I}_b = \frac{|\mathbf{V}_{bn}|\angle 120^o}{|\mathbf{Z}_L|\angle\phi} = |\mathbf{I}_b|\angle(120 - \phi)$$

$$\mathbf{I}_c = \frac{|\mathbf{V}_{cn}|\angle 240^o}{|\mathbf{Z}_L|\angle\phi} = |\mathbf{I}_a|\angle(240 - \phi) \tag{6.24}$$

The phase currents also have the same magnitude and they are shifted by 120° in respect to each other, so the vector sum of three-phase currents is equal to zero. Therefore, in a balanced system, no current flows through the dashed line in Figure 6.7 connecting the neutral of the supply and load circuits. Since the electric generators are designed to produce voltages with equal magnitude on their terminals, three-wire overhead lines without a return wire are sufficient for long distance transmission of three-phase power.

6.3.6 Per-Phase Analysis

If the three-phase system is balanced, it may be analyzed using a single-phase equivalent circuit. For a star-connected source supplying a star-connected load as shown in Figure 6.7, the voltages and currents in a particular phase are identical to the corresponding voltages and currents in the other phases except the 120° phase shift. Therefore, a single circuit for one of the phases and the neutral line completing the current loop may be analyzed and the results translated to the other phases considering their corresponding phase shifts.

When either the source or the load is delta connected, the delta side is transformed into wye configuration before applying per-phase analysis. Then, the results are converted back to the actual equivalents.

6.4 AC Power

An AC power source can be represented by Thevenin equivalent consisting of a single ideal voltage source and a series impedance as depicted in Figure 6.10. If the power supply is three-phase, the equivalent circuit corresponds to one of the phases as explained in Section 6.3.6 for per-phase analysis.

6.4.1 Power in Single-Phase Circuits

A simple AC power supply that consists of a power supply and a load is shown in Figure 6.11. The dashed rectangle on the left contains the equivalent circuit of the power source. The load is represented by its equivalent impedance $\mathbf{Z}_L$. The voltage across the terminals and the current through the load are characterized by the phasors $\mathbf{V}_L$ and $\mathbf{I}_L$ respectively. The vector diagram on the right side of Figure 6.11 shows the relative positions of the phasors. It is generally convenient to select the phase angle of the supply voltage ($\mathbf{V}_S$ in this circuit) zero to define the reference direction. The angles of all other phasors are expressed in respect to this direction.

Time variations of the voltage across the load terminals and the current flowing through the load are expressed in Eq. (6.25) in terms of the effective values.

$$v(t) = \sqrt{2}V \cos(\omega t + \theta_v)$$
$$i(t) = \sqrt{2}I \cos(\omega t + \theta_i) \tag{6.25}$$

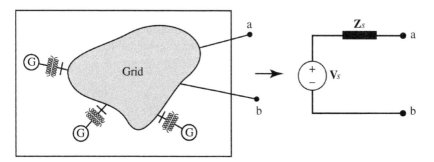

Figure 6.10 Thevenin equivalent of an electric grid.

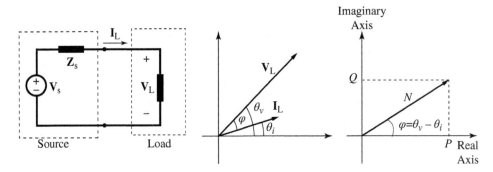

Figure 6.11 A simple AC power supply circuit and phasors at the load terminals.

Instantaneous power is the product of voltage and current at any instant.

$$p(t) = v(t) \cdot i(t) = \left[\sqrt{2}V \cos(\omega t + \theta_v) \right] \cdot \left[\sqrt{2}I \cos(\omega t + \theta_i) \right] \tag{6.26}$$

Using the trigonometric identity

$$\cos A \cos B = \frac{1}{2}[\cos(A - B) + \cos(A + B)] \tag{6.27}$$

the instantaneous power expression reduces to

$$p(t) = VI \cos(\theta_v - \theta_i) + VI \cos(2\omega t + \theta_v + \theta_i) \tag{6.28}$$

Average power is the average of the instantaneous power over one period. The first term on the right side of this equation is constant, however the second term is sinusoidal. The average of a sinusoidal function is zero. Therefore, the average power equals the first term, as shown in Eq. (6.29).

$$P = VI \cos \varphi \qquad \varphi = \theta_v - \theta_i \tag{6.29}$$

In a single-phase circuit, real power absorbed by a load impedance is the product of the voltage, current, and cosine of the phase angle between the voltage and current phasors. The cosine of the angle between the voltage and current phasors is called the power factor. In the case of a pure resistance, the phase angle between voltage and current phasors is zero, therefore the power factor is unity. That means power is simply equal to the product of the effective values of current and voltage, like in a DC resistor. Real power, also referred to as active power is converted to heat, mechanical power, or other forms of energy in unit time. The unit for active power is the watt (W).

As seen in the vector diagrams of Figure 6.6, active power is the real part of the vector called complex power, which is calculated by multiplying the voltage phasor and the complex conjugate of the current phasor.

$$S = VI^* = P + jQ \tag{6.30}$$

The imaginary part of the complex power is called reactive power. An ideal inductor or capacitor exchange only reactive power with the external circuit. The unit of reactive power is derived from voltage and current units and denoted as VAr, signifying volt-ampere reactive.

The magnitude of the complex power is simply the product of the effective values of voltage and current, similar to DC power. However, since it does not include the power factor, apparent power has no physical meaning. In practice, AC equipment is usually specified with apparent power since the active and reactive powers can be determined if the power factor is known. Apparent power is expressed in volt-amps (VA). Equations (6.31) summarize three AC power definitions.

$$P = VI \cos \varphi = N \cos \varphi \quad [\text{W}]$$
$$Q = VI \sin \varphi = N \sin \varphi \quad [\text{VAr}]$$
$$N = \sqrt{P^2 + Q^2} = VI \quad [\text{VA}] \tag{6.31}$$

Power absorbed by an impedance can be derived by substituting either the voltage or current obtained from Eq. (6.18) into (6.30).

$$\mathbf{S} = \mathbf{V}\mathbf{I}^* = (\mathbf{Z}\mathbf{I})\mathbf{I}^* = \mathbf{V}\left[\frac{\mathbf{V}}{\mathbf{Z}}\right]^* \tag{6.32}$$

Since the product of a complex number with its conjugate gives the square of its magnitude, power absorbed by an impedance is proportional to the square of the effective value of the current or voltage.

$$\mathbf{S} = I^2 \mathbf{Z} = \frac{V^2}{\mathbf{Z}^*} \tag{6.33}$$

If the impedance is only resistive, then complex power does not have an imaginary part, hence it is purely active power. In the case of an ideal inductor or capacitor the real part of the complex power is zero. Therefore, an ideal inductor or capacitor does not consume active power.

6.4.2 Power Factor Considerations

From Eqs. (6.31) power factor can be defined as the ratio of the active power to the apparent power, which is simply the product of the effective values of the voltage and current. For a given supply voltage and active power of the load, the current is inversely proportional to the power factor.

Most electric equipment contains electric motors, transformers, and other inductive elements that absorb reactive power for their operation. In an energy conversion system, only the active power is converted into mechanical work, heat, light, or any other useful form of energy. Reactive power is exchanged between the electric supply system and energy storing elements (that is inductors and capacitors) without changing the real power used. Reactive power absorbed by the load has significant adverse effects on the cost and operation of electric power supply systems.

Consider the AC supply system shown in Figure 6.12. The equivalent impedance of the transmission system, including all transformers and transmission lines from the source to the load, is represented by $\mathbf{Z}_T$ composed of a resistance R_T and a reactance jX_T. An electric utility supplies the consumers with a standard grid voltage V.

When a consumer uses P watt active power with an inductive power factor of $\cos\varphi$, the load current with magnitude I_L is drawn from the transmission line with φ-degree phase

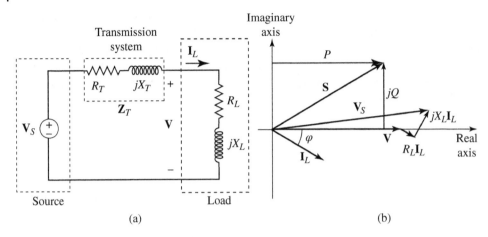

Figure 6.12 Phasor diagram of an AC supply system.

angle lagging the supply voltage.

$$I_L = \frac{P}{V \cos \varphi} \tag{6.34}$$

For the same active power and supply voltage, the magnitude of the load current increases if the power factor is smaller. The load current causes transmission loss of

$$P_{loss} = R_T I_L^2 \tag{6.35}$$

If the consumers' power factor is consistently low, the power company must build the transmission lines and transformers with higher current ratings to supply the load current increased by the reactive power. Consequently, a lower power factor causes a significant increase of the supply grid cost.

The source voltage V_S necessary to supply the standard grid voltage V at the receiving end of the transmission system is calculated by applying the circuit equations in the phasor domain.

$$V_s = V + (R_T + jX_T)I_L = V_s = V + (|I_L|\angle\varphi)(|Z_T|\angle\theta) \tag{6.36}$$

In Eq. (6.36), θ is the angle of the impedance and φ is the power factor angle. The impact of the load current on the source voltage can be visualized better by examining the vector diagram shown in Figure 6.12(b). Load current with a lagging phase angle causes larger voltage variation on the transmission line, therefore the source voltage must be increased to supply the lower power factor load. This causes the generators to operate at higher current to supply the same active power.

Consequently, it is more advantageous to increase the power factor at the consumer end. Since the power factor depends on the circuit elements necessary for the operation of electrical devices, reactive power be compensated at the point of use rather than transmitted all the way from electric generation facilities to the consumers. The process of increasing the power factor without changing the voltage or current on the load terminals is known as power factor correction or reactive power compensation. Since most loads are inductive, one common approach is to place capacitors in parallel with the load to improve the power factor.

A capacitor bank is sized to compensate the reactive power such that the power factor is closer to but less than unity. Overcompensation results in a leading phase angle of the current, which causes a voltage increase across the load. In practice, a power factor around 0.9 is desirable. Suppose that the power factor is desired to increase from $\cos\varphi_1$ to $\cos\varphi_2$ by connecting a shunt (parallel) capacitor across the load. The capacitor must provide the reactive power

$$Q_C = Q_1 - Q_2 = P(\tan\varphi_1 - \tan\varphi_2) \tag{6.37}$$

Using Eq. (6.33) for the power on the capacitive reactance $X_C = 1/j\omega C$ we obtain

$$C = \frac{Q_C}{\omega V^2} = \frac{P(\tan\varphi_1 - \tan\varphi_2)}{\omega V^2} \tag{6.38}$$

Some large industrial consumers may draw from the supply system large amounts of power at variable power factors, depending on their process cycles. In the steel industry, where large arc furnaces draw megawatts of power at continuously changing power factors, is a typical example. As the power factor changes in a wide range, the voltage at the consumers' electric network supplied from the same grid fluctuates in time, causing disturbing effects such as flickering of lights. A utility company, which is responsible for the power quality of the supplied electricity, requires industrial consumers that draw electric power with variable power factor to install dynamic power factor compensation systems to eliminate the fluctuations. Synchronous condensers or "Static VAr Compensation Systems (SVC)" can be used in such situations. A dynamic (or synchronous) condenser is a synchronous motor operated without mechanical load that can supply the required reactive power to the system by adjusting the field current (Nasar 1996). SVC includes a combination of thyristor switched capacitors, reactors, and mechanically switched capacitor banks. Development of power electronics systems has reduced the initial and maintenance costs of dynamic compensation units.

6.4.3 Power in Three-Phase Systems

Since the magnitudes of phase voltages and currents are equal in a balanced multiphase system, power calculations can be done for one phase as explained in Section 6.4 multiplied by the number of phases. In practice, it is generally more convenient to work with phase-to-phase voltages rather than phase-to-neutral. Hence, power in a balanced three-phase circuit is obtained as:

$$P_{3phase} = 3\mathbf{VI}^* = 3\left(\frac{\mathbf{U}}{\sqrt{3}}\right)\mathbf{I}^* = \sqrt{3}\,\mathbf{UI}^* \tag{6.39}$$

Active, reactive, and apparent powers are defined in the same manner as in single-phase circuits.

$$P_{3phase} = 3VI\cos\phi = \sqrt{3}UI\,\cos\phi$$

$$Q_{3phase} = 3VI\sin\phi = \sqrt{3}UI\,\sin\phi$$

$$N_{3phase} = 3VI = \sqrt{3}\,UI \tag{6.40}$$

6.5 Electromagnetic Field

Electric generators, transformers, and motors rely on interactions between electric circuits and magnetic fields. Electric current flowing on a conductor creates a magnetic field in the surrounding space. A magnetic field changing in time produces electric potential across the conductors placed around it. This section provides a review of the fundamental concepts of electromagnetic theory pertinent to electric power generation, voltage transformation, and electromechanical energy conversion. Detailed information can be found in fundamental textbooks such as J.D. Kraus (1992), S.J. Chapman (1991), and Fitzgerald, Jingsley, and Umans (1992).

6.5.1 Ampere's Law

Ampere's circuital law states that *the line integral of the magnetic field intensity on a closed path is equal to the total electric current enclosed in the path.* A mathematical expression of the Ampere's law can be written as shown in Eq. (6.41).

$$\oint_C \vec{H} \cdot d\vec{l} = \iint_S \vec{J} \cdot d\vec{S} \tag{6.41}$$

In Eq. (6.41) H represents the intensity of the magnetic field vector, and dl is the differential length vector on a closed path C. The line integral of the scalar (dot) product of the magnetic field intensity and line segment vectors equals the integral of the current density on a surface S bordered by the closed path C. In most practical electromagnetic devices, the magnetic field vectors are parallel to the differential length and constant on separate sections of the path. Also, all currents flow through distinct wires orthogonal to the surface surrounded by the magnetic field lines. In practice, Ampere's law can be further simplified to simple additions on both sides of Eq. (6.41), as in Eq. (6.42).

$$\sum_j H_j \cdot l_j = \sum_k I_k \tag{6.42}$$

For example, Ampere's law can be expressed as follows for the electromagnetic device shown in Figure 6.13.

$$H_1 l_1 + H_2 l_2 = I_1 + I_2 - I_3$$

The standard unit for magnetic field intensity is amperes per meter (A/m) as the expression implies. Note that H depends only on the length of the magnetic path and the total current enclosed in the path, not the magnetic properties of the material in which the magnetic field occurs.

6.5.2 Magnetic Flux

In early years during the development of the magnetic field theory, scientists were visualizing the magnetism as field lines that form a magnetic flux. Although magnetic field does not actually flow through matter, the concept of magnetic flux has been helpful in analysis and design of electromagnetic devices. Magnetic flux is expressed in Weber in the international unit system (SI).

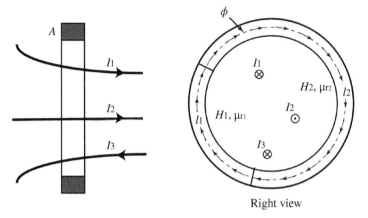

Figure 6.13 An electromagnetic structure to illustrate Ampere's law.

Magnetic flux per unit area is defined as flux density. In modern electromagnetic theory, the field line analogy is no longer used, and the magnetic flux is defined as the surface integral of the normal component of the flux density on a surface. If the magnetic field vectors are in different directions at every point of the surface, magnetic flux at each point is obtained from the scalar product of the flux density vector at that point and the normal vector perpendicular to the infinitesimal surface element with magnitude dA.

$$\phi = \iint_S \vec{B} \cdot d\vec{A} \tag{6.43}$$

If the flux density is constant at every point in the same angle α in respect to the normal direction of the surface, the expression of flux is reduced to a simple multiplication.

$$\phi = BA \cos \alpha \tag{6.44}$$

In most practical magnetic devices, the induction vector is perpendicular to the surface and the expression is further simplified to $\phi = BA$. In the international unit system (SI) flux density is in the dimension of Weber per square meter (Wb/m^3). The General Conference on Weights and Measures named the standard unit of flux density Tesla (T) in 1960 in honor of Nikola Tesla, who made enormous contributions to the development of electric generation and transmission systems. One T is equal to 10,000 Gauss (G), the older unit in the cgs unit system. Another term for flux density is *magnetic induction* and both terms are used interchangeably.

6.5.3 Magnetic Properties of Substances

Flux density depends on the magnetic properties of substances. In a vacuum, the flux density is obtained by multiplying the field intensity with a constant known as the permeability of vacuum.

$$B_{vacuum} = \mu_0 H \quad (\mu_0 = 4\pi 10^{-7}) \tag{6.45}$$

The permeability of most substances with weak magnetic properties is expressed relative to the permeability of vacuum.

$$\mu = \mu_r \mu_0 \tag{6.46}$$

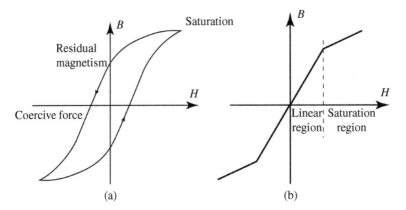

Figure 6.14 Hysteresis loop of a ferromagnetic material and its simplified representation.

Air, water, wood, plastics, and any metal that is not attracted by a magnet has a relative permeability around unity. However, metals such as iron, steel, chromium, cobalt, nickel, and many iron alloys have strong magnetic properties because of the way the atoms are aligned in their crystals. Such metals, commonly called ferromagnetic materials, display nonlinear magnetic properties represented by a hysteresis loop shown in Figure 6.14(a). The flux density versus field intensity (B–H) characteristic is different as the field intensity H increases or decreases, indicated by an arrow on the figure. Once the material is magnetized, the flux density remains at a certain value called residual magnetism, even if the field intensity becomes zero. Since the permeability changes at every point of the curve, the slope of the tangent to the hysteresis curve at the operating point is defined as incremental permeability.

A hysteresis curve is not convenient for practical calculations; for simplification, the two branches of the curve are unified on average straight lines, as shown in Figure 6.14(b). The first line segment to cross the origin represents the linear region of the characteristic. The slope of the first segment defines the relative permeability in the linear region. Ferromagnetic materials used in electric machines have relative permeability in the order of thousands. The second segment, which has a smaller slope is called the saturation region. For better efficiency and controllability, electromagnetic devices are generally operated in the linear region.

6.5.4 Magnetic Circuits

Electromechanical systems used as transformers, machines, actuators, and transducers have rather complex structures magnetically linking several coils. Analysis of the magnetic flux distribution in such devices is generally simplified by modeling the structure similar to electric circuits and applying circuit analysis methods.

Combining Eqs. (6.42–6.46), we can write:

$$\sum_{j=1}^{n} \frac{l_j}{A_j \mu_j} \phi_j = \sum_{k=1}^{m} N_k I_k \tag{6.47}$$

The terms of the summation on the left side of the equation are proportional to the magnetic flux, similar to the Ohm's law in electric circuits. By analogy, the proportionality constant is defined as reluctance.

$$R = \frac{l}{\mu A} \tag{6.48}$$

The terms in the summation on the right side are called magnetomotive force (MMF), similar to the EMF in electric circuits. A magnetic device can be, therefore, converted to a resistive circuit where the resistances are replaced with reluctances, and voltage sources with MMFs. Generally, θ is used to represent MMF. Since the number of conductors N is defined by the number of turns in a coil wrapped around the magnetic core, the unit for MMF is Ampere-turns.

$$\theta = NI \tag{6.49}$$

Example 6.1 Obtain an expression in terms of the currents for magnetic flux in the electromagnetic structure shown in Figure 6.13.

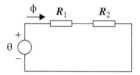

Solution
The structure has only one path for magnetic flux. The equivalent magnetic circuit can be drawn as shown Figure 6.13. Note that the reference directions for the MMF and flux are selected similar to the passive notation we used for electric circuits. That is, the flux arrow leaves the MMF from the positive sign.

$$(R_1 + R_2)\phi = \theta \quad \Rightarrow \quad \phi = \frac{\theta}{R_1 + R_2}$$

$$R_1 = \frac{l_1}{\mu_{r1}\mu_0 A}; \quad R_2 = \frac{l_2}{\mu_{r2}\mu_0 A}; \quad \theta = I_1 + I_2 - I_3$$

$$\phi = A\mu_0 \left[\frac{\mu_{r1}}{l_1} + \frac{\mu_{r2}}{l_2} \right] (I_1 + I_2 - I_3)$$

Magnetic circuits of all electromechanical devices consist of a core made from a ferromagnetic material and air space so that the part generally called rotor or armature can move. Because the relative permeability of the iron core is much larger compared to the permeability of air, the reluctance of the iron parts may be negligible in practical calculations.

Example 6.2 The electromagnetic device shown in Figure 6.15 has a toroidal ferromagnetic core of radius $R_1 = 45$ mm and $R_2 = 55$ mm, with an air gap of 1 mm. Thickness of the core is $t = 20$ mm, and relative permeability of the ferromagnetic material is 3000. A DC current of 2 A flows through the 100-turn coils. Determine the flux density in the core.

Solution
The core structure is similar to the one in Example 6.1. Segment 1 of the structure is made from ferromagnetic material and segment 2 is an air gap. The same current flows on N

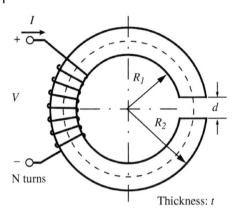

Figure 6.15 A toroidal magnetic core with an air gap described in Example 6.2.

I

+ ○

V

N turns

R_1

R_2

d

Thickness: *t*

turns of the coil. We can therefore substitute the corresponding numbers in the expression obtained in the previous example. The average radius of the magnetic path is

$$l_m = 2\pi \frac{R_1 + R_2}{2} = \pi \times (45 + 55) \cdot 10^{-3} = 0.314 \text{ m}$$

Average length of the core is 0.313 m and the length of the air gap is 0.001 m. The area of the core is $A = 10 \times 20 \times 10^{-6} = 2 \cdot 10^{-4}$ m². Reluctances of the core and air gap are:

$$R_1 = \frac{0.313}{3000 \times 4\pi 10^{-7} \times 2 \cdot 10^{-4}} = 0.415 \cdot 10^6$$

$$R_2 = \frac{0.001}{1 \times 4\pi 10^{-7} \times 2 \cdot 10^{-4}} = 3.979 \cdot 10^6$$

Note that although the length of the magnetic field lines in the core is 314 times longer than the air gap, the reluctance of the core is about one-tenth of the reluctance of the air gap. In many engineering calculations, 10% tolerance is acceptable. In this example, without neglecting the reluctance of the iron core, the flux is obtained

$$\phi = \frac{NI}{R_1 + R_2} = \frac{100 \times 2}{(0.415 + 3{,}979) \times 10^6} = 0.45 \cdot 10^{-6} \text{ Wb}$$

6.5.5 Faraday's Law

In the early 1830s Michael Faraday experimentally discovered that a magnet moving inside of a coil of copper wire produces electric voltage across the terminals of the coil. Faraday explained his observations in terms of magnetic flux lines. In addition, he discovered that changing current in a wire could induce a current in the neighboring wires. Faraday demonstrated his findings in public experiments and concluded that the EMF across the terminals of a coil was directly proportional to the rate of change of the magnetic flux lines that cut the coil conductors.

A common statement of the Faraday's law is: *"The induced emf in a closed loop is equal to the time rate of change of the magnetic flux through the loop."* A few years after Faraday published the magnetic induction law, Emile Lenz added a negative sign stating that the current induced in a conductor creates a magnetic field in the opposite direction to the changing magnetic field which creates it.

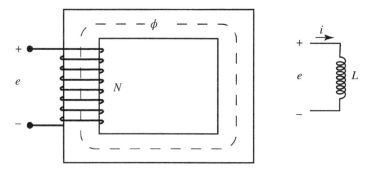

Figure 6.16 Illustration of Faraday's law and circuit representation of a self-inductance.

Practical electromechanical devices have coils rather than a single conductor. The magnetic flux that links all conductors of an N turn coil is called *flux linkage* and expressed as $\lambda = N\phi$. According to the Faraday's law, the EMF e across the terminals of an N turn coil is expressed as the time derivative of the flux linkage with a negative sign due to Lenz's law.

$$e = -N\frac{d(\phi)}{dt} = -\frac{d(N\phi)}{dt} = -\frac{d\lambda}{dt} \tag{6.50}$$

Faraday's law is the mathematical expression of the magnetic induction concept, which is the basis of transformers, motors, and generators, and electromagnetic transducers that have moving coils.

If the $B–H$ characteristic of the magnetic core is linear or assumed to be in the linear region, the flux linkage is proportional to the current. The proportionality coefficient is named self-inductance and represented by the symbol L. The unit of self-inductance is the Henry, named in the honor of Joseph Henry, who discovered magnetic induction independently at the same time as Faraday.

In electric circuits, an ideal inductance represents a coil with zero resistance (Figure 6.16). The following equation describes the voltage current relationship of an inductance.

$$\lambda = Li \implies e = L\frac{di}{dt} \tag{6.51}$$

In Eq. (6.51) the negative sign is dropped because of the selected reference directions of the current and EMF. The operation principles of all practical electric machines are based on Ampere's and Faraday's laws. Electromechanical energy conversion occurs in the magnetic field created by the coils of electric machines. An expression of magnetic energy can be obtained by substituting the Faraday's equation given in Eq. (6.50) into the mathematical expression of electric energy.

$$W = \int_{\lambda_1}^{\lambda_2} p(t)dt = \int_{t_1}^{t_2} -i\frac{d\lambda}{dt}dt = -\int_{\lambda_1}^{\lambda_2} id\lambda \tag{6.52}$$

The negative sign in this expression implies that if the flux linkage λ increases for constant current, and the magnetic system delivers energy. In a linear system, where flux is proportional to the electric current that creates it, magnetic energy stored in or delivered by an inductor is proportional to the square of the current.

$$\int_{\lambda_1}^{\lambda_2} \frac{\lambda}{L}d\lambda = \frac{1}{2}\frac{\lambda_2^2 - \lambda_1^2}{L} = \frac{1}{2}L\left(i_2^2 - i_1^2\right) \tag{6.53}$$

6.6 Transformers

Transformers are essential elements of electric transmission and distribution networks. In energy systems, their main role is to change the voltage levels at different points of the energy system from electric generation units to consumers to reduce the transmission losses.

6.6.1 Operation Principle

A transformer consists of a magnetic circuit generally called a transformer core, and two or more coils linked by the time-varying magnetic flux. In practical transformers used in AC power systems, the core is made by stacking laminations of special steel with superior magnetic properties. The coils supplied from a source to create the magnetic flux are referred to as primary windings and the coils that supply an electric load are referred to as secondary windings. Figure 6.17(a) shows a basic transformer with two windings placed on a magnetic core.

In an ideal transformer, the same magnetic flux encircles both windings. EMFs induced across the terminals of the coils with N_1 and N_2 turns is obtained by applying Faraday's law.

$$e_1 = N_1 \frac{d\phi}{dt}; \quad e_2 = N_2 \frac{d\phi}{dt} \tag{6.54}$$

Since the variation of the magnetic flux in both windings is the same, the ratio of e_1 and e_2 is simply the ratio of the number of turns, defined as the transformation ratio.

$$\frac{e_1}{e_2} = \frac{N_1}{N_2} = a \tag{6.55}$$

A step-down transformer decreases the voltage from its primary to secondary, hence the transformation ratio is larger than unity. Step-up transformers increase the voltage; therefore, their transformation ratio is less than unity. An ideal transformer transfers the energy absorbed by the primary winding to the secondary winding without any loss. Hence, electric power at both sides are equal. The relationship between the primary and secondary

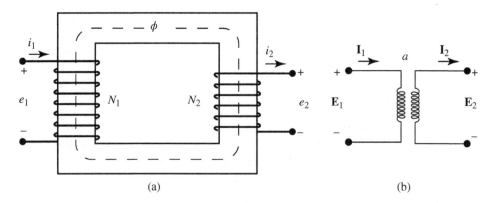

(a) (b)

Figure 6.17 Ideal transformer and its equivalent circuit.

currents is obtained by applying the invariance of power property.

$$e_1 i_1 = e_2 i_2 \quad \Rightarrow \quad \frac{i_1}{i_2} = \frac{e_2}{e_1} = \frac{1}{a} \tag{6.56}$$

The equivalent circuit of an ideal transformer is shown in Figure 6.17(b). In power systems the mutual flux linking both windings is represented by a sinusoidal function changing at time with grid frequency.

$$\phi = \phi_m \sin \omega t \tag{6.57}$$

According to Faraday's law, the induced EMFs are expressed as

$$e_1 = \frac{d\lambda_1}{dt} = N_1 \frac{d\phi}{dt} = \omega \phi_m N_1 \cos \omega t$$

$$e_2 = \frac{d\lambda_2}{dt} = N_2 \frac{d\phi}{dt} = \omega \phi_m N_2 \cos \omega t \tag{6.58}$$

Then, the effective (rms) values of the induced voltages are

$$E_1 = \frac{1}{\sqrt{2}}(2\pi f)\phi_m N_1 = 4.44 f \phi_m N_1$$

$$E_2 = \frac{1}{\sqrt{2}}(2\pi f)\phi_m N_2 = 4.44 f \phi_m N_2 \tag{6.59}$$

The relationship between the effective values of the voltages and currents across the primary and secondary terminals are similar to their time domain functions.

$$\frac{E_1}{E_2} = a \quad \frac{I_1}{I_2} = \frac{1}{a} \tag{6.60}$$

Note that the primary winding creates magnetic flux that closes its path only through the core, and there is no flux line that closes its path in the air. When the secondary winding supplies a load, the current flowing through it creates a magnetic flux that opposes the flux created by the primary winding. As power is transferred from the primary to the secondary side of an ideal transformer magnetic fluxes created by the windings cancel each other.

An actual transformer winding has conductor resistance which causes resistive voltage drop and Joule losses. Also, some part of the magnetic fluxes considered as "leakage flux" close their paths through air instead of the core. Consequently, the voltage and current relationships between the primary and secondary sides of a practical transformer differ from ideal transformers. In addition, during the energy transfer from the primary to the secondary terminals heat losses occur on the winding resistances and the magnetic core. Figure 6.18(a) illustrates the outline of a practical transformer and (b) shows the corresponding equivalent circuit.

In the equivalent circuit the leakage flux of each winding corresponds to a linear inductance jX_σ connected in series with the winding resistance R. The non-linear reactance due to the core flux jX_μ and a resistance that corresponds to magnetic losses in the core R_μ are represented by a parallel RL circuit lumped at the primary circuit. Transformation of the voltage and current values is represented by an ideal transformer between the primary and secondary sides (Figure 6.19).

Analysis of the equivalent circuit can be simplified by referring the parameters on the secondary circuit to primary and moving the ideal transformer to either the load or source

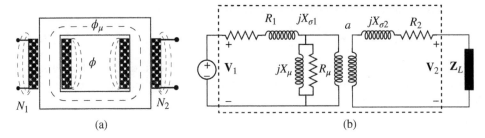

Figure 6.18 Non-ideal transformer outline and equivalent circuit.

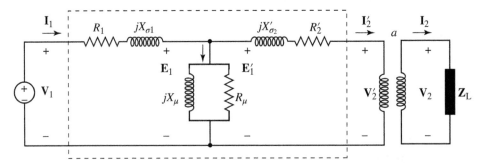

Figure 6.19 Equivalent circuit of a non-ideal transformer referred to the primary side.

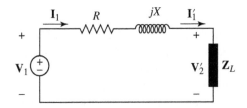

Figure 6.20 Approximate equivalent circuit of a practical transformer.

terminals. An impedance $\mathbf{Z}_2$ on the secondary circuit is referred to the primary winding $\mathbf{Z}_2'$ using the terminal relationship and the transformation ratio.

$$\mathbf{Z}_1 = \frac{\mathbf{V}_1}{\mathbf{I}_1} = \frac{a\mathbf{V}_2}{{}^1\!/_a\mathbf{I}_2} \quad \Rightarrow \quad \mathbf{Z}_2' = a^2\mathbf{Z}_2 \tag{6.61}$$

Resistances, inductances, capacitances, and corresponding reactances can be referred to the primary side using corresponding expressions in the form of Eq. (6.61). In practice, the magnetizing current $\mathbf{I}_\mu$ is negligible when the transformer operates around its rated voltage and power. Hence, the equivalent circuit can be further simplified as shown in Figure 6.20. In the approximate equivalent circuit, the resistance and reactance are the total values referred to the primary circuit.

$$R = R_1 + R_2' = R_1 + a^2 R_2$$
$$X = X_1 + X_2' = X_1 + a^2 X_2 \tag{6.62}$$

6.6.2 Industrial Transformer Tests

A series of routine tests are done on every industrial size transformer throughout the manu-facturing process. Tests of the turn ratio, polarity, resistance measurements, and open- and short-circuit tests are standard procedures before a power transformer is delivered to the customer.

Turn ratio and polarity tests are based on accurate voltage measurements. Winding resistances are directly measured by sensitivity resistance test equipment or a Wheat-stone bridge. Other equivalent circuit parameters and losses of a power transformer are determined from *open-circuit* and *short-circuit tests*.

6.6.2.1 Open-circuit (No-load) Test

The purpose of the open-circuit test is to determine the core losses and corresponding mag-netization parameters of a transformer. The rated voltage is applied to one of the windings while the terminals of the other are open. In utility-size transformers, it is more convenient to apply voltage to the low-voltage winding and leave the high-voltage (HV) terminals open.

Power measured when the transformer is not supplying any external load is mainly the core losses, since the small magnetizing current dissipates negligible resistive power on the series resistance R_1. The resistance that represents core losses R_μ is calculated from the measured open circuit power and voltage.

$$R_\mu = \frac{V_{oc}^2}{P_{oc}} \tag{6.63}$$

Since the resistance and reactance corresponding to the magnetization properties of the core are represented in a parallel circuit, the resistive component of the open-circuit current is directly determined from the open-circuit voltage and the resistance calculated from the open-circuit power.

$$I_{R\mu} = \frac{V_{oc}}{R_\mu} \tag{6.64}$$

The reactance X_μ corresponding to the magnetic flux is obtained from the measured open-circuit current and its calculated resistive component.

$$X_\mu = \frac{V_{oc}}{\sqrt{I_{oc}^2 - I_{R\mu}^2}} \tag{6.65}$$

6.6.2.2 Short-circuit Test

Losses on the winding resistances and leakage inductances are determined in a short-circuit test. The terminals of one winding are short-circuited, and a reduced voltage is applied to the other winding such that the rated current flows through the short-circuited winding. In a utility-size transformer, it is more convenient to connect together the low-voltage terminals and apply reduced voltage to the high-voltage terminals.

The magnetic flux created by the secondary winding supplying the rated current cancels the flux created by the primary winding since it is in the opposite direction. Therefore, the power measured in the short-circuit test represents the Joule losses on the coil resistance. In practice the term *copper loss* refers to the power dissipated on the winding resistances.

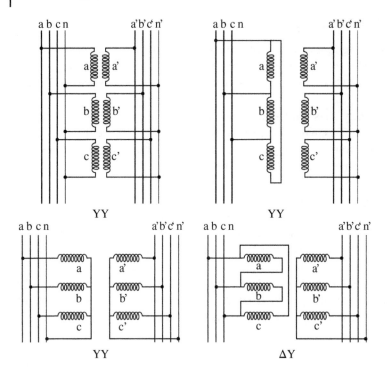

Figure 6.21 Three phase transformer combinations.

The magnetizing current $\mathbf{I}_\mu$ is negligible compared to the rated current, therefore the voltage mainly drops across the total series impedance. The magnitude of the series impedance can be obtained from the measured short-circuit current.

$$|\mathbf{Z}| = Z = \frac{V_{sc}}{I_{sc}} \tag{6.66}$$

The impedance is a series combination of the total winding resistances measured directly and the total leakage reactance expressed in (6.62).

$$X = \sqrt{Z^2 - R^2} \tag{6.67}$$

6.6.3 Three-phase Transformers

Transformers used in three-phase circuits can be formed either by installing three separate transformers for each phase or building a single core to transform three-phase voltages on the same magnetic circuit. In either case the phase windings can be connected in wye (Y) or delta (Δ) configuration. Figure 6.21 shows separate transformers for each phase (top) and single three-phase transformer (bottom). In practice, various combinations of primary and secondary connections (Y-Y, Y-Δ, ΔY, or Δ-Δ) are preferred depending on supply and source conditions.

In practice, a delta connection is preferred to reduce transmission line cost by eliminating the neutral wire if the voltages and currents are balanced. Unbalanced loads cause

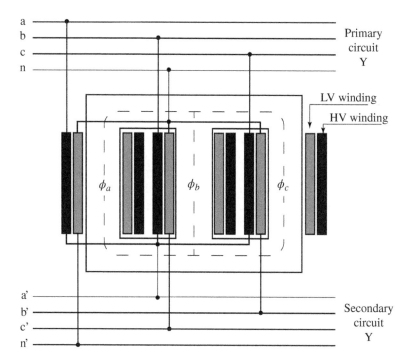

Figure 6.22 Schematic diagram of a Y–Y-connected three-phase transformer.

undesired circulation current through the phase windings. A wye connection is similar to three separate single-phase systems and the neutral wire can conduct a current if the load is unbalanced. A Y-Y configuration without a neutral wire is also possible if the three-phase system is perfectly balanced. Installing three separate transformers is more expensive but provides higher reliability to the electric system since it is easier to replace one transformer if it fails. Large extra-high voltage (EHV) transformers are often manufactured as single-phase because of transportation limitations.

A three-phase transformer core has three vertical legs joined by an upper and lower yoke as illustrated in Figure 6.22. Balanced three-phase voltages applied to the primary windings create fluxes with the same magnitude, but shifted by 120° on each leg end portion of the yoke. The sum of these fluxes is zero at any instant.

$$\phi_a = \phi_m \sin \omega t$$
$$\phi_b = \phi_m \sin(\omega t + 120°)$$
$$\phi_c = \phi_m \sin(\omega t + 240°)$$
$$\phi_a + \phi_b + \phi_c = 0 \tag{6.68}$$

The EMFs induced by these magnetic fluxes form a balanced three-phase voltage system with equal effective values at each phase that may be calculated using Eq. (6.59). In a three-phase transformer, primary and secondary windings of each phase are placed on the same leg. In general, for practical reasons the inner coil is lower voltage and the outer coil is higher voltage. Utility-size transformers are typically enclosed in a tank filled with a special mineral oil called transformer oil. The purpose of the transformer oil is both electrical insulation and transfer of the heat dissipated from the coils and the core to the external radiators or heat exchangers.

6.7 Electromechanical Energy Conversion

Electrical and mechanical energy can be converted into each other using electric machines. Energy flow in all electric machines is bidirectional. In other words, the same electric machine can operate as a generator to convert mechanical energy into electrical form or as a motor to convert electrical energy into mechanical form. In generator operation a prime mover like a turbine, combustion engine, or a mechanical energy storage device drives the generator, which delivers electric power to the external circuit. In the motor operation, the electric machine converts electrical energy into rotational work to drive a passive mechanical load or a mechanical energy storage device.

Most electric machines used in energy systems perform rotational motion. Figure 6.23(a) shows a schematic diagram of a rotational electromechanical energy conversion system. All reference directions in the figure are chosen according to the passive convention that the absorbed power is positive and delivered power is negative. At the electrical side, current enters from the positive sign of the voltage reference.

In motor operation both voltage and current are either positive or negative, thus the electrical input power is positive. At the mechanical side, ω represents the angular velocity and T represents the torque. The angular velocity and torque must have opposite signs so that the mechanical power is negative, and thus becomes output power.

In generator operation, torque and rotation are in the same direction, therefore the mechanical power is negative, meaning that it is the input power. The current becomes opposite to the reference direction making the electrical input negative, or output.

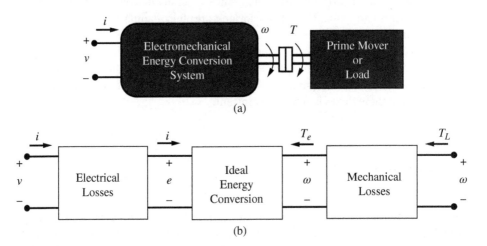

Figure 6.23 Schematic representation of an electromechanical system.

A generator receives mechanical power and delivers electric power. The expressions below summarize the motor and generator operation modes.

Motor operation	$v > 0; i > 0$ or $v < 0; i < 0 \Rightarrow P_{elec} > 0$
	$\omega < 0; T > 0$ or $\omega > 0; T < 0 \Rightarrow P_{mec} < 0$
Generator operation	$v > 0; i < 0$ or $v < 0; i > 0 \Rightarrow P_{elec} < 0$
	$\omega < 0; T > 0$ or $\omega < 0; T < 0 \Rightarrow P_{mec} > 0$

In Figure 6.23(b) electrical and mechanical losses are separated. The middle block performs ideal energy conversion where the sum of the mechanical and electrical input powers is zero.

$$ei + \omega T = 0 \tag{6.69}$$

The torque is often expressed in terms of the electrical variables and angular velocity using Eq. (6.69).

$$T_e = -\frac{e \cdot i}{\omega} \tag{6.70}$$

6.7.1 Basic Motor and Generator

Consider the classical example of a segment of rigid conductor of length l sliding freely on two parallel conducting rails connected to a galvanometer as shown in Figure 6.24(a). Any friction and contact resistance between the conductor and rails is negligible. The conductor is placed in a magnetic induction B perpendicular at all points to the conductor. For a differential movement of the conductor from the original position, the magnetic flux will change by

$$d\lambda = BA = Bldx \tag{6.71}$$

The voltage across the moving conductor can be obtained by applying Faraday's law.

$$e = -\frac{d}{dt}(Bldx) = -Bl\frac{dx}{dt} = -Blu \tag{6.72}$$

In Eq. (6.72), the derivative of the displacement is replaced by the linear velocity u of the conductor. Due to the Lenz law, as the conductor moves from x_0 to the right in the direction

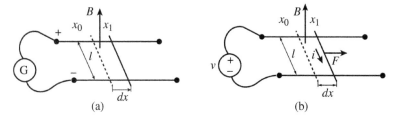

Figure 6.24 Electromechanical energy conversion on a moving conductor.

to increase the initial area, the induced EMF is negative. If the initial area decreases with the motion, the EMF becomes positive.

Now consider Figure 6.24(b), where a voltage source is connected to the rails such that a constant current i passes through the moving conductor. The differential magnetic energy as the conductor moves an infinitesimal distance dx in the direction of F will be:

$$dW = id\lambda = iBldx \tag{6.73}$$

Since the work done by the force F as the conductor moves over the distance dx is $W = Fdx$, Eq. (6.74) is therefore obtained for the magnitude of the force developed.

$$F = Bli \tag{6.74}$$

If the magnetic flux is not perpendicular to the moving conductor, the force is obtained by the vector (cross) product of the vectors $\vec{B}$ and $\vec{l}$.

$$\vec{F} = (\vec{B} \times \vec{l})i \tag{6.75}$$

The current i is a scalar quantity since it is the time rate of change of electric charge. The magnitude of the vector $\vec{l}$ is equal to the length of the conductor in the magnetic field, and its direction is the same as the current flowing on the conductor.

Some machines are designed to operate as a generator and some machines as a motor under normal steady state conditions. All machines, however, exchange energy in either direction under transient operation. Electric machines used in hybrid or electric vehicles operate as a generator when the vehicle slows down, or elevation decreases. Sudden change of load or certain fault conditions may cause electromechanical oscillations of generators installed in power plants.

Obviously, at any instant if the power at one side is positive, power at the other side must be negative. According to the passive notation, the shaft power is positive when the machine absorbs mechanical power, therefore shaft rotation and torque developed on the rotor are in the same direction in generator mode. In motor operation, the machine delivers mechanical power, therefore torque developed on the rotor is opposite to the shaft rotation to balance external mechanical load.

6.7.2 Efficiency of Electromechanical Energy Conversion

Efficiency of an energy conversion system is always defined as the ratio of the output energy (or power) to the input energy (or power). If the generator operation input power is mechanical, output power is electrical. In motor operation, the input and output variables are simply reversed. Efficiency is determined using Eqs. (6.76). In practice efficiency is often given in percent value, which is obtained by simply multiplying the decimal figure with 100.

$$\eta = \frac{P_{out}}{P_{in}} = \frac{P_{in} - P_{loss}}{P_{in}} = 1 - \frac{P_{loss}}{P_{in}} \tag{6.76}$$

Total energy dissipated to the surroundings of a machine as unused heat is called losses. During a practical conversion process, electrical and mechanical losses occur. Figure 6.25 illustrates energy flow in generator and motor operation.

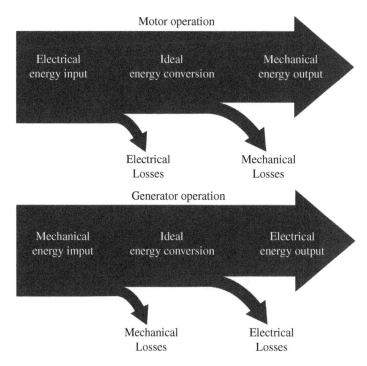

Figure 6.25 Energy flow in generator and motor operation of en electric machine.

In motor operation, part of the electrical input energy is converted into heat on the conductors and magnetic materials. Electrical losses include Joule heating effect RI^2 and magnetic losses in the ferromagnetic core. The remaining electric energy is completely converted into mechanical form in the imaginary ideal part. Before leaving the non-ideal machine, part of the mechanical energy developed in the rotor is dissipated as heat due to friction on the bearings. Larger machines have an internal fan and sometimes water or oil circulation pumps to cool the windings and magnetic core. Energy consumed by the cooling elements are also part of the mechanical losses.

In a generator the energy flow occurs in the reverse order. Part of the mechanical energy applied to the shaft is first lost as friction in the bearings and energy consumed by cooling elements such as fans and pumps. The remaining energy is converted into electricity on the armature windings. Before reaching the external load, part of the induced electric energy is dissipated as heat on the conductors and magnetic materials.

In an electromechanical energy conversion process the changes of entropy and enthalpy are negligible. Unlike thermodynamic cycles, in ideal electromechanical conversion, all mechanical energy is available for conversion to electrical and vice versa. This property makes electromechanical conversion much more efficient than thermodynamic cycles. The efficiency of most electric machines exceeds 90% compared to typical thermodynamic efficiency in the range of 30–40%.

Example 6.3 An electric generator operates at 90% efficiency when it produces 50 MW electric power. Calculate the mechanical power it absorbs, and total energy lost in one hour.

Solution

For 50 MW output power and 90% efficiency, input power is:

$$P_{in} = \frac{50}{0.90} = 55.6 \text{ MW}.$$

Since 5.6 MW energy is lost per second, energy lost in one hour is therefore 5.6 MWh.

6.8 Electric Generation

While electric power can be generated by many conversion techniques, the largest part of electric generation over the world is achieved by generators that convert mechanical energy into electricity. In a generator a magnetic field changing in time by motion induces an EMF across the terminals of coils, according to Faraday's law. A magnetic flux that changes periodically by rotation of the shaft induces an AC voltage across the coil terminals.

There are three practical types of generators named synchronous, asynchronous, and DC generators. Since interconnected transmission and distribution systems are three-phase, generators in all electric power stations deliver three-phase AC power. When DC generation is needed for some special industrial applications or transportation, either a mechanical system consisting of a commutator and brushes or an electronic circuit is used to convert AC into a unidirectional voltage.

The set of coils which deliver electric power is called armature windings. Because fixed contacts are more convenient and reliable to connect external circuit directly to non-moving coils, the armature of practical AC generators is usually on the stationary part called stator.

In AC generators, the magnetic field is produced on the rotor by permanent magnets, or a set of field windings. The rotor may be the internal or external part of the generator, depending on the mechanical design. Since permanent magnet generators do not need electric conduction to the rotor, they are more cost-effective and reliable. Strong permanent magnets were developed after the 1970s allowing commercial production of AC generators with higher output powers up to kilowatts. Such generators are especially used in small size wind turbines.

Larger synchronous generators, such as those used in power stations, have coils rotating with the shaft to produce the magnetic field. A DC field current much smaller than the armature current must be conducted to the rotor windings through slip rings. In asynchronous machines, a magnetic field is produced by currents induced on the rotor windings, therefore conduction of a current to the rotor windings is not needed. In a conventional DC generator, however, the armature windings must be on the rotor to use the commutator. Brushless DC generators developed later have an electronic rectifier to convert AC to DC.

6.8.1 Synchronous Generators

Synchronous generators are the most common type of AC generators, and are called "alternators" in practice. Frequency of the voltage generated by a synchronous machine is

Figure 6.26 Simplified diagram of a permanent magnet generator.

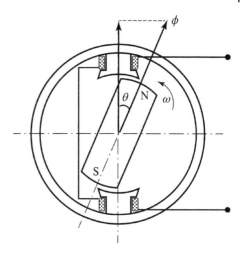

proportional to the rotation speed of the shaft. All generators operating in hydroelectric and thermoelectric power plants, as well as smaller backup generation units, are of this type.

6.8.1.1 Single-Phase Generation

Most synchronous generators are designed to generate three-phase power. To explain the operation principle, we will first consider the simplified structure of a single-phase generator shown in Figure 6.26. The two-pole magnet turns with the angular velocity ω, thus the angle between the flux and the stator winding changes in time as $\theta = \omega t$. The voltage induced on the stator winding is obtained by taking the derivative of the rotating flux component on the stator winding axis according to the Faraday's law

$$v(t) = \frac{d}{dt}(\phi \cos \omega t) = \omega \phi \sin \omega t = V_m \sin \omega t \qquad (6.77)$$

A two-pole generator produces one electric cycle at each rotation. Practical generators usually have multiple pairs of poles. Since the number of cycles per revolution equals half of the total number of poles, frequency of the voltage generated by a synchronous generator that has p poles and is turning at N revolutions per minute (rpm) is

$$f = \frac{p}{2} \frac{N}{60} \qquad (6.78)$$

In a multiple pole generator, it is important to distinguish mechanical and electrical angles. When a p pole rotor completes one revolution, the generated voltage completes $p/2$ electrical cycles. Therefore, the number of electrical degrees equals $p/2$ times the number of mechanical degrees in any electric machine.

6.8.1.2 Three-phase Generation

All commercial-sized synchronous generators operating in electric power stations have three-phase windings distributed around the stator as shown in Figure 6.27. Since stator windings are positioned with equal angles of 120°, as the rotor turns at the angular speed of ω radians per second, voltages with the same magnitude and frequency but shifted by the phase angle of 120° ($2\pi/3$ rad) are induced across their terminals.

$$v_a = V_m \sin(\omega t)$$
$$v_b = V_m \sin\left(\omega t - \frac{2\pi}{3}\right)$$
$$v_c = V_m \sin\left(\omega t - \frac{4\pi}{3}\right) \tag{6.79}$$

In practice, there are two distinct rotor structures. In a round rotor [Figure 6.27(a)], the field windings are inserted into slots around the cylindrical rotor. The air gap between the stator and rotor is practically constant at all points. Such generators are typically used with higher speed gas turbines, and are hence called turbo-generators. To reduce the moment of inertia, the rotor of turbo-generators is designed with a smaller diameter and longer length. In slower systems such as hydroelectric generation, the rotor diameter is larger, and the length is shorter. The poles are mounted around the rotor as shown in Figure 6.27(b), hence the air gap varies around the rotor. Such a structure is specified as hydro-generator because they are common in hydroelectric power plants.

Because the induced voltage is proportional to the time derivative of the magnetic field according to the Faraday's law, in all types of generators the voltage produced across the windings is proportional to the rotation speed if no power is delivered to an external circuit, a special operating condition described by the term "no-load operation" in practice.

6.8.1.3 Motor Operation

Like all electromechanical conversion systems, synchronous machines can also work either as a motor or generator. Although synchronous machines used in electric power systems are designed to operate as generators under normal conditions, under abnormal transient conditions they may temporarily turn into a motor producing torque on the shaft, which would accelerate the prime mover.

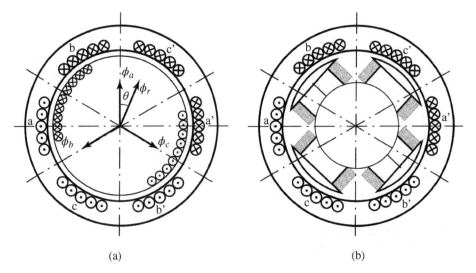

(a)   (b)

Figure 6.27 Simplified structure of three-phase synchronous generators with (a) round-rotor and (b) salient-pole rotor.

In the motor operation mode, the shaft of a synchronous machine must turn at a speed precisely imposed by the supply frequency. The expression below defines the synchronous speed (in rpm) in terms of the supply voltage frequency and total number of poles.

$$n_s = \frac{120f}{p} \tag{6.80}$$

A synchronous motor rotates at the synchronous speed at any load within its range of operation. If overloaded, a synchronous motor becomes unstable and cannot continue to deliver mechanical power. In early years of electrical engineering, synchronous motors were used in applications where constant speed was needed. With the evolution of feedback control systems, they have been replaced with other types of machines with electronic speed controllers.

6.8.1.4   Rotating Magnetic Field

When three stator windings placed 120° apart are supplied from a balanced three-phase source, the resultant magnetic field with constant magnitude rotates at the synchronous speed around the rotor axis. To illustrate the rotating magnetic field concept, an empty stator containing just three coils is shown in Figure 6.28(a).

The expression of the flux densities created by phase windings can be written as

$$\mathbf{B}_{aa'}(t) = [B_m \sin \omega t]\angle 0°$$
$$\mathbf{B}_{bb'}(t) = [B_m \sin(\omega t - 120)]\angle 120°$$
$$\mathbf{B}_{cc'}(t) = [B_m \sin(\omega t - 240)]\angle 240° \tag{6.81}$$

Note that the magnitudes of the flux density vectors are sinusoidal time functions, and at the same time the directions of the vectors are rotated by 120° with respect to each other.

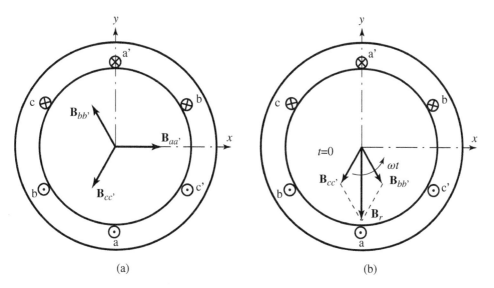

(a)   (b)

Figure 6.28 Production of a rotating magnetic field.

For example, at $t = 0°$, the flux density vectors will be

$$|\mathbf{B}_{aa'}| = 0$$

$$|\mathbf{B}_{bb'}| = B_m \sin(-120) = -\frac{\sqrt{3}}{2} B_m$$

$$|\mathbf{B}_{cc'}| = B_m \sin(-240) = \frac{\sqrt{3}}{2} B_m \qquad (6.82)$$

As shown in Figure 6.28(b), the sum of three flux density vectors at $t = 0$ yields

$$B_r = 0 + \left[-\frac{\sqrt{3}}{2} B_m \angle 120° \right] + \left[\frac{\sqrt{3}}{2} B_m \angle 240° \right] = 1.5 B_m \angle 270° \qquad (6.83)$$

At all instants, the resultant flux density B_r will have the same magnitude of 1.5 B_m and the angle ωt. Therefore, the total magnetic flux created by three stator coils shifted in space by 120° and supplied by equal magnitude currents with angles shifted in time by 120° produce a magnetic field equivalent to a magnet rotating at the speed ω.

6.8.2 Induction Machines

Induction machines are also known as asynchronous machines because the magnetic flux necessary for electromechanical energy conversion is created on the rotor conductors by the stator windings through electromagnetic induction. The stators of induction machines are similar to synchronous machines. In practice, induction machines are mostly used as motors, but their operation as generators became more important with the development of commercial-size wind turbines and small-scale hydroelectric units where the turbine speed changes in a wide range.

There are two types of induction (or asynchronous) machines. One type, called "short-circuit" or "squirrel cage" induction machines, have conductor rods embedded in the rotor and are short-circuited at both ends by rings. The name squirrel cage refers to the similarity of the rotor conductor assembly to rotating cages in which small pets like squirrel or hamster can run. The other type is known as wound-rotor or slip-ring induction machine. In this type of machine, the rotor holds windings connected to an external circuit for control purposes via brushes pressing on slip rings mounted on the shaft.

6.8.2.1 Induction Motor

In motor operation of a three-phase induction machine, the stator windings produce a magnetic field revolving at the synchronous speed defined by Eq. (6.80). When the rotor is at rest, the rotating stator magnetic field $\mathbf{B}_s$ sweeps across the rotor conductors at the maximum rate and induces an EMF on the rotor depending on the stator voltage, number of turns in the stator coil, and number of conductors in the rotor coil. The induced EMF creates a current through the closed loop of the rotor coils, producing a counter EMF on the stator coils.

As the rotor accelerates, its relative speed with respect to the stator rotating field decreases. Hypothetically, if the rotor could turn at the same speed as the stator field, there would not be any induced voltage to supply the rotor magnetic field; therefore, it would slow down. In other words, an induction motor can never reach exactly the synchronous

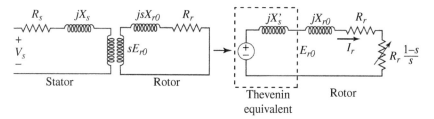

Figure 6.29 Induction machine equivalent circuit.

speed, which explains why it is also called an asynchronous motor. The difference between the synchronous speed and the actual rotor speed is called the "slip speed." The variable named "slip" refers to the ratio of the slip speed to the synchronous speed. In practice, slip may be expressed in percentage.

$$s = \frac{n_s - n_r}{n_s} \tag{6.84}$$

When the rotor is at rest slip is 1, and at synchronous speed slip is 0. Therefore, in the normal operation of an induction motor, slip varies between zero and one.

$$0 < s < 1 \tag{6.85}$$

Operation of an induction motor can be analyzed using the equivalent circuit shown in Figure 6.29. The equivalent circuit of a squirrel cage induction machine is similar to the equivalent circuit of a transformer. Magnetic coupling between the stator and rotor windings is represented by an ideal transformer. When the rotor is blocked (standstill), the magnitude of the EMF induced on the rotor coil E_{r0} depends on the ratio of the rotor to stator turns per phase. The frequency of the induced EMF is the same as the frequency of the stator supply voltage. When the rotor turns due to the relative speed between the rotating stator field and moving rotor coils, the magnitude and frequency of the induced EMF decreases proportionally to the slip. Therefore,

$$E_r = sE_{r0}$$
$$f_r = sf_{r0} \tag{6.86}$$

Since the reactance of an inductor is $X = j\omega L = j(2\pi f)L$, the reactance of the rotor winding is obtained by multiplying the rotor reactance at rest X_{r0} with the slip. The rotor current is

$$I_r = \frac{sE_{r0}}{R_r + jsX_{r0}} = \frac{E_{r0}}{\frac{R_r}{s} + jX_{r0}} \tag{6.87}$$

On the right side of Figure 6.29, the rotor circuit and the ideal transformer is replaced with the Thevenin equivalent seen from the rotor. By dividing the numerator and denominator of the current to s, the circuit parameters are transformed to E_{r0}, jX_{r0}, and R_r/s. In addition, the actual rotor resistance is separated from the variable load resistance.

$$\frac{R_s}{s} = R + \left(\frac{1}{s} - 1\right)R_s = R_s + \frac{1-s}{s}R_s \tag{6.88}$$

A small part of the electric power transferred to the rotor is converted to heat on the actual rotor resistance as $I^2 R_r$ and a larger part is converted to mechanical power on the load resistance $R_s(1-s)/s$, which corresponds to the mechanical load. Mechanical power developed on the rotor is therefore

$$P = 3I_r^2 \frac{1-s}{s} R_r \tag{6.89}$$

The torque developed on the rotor shaft is obtained by multiplying the rotational speed by the mechanical power.

$$T = \frac{P}{\omega} = \frac{3I_r^2}{\omega} \frac{1-s}{s} R_r \tag{6.90}$$

In practice, rated power of a motor is expressed in horsepower (1 HP = 746 W_e) and the rotation speed of the shaft in rotations per minute (rpm). Then the radial speed is found from the expression $\omega = 2\pi N/60$.

Detailed analysis of induction machine performance is beyond the scope of this text. The reader may refer to specific texts on electric machinery such as Chapman (1991), and Fitzgerald et al. (1992). The National Electrical Manufacturers Association (NEMA) has published standards on induction motor parameters and characteristics. A typical torque-speed characteristic of NEMA B-type induction motor is shown in Figure 6.30. Significant points are marked with letters on the figure. Point "a" is the starting torque. This torque also corresponds to breakaway torque, which occurs when the rotor is blocked while the stator is energized. NEMA defines the locked-rotor torque as "*the minimum torque developed by the motor at rest for all angular positions of the rotor, with rated voltage applied at rated frequency*" (NEMA 2014).

As the rotor accelerates torque reaches a maximum value referred to as breakdown torque (point b). The minimum torque developed by the motor during the period of acceleration from rest to the speed at which breakdown torque occurs is called pull-up torque. In motor operation, breakdown torque corresponds to the maximum mechanical load that an induction machine can handle. Rated torque (point c) the motor can deliver under normal load conditions at the rated voltage and frequency. At no-load conditions, the speed approaches the synchronous speed n_s. Between the points b and d, the torque-speed characteristic is almost linear.

6.8.2.2 Induction Generator

If an external prime mover forces the rotor to turn above the synchronous speed while the stator is connected to a power source, the induction machine operates as a generator. Suppose that an induction machine is supplied from a three-phase source and its shaft is mechanically coupled with a prime mover. While the machine is operating as a no-load motor, the mechanical prime mover starts to turn the shaft in the same direction to increase the speed above the synchronous speed imposed by the supply circuit. In this situation, slip becomes negative and makes the power developed in the air gap negative. Hence, the induction machine delivers electrical energy to the source. Torque-speed characteristic of an induction generator operating in the generation mode is shown in the right section of Figure 6.30. The maximum torque in generator operation mode is called pushover torque, which may be a different value than the breakdown torque in motor operation.

As a generator, an induction machine has limitations. First, because it has no separate field circuit, it requires reactive power supplied to the stator to maintain the magnetic

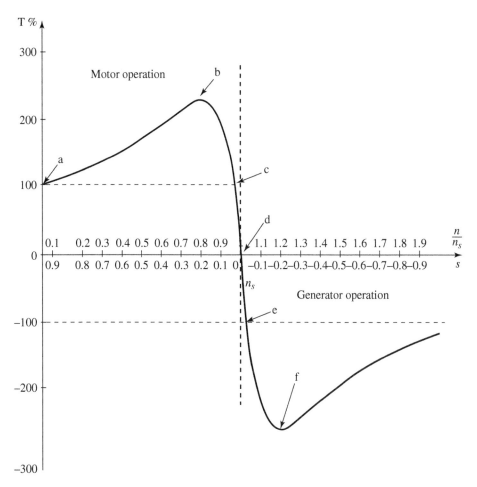

Figure 6.30 Torque-speed characteristic of an induction machine.

field. The output voltage and frequency cannot be adjusted by changing its speed and the magnetic field as in synchronous generators. Also, it cannot produce, but rather consumes reactive power.

When the stator is connected to a strong three-phase grid, the reactive power needed to create the magnetic field is drawn from the supply source. The stator voltage and frequency are determined by the grid. Variations of the generator speed change the generated power delivered to the grid.

However, if the induction generator operates alone, the reactive power must be supplied from a capacitor bank. In order to start the self-excitation process, there must be a residual magnetic flux from previous operation of the machine. As the generator starts to turn, the residual magnetism produces a small voltage on its terminals. That voltage causes a capacitive current flow, which increases the terminal voltage. Consequently, the capacitive current increases until the straight line defined by the capacitance value and the nonlinear current defined by the magnetization curve intercept. Thus, the terminal voltage and frequency of an induction generator operating independently from an external source depend on the capacitance value, rotation speed, and load conditions. In fact, the terminal voltage varies in a wide range as the electrical load changes. In the case of an inductive load, the voltage may collapse rapidly since the reactive power of the capacitors will be also supplying the reactive power of the load. The frequency also changes with variable load due to the torque-speed characteristic; but since the characteristic is steep within the normal operation range, the frequency variation remains limited.

In practice, induction generators are not preferred for standalone operation. However, application of induction generators has gained particular importance with the increase of commercial-scale wind power generation. Currently, induction generators are the most common generator type used in commercial-size wind turbines (Hansen 2005).

Both squirrel cage (short-circuit) and wound rotor (slip-ring) type induction generators are used in large wind turbines. Squirrel cage induction generators offer the advantages of robustness, mechanical simplicity, and lower cost due to series production. Wound-rotor type generators offer more control flexibility in wind power applications. In either case, the generator is coupled to the wind turbine rotor through a gearbox since the optimal turbine and generator speeds are different. Induction generators driven by wind turbines are equipped with a soft-starter mechanism and a capacitor bank for reactive power compensation.

When a squirrel cage generator is directly coupled to the grid, its speed changes only a few percent around the synchronous speed dictated by the grid. Hence, this type of generator is more suitable for constant-speed wind turbines. Active power, reactive power, terminal voltage, and rotor speed of a squirrel cage generator are interrelated by a unique equation. The amount of absorbed reactive power is not controllable and speed changes directly increase the delivered active power as well as the consumed reactive power. While capacitor banks are not needed for grid-connected operation of a squirrel cage generator, proper reactive power compensation reduces transmission losses. The main drawback of the capacitors used to compensate reactive power is electrical transients during switching instant.

Wound-rotor generators are more expensive and less robust compared to squirrel-cage generators. However, their voltage-speed characteristic can be controlled. Rotor windings are either connected to an external circuit by sliprings and brushes or to a power electronics converter system which may or may not require connection to the external circuit. The rotor windings may be connected to a set of resistors and a converter circuit moving with the rotor. The converter optically linked to an external controller modifies the slip of the generator by changing the current that flows through the resistance. Such an arrangement does not need sliprings. Doubly fed induction generators have stator windings directly connected to the constant frequency power grid and the rotor windings supplied through a bidirectional electronic converter. This system allows a variable speed operation over a wide range. The market for doubly fed induction generators has been significantly growing.

6.9 Electric Transmission and Distribution

Electric transmission and distribution systems consist of transformers, transmission lines, cables, switchgear, protection relays, supervisory control and data acquisition (SCADA) devices, and reactive power compensation units. Step-up and step-down transformers change the voltage to the appropriate level depending on the transmission distance and amount of power transmitted. Long distance transmission is achieved by high-voltage (HV) or EHV overhead lines. Distribution of electric power in populated areas is typically at medium and low voltage levels. Electric power is delivered to consumers via low-voltage overhead lines or underground cables. Transformers are covered in Section 6.6. This section focuses on the electrical characteristics, equivalent circuits, and voltage-current relationships of transmission lines.

Transmission lines are designed for diverse nominal system voltages depending on many factors including the transferred power, distance, generation siting, terrain types, ground cover, population density, land usage, climate, visual impact, communication interference, and possible biologic effects. In general, below 100 kV is considered medium voltage and used for secondary transmission. High voltage lines in the 115–230-kV range are used for primary transmission. EHV lines above 345 kV, are used for long distance transmission. In the US, 138-kV is the most common system voltage with total 16,186 circuit-miles. Transmission voltages between 230 and 500 kV became common in long distance power lines. For instance, hydropower from the Hoover Dam is transmitted to southern California at 500 kV. Except for some highly populated areas and underwater connections, all high voltage transmission is done by three-phase overhead lines.

Electric distribution systems deliver electric power to consumers via relatively small transformers, insulated or uninsulated overhead lines, and cables. Distribution transformers may be a bank of three single-phase transformers mounted on a pole or a three-phase transformer placed in a cabinet near residences. Electric power is distributed to consumers at a standard low-voltage level for safety. Some larger buildings or compounds may be supplied directly in three-phase and step-down the voltage to the standard low level by a transformer installed in a building or secured area.

In North America most residential and small commercial consumers are supplied with 120/240-V at the grid frequency of 60 Hz. Three-wire single-phase wiring where one of the wires is called neutral and grounded is common in commercial and residential buildings. The voltage between two live lines is 240 V and between line and neutral is 120 V. In Europe and most parts of the world, groups of houses and larger buildings are supplied through three-phase circuits where the phase to neutral voltage is 220 or 240 V depending on the country. Equipment that needs relatively higher power may be supplied directly in three-phase. Three-phase electric power is distributed as equally as possible to supply smaller appliances and lighting loads.

6.9.1 Transmission Line Parameters

Electrical behavior of transmission lines depends on their resistive, inductive, and capacitive parameters distributed along the wires. Analysis of transients that occur due to lightning strikes and switching surges require distributed parameter analysis either using comprehensive computer models or analysis of electromagnetic wave propagation along the line. Low frequency transients and steady-state operation can be performed using equivalent circuits where the total resistance and reactances of the wires are assumed to be lumped at certain points, mostly the ends and/or center of the line. Since electrical parameters depend on the wire diameters and distances to each other and ground, the parameters can be assumed constant along the line and determined per unit length. In balanced three-phase systems, parameters are generally defined per phase.

6.9.1.1 Line Resistance

Resistance of a long wire is proportional to its length and inversely proportional to its cross-sectional area. The resistance of a long wire with cross sectional area of A-mm^2 made from a material with σ [m/($\Omega \cdot$mm^2)] is calculated using Eq. (6.91).

$$R_T = \frac{1}{\sigma A} \; \Omega/m \tag{6.91}$$

Among all metals listed in Table 6.1, copper is the best conductor after silver. Because of the high cost, use of silver as a conductor is limited to internal wiring of some semiconductor devices including PV modules. While gold is not a better conductor than silver and copper, it is preferred in high-quality connectors because it is not affected by corrosion. Aluminum conductor is an economical choice for some electric equipment, cables, and transmission lines. Its electrical conductivity is lower than copper but since its density is also lower, aluminum wire with the same resistance as copper wire has lower cost and weight. Aluminum conductors suffer from two major practical limitations: first, soldering and welding aluminum conductors is not as easy as copper; and second, when aluminum is in contact with copper, electro corrosion effect causes oxidation of both surfaces and creates additional resistance at the point of connection.

Overhead lines and cables are generally made from annealed copper or electrical grade aluminum. In practice, transmission line conductors are made from a bundle of stranded steel core to increase mechanical strength surrounded by aluminum conductors for better electrical conduction. Such wires designated with the acronym ACSR (Aluminum Conductor Steel Reinforced) is the most common conductor material in high-voltage overhead lines.

Table 6.1 Electrical conductivity and temperature coefficient of common metals.

Material	Electrical Conductivity σm/(Ωmm^2)	Temperature Coefficient α (1/°C)
Silver	63.0	6.1×10^{-3}
Copper	59.6	4.29×10^{-3}
Copper – annealed	58.0	3.9×10^{-3}
Gold	45.2	—
Aluminum	37.7	3.8×10^{-3}
Tungsten	18.9	4.5×10^{-3}
Zinc	16.6	3.7×10^{-3}
Nickel	14.3	6.4×10^{-3}
Iron	9.93	6.4×10^{-3}
Platinum	9.66	3.9×10^{-3}
Brass	14–17	1.5×10^{-3}

Source: https://www.engineeringtoolbox.com/ (accessed on 8/9/2019).

Resistance of a wire is temperature-dependent. As current passes through a transmission line, the wire temperature rises by the temperature coefficient specific to the conductor material. A resistance R at the temperature T increases to R' at a higher temperature T' as described by Eq. (6.92).

$$R' = R[1 + \alpha(T' - T)] \tag{6.92}$$

Electrical conductivity at 20 °C and temperature coefficient of some common metals is listed in Table 6.1.

6.9.1.2 Line Inductance

The current flowing through a transmission line creates a magnetic field inside the conductor and in the space around the wires. Inductive property of transmission lines does not have any effect in DC transmission. In AC transmission, however, inductive reactance is a significant part of the wire impedance.

Consider a single-phase line formed by two wires with the same radius r separated by a distance D. The total line inductance per unit length can be calculated by Eq. (6.93) (Faulkenberry and Coffer 1996):

$$L = (4 \times 10^{-7}) \ln \left[\frac{D}{r'}\right] \quad \text{H/m}$$
$$r' = re^{1/4} = 0.7788r \tag{6.93}$$

The factor 4×10^{-7} comes from the permeability of vacuum divided by the number pi (3.14). The relative permeability of air and the wire conductors are assumed to be unity. The factor 0.7788 is the result of mathematical simplification.

In three-phase overhead lines supplying a balanced load the sum of phase currents of a transmission line is zero, therefore no return current flows through the neutral. In most three-phase overhead lines no neutral wire is needed, but a small-diameter grounded wire is installed above the phase wires for lightning protection. If the phase wires are a symmetrically spaced, the inductance per phase would be half of the inductance of a two-wire single-phase line.

$$L_{3\phi} = (2 \times 10^{-7}) \ln \left[\frac{D}{r'}\right] \quad \text{H/m per phase} \tag{6.94}$$

The three conductors of a three-phase line may not be equally spaced for some practical tower structures. An unsymmetrical spacing results in unequal inductances for phase conductors causing a different voltage drop on each phase. To resolve this issue, phase wires are transposed at equal intervals along the transmission line. If the wires are not equally spaced, the geometric mean distance D_e should be used for the wire distance in Eq. (6.94).

$$D_e = \sqrt[3]{D_{ab}D_{bc}D_{ca}} \tag{6.95}$$

6.9.1.3 Line Capacitance

The wires of a transmission line can be considered as a capacitor formed by two or more parallel cylindric conductors in a non-conducting medium, which is air. The potential difference between the wires of a transmission line creates electric charges on each conductor at a different polarity in respect to each other and the ground. The accumulated charge is proportional to the voltage between the conductors. Therefore, the capacitance depends on the radius and separation distance of the conductors.

The conductors of a single-phase transmission line can be considered like two parallel axis cylinders separated by the distance of the wires. The following expression can be derived for the capacitance per unit length of a single-phase, two-wire transmission line with wires of equal radius r, separated by the distance D.

$$C = \frac{\pi \varepsilon_0}{\ln\left[\frac{D}{r}\right]} \text{ F/m} \tag{6.96}$$

where ε is the permittivity of air, approximately equal to the permittivity of free space (vacuum), 8.854×10^{-12} F/m.

For a three-phase line that has equilaterally spaced conductors and is carrying balanced voltages, the per-phase (line-to-neutral) capacitance can be expressed as (Faulkenberry and Coffer 1996):

$$C = \frac{2\pi \varepsilon_0}{\ln\left[\frac{D}{r}\right]} \text{ F/m} \tag{6.97}$$

If the distances between wires are not equal, the geometric mean distance given in Eq. (6.95) should be used for the distance D.

In DC transmission, line capacitance is equivalent to an open circuit, therefore does not have any effect on the voltages and currents in steady-state operation. In AC, however, the line capacitance produces a reactance, which results in a capacitive current. Capacitive current on a transmission line affects the power transmitted, power factor, and the voltage drop along the transmission line.

6.9.2 Representation of Transmission Lines

A transmission line can be represented by an equivalent circuit comprising parameters calculated for the full length of the wire. In DC transmission, the equivalent circuit does not contain the line inductance and capacitance. In short overhead lines (less than 80 km or 30 mi), the capacitances between lines and ground are negligible. Figure 6.31 shows the

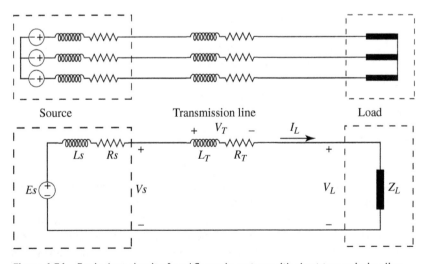

Figure 6.31 Equivalent circuit of an AC supply system with short transmission line.

per-phase equivalent circuit of a three-phase balanced AC supply system. The capacitive reactance is considered in the analysis of a medium length transmission line (80–240 km or 30–150 mi). Long transmission lines are analyzed by solving differential equations written for distributed parameters. Distributed parameter representation and analysis are beyond the scope of this text.

6.9.3 Short Transmission Lines

In an overhead line shorter than 80 km (30 mi) the effect of the capacitance between the wires and ground is negligible. Therefore, the line can be represented as shown in Figure 6.31. The voltage variation between the sending end (source side) and the receiving end (load side) of the line can be obtained in the phasor domain as

$$\mathbf{V}_s - \mathbf{V}_L = \mathbf{Z}_T \mathbf{I}_L = (R + j\omega L)\mathbf{I}_L \tag{6.98}$$

The voltage variation at the receiving end of the line between the no-load and full-load conditions relative to the no-load voltage expressed in percent is defined as voltage regulation.

$$\varepsilon_\% = \frac{V_{L,no-load} - V_{L,full-load}}{V_{L,no-load}} \times 100 \tag{6.99}$$

Voltage regulation depends on the line impedance, full-load current, and the power factor of the load. Effects of the load power factor on voltage variation can be visualized from the phasor diagrams shown in Figure 6.32. Inductive load causes a voltage decrease (a) and a capacitive load increases the voltage (c) for the same no-load voltage at the receiving end and same current magnitude. Pure resistive load (b) causes the minimum voltage variation from no-load to full-load.

6.9.3.1 Resistive Losses

The power consumed on the transmission line is dissipated to the surrounding area in the form of unused heat, referred to as transmission loss. Power dissipated on a transmission line can be expressed in terms of the load current I_L and total line resistance R_T.

$$P_T = R_T I_L^2 \tag{6.100}$$

Efficiency of a transmission line can be defined similar to the efficiency of converters.

$$\eta_T = \frac{P_T}{V_S I_L} = \frac{V_S I_L - R_T I_L^2}{V_S I_L} = 1 - \frac{R_T I_L}{V_S} \tag{6.101}$$

where V_S is the voltage at the sending end (source side) of the line.

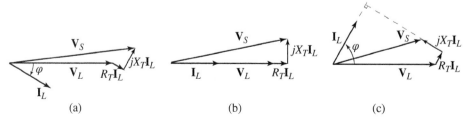

Figure 6.32 Short transmission line phasor diagrams for inductive, resistive, and capacitive loads.

The efficiency of a transmission line can be increased either by reducing the line resistance or increasing the supply voltage. Since transformers do not work on DC, the supply voltage is slightly higher than the load voltage by the amount of voltage drop on the transmission line. The major practical challenges of low voltage DC power systems are the increased voltage drop and transmission losses due to higher current.

Although copper is a better electric conductor, aluminum is more economical in longer overhead lines because of its lower specific mass and price. In shorter distance underground transmission and buildings wiring, plastic insulated copper wires are used. For safety, the US National Electric Code (NEC) and similar codes require that DC cables are protected in metal conduits.

As indicated in the previous section, in early DC systems the transmitted power was limited by the highest voltage level safe to be used by consumers. Increasing the conductor diameter to reduce the heat losses on the wires increases the cost significantly.

Electric loss on electric wires is dissipated as heat around the conductors. Overheating may damage the cable insulation, ultimately turning the insulation materials into carbon, which, in fact, conducts electric current and results in short circuits. The NEC and similar electric codes around the world specify the current carrying capability of insulated conductors (see Table 6.2).

Table 6.2 Standard cable sizes and recommended current carrying capacities.

| | | | | Max load (A) | | |
| | | | | Number of cores | | |
AWG	Diameter (mm)	Area (mm^2)	Single core	1	up to 3	4 to 6
24	0.51	0.02	0.2	3.5	2	1.6
22	0.64	0.025	0.33	5.0	3	2.4
20	0.81	0.032	0.5	6.0	5	4.0
18	1.0	0.04	0.82	9.5	7	5.6
16	1.3	0.051	1.3	15	10	8.0
14	1.6	0.064	2.1	24	15	12
13	1.8	0.072	2.6	—	—	—
12	2.1	0.081	3.3	34	20	16
10	2.6	0.1	5.3	52	30	24
8	3.3	0.13	8.3	75	40	32
6	4.1	0.17	13.3	95	55	44
4	5.2	0.2	21.2	120	70	56
3	—	—	26.7	154	80	64
2	6.5	0.26	33.6	170	95	76
1	7.4	0.29	42.4	180	110	88

Source: https://www.engineeringtoolbox.com/ (accessed on 9/8/2019).

6.9.4 DC Transmission and Distribution

DC transmission has practical applications at both very high-voltage or low-voltage levels. HVDC systems serve in long distance power transmission or asynchronous connection of separate grids. Consumer electronics, computers, data storage systems, telecommunication systems, and plug-in vehicles are major electric loads supplied by DC. Since electric power is delivered to consumers in AC, low-voltage DC to supply such devices is produced near or at the point of use.

Figure 6.31 shows the equivalent circuit of a basic DC transmission system. The power source (DC generator, PV array, battery bank, etc.) is represented by an ideal voltage source and a series resistance. The ideal voltage source E_s is the induced electromotive force (EMF) or open-circuit (no-load) voltage of the physical DC converter. R_s corresponds to the equivalent internal resistance of the non-ideal source. Output voltage of the DC source drops to V_s when the load current I_L is drawn. The resistor R_T is the total resistance of the transmission line. In DC, two wires are needed to complete the current flow between the source and the load. Therefore, R_T is equal to the sum of the resistance of the two wires that constitute the transmission line. R_L is the total equivalent resistance of the supplied electric load.

In general, the rated voltage of a power source is given for continuous operation when the source is supplying the rated current. At the rated operation conditions, the voltage across the load terminals can be obtained by applying Kirchhoff's voltage law (KVL) to the current loop.

$$V_L = V_S - V_T \tag{6.102}$$

Hence, the load current is calculated by using the Ohm's Law to the equivalent resistance of the DC load.

$$I_L = \frac{V_L}{R_L} \tag{6.103}$$

Power delivered by the source is the product of the terminal voltage and load current.

$$P_S = V_S I_L = (V_T + V_L) I_L \tag{6.104}$$

Electric power produced by ideal energy conversion is the product of the induced EMF and the delivered current.

$$P_S = (E_S - R_S I_L) I_L = E_S I_L - R_S I_L{}^2 \tag{6.105}$$

The term $R_S I_L{}^2$ corresponds to the heat losses in the conversion system. The efficiency of an electric transmission system is the ratio of the active power it delivers (P_s) to the load to the active power drawn from the source ($E_s I_L$).

$$\eta_S = \frac{P_S}{E_S I_L} = \frac{E_S I_L - R_S I_L{}^2}{E_S I_L} = 1 - \frac{R_S I_L}{E_S} \tag{6.106}$$

Electronic devices used to produce DC from AC (rectifiers) are the most efficient converters. Electric generators have additional losses due to magnetic core and friction; they operate typically at an efficiency above 90%. Electrochemical conversion in batteries and fuel cells is less efficient and depends on the electrodes and composition of the electrolyte. PV conversion has a lower efficiency but the advanced cell technologies have improved the efficiency to the range of 20–25%.

6.9.4.1 Voltage Regulation

In addition to transmission losses, voltage drop from the generation unit to the consumer is a major challenge in low voltage DC transmission. In practice, voltage regulation is defined as the deviation of the voltage across the transmission line relative to the voltage at the sending end. Voltage regulation is generally expressed in percent by multiplying its decimal value by 100.

$$\varepsilon = \frac{V_S - V_L}{V_S} = \frac{R_T I_L}{V_S} \tag{6.107}$$

In DC transmission, voltage regulation is proportional to the product of transmission line resistance and load current. Terminal voltages and powers are more often used in the analysis of electric energy systems. Power consumed by the DC load is

$$P_L = V_L I_L = R_L I_L^2 = \frac{V_L^2}{R_L} \tag{6.108}$$

Transmission loss is the power dissipated as heat on transmission lines. It can be expressed in terms of supply voltage and transmitted power.

$$P_{Loss} = R_T I_L^2 = R_T \left[\frac{P_L}{(1 - \varepsilon)V_S} \right]^2 \tag{6.109}$$

Both the voltage regulation and transmission losses must be minimized in a practical power supply system.

Because of the voltage regulation and transmission loss issues, sources that produce direct DC power are connected to a distant load through DC-AC-DC conversion systems. Power electronic converters have become cost effective for such conversions. For example, a microinverter converts the DC output of a PV module at its terminals to AC and combines the power generated by each module in a PV unit. Then the total generated power is transmitted into AC to the connection point with the utility grid. Uninterruptible power supply (UPS) modules have an internal backup battery which is charged by a rectifier. When the grid power becomes unavailable, the UPS supplies the load as AC through an inverter.

6.10 Electric Loads

At the consumer side, electrochemical processes such as electrolysis, metal plating using galvanoplasty methods, charging batteries, some transportation systems, computers, data centers, and consumer electronics use DC power.

DC power converted to other forms at the consumer side is calculated similarly using the equivalent resistance of the load R_L. Only heating equipment convert all electric power consumed to useful power if the goal is heating. Losses are inevitable in practical electric equipment. In electromechanical devices, a fraction of the equivalent load resistance corresponds to losses dissipated as unused heat. The remaining fraction of the resistance represents the electric power converted into mechanical power. In lighting equipment, the desired output is the radiant power, usually known as light intensity. Incandescent lightbulbs are based on heating a filament; therefore, a big proportion of the electric power is dissipated as heat to produce a certain radiant power. Conventional fluorescent lamps and compact fluorescent lamps (CFL) are more efficient since most of the electric power is converted to light by discharge in a low pressure gas rather than electric resistance. LED lamps are basically solid-state elements without a heating element that produce light directly from DC power. Although a small part of the electric power is still converted to heat on the electronic elements, LED lights are much more efficient than incandescent lightbulbs.

Power electronics circuits allow efficient conversion of AC to DC and DC to AC. Sources that deliver direct DC (PV modules, batteries, fuel cells) and equipment that operate on DC (LED lights, consumer electronics, etc.) are integrated in the utility grid through electronic converters that perform the conversion between AC and DC at the point of use.

6.11 Chapter Review

This chapter presents an overview of electrical energy systems. Supply of electric power to the general public started during the second half of the nineteenth century with DC generation units. Because DC voltage could not be stepped up or down directly, the distance over which electric power could be transmitted was limited, and DC generation units had to be installed near the consumers. Development of transformers and three-phase AC systems resolved this issue at the turn of the century.

Typical output voltage of an AC generator is in the 11–25 kV range due to practical and economic limitations. Transformers step up the generators' output voltage to a much higher level for long distance transmission, and step down to a safe and convenient level for distribution to consumers. High voltage and EHV lines can transmit larger amounts of power over longer distances. From the generation units to consumers, the voltage level is increased or decreased in several substations. Local and regional power grids are interconnected to optimize the capacity factor of generation units while increasing the system reliability. Currently, wide-area interconnected grids formed by connecting national grids cover most part of the world.

HVDC transmission systems have evolved to transmit higher voltages over longer distances. In an HVDC system, generator output voltage is increased to high voltage levels ranging between 500 and 765 kV (or even higher), then converted to DC by power electronic converters. DC transmission has the benefit of not being affected by inductive and capacitive transmission line reactances. At the receiving end, HVDC is converted back to AC through power electronic inverters and stepped down to a voltage level convenient to regional transmission. HVDC systems are also used for underwater power transmission to and between islands, and connect two electric grids operating at different frequencies, by the method called asynchronous interconnection.

Major elements of an electric power system are generators, transformers, transmission lines, cables, switchgear, and equipment used for protection, control, and data collection. Three-phase AC power dominates the electric generation, transmission, and distribution systems. DC is used in HVDC transmission, some rail transportation, autonomous vehicles, and electronic equipment including telecommunication, consumer electronics, and computers.

Voltage, current, and power distribution in an electric energy system are studied using electric circuit analysis methods. In AC circuits, voltages and currents are represented by complex numbers, or vectors, called phasor quantities. The magnitude of a phasor is equal to the effective (or rms) value of the corresponding sinusoid and angle equal to the phase angle from a predefined reference. Impedance is a complex number found by dividing the voltage phasor with the current phasor. In AC networks, complex power is calculated by multiplying the voltage phasor with the conjugate of the current phasor. Magnitude of the complex power is the product of the effective (rms) values of the voltage and current, defined as apparent power. Real part of the complex power is active power, which is converted to heat, mechanical work, or other forms of energy. Power factor is the ratio of the active and apparent powers. Active power is also obtained by multiplying the apparent power with the power factor. Reactive power is the imaginary part of the complex power, which is absorbed by inductive and capacitive elements of an AC circuit. If a load absorbs excessive reactive power, the supply current increases resulting in additional heat losses in generators, transformers, and transmission lines. Reactive power causes inefficient operation of the energy system, hence it must be generated at the point of use by connecting capacitors and reactors in parallel to the load. Increasing the power factor without changing the supplied active power and the voltage across the load is called power factor correction or reactive power compensation.

The largest amount of electric power is generated by electromechanical energy conversion in synchronous generators driven by thermal or hydraulic turbines. A smaller fraction of electric power is generated by asynchronous generators and permanent magnet generators in wind turbines and direct PV conversion in solar generation units.

In all electric generators, a magnetic field changing in time due to the motion of the rotor induces voltages across the terminals of armature windings, according to Faraday's electromagnetic induction law. In larger synchronous generators, the rotor windings create a rotating magnetic field sweeping the stator windings where voltage is induced. Frequency of the voltage generated by synchronous generators depends only on the product of rotor speed and the number of poles ($f_s = pN_s/120$). In small synchronous generators, permanent magnets rotating with the shaft create the magnetic field. The speed of the rotating magnetic field N_s is called synchronous speed.

In induction generators, also known as asynchronous generators, a rotating field created by stator windings induces voltages across the rotor conductors by electromagnetic induction effect. Induction machines are the most common type of AC motors used in industrial equipment and domestic appliances. In motor operation mode, an induction machine always rotates at a speed lower than the synchronous speed. The difference between the synchronous speed and the rotor speed determines the variable defined as slip ($s = [N_s - N]/N_s$), which is unity when the rotor is blocked and approaches zero at no-load in motor operation. When the induction machine is forced to turn above the synchronous speed by a prime mover, it operates as generator. Most commercial-size wind turbines operate asynchronous machines in the generator operation mode. Smaller wind turbines utilize permanent magnet generators.

Electric power is delivered to consumers through overhead lines or underground cables. A transmission line can be represented by an RLC circuit where R corresponds to the resistance of the wire conductor, L is the inductance due to the magnetic field around and between the conductors, and C is the equivalent capacitance between the phase lines and ground. In DC transmission, line reactances do not affect the power flow. Inductances and capacitances cause reactive currents in AC, which affect the transmitted power and voltage variations between the sending and receiving ends of a transmission line.

Review Quiz

1 The largest fraction of electric power is generated by
  a. DC generators.
  b. induction generators.
  c. direct conversion from heat.
  d. synchronous generators.

2 Asynchronous machines operate as generators if
  a. they are rotated in the opposite direction to the rotating field.
  b. they are driven to turn faster than the synchronous speed.
  c. they turn slower than the synchronous speed.
  d. their stator voltage is reversed.

3 Faraday's law states that the voltage induced across a coil is proportional to the time rate of change of
  a. magnetic flux.
  b. electrostatic field.
  c. electric charge.
  d. current flowing through it.

4 Which type of machines are dominant in commercial-scale wind powered generation?
  a. Asynchronous generators
  b. DC generators
  c. Permanent magnet generators
  d. Synchronous generators

5 Which type of machines are dominant in small-scale wind powered generation?
  a. Asynchronous generators
  b. Permanent magnet generators
  c. DC generators
  d. Synchronous generators

6 In electrical energy systems transformers are used to
  a. convert AC to DC.
  b. convert DC to AC.
  c. change the frequency from one side to the other.
  d. change the voltage from one side to the other.

7 High voltage DC (HVDC) transmission is more suitable for
  a. long distance transmission of larger amounts of power.

b. short distance transmission from a PV generation unit to the grid.

c. battery backup distribution systems.

d. synchronous interconnection of electric grids.

8 An HVDC system consists of

   a. a high voltage DC generator, transmission line, and transformer.

   b. a step-up transformer, AC-DC converter, DC transmission line, DC-AC converter, and a step-down transformer.

   c. a step-up transformer, AC-DC converter, DC transmission line, and a step-down transformer.

   d. a step-down transformer, AC-DC converter, DC transmission line, DC-AC converter, and a step-up transformer.

9 Power factor is

   a. the ratio of active power and reactive power.

   b. the ratio of active power and complex power.

   c. the ratio of reactive power and apparent power.

   d. the ratio of active power and apparent power.

10 Lower power factor results in

   a. higher supply current for the same active power.

   b. lower supply current for the same active power.

   c. higher supply voltage for the same current.

   d. frequency increase for the same voltage.

Answers: 1-d, 2-b, 3-a, 4-a, 5-b, 6-d, 7-a, 8-b, 9-d, 10-a

Research Topics and Problems

Research and Discussion Topics

1 What are the possible impacts of generating more electric power than needed by consumers?

2 What are the possible impacts of a sudden drop of the supplied electric power due to a failure in the transmission system?

3 What are the environmental effects of extra high voltage (EHV) transmission systems?

4 Discuss the possible effects of a growing electric vehicle (EV) market on the electric power system.

5 Compose a white paper to present the benefits and drawbacks of distributed electric generation from PV arrays and small wind turbines on the interconnected electric grid.

Problems

1 A generator rated 60 MVA and 15 kV delivers 40 MW with 0.8 inductive power factor. Determine the load current.

2 An electric load draws 10 MW with 0.7 inductive power factor. Determine the value of the capacitance to be connected in parallel to raise the power factor to 0.95.

3 Rated speed of a hydro-turbine is 200 revolution per minute (rpm). How many poles must a generator driven by this turbine have to generate 60 Hz output voltage?

4 Rated speed of a gas turbine is 1800 revolution per minute (rpm). How many poles must a generator driven by this turbine have to generate 60-Hz output voltage?

5 The resistance of a 69-kV, three-phase transmission line is 0.3 Ω/mi. Calculate the efficiency as the ratio of the resistive loss to the transmitted power when each wire of this line conducts 100 A to a distance of 100 mi.

6 Repeat Problem 5 for the same current and distance for a 345-kV transmission line with conductor resistance of 0.15 Ω/mi.

Recommended Web Sites

- Engineering Toolbox: https://www.engineeringtoolbox.com/
- History.com: https://www.history.com/news/what-was-the-war-of-the-currents
- Scientific American: https://www.scientificamerican.com/
- US Department of Energy: https://www.energy.gov/

References

Chapman, S.J. (1991). *Electric Machinery Fundamentals*, 2e. New York, NY: McGraw-Hill.

Ellert, F.J., Miske, S.A., and Truax, C.J. (1982). EHV-UHV transmission systems. In: *Transmission Line Reference Book – 345 kV and Above* (ed. J.J. LaForest), 11–62. Palo Alto, CA: Electric Power Research Institute (EPRI).

Faulkenberry, L.M. and Coffer, W. (1996). *Electrical Power Distribution and Transmission*. Upper Saddle River, NJ: Prentice-Hall.

Fitzgerald, A.E., Jingsley, C.J., and Umans, S.D. (1992). *Electric Machinery*, 5e. New York, NY: McGraw-Hill.

Gross, C.A. (1979). *Power System Analysis*. New York, NY: Wiley.

Hansen, A.D. (2005). Generators and power electronics for wind turbines. In: *Wind Power in Power Systems* (ed. T. Ackerman), 54–78. West Sussex, England: Wiley.

Jarvis, M. C., 1955. The History of Electrical Engineering: 4. Machinery for the New Light: Part 2. *IEE Journal*, September Issue, pp. 566–574.

Josephson, M. and Conot, R. E., 2019. *Thomas Edison: American Inventor*. s.l.: s.n.

Kraus, J.D. (1992). *Electromagnetics*, 4e. New York, NY: McGraw-Hill.

Lantero, A., 2014. *War of the Currents: AC versus DC power*. [Online] Available at: https://www .energy.gov/articles/war-currents-ac-vs-dc-power [Accessed 29 August 2019].

Nasar, S.A. (1996). *Electric Energy Systems*. Upper Saddle River, NJ: Prentice-Hall.

NEMA (2014). *NEMA Condensed MG 1-2011: Information Guide for General Purpose Industrial AC Small and Medium Squirrel-cage Motor Standards*. Rosslyn, VA: National Electrical Manufacturers Association.

Stevenson, W.D. (1982). *Elements of Power System Analysis*. New York, NY: McGraw-Hill.

7

Thermal Power Generation

Mount Storm coal fired power plant located in Grant County West Virginia, near Bismarck. The facility, with 1.629-MW generation capacity, consumes on average 15,000 tons of bituminous coal and 700 tons of limestone per day.

The facility is owned and operated by Dominion Energy, has three units which started commercial operation in 1965, 1966, and 1973, and is located on the west bank of the 1200-acre Mount Storm lake which was built to provide cooling water for the station. The lake, which remains at significantly high temperatures due to the energy discharged from the power station, is also used for recreational activities such as fishing and scuba diving.

Source: Dominion Energy web site https://www.dominionenergy.com (accessed on 9/28/2019).

Source: © H&O Soysal

Energy for Sustainable Society: From Resources to Users, First Edition. Oguz A. Soysal and Hilkat S. Soysal.
© 2020 John Wiley & Sons Ltd. Published 2020 by John Wiley & Sons Ltd.

7.1 Introduction

Thermal power generation involves conversion of heat into mechanical work, which may be used directly or converted into electric energy. Many industrial plants produce mechanical work for their operations directly by using steam and gas turbines. Electrothermal power generation is based on converting heat into electricity using an electric generator driven by a heat engine, which converts thermal energy (heat) into useful mechanical work. All electric generation facilities using fossil fuels, biomass, waste, geothermal energy, nuclear energy, and concentrated solar energy as their primary sources convert heat into electric power using heat engines. Electrothermal power stations produce approximately 75% of the total electricity worldwide.

In electric power stations, steam and gas turbines are commonly used to produce mechanical energy needed to turn the shaft of an electric generator. Whereas reciprocating engines are used in small backup generators, this Chapter will focus on turbines that are an essential element of thermal power plants. Heat engines have certain common properties; they all receive heat from a high-temperature source, produce useful mechanical work, reject part of the heat to a low-temperature sink, and operate on a thermodynamic cycle.

In steam turbines, heat obtained from combustion of a fuel, nuclear reaction, geothermal fluid, or concentrated solar energy heats pressurized water in an external boiler. Saturated steam turns the turbine blades, where heat is converted into mechanical work. The steam-water mixture leaving the turbine is cooled in a condenser. A pump or compressor circulates the cool working water back to the boiler at high pressure. The turbine converts heat into useful work, turning the shaft of an electric generator. The generator converts mechanical energy into electrical to supply the electric network through the transmission

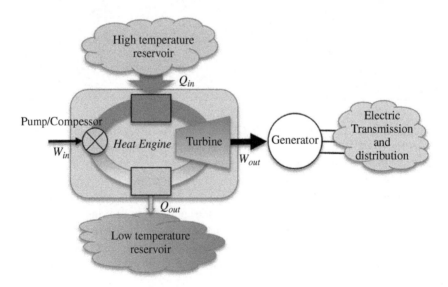

Figure 7.1 Outline of an electrothermal power plant.

and distribution system. Figure 7.1 outlines the structure of a thermoelectric energy conversion network.

In gas turbines, the working fluid is generally hot air. Operation of gas turbines is similar to jet engines; heated atmospheric air turns the turbine blades and is discharged back into the open air where the thermodynamic cycle is completed.

7.2 Principles of Thermodynamics

Thermodynamic processes are interactions between heat and mechanical work, consequently mechanical energy. Chemistry uses thermodynamic principles to explain molecular level interactions and energy exchange in chemical reactions. In physics, thermal processes are explained and analyzed using thermodynamics.

Heat engines, which are among the essential components of an energy system, convert heat into mechanical energy through thermodynamic cycles. In a thermodynamic process, internal energy of the working fluid changes due to temperature differences between the heat source and sink, where part of the heat is released as part of the thermodynamic cycle that converts heat to mechanical work or vice versa. Steam and gas turbines, as well as internal combustion engines, operate on thermodynamic cycles to convert heat to mechanical energy. Combined heat and power (CHP) systems return part of the released heat back to increase efficiency or use in industrial processes. Heat pumps use mechanical energy to transfer heat from lower temperatures to a higher temperature reservoir (Cengel and Boles 2006).

Physical systems are classified in three categories based on their interactions with the environment. A system that exchanges both energy and matter with its surrounding is called an *open system*; if it exchanges only energy, not matter, it is a *closed system*. A system that does not exchange any energy or matter with the environment is called an *isolated system*. Since the Earth receives energy from the Sun but does not exchange matter with outer space (neglecting the mass of meteors that may hit the earth), it can be considered a closed system.

7.2.1 Heat and Temperature

In the sense of thermodynamics, heat is an energy interaction solely due to a temperature difference. Temperature is a measure of hotness or coldness of an object. Two bodies exchange energy in the form of heat only if they are at different temperatures. When two objects are at the same temperature, there is no heat interaction and the objects are in thermal equilibrium.

7.2.1.1 Common Temperature Scales

In the metric unit system, temperature is measured in degrees Celsius (°C), where freezing temperature of pure water is 0 °C and the boiling temperature of pure water at normal atmospheric pressure at sea level (1 atm) is 100 °C. The Celsius scale is used in most parts of the world.

In the older Fahrenheit scale water freezes at 32 °F and boils at 212 °F under normal pressure. Equation (7.1) converts the temperature scale between Celsius and Fahrenheit.

$$[°C] = ([°F] - 32) \times \frac{5}{9}$$
$$[°F] = [°C] \times \frac{9}{5} + 32 \tag{7.1}$$

Until the 1960s, Fahrenheit was the principal temperature unit in English-speaking countries. Today it is used in the US and a few other countries. Some parts of Europe used the Réaumur scale before the Celsius scale was adopted as metric standard. In the Réaumur scale, which is represented by the symbol "Ré," water freezes at 0° and boils at 80°.

7.2.1.2 Absolute Temperature Scale

On absolute temperature scales, only one basis temperature is defined instead of two, as suggested originally by James Prescott Joule and William Thompson (Lord Kelvin). This idea stems from the ideal gas law and development of gas thermometers.

French physicists J.A.C. Charles and J.L. Gay-Lussac observed experimentally that if the volume of a gas is kept constant, its pressure increases proportionally as temperature increases. This was the foundation for the ideal gas law. A gas thermometer, a container that has a constant volume v, is filled with an ideal gas at a low-pressure p at 0 °C. The ratio of pressures at two different temperatures is equal to the ratio of the pressures. Let us represent the temperature of ice-water mixture with T_i and the corresponding gas pressure with p_i. At the temperature of boiling water T_b, the pressure of gas in the container will increase to p_b. Then, we can write Eq. (7.2) using the ideal gas law for constant volume.

$$\frac{T_b}{T_i} = \frac{p_b}{p_i} \tag{7.2}$$

Many scientists experimentally demonstrated that as the lower pressure is reduced enough, the ratio tends to be constant for all volumes and all ideal gases.

$$\lim_{p_i \to 0} \left[\frac{p_b}{p_i} \right] = 1.36609 \tag{7.3}$$

Using the reference temperatures of the Celsius scale we can write a second equation

$$T_b - T_i = 100 \tag{7.4}$$

The value of T_i and T_b can be obtained by solving Eqs. (7.3) and (7.4) together.

$$T_i = 273.16 \quad T_b = 373.16 \tag{7.5}$$

The absolute temperature scale measures temperatures in degree Kelvin (K), obtained by adding 273.16 to the Celsius scale. If temperature is expressed in Fahrenheit, absolute temperature is obtained in Rankine by shifting the Fahrenheit scale by 459.67 °F.

$$[K] = [°C] + 273.16$$
$$[R] = [°F] + 459.67 \tag{7.6}$$

The Rankine scale should not be confused with the Réaumur scale that was used in some parts of Europe.

Heat and temperature are sometimes confused in everyday conversations. One should always keep in mind that heat is an energy interaction resulting from a temperature difference. In addition, many expressions or phrases used in everyday language in technical or non-technical conversations that associate heat interactions to fluid movements are not consistent with the thermodynamic principles. Heat flow, heat addition, heat loss, heat rejection, wasted heat, and many other terms are examples. Even terms like latent heat and specific heat contradict with the definition of heat. Such sentences and terms, however, have been commonly used by ordinary people and scientists without causing a misunderstanding. For practical reasons, this text will also use the common vocabulary.

Box 7.1 The Story of Practical Temperature Scales

Although heat and cold were one of the first physical effects early humans perceived, it took millions of years to quantify how hot or how cold air, water, or any object is. Earlier devices developed to measure temperatures were based on the expansion of a liquid with temperature similar to the alcohol or mercury thermometers we still use. A liquid filled in the bulb expands in a bore marked with linear intervals between two well-known standard temperatures. Originally, Isaac Newton proposed in 1701 to setting the lower reference point to the freezing temperature of water, the upper reference point to the body temperature of a healthy male, and dividing the range by 12 (Fenn 1982). The zero reference of Newton's scale depends on the purity of water, the higher reference can be slightly different from one person to another.

In 1724, Dutch-German physicist and instrument maker Gabriel Fahrenheit proposed that the zero point be taken as the freezing temperature of a solution made by mixing equal amounts of ice, water, and salt. The body temperature would still be 12, but to increase the precision, the intervals would be divided by 8. Hence, there would be 96 marks between lower and the upper temperatures. Fahrenheit was the first to use mercury as fluid in thermometers to reduce the dependence on the expanding fluid, and he demonstrated experimentally that the temperature of the ice-water mixture and the boiling temperature of water were constant and reproducible. The Fahrenheit scale was later adjusted by setting the melting point of ice to 32° and the boiling point of water to 212° at standard atmospheric pressure at sea level. Based on the readjusted scale, the normal body temperature was determined as 98.6 °F, slightly different from the value Fahrenheit originally estimated. This scale was used in English-speaking countries until 1948, and it is still used in the US and a few other countries.

Several scientists proposed using the freezing and boiling temperatures of water as lower and higher references. French scientist R.A.F de Réaumur proposed dividing this interval into 80°. The metric temperature scale, which divides the temperature range between the temperature of melting ice and boiling water by 100° is credited to Swedish astronomer Anders Celsius. In 1954, the Tenth International Conference on Weights and Measures officially accepted the Celsius scale represented as °C (Fenn 1982).

The temperature scales explained above had two major weaknesses. First, each liquid has a different expansion characteristic in respect to temperature variations. For example, if water were used in a thermometer, the error for temperatures closer to zero would be large because the volume of water at 4 °C is smaller than the volume of ice. The second issue is the use of

---

(Continued)

Box 7.1 (Continued)

two reference temperatures. Alcohol and mercury thermometers resolved the first issue but not the second. Development of thermometers using an ideal gas, such as hydrogen, helium, or argon allowed the scientists to define an absolute temperature scale that had only one well-defined reference. The triple point temperature of water, where the solid, liquid, and gas states of water coexist at a low pressure, is selected as the single reference temperature.

The triple point temperature of water, where water, ice, and vapor coexist at 0.61 kPa pressure, is 273.16 °C (Rossini, Wagman, Evans et al. 1952). In the thermodynamic temperature scale named after Lord Kelvin, the absolute temperature is obtained by adding 273.16 to the temperature measured in the Celsius scale. The unit of absolute temperature is K (not °K since the symbol ° was officially dropped in 1976). The international temperature scale of 1990 adopted by the International Committee of Weights and Measures defines degree Kelvin (K) as 1/273.16 of the thermodynamic temperature of the triple point of water.

In the imperial unit system, the Rankine scale defines the absolute temperature by adding 491.68 to the temperature measured in Fahrenheit, which is the triple point temperature of water in Fahrenheit scale.

7.2.2 Internal Energy

Internal energy is the total kinetic and potential energy of atomic particles that form matter. Internal energy changes when the temperature, volume, or pressure changes. As the temperature changes, the energy that holds atomic particles together changes and they move closer or farther. A change of internal energy can change the phase of a substance from solid to liquid, liquid to gas, or vice versa. Changing the volume of a gas by pressure or temperature also changes the internal energy. In a thermodynamic process, interaction of heat and mechanical work cause significant changes of the internal energy of the working fluid. However, in practical processes that operate in steady-state cycles, the change of internal energy during each cycle is zero.

7.2.3 Laws of Thermodynamics

7.2.3.1 Thermal Equilibrium: Zeroth Law of Thermodynamics

Since no heat interaction occurs between two objects at the same temperature, these objects are said to be in thermal equilibrium. R.H. Fowler indicated the thermal equilibrium conditions in 1931; if two objects are in thermal equilibrium with a third object, they are in thermal equilibrium with each other. Although this statement seems trivial, it is the basis of temperature measurements. Because the first and second laws of thermodynamics were already established years before thermal equilibrium was revealed, the formal statement of this concept is known as the zeroth law of thermodynamics:

"*If two objects are in thermal equilibrium with a third object, they are in thermal equilibrium with each other.*"

7.2.3.2 First Law of Thermodynamics: Conservation of Energy

The conservation of energy principle is also known as the first law of thermodynamics, which states that *in an isolated system, the total amount of energy is constant.*

A thermodynamic process is a conversion between heat and mechanical energy. Suppose that a thermodynamic system receives the thermal energy Q_H from a higher temperature source. Part of this heat is converted to mechanical work while the internal energy of the system changes. The remaining thermal energy Q_H is rejected to a lower temperature sink. According to the first law of thermodynamics, the difference between Q_H and Q_L is equal to the sum of the change in internal energy ΔU and the mechanical work W_m. Energy balance in such thermodynamic process can be expressed with Eq. (7.7).

$$Q_H - Q_L = \Delta U + W_m \tag{7.7}$$

Internal energy of a matter is the sum of randomly distributed kinetic and potential energy of the atomic particles. The principle of energy conservation includes the work done by microscopic forces on atomic particles. Part of the input energy is transferred to the internal energy changing system variables such as temperature, volume, pressure, phase, and dimensions. Input energy may also increase or decrease the amount of energy stored within the system in different forms.

7.2.3.3 Second Law of Thermodynamics: Direction of Heat Flow

Energy can flow only from a higher temperature reservoir (source) to a lower temperature reservoir (sink) if no external work is added to the system. This principle leads to the second law of thermodynamics, which states that *it is impossible for any device that operates on a cycle to receive heat from a single reservoir and produce a net amount of work.* This statement is quoted as the *Kelvin-Planck statement.* Rudolf Clausius, who introduced the concept of entropy in 1865, made an equivalent statement of the second law of thermodynamics: *It is impossible to construct a device that operates in a cycle and produces no effect other than the transfer of heat from a lower temperature body to a higher temperature body.*

7.2.4 Entropy

Entropy is a fundamental concept to understand the physical limitations of energy conversion. To explain the physical meaning of entropy, suppose that heat is added to an ideal gas so that it expands freely at constant temperature. Such expansion is called *isothermal.* Internal energy of an ideal gas depends on its absolute temperature. The expansion process is therefore reversible because if the gas is compressed back to the initial volume, it gives out the same heat to maintain the temperature constant. Differential work done by the internal forces on gas atoms is equal to the product of the pressure p and differential change of volume dV. If all frictions are neglected and no external forces are present, the work done is equal to the heat input according to the principle of conservation of energy principle.

$$dQ = dW = p \cdot dV \tag{7.8}$$

We use the ideal gas equation $pV = kT$, where T is the absolute temperature in Kelvin and k is constant. By substituting in Eq. (7.8) we can write

$$\frac{dQ}{T} = k\frac{dV}{V} \tag{7.9}$$

The change of heat per absolute temperature is defined as differential entropy for an infinitesimal reversible process. We clearly see that as the gas expands at constant temperature, entropy increases.

$$S = \frac{dQ}{T} \tag{7.10}$$

The fractional volume change in Eq. (7.9) reflects the disorder of the gas molecules since they can move more randomly in increased volume. Differential entropy of a system over an *internally reversible* path is proportional to the randomness of the gas molecules. The total amount of heat added during a reversible process at absolute temperature T yields the total entropy change given by

$$\Delta S = S_2 - S_1 = \frac{Q}{T} \tag{7.11}$$

The unit of entropy is unit of energy divided by unit of temperature. In SI, the unit of entropy is Joules per Kelvin (J/K).

In practical systems where processes are irreversible because of frictions and unavoidable external effects, entropy always increases and affects the performance of the system adversely.

7.2.5 Enthalpy

Consider a thermodynamic process where a gas is heated at constant pressure p. Internal energy of the system changes by dU, and the change of volume by dV results in a work $p \cdot dV$. Let us assume that the potential, kinetic, or electromagnetic energy of the system does not change. According to the conservation of energy principle, the small energy change δq equals the sum of differential internal energy and work.

$$\delta q = dU + p \cdot dV \tag{7.12}$$

Since the pressure is constant, Eq. (7.12) can be arranged to combine the internal energy and work in a single differential.

$$\delta q = d(U + pV) = dH \tag{7.13}$$

The term $H = U + pV$ is defined as *enthalpy*. In other words, enthalpy is the sum of the internal energy of a body and the product of its volume and the pressure exerted on it. Note that dimensional analysis of $p \cdot V$ shows that it has the unit of energy. In analysis of thermodynamic processes, specific enthalpy is a useful parameter to simplify calculations. Specific enthalpy of a body is expressed in terms of the internal energy per unit mass u, specific volume v, and pressure p.

$$h = u + pv \tag{7.14}$$

The SI unit of specific enthalpy is J/kg. During heat transfer in boilers and condensers where the pressure is constant, the change of enthalpy is equal to the heat input. When there is no heat transfer in the process (adiabatic), the change in enthalpy is equal to the work done.

7.2.6 Reversibility of Energy Flow

The principle of energy conservation deals only with the total quantity of energy, which has to remain constant in a closed system. This property, however, becomes insufficient when a question arises about the direction of the energy flow.

Processes where energy can flow in either direction are called reversible. Many practical systems, however, involve irreversible processes. For example, consider a hot object in a cooler environment. The temperature of the object continuously drops as the heat is dissipated to the environment. This process is irreversible since the object cannot take back the energy that exists in the cooler environment and warm up. Heat always flows from higher temperature to cooler. In order to reverse this flow, some external energy must be given to the system.

A large group of electric power plants burn some type of fuel to produce heat and generate electricity. When the fuel burns, the molecules decompose allowing carbon and hydrogen atoms to combine with the oxygen and produce heat. There is no known practical process that can recombine carbon and hydrogen atoms to regenerate the fuel. In addition, the power plant must be cooled down to continue the energy conversion process. The heated water or air does not return to the power plant to generate electricity unless an external work is done. Similarly, if no external work is added to the system, potential energy changes from higher to lower value. In order to increase the potential energy of an object, an external work must be done to lift the object to a higher elevation.

7.2.7 State of a System

In physical sciences and engineering the term *state* is used to describe the particular condition of a substance, object, or system at a specific time. State variables are a set of physical properties that completely describe the condition of a system. In thermodynamic systems, temperature, volume, pressure, mass, internal energy, and entropy are the key variables to describe the state of a system. A system is said to be in equilibrium if it does not experience any changes and does not interact with its surroundings.

The state of a system changes if one or several of the state variables change. In thermodynamic systems involving ideal gases, pressure, volume, and temperature are the essential state variables. If the pressure is not exceedingly high, Eq. (7.15) describes the relationship between these state variables in an ideal gas.

$$\frac{V \cdot P}{T} = nR = \text{constant} \tag{7.15}$$

In this equation, n is the number of moles, T is the absolute temperature in Kelvin (K), and R is called universal gas constant. The numerical value of R depends on the units used for p and V. If p is in atmospheres and V is in cubic centimeters, R is $82.07\,\text{cm}^3\,\text{atm/K}$.

7.3 Thermodynamic Processes

Thermodynamic processes are the transition of an object or a system from one state to another change of state variables. Thermodynamic processes of an ideal gas are important to understand the operation of heat engines. Such engines operate in thermodynamic cycles such that in the end of each cycle the system returns to the initial state. Three particular processes described below are often used in the study of a thermodynamic cycles (Sonmtag, Borgnakke, and Van Wylen 1998)

7.3.1 Isothermal Process

A thermodynamic process where the temperature remains constant is called an *isothermal process*. If a volume of ideal gas is heated, the product of the pressure and volume must remain constant to satisfy the ideal gas Eq. (7.15). Consider heating an ideal gas at constant temperature. According to the conservation of energy principle, for a small amount of heat entering the system we can write

$$\delta q = dU + p \, dV \tag{7.16}$$

Since temperature is constant, internal energy of the gas will not change and all heat will be converted to work. By substituting $p = RT/V$ for 1 mol of gas and integrating both sides of the equation, we obtain

$$\int_{q_1}^{q_2} dq = \int_{V_1}^{V_2} \frac{RT}{V} dv \quad \Rightarrow \quad Q = RT \ln \frac{V_2}{V_1} = W \tag{7.17}$$

We see that the heat absorbed by the gas in an isothermal process is converted to mechanical work. In addition, the work is proportional to the absolute temperature at which the expansion occurs.

7.3.2 Adiabatic Process

Thermodynamic processes that occur without any heat interaction are called adiabatic. Again, applying the conservation of energy principle, we can write

$$dU + p \, dV = 0 \tag{7.18}$$

Internal energy is proportional to the temperature difference. If the specific heat at constant volume is represented by C_V, Eq. (7.18) becomes

$$C_V dT = -p \, dV \tag{7.19}$$

Substituting the ideal gas equation and integrating both sides, we obtain Eq. (7.20).

$$C_V \int_{T_1}^{T_2} \frac{dT}{T} = -R \int_{V_1}^{V_2} \frac{dV}{V} \quad \Rightarrow \quad C_V \ln(T_2/T_1) = -R \ln(V_2/V_1) \tag{7.20}$$

The difference of specific heat of an ideal gas at constant volume (C_V) and at constant pressure (C_p) is equal to the universal gas constant R. By substituting $\gamma = C_p/C_V$, the relationship between temperature and pressure in adiabatic process can be simplified to

$$\ln(T_2/T_1) = \frac{\gamma - 1}{\gamma} \ln(p_2/p_1) \tag{7.21}$$

Since no heat interaction occurs, or in other words $Q = 0$ in an adiabatic process, we can see from Eq. (7.11) that the entropy of the ideal gas remains constant.

7.3.3 Carnot Cycle

Heat engines convert thermal energy to mechanical work through a cycle such that internal energy of the working fluid remains unchanged at the end of each cycle. A thermodynamic cycle is completed in four separate processes. In the ideal case when the state of the system can be completely restored by the process, it is called reversible. In practice all thermodynamic processes are irreversible because of heat losses, frictions, and non-ideal fluid flow. Several thermodynamic cycles were proposed to simplify the study of heat engine operations. In this chapter we will focus on Carnot cycle, which is essential to understand the maximum efficiency of heat engines.

The first commercially successful steam engine developed by Thomas Newcomen in 1712 was an extremely low-efficiency reciprocal engine working on cycles. In this engine, the cylinder where a piston travels needed to be cooled at each cycle. James Watt made several major modifications on the Newcomen's engine to improve the efficiency. While James Watt's invention was successful enough to be mass produced, the power and efficiency limitations motivated other inventors to try new design ideas.

Inspired by the steam engine designs available at the time, French engineer and physicist Sadi Carnot proposed in 1724 an ideal and reversible thermodynamic cycle to determine the maximum efficiency attainable by heat engines. A Carnot cycle consists of four reversible processes to produce useful work by extracting heat from a high temperature source (reservoir) and rejecting heat to a lower temperature sink.

A graphical representation of the Carnot cycle is shown on the *V-T* and *S-T* diagrams in Figure 7.2. The cycle starts at point 1, where the gas is at a temperature T_H and pressure P_H. On the segment 1–2, external heat is added, and the volume increases such that the temperature remains unchanged. Since the state changes at constant temperature, the process is *isothermal expansion*. During this process, heat increases the entropy of the ideal gas since no mechanical work is involved. The addition of heat stops at point 2. On the segment 2–3, temperature drops from T_H to T_L. Because no heat is added or removed

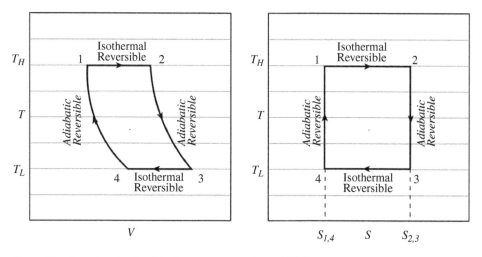

Figure 7.2 Carnot cycle of an ideal gas represented on a V-T diagram.

during this expansion, the process is an *adiabatic expansion*. The internal energy of the gas is therefore converted to external work according to the first law of thermodynamics while entropy remains constant. On the segment 3–4, heat is rejected to the environment and the volume is decreased to keep the temperature constant. During the isothermal process, entropy changes from s_3 to s_4. During the last sequence of the cycle, the gas is compressed to return to the initial state, temperature increases but entropy does not change. The entropy-temperature diagram of a Carnot cycle is a perfect rectangle.

7.3.4 Carnot Heat Engine

Carnot developed his model to analyze the operation of reciprocating steam engines as explained the previous paragraph. The same cycle can also be executed in a steady flow system. In fact, Steam turbines operate on steady flow of pressurized water and superheated vapor. The hypothetical engine that operates on Carnot cycle is known as a Carnot heat engine. Modern steam turbines differ in many aspects from a Carnot heat engine. First, steam is not an ideal gas and the actual processes in steam turbines are not completely reversible because of frictions, losses, and other non-ideal effects. However, the Carnot heat engine is an idealized and simplified model to understand the operation of steam turbines. A Carnot engine, although impractical, provides the highest possible thermal efficiency for any heat engine.

7.4 Efficiency and Heat Rate

Because the components of a heat engine involve mass flow in and out, they should be considered as an open system. However, if the leakages are disregarded, the heat engine is a closed system because the amount of working fluid remains the same at each cycle, while only work and heat pass across boundaries. The efficiency of a heat engine is the ratio of useful net output work to the net heat input.

$$\eta = \frac{W_{net}}{Q_{in}} \tag{7.22}$$

Net output work is the difference between the output work and mechanical work input of mechanical devices such as pumps and compressors.

$$W_{net} = W_{out} - W_{in} = Q_{in} - Q_{out} \tag{7.23}$$

Substituting Eq. (7.23) into Eq. (7.22), we can express the thermal efficiency in terms of the heat input and output.

$$\eta = 1 - \frac{Q_{out}}{Q_{in}} \tag{7.24}$$

7.4.1 Carnot Efficiency

Thermal efficiency of a thermodynamic cycle is defined as the ratio of the total mechanical work done W_{net} to the heat received from a high temperature reservoir Q_H. Since during the cycle some part of the heat is rejected to a lower temperature reservoir the net work done is equal to the difference of received and rejected heats due to the first law of thermodynamics (conservation of energy principle).

$$\eta = \frac{W}{Q_H} = \frac{Q_H - Q_L}{Q_H} = 1 - \frac{Q_L}{Q_H} \qquad (7.25)$$

Since the change of entropy is $\Delta S = Q/T$, the amounts of heat received from the source and rejected to the sink are expressed in terms of the temperatures of isothermal processes and the change of entropy from beginning to end states.

$$Q_H = T_H(s_2 - s_1)$$
$$Q_L = T_L(s_3 - s_4) \qquad (7.26)$$

The adiabatic processes during expansion and compression do not change the entropy; hence, $s_1 = s_4$ and $s_2 = s_3$. The ratio of the rejected to received heat is therefore equal to the ratio of lower and higher temperatures. By substituting this property in Eq. (7.25) efficiency of a Carnot cycle can be written in terms of temperatures.

$$\eta = 1 - \frac{T_L}{T_H} \qquad (7.27)$$

Thermal efficiency of Carnot heat engine is the maximum efficiency that can be obtained by any heat engine because it is derived for reversible ideal processes. Practical heat engines always involve frictions, heat losses, turbulences, and steam leakages. Note that the maximum thermal efficiency depends only on the higher temperature and the lower temperature of the process.

7.4.2 Heat Rate of Thermoelectric Generation Units

Most electric generators worldwide are driven by heat engines that convert heat into mechanical energy. A statistical measure named "heat rate" is useful to compare the efficiency of fossil fuel fired and nuclear electric generation plants. The US Energy Information Administration (EIA) defines heat rate as the amount of thermal energy in British Thermal Units (Btu) used by an electric power plant to generate 1 kilowatt-hour (kWh) of net electricity supplied to the power transmission network (EIA 2019).

$$HR = \frac{Q}{W_{elec}} \ [\text{Btu/kWh}] \qquad (7.28)$$

Heat rate of electric generation plants should not be confused with the heat content of fuels. Heat rate is a measure of the efficiency of a power plant that converts into electricity. The percent efficiency of a power plant can be calculated by dividing the equivalent heat content of electricity in Btu by the heat rate. Since 1 kWh (3600 J) is equivalent to 3412 Btu, the expression below gives efficiency in percent.

$$\%\eta = \frac{3412}{HR} \times 100 \qquad (7.29)$$

Heat rate of a power plant depends on the quality of the fuel used to generate heat in addition to the thermal and electrical efficiency of the components. The US EIA publishes heat rates for fossil fuel-fired and nuclear power plants. Table 7.1 shows the operating heat

Table 7.1 Average operating heat rate and efficiency of thermal power plants in the US as of 2017.

Primary source	Heat rate (Btu/kWh)	Efficiency (%)
Coal	10,465	32.6
Petroleum	10,834	31.5
Natural gas	7812	43.7
Nuclear	10,459	32.6

Source: DOE/EIA MER (2017).

rates and efficiencies of thermal and nuclear power plants in the US in 2017 (DOE/EIA MER 2017). Note that the overall efficiencies of thermal power plants listed in Table 7.1 vary between 31.5% and 43.7%, depending on the fuel. Petroleum is the least efficient fuel for electric generation. Efficiency of coal fired and nuclear power stations are equal. Natural gas burning power stations are the most efficient among different types of thermal power plants.

Example 7.1 A coal-burning power plant consumes 35 short tons of coal per hour when it operates at 60 MW output power capacity. Given the heat content of the coal used in this power plant is 19.61 million Btu per short ton, calculate the heat rate and efficiency of this facility.

Solution
Heat produced by burning 35 short tons of coal in one hour to generate 60 MWh electric energy is:

$$W_m = 35 \times 19.61 = 679.35 \cdot 10^6 \text{ Btu}$$

Heat rate calculated from (7.28) is:

$$HR = \frac{679.35 \cdot 10^6}{60 \cdot 10^6} = 11,300 \text{ Btu/kWh}$$

Using Eq. (7.29) the efficiency is obtained as:

$$\eta = \frac{3,412}{11,300} 100 = 30\% \therefore$$

7.5 Steam Turbines

In 1781 James Watt patented a piston-type steam engine that could produce mechanical power. The invention and development of steam engines replaced manpower and animal power in manufacturing and led to the industrial revolution. The steam turbine, developed by Sir Charles Parsons about 100 years ago, became the favorite prime mover for electric generation. Today steam turbines are widely used in energy systems; particularly to drive large electric generators in fossil fuel, nuclear, geothermal, and concentrated solar power plants. In 2013 about 48% of installed capacity of all electric generation plants in the US was powered by steam turbines. Steam turbine-driven generators, on the other hand, generated 60% of the total electricity (EIA 2013).

Steam turbines are heat engines that produce direct rotary motion on a steady flow of water and steam. Thermodynamic processes implemented in modern steam turbines are variations of the Carnot cycle.

7.5.1 Evaporation Properties of Water

As we know, a substance can be in solid, liquid, or gas phase depending on the pressure and temperature. These phases may also coexist under certain physical conditions. Different phases of water; ice, liquid water, and vapor are part of our everyday life in many situations. Since water is used as the working fluid of steam turbines, particularly its change of phase from liquid to gas and from gas to liquid, it has importance in energy engineering.

In the solid phase, the molecules are strongly tied together with microscopic forces that restrict their motion in respect to each other. By increasing the temperature, energy is added to the molecules so that molecules gain freedom of motion. That is, some part of ice melts down and turns to liquid. Energy required to change phase is called *latent heat*. During the phase change from solid to liquid the temperature of the melting mixture remains constant. At atmospheric pressure, this temperature is by definition 0 °C. As soon as all ice is melted, the water temperature can increase if external energy is added or decrease if energy is removed. In the liquid phase, the molecules have more freedom than the solid phase so that they can displace easier.

When the temperature reaches 100 °C under normal atmospheric pressure at sea level (1 atm or 760 mmHg), water starts to change its phase from liquid to gas. During the phase change the temperature remains constant until all liquid water drops turn to vapor. After that point, the temperature of the vapor can increase. The molecules of water in the vapor phase have the highest energy, which allows them to move freely, and to collide with each other and the boundaries of the container.

Needless to say, water has the same chemical composition in all phases. The only difference is its internal energy, which allows random movement of its molecules. Entropy represents the degree of this randomness, therefore adding external energy increases the entropy and removing energy decreases entropy.

Evaporation curves of water at different pressures are plotted on a temperature-entropy plane in Figure 7.3 (Lemmon et al. 2015). While water vapor and steam are synonyms, in energy engineering the word steam is used more frequently. The bell-shaped *saturation curve* is the envelope of saturation points at different pressures and represents the transition from water to steam. The left side of the curve is the boundary between water and vapor. As soon as the boundary is crossed water starts to boil. In the region encountered by the saturation curve, water and vapor coexist. The temperature of the boiling water remains the same until all water particles completely disappear at a value that depends on the pressure. The two-phase mixture under the saturation curve is called *wet steam*. The right branch of the curve corresponds to the points where water particles completely disappear. The right side of the bell-shaped saturation diagram is the *dry steam* region where the temperature increases again.

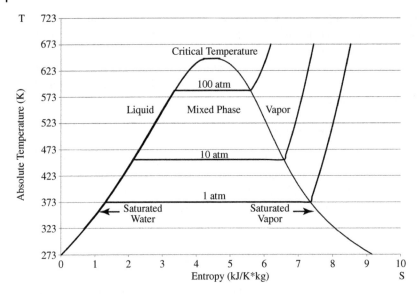

Figure 7.3 Saturation diagram for water. Source: Lemmon et al. (2015).

In the wet steam region, the total mass of the mixture is the sum of the masses of water and gas

$$m = m_f + m_g \tag{7.30}$$

The index f is used for liquid water and g is used for the water vapor. The quality of steam is defined as the ratio of the mass of gas to the total mass m of the mixture.

$$x = \frac{m_g}{m} \tag{7.31}$$

Using the specific volumes for water and steam we can write:

$$v = \frac{V}{m} = \frac{m_f v_f + m_g v_g}{m} = \frac{(m - m_g)v_f + m_g v_g}{m} \tag{7.32}$$

Substituting the quality of steam ratio x, the following expression is obtained for the specific volume of the mixture in terms of the specific volumes of water and vapor.

$$v = (1 - x)v_f + xv_g \tag{7.33}$$

The expressions of internal energy, entropy, and enthalpy have the same form for a mixture with a steam quality x:

$$u = (1 - x)u_f + xu_g$$
$$s = (1 - x)s_f + xs_g$$
$$h = (1 - x)h_f + xh_g \tag{7.34}$$

Steam tables giving thermal properties of water-steam mixtures are available in reference publications and online databases (Lemmon et al. 2015). These tables usually include the values of specific volume, internal energy u, entropy S, and enthalpy u for liquid

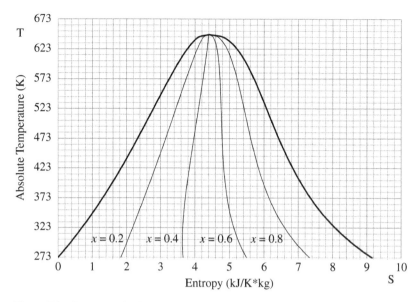

Figure 7.4 Steam quality in the two-phase region.

and gas components of the mixture by temperature and pressure. Dashed lines on the entropy-temperature (S-T) diagram of Figure 7.4 represent states for the same steam quality.

In the analysis of steam engines, we often need to determine internal energy or enthalpy of water-steam mixtures. If the entropy of a mixture is known, then the steam quality can be determined solving the second equation in Eq. (7.34) for x_{mix}.

$$x_{mix} \frac{s_{mix} - s_f}{s_g - s_f} \tag{7.35}$$

Knowing the steam quality, the internal energy and enthalpy of the mixture can be found using respective equations in (7.34).

Example 7.2 Entropy and enthalpy data for water-steam mixture at 100 and 300 °C are given below. A water-steam mixture has a steam quality of 0.6. Determine:

a) Entropy
b) Enthalpy
c) Steam quality when the mixture is heated to increase the entropy to 6 kJ/kg·K
   Steam data

T (°C)	h_f (kJ/kg)	h_g (kJ/kg)	s_f (kJ/kg·K)	s_g (kJ/kg·K)
100	419.2	2676	1.3072	7.3541
300	1345	2750	3.2552	5.7059

Solution

a) The saturation entropies at $100\,°C$ are given in the steam table:

$$s_f = 1.3072 \quad \text{and} \quad s_g = 7.3541$$

Using the second equation in Eq. (7.34) for $x = 0.6$, we obtain:

$$s = (1 - 0.6)1.3072 + 0.6 \times 7.3541 = 4.9353$$

b) The saturation enthalpies at $100\,°C$ are given in the table:

$$h_f = 419.2 \quad \text{and} \quad h_g = 2676$$

Using Eq. (7.34) for $x = 0.6$, we obtain:

$$s = (1 - 0.6)419.2 + 0.6 \times 2676 = 1773.28$$

c) Using Eq. (7.35) for $s = 6\,\text{kJ/kg·K}$.

$$x = \frac{s - s_f}{s_g - s_f} = \frac{6 - 1.3072}{7.3541 - 1.3072} = 0.78.\therefore$$

Note that when the mixture is heated the steam quality improves. On the contrary, as heat is taken out, or the mixture is cooled, the steam quality degrades.

7.6 Carnot Heat Engine

Although we explained Carnot cycle using a reciprocating system, the same cycle can also be executed in a steady flow system. In fact, Carnot developed his model to analyze the operation of reciprocating steam engines. Steam turbines operate on a steady flow of pressurized water and superheated vapor.

Modern steam turbines differ in many aspects from a Carnot heat engine. First, steam is not an ideal gas and the actual processes in steam turbines are not completely reversible because of frictions, losses, and other non-ideal effects. However, the Carnot heat engine is an idealized and simplified model to understand the operation of steam turbines.

A simplified diagram of steam turbines is shown in Figure 7.5. A heat source, for example combustion of a fuel, nuclear reaction, or concentrated solar energy is used to produce high-pressure steam in the boiler (1–2). Water evaporates at a constant temperature and the saturation temperature is a function of the pressure. High-pressure steam at a high temperature expands while flowing through the blades of a turbine. Rotation of the turbine shaft produces useful output work. At the same time, the temperature and pressure decrease during the expansion starting to condense steam back to the liquid water (2–3). The mixture of steam and water is further cooled in a condenser by releasing energy to the environment, for example a lake, river, sea, or air (3–4). A compressor pump feeds high-pressure water into a boiler (4–1). External work is done to operate the compressor. Assuming that the system is well insulated and there is no heat exchange with the environment, this part of the process is adiabatic.

In our first discussion of the Carnot cycle, a volume-temperature (V-T) diagram was convenient for visualization of the physical process. However, a pressure-volume (P-V) or

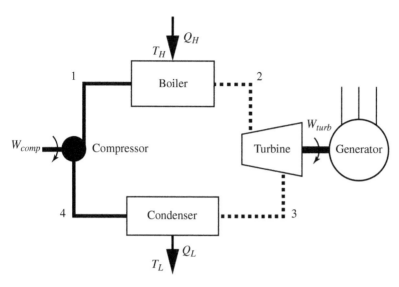

Figure 7.5 A simplified diagram of steam engine.

temperature-entropy diagram allows direct calculation of energy using the areas enclosed by the segments of the cycle. Since no heat exchange occurs during the adiabatic process, according to the energy conservation principle, the external work applied to the compressor shaft is converted to internal energy of the water and the internal energy of the steam is converted to output work in the turbine. During the adiabatic compression and expansion processes, the entropy of the system remains constant. As external heat is injected to the system in the boiler, the entropy increases. On the other hand, taking heat out of the fluid decreases the entropy. Therefore, the Carnot cycle is represented by a rectangle on the T-s diagram. Equation (7.38) shows a Carnot cycle superposed on the saturation curve of water on the T-s diagram. A numerical example for this Carnot cycle is analyzed in 7.3.

As the diagram shows, the adiabatic compression and expansion take place on the regions where the working fluid is a mixture of water and steam. Two-phase fluid (liquid and gas) presents practical challenges in the compressor design. In addition, the presence of humidity in steam causes corrosion on turbine blades.

A Carnot heat engine, although impractical, provides the highest possible thermal efficiency for any heat engine, see Example 7.3. The thermal efficiency is defined as the ratio of the total work done by the system to the input heat energy.

$$\eta_{TH} = \frac{W_{net}}{Q_H} = \frac{Q_H - Q_L}{Q_H} = 1 - \frac{Q_L}{Q_H} \tag{7.36}$$

Since the change of entropy in isothermal process is obtained by dividing the change of heat to the absolute temperature,

$$\Delta s = \frac{\Delta Q}{T}, \tag{7.37}$$

the heat input to the boiler (b-c) and the heat dissipated by the condenser (d-a) can be expressed in terms of entropy and temperatures.

$$Q_H = T_H(s_2 - s_1)$$
$$Q_L = T_L(s_3 - s_4) = T_L(s_2 - s_1) \tag{7.38}$$

By substituting Eq. (7.38) into Eq. (7.36) we obtain:

$$\eta_{TH} = 1 - \frac{T_L}{T_H} \tag{7.39}$$

Thermal efficiency of a Carnot heat engine is the maximum efficiency that can be obtained by any heat engine because practical heat engines always involve frictions, heat losses, turbulences, and steam leakages. Note that the maximum thermal efficiency depends only on the highest and the lowest temperature of the process. In the early eighteenth century, engineers were trying to improve the efficiency of steam engines by replacing water with another working fluid. Carnot proved that the physical properties of the working fluid do not affect the maximum thermal efficiency.

Example 7.3   A steam engine operates in the Carnot cycle illustrated in Figure 7.6. The boiler temperature is 340 °C. The condenser lowers the temperature to 50 °C. Using the steam data provided below, determine

a) the thermal efficiency of the cycle.
b) the heat input of the boiler.
c) the work done by the turbine.
d) the heat dissipated by the condenser.
e) the proportion of the heat converted to work to the input heat.

Solution

a) Thermal efficiency for the Carnot cycle is given in Eq. (7.39). Using absolute temperatures corresponding to the given higher and lower temperature values, we obtain

$$\eta_{TH} = 1 - \frac{T_H}{T_L} = \left(1 - \frac{273 + 50}{273 + 340}\right) \times 100 = 47\%$$

b) As shown in Figure 7.6 the boiler expands the gas from 1 to 2 at a constant pressure. Therefore, the heat input is the difference of enthalpy at these points (Table 7.2).

$$Q_H = h_2 - h_1 = 2621.8 - 1594.5 = 1027.3 \text{ kJ/kg}$$

c) The turbine converts thermal energy to useful work through an adiabatic expansion from 2 to 3 as shown in Figure 7.6.

$$W_{out} = h_2 - h_3 = 2621.8 - h_3$$

We can obtain h_3 by using Eq. (7.34) but we must first determine the steam quality x. Since the expansion is adiabatic, the entropy remains 5.3356 kJ/kg from 2 to 3. On the isothermal process between 3 and 4 the entropy changes from $s_g = 8.0748$ to $s_f = 0.7038$.

$$s_3 = (1 - x_3)0.7038 + x_3 8.0748 = 5.3356$$

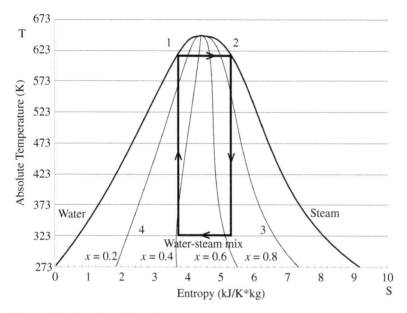

Figure 7.6 Carnot cycle for steam-water mixture.

Table 7.2 Steam table data for Example 7.3.

T (°C)	P (MPa)	h_w (kJ/kg)	h_v (kJ/kg)	s_w (kJ/kg)	s_v (kJ/kg)
50	0.012 35	209.34	2591.3	0.7038	8.0748
340	14.601	1594.5	2621.8	3.6601	5.3356

Source: Lemmon et al. (2015).

The steam quality ratio is obtained by solving the equation.

$$x_3 = \frac{5.3356 - 0.7038}{8.0748 - 0.7038} = 0.628$$

Hence, the enthalpy at 3 is

$$h_3 = (1 - x_3)h_f + x_3 h_g = (1 - 0.628)209.34 + 0.628 \times 2591.3$$
$$h_3 = 1706 \text{ kJ/kg}$$

The useful work delivered by the turbine is

$$W_{out} = h_2 - h_3 = 2621.8 - 1706.1 = 915.7 \text{ kJ/kg}$$

d) Heat exchanged between the compressor and the environment during the isothermal process is the difference of enthalpies between 3 and 4. To determine the enthalpy at point 4, we need to know the steam quality at there. Again, since the process from 4 to 1 is adiabatic, the entropy of the fluid at 4 is 3.660 kJ/kg, the same as at 1. On segment 3–4 the entropy is reduced from 8.0748 kJ/kg to 0.7038 kJ/kg.

$$(1 - x_4)0.7038 + 8.0748 \cdot x_4 = 3.660$$

Solving this equation, the steam quality at 4 is obtained $x_4 = 0.4$. Enthalpy at the point 4 is

$$h_4 = (1 - 0.4)209.34 + 0.4 \times 2591.3 = 1165 \text{ kJ/kg}$$

The condenser operates at constant pressure. Therefore, the heat lost to the environment is

$$Q_L = h_3 - h_4 = 1706 - 1164 = 541.4 \text{ kJ/kg}$$

e) Heat converted to work is the difference of the input heat and the heat lost to the environment in the boiler.

$$Q_w = Q_H - Q_L = 1027.3 - 541.4 = 486 \text{ kJ/kg}$$

The proportion of the heat converted to work to the input heat is:

$$\frac{Q_w}{Q_H} = \frac{486}{1027} = 0.47 \therefore$$

Note that the proportion of the heat converted to useful work to the input heat is equal to the thermal efficiency obtained in part a.

7.7 Rankine Cycle

Condensing the steam completely until the liquid phase is reached and operating the compressor with liquid water can eliminate the problems associated with the two-phase compressor of the Carnot heat engine. In addition, operating the turbine with superheated steam instead of a mixture of water and steam reduces the risk of corrosion on the turbine blades. A thermodynamic cycle named after the Scottish physicist and engineer William Rankine allows such operation. An idealized Rankine cycle is plotted in Figure 7.7 for water, which is the working fluid for modern steam turbines. Sequences of the Rankine cycle are described below:

- a–b: Saturated water is compressed. As the fluid is at liquid phase (not a two-phase mixture), the pump requires less input energy compared to the Carnot heat engine.
- b–c: Water is heated at high pressure up to the boiling temperature (liquid saturation point). This stage of the cycle is called *economizer*.
- c–d: Pressurized water entering the boiler (c) is heated at constant pressure and temperature up to the saturation point.
- d–e: The vapor is further heated to obtain superheated steam, typically around 600 °C.
- e–f: The superheated steam expands while it passes through the blades of the turbine producing mechanical work on the shaft. In the meantime, the temperature and pressure decrease, and some condensation may occur in the turbine. If the heat loss in the turbine is neglected, this process can be considered adiabatic since there is no external heat added.
- f–a: The wet vapor enters a condenser where it is condensed at a constant pressure to become saturated water.

The method used to determine the efficiency of a Carnot cycle cannot be applied to a Rankine cycle. We can instead use the thermal efficiency defined in Eq. (7.36).

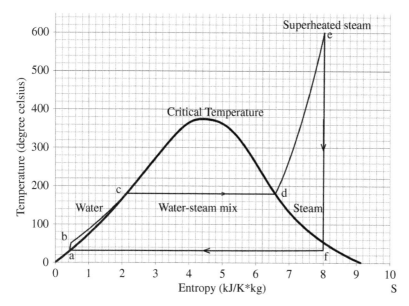

Figure 7.7 Idealized Rankine cycle for water.

The compressor increases the pressure at constant entropy. Assuming this process is adiabatic, the work done on the compressor can be calculated using the energy conservation principle. The work done on the fluid by the compressor is the product of the fluid volume and the difference of the input and output pressures.

$$W_{comp} = v(p_b - p_a) \tag{7.40}$$

If the specific volume in m³/kg is used in this equation, the work is obtained in J/kg. On the other hand, the compressor increases the enthalpy of the working fluid.

$$W_{comp} = h_b - h_a \quad \Rightarrow \quad h_b = W_{comp} + h_a \tag{7.41}$$

Water entering the boiler has the enthalpy h_b. The value of h_a is determined using steam data for saturated water at the condenser temperature. In modern steam power plants water is boiled in three separate sections named *economizer*, *boiler*, and *superheater*. Throughout the boiling process (b–e) the pressure remains constant (isobaric). Therefore, the heat obtained from external combustion of a fuel increases the enthalpy in all sections of the boiler.

$$Q_H = h_e - h_b \tag{7.42}$$

The overheated steam expands in the turbine producing work on the shaft, then exits the turbine at a partially condensed state with much lower temperature. If the process is assumed adiabatic, the work done by the steam is equal to the difference of enthalpies at the entrance and exit of the turbine.

$$W_t = h_e - h_f \tag{7.43}$$

The condenser dissipates the heat to the environment while the water-steam mixture condenses to saturated water. The heat expelled is the difference of the enthalpies at

Table 7.3 Steam data for Example 7.4.

T (°C)	P (MPa)	h_w (kJ/kg)	h_v (kJ/kg)	s_w (kJ/kg)	s_v (kJ/kg)
25	0.0032	104.83	2546.5	0.3672	8.5566
180	8.5879	1345.00	2749.6	3.2552	5.7059
600	8.5879	—	3637.5	—	6.9851

Source: Lemmon et al. (2015).

states f and a.

$$Q_L = h_f - h_a \tag{7.44}$$

Since the process in the turbine is isentropic, the steam at the exit has the same entropy as the state e. Steam quality at the state f is determined by using Eq. (7.35).

The efficiency of the cycle is the ratio of the net mechanical work done by the system to the heat input (Table 7.3).

$$\eta = \frac{W_{net}}{Q_H} = \frac{W_t - W_{comp}}{Q_H} \tag{7.45}$$

The efficiency obtained from Eq. (7.45) is smaller than the thermal efficiency for the same higher and lower temperature reservoirs given by the Carnot expression.

$$\eta_R < 1 - \frac{T_L}{T_H} \tag{7.46}$$

Example 7.4 A heat engine operating on an ideal Rankine cycle like the one shown in Figure 7.7 has the specifications listed below. Using the steam data given in Table 7.3, determine the following for each cycle.

a) The work done by the compressor
b) The heat input in the boiler
c) Work done on the turbine
d) The heat rejected by the condenser
e) The efficiency of the engine ignoring frictions and non-idealities
   Engine specifications

Boiler pressure	8.5879 MPa
Condenser pressure	0.0032 MPa
Boiler temperature	180 °C
Super heater temperature	600 °C

Solution
The processes will be identified with the same numbers as in Figure 7.7.

a) Since water can be assumed incompressible, the work done by the compressor is the product of volume and the difference of input and output pressures. The specific volume of water at 25 °C is 10^{-3} m³/kg.

$$W_{com} = v(p_b - p_a) = (8.5879 - 0.0032) \times 10^6 \times 10^{-3} = 8.6 \ \text{kJ/kg}$$

b) Assuming that the process in the compressor is adiabatic (no heat input), the work done on by the compressor changes the enthalpy.

$$W_{comp} = h_b - h_a$$

Enthalpy at the entrance of the boiler is

$$h_a = W_{comp} + h_b = 8.6 + 104.8 = 113.4 \text{ kJ/kg}$$

Super-heated dry steam leaves the boiler at 600 °C with the enthalpy of 3637.5 kJ/kg. Since the process is isobaric, the heat input to the boiler is the difference of the enthalpies at the input and output.

$$Q_{in} = h_e - h_b = 3637.5 - 113.4 = 3524 \text{ kJ/kg}$$

c) The process in the turbine is adiabatic. Therefore

$$W_t = h_e - h_f$$

Steam quality at the entrance of the condenser (point f) can be determined by using the entropy of the steam at 600 °C and the saturation entropies at 25 °C.

$$x_f = \frac{S_e - S_{w,25°C}}{S_{v,25°C} - S_{g,25°C}} = \frac{6.9851 - 0.3672}{8.5566 - 0.3672} = 0.81$$

Enthalpy of the mixture at the exit of the turbine is obtained for the steam quality:

$$h_f = (1 - x_f)h_w + x_f h_v =$$
$$= (1 - 0.81) \times 104.8 + 081 \times 2546.5 = 2078 \text{ kJ/kg}$$

Work done by the turbine is

$$W_t = 2078 - 3637.5 = 1560 \text{ kJ/kg}$$

d) Since the process in the condenser is isobaric, the heat dissipated is equal to the difference of the enthalpies:

$$Q_{out} = h_f - h_{w,25} = 2078 - 105 = 1973 \text{ kJ/kg}$$

e) Efficiency of the engine is

$$\eta = \frac{W_{net}}{Q_{in}} = \frac{W_t - W_{comp}}{Q_{in}} = \frac{1560 - 8.6}{3524} = 0.44 \text{ or } 44\%$$

7.8 Improved Efficiency Steam Turbines

In modern steam turbines the thermal efficiency is improved in several ways by modifying the Rankine cycle. One method is lowering the condenser pressure to lower the temperature of the working fluid T_L exiting the condenser. Another method is increasing the boiler pressure such that steam exits the boiler at a higher temperature. The optimal way is to increase the boiler pressure without increasing the moisture content of the steam that enters the turbine. For maximum efficiency, it is desirable to operate the turbine at a temperature as

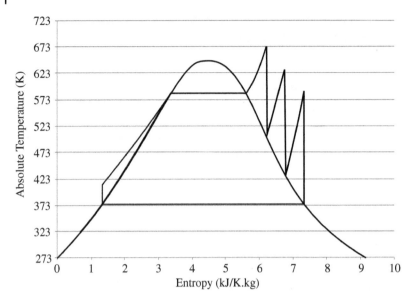

Figure 7.8 Rankine cycle with multiple reheating.

high as possible. However, extreme temperatures will fatigue boiler and turbine materials. Corrosion and erosion of boiler tubes and turbine blades become more significant at higher temperatures.

Modern thermal power plants use multiple stage turbines with reheat. Steam exiting the first turbine is reheated and passed through other turbines several times before entering the condenser. Figure 7.8 shows a Rankine cycle three-stage reheating. The first turbine operates at higher pressure and higher temperature. Steam exiting this turbine is reheated and enters an intermediate stage pressure turbine, then is reheated before entering a low-pressure turbine. A multiple reheating process increases the overall efficiency and reduces the formation of water droplets in the steam.

7.9 Gas Turbines

Like steam turbines, gas turbines also work on a cycle, but there are several major differences: first, the working fluid is air; the boiler is eliminated, and the fuel and air mixture are fired in a combustion chamber; finally, there is no need for a condenser since the exhaust gas is released into the atmosphere. While different types of gas turbines have been developed for various applications, in this section we will focus on continuous combustion, stationary, and open-cycle plants, which are more commonly used in energy systems.

Gas turbines were developed in several places at the beginning of the twentieth century. Progress in the field of aerodynamics, development of new materials that can withstand higher temperatures, and the invention of turbo-jet engines for aircrafts and guided missiles helped the development of modern gas turbines used in electric power generation plants. Gas turbine plants are reliable and free from vibrations. They can utilize grades of fuel that

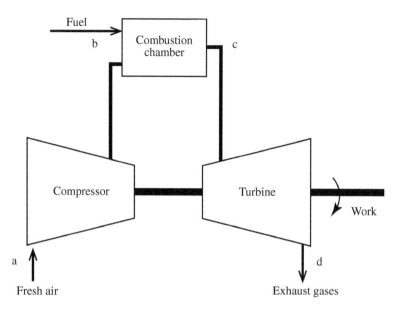

Figure 7.9 Schematic structure of a gas turbine.

are not suitable for spark-ignition engines and they can produce large amount of power from units relatively lightweight and smaller in size (Patel and Karamchandani 1996).

Figure 7.9 outlines the structure of an open cycle continuous combustion gas turbine plant. The main components of a gas turbine are compressor, combustion chamber, and turbine blades. Most gas turbines combine these sections in a single unit, where the turbine shaft also drives the compressor. Fresh air from the atmosphere enters the compressor and is pressurized to about 1–2 MPa. The compressed air is mixed with the fuel in the combustion chamber. The fuel burning in the combustion chamber produces combustion gases and further increases the pressure of the air. The air and combustion gas mixture expand through the turbine blades producing mechanical work on the shaft. The exhaust gas is released to the atmosphere at lower temperatures. While the process seems to be open, the atmosphere can be considered as a heat sink like the condenser of the steam turbines. A separate starter unit is used to provide the first rotor motion, until the turbine's rotation reaches the rated speed and the combustion sustains. The thermodynamic cycle of simple gas turbines is named the Brayton or Joule cycle.

7.9.1 Brayton (Joule) Cycle

A Brayton cycle is depicted in Figure 7.10 on both volume-pressure and entropy-temperature planes. During the compression process (a–b) the incoming air ideally does not exchange heat with the environment, so compression can be considered adiabatic. Therefore, according to the first law of thermodynamics, the external work done on the shaft increases the internal energy of the air.

$$Q = \Delta u + W_{comp} = 0 \quad \Rightarrow \quad \Delta u = -W_{comp} \tag{7.47}$$

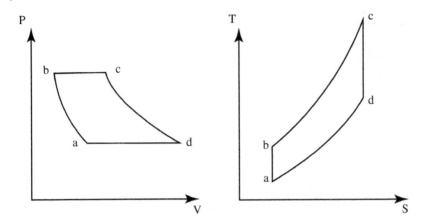

Figure 7.10 Thermodynamic diagrams for a Brayton (Joule) cycle.

At the same time, the compressor increases the enthalpy from h_a to h_b. Since specific heat at constant pressure c_p is approximately constant at this pressure level, the change of enthalpy can be expressed in terms of the temperature difference.

$$W_{comp} = h_b - h_a = c_p(T_b - T_a) \tag{7.48}$$

The work done by the turbine is obtained from a similar expression.

$$W_T = h_c - h_d = c_p(T_c - T_d) \tag{7.49}$$

The combustion process does not involve external work. Therefore, the input heat is solely responsible for the increase of enthalpy.

$$Q = h_c - h_b = c_p(T_c - T_b) \tag{7.50}$$

The efficiency of the cycle is the ratio of the net shaft work to the heat input. The expression of efficiency contains only the temperatures since the specific heat is simplified.

$$\eta = \frac{(T_c - T_d) - (T_b - T_a)}{T_c - T_b} \tag{7.51}$$

It is sometimes more convenient to use the pressures instead of the temperatures to calculate the efficiency of a gas turbine. In an adiabatic expansion of an ideal gas the differential heat is zero. Therefore, for a differential change of volume the first law of thermodynamics can be written as Eq. (7.52).

$$du + p \, dv = 0 \quad \Rightarrow \quad c_v dT + p \, dv = 0 \tag{7.52}$$

On the other hand, differentiating the equation of state for ideal gas, we obtain

$$d(pv) = d(RT) \quad \Rightarrow \quad p \, dv + v \, dp = R \, dT \tag{7.53}$$

Substituting $R = c_p - c_v$, $\gamma = c_p/c_v$ and $v = RT/p$, Eq. (7.53) becomes

$$\frac{dT}{T} = \left(\frac{\gamma - 1}{\gamma}\right) \frac{dp}{p} \tag{7.54}$$

Integration of both side yields:

$$\ln T = \left(\frac{\gamma - 1}{\gamma}\right) \ln p + C \tag{7.55}$$

C is integration constant so Eq. (7.55) can be written alternatively as in Eq. (7.56), where A is another constant equal to $\ln C$.

$$T = A \cdot p^{\frac{\gamma-1}{\gamma}} \tag{7.56}$$

Using this expression for T_a and T_b and dividing both sides we can write the following ratios.

$$\frac{T_a}{T_b} = \left(\frac{p_a}{p_b}\right)^{\frac{\gamma-1}{\gamma}} \quad \text{and} \quad \frac{T_c}{T_d} = \left(\frac{p_c}{p_d}\right)^{\frac{\gamma-1}{\gamma}} \tag{7.57}$$

By arranging and substituting $r = p_b/p_a = p_c/p_d$, we can write:

$$\eta = 1 - r^{\frac{1-\gamma}{\gamma}} \tag{7.58}$$

The efficiency of a Brayton cycle depends only on the ratio of the pressure of the incoming and the compressed air. The thermal efficiency is independent on the temperatures. The ratio of the specific heats at constant pressure and constant volume $\gamma = c_p/c_v$ is 1.4 for air.

Example 7.5 An ideal continuous combustion gas turbine plant has a compressor that raises the pressure of the incoming air five times and the maximum temperature in the combustion chamber is 1200 °C. Determine

a) the thermal efficiency of this plant.
b) the temperature of the exhaust gas.

Solution

a) Given the pressure of the air supplied to the combustion chamber (p_b) is five times the pressure of the incoming air (r_a) and $\gamma = 1.4$; using Eq. (7.58) we obtain:

$$\eta = 1 - (5)^{\frac{1-1.4}{1.4}} = 0.37 \quad \text{or} \quad 37\%$$

b) To obtain the temperature of the exhaust gas, we can use Eq. (7.57).

$$T_d = T_c r^{-\left(\frac{\gamma-1}{\gamma}\right)} = (1200 + 273)5^{-\left(\frac{1.4-1}{1.4}\right)} = 930 \text{ K}$$

The exhaust gas temperature in degree Celsius is therefore $930 - 273 = 657°C$∴.

7.10 Improved Efficiency Thermal Systems

Efficiency of thermal power systems is limited with the Carnot efficiency, which is determined by the upper and lower temperatures. In steam turbines, the upper temperature is constrained by practical limitations of boiler and turbine pressures. Gas turbines can operate at higher temperatures since the working fluid is air. In both turbine types a considerable amount of heat is rejected to the environment. Various combinations have been developed to increase the overall efficiency of thermal power systems.

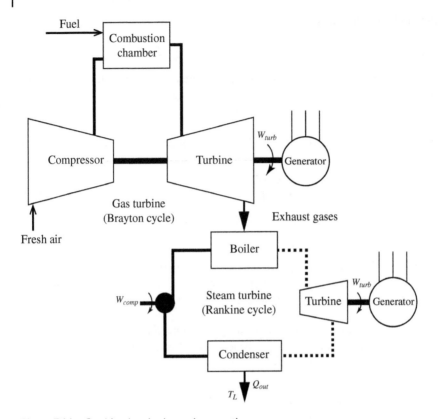

Figure 7.11 Combined cycle thermal generation.

7.10.1 Combined Cycle Gas Turbine (CCGT)

The exhaust gas leaving a gas turbine contains a significant amount of energy that can be reused to heat working fluid of a steam turbine. A system formed by combining gas and steam turbines is known as a combined cycle turbine. Such a system shown in Figure 7.11 achieves a higher thermal efficiency by using the exhaust gas from Brayton cycle to heat the working fluid in Rankine cycle. The net efficiency of a combined cycle turbine system is equivalent to a thermodynamic cycle operating between the higher temperature of the Brayton cycle and the lower temperature of the Rankine cycle. Efficiencies around 60% are common for combined cycle turbine systems.

7.10.2 Combined Heat and Power (CHP) Systems

In CHP systems, the released heat from the fluid exiting the turbine is used for heating of industrial process, commercial buildings, building complexes, or public services. In a CHP power plant using steam turbines, the condenser operates at a higher temperature than a conventional power plant. CHP systems are common in waste-to-energy and geothermal energy systems. Because of the high cost of thermally insulated pipelines and other infrastructure to convey steam to a long distance, such systems are more convenient for industrial process heating and district heating in densely populated areas.

7.11 Chapter Review

Thermodynamics is the study of interactions between heat and mechanical work. In the sense of thermodynamics, heat is an energy interaction due to temperature difference. Temperature is the measure of hotness or coldness. When two objects are at the same temperature, they are said to be in thermal equilibrium and no thermodynamic interaction occurs. Celsius and Fahrenheit scales are commonly used to express temperature. Both temperature scales are standardized using the freezing and boiling temperatures of pure water. Absolute temperature scales use a single reference temperature and use the unit Kelvin (K).

Internal energy is the total kinetic and potential energy of atomic particles in a substance or object. Thermodynamic processes cause changes of internal energy.

The conservation of energy principle is known as the first law of thermodynamics. According to the first law of thermodynamics, *the net heat processed in a system is equal to the sum of the internal energy change and mechanical work.*

The second law of thermodynamics states that energy can flow only from a higher temperature reservoir to a lower temperature reservoir if no external work is added to the system. This principle led to the concept of entropy. Differential entropy is obtained by dividing the change of heat with the absolute temperature. Over an internally reversible process, differential entropy of a system corresponds to the randomness of molecules. The total entropy change is obtained by dividing the heat by absolute temperature. The unit of entropy is Joules per Kelvin (J/K).

Enthalpy is a combined property defined as the sum of internal energy u and flow energy expressed as the product of volume and pressure pv. A thermodynamic process is a transition of a gas from one state to another. An ideal process is reversible but in practice most thermodynamic processes are irreversible due to frictions and heat losses. In an isothermal process the temperature is constant, and the internal energy does not change. In an adiabatic process no heat interaction occurs, therefore the internal energy of the system changes due to the temperature change. Since no heat is added or taken out, the entropy remains constant during an ideal adiabatic process.

Carnot cycle is an ideal reversible thermodynamic process consisting of two isothermal and two adiabatic processes. Efficiency is defined as the ratio of the output energy to input energy. Efficiency of Carnot cycle is the maximum efficiency attainable by a thermodynamic process. Carnot efficiency depends only on the absolute temperature of the heat source and the absolute temperature of the heat sink. The expression $\eta = 1 - T_H/T_L$ gives the Carnot efficiency.

In the Carnot heat engine, compression and expansion take place in the regions where the working fluid is a mixture of water and steam. Two-phase fluid (liquid and gas) presents practical challenges in the compressor design. In addition, the presence of humidity in steam causes corrosion on turbine blades.

In the Rankine cycle, steam is completely condensed until the liquid phase is reached to operate the compressor with liquid water and eliminate the problems associated with the two-phase compressor of the Carnot heat engine. In addition, operating the turbine with superheated steam instead of a mixture of water and steam reduces the risk of corrosion on the turbine blades.

Efficiency of the Rankine cycle is improved by reheating the working fluid several times before entering the condenser.

Gas turbines work on a Brayton cycle, also known as Joule cycle. The working fluid of gas turbines is air. The boiler is eliminated, and the fuel and air mixture are fired in a combustion chamber. There is no need to use a condenser in a gas turbine since the exhaust gas is released to the atmosphere.

The efficiency of a Brayton cycle depends only on the ratio of the pressure of the incoming and the compressed air; the thermal efficiency is independent of the temperatures.

Combined cycle turbine systems use the exhaust gas from a gas turbine to heat the working fluid of a steam turbine to increase the overall thermal efficiency.

Review Quiz

1 Approximately what percentage of the electric worldwide is generated from thermal energy?
   a. 10%
   b. 50%
   c. 75%
   d. 95%

2 Heat engines convert
   a. mechanical work to useful heat.
   b. thermal energy to useful work.
   c. mechanical work to electricity.
   d. temperature to mechanical work.

3 Which one of the following is/are not a primary source of energy for a thermal power plant?
   a. Coal
   b. Natural gas
   c. Nuclear reactions
   d. Tidal energy

4 During isothermal expansion of an ideal gas
   a. the volume changes with heat input while the temperature remains constant.
   b. the temperature changes with heat input while the volume remains constant.
   c. the volume changes without heat input while the temperature remains constant.
   d. the volume and temperature change without heat input.

5 During isothermal compression of an ideal gas
   a. the temperature changes with heat input while the volume remains constant.
   b. the volume changes without heat input while the temperature remains constant.
   c. the volume decreases while the temperature remains constant and heat is released.
   d. the volume and temperature change without heat input.

6 During adiabatic expansion of an ideal gas
   a. the volume changes with heat input while the temperature remains constant.
   b. the temperature changes with heat input while the volume remains constant.
   c. the volume changes without heat input while the temperature remains constant.
   d. the volume and temperature change with heat input.

7 The thermodynamic process performed by the pump/compressor in a Carnot engine is
   a. adiabatic compression.
   b. adiabatic expansion.
   c. isothermal compression.
   d. isothermal expansion.

8 The thermodynamic process performed by the boiler in a Carnot engine is
   a. adiabatic compression.
   b. adiabatic expansion.
   c. isothermal compression.
   d. isothermal expansion.

9 The thermodynamic process performed by the turbine in a Carnot engine is
   a. adiabatic compression.
   b. adiabatic expansion.
   c. isothermal compression.
   d. isothermal expansion.

10 Which element in a steam engine performs isothermal process?
   a. Compressor
   b. Turbine
   c. Boiler
   d. None of them

11 Which part of a thermal power plant performs adiabatic process?
   a. Condenser
   b. Turbine
   c. Boiler
   d. None of them

12 Which one of the following statements is true for a heat engine?
   a. The efficiency of a heat engine depends on the working fluid.
   b. Some heat engines can produce useful mechanical work by receiving thermal energy from only one reservoir.

   c. Heat engines convert mechanical work into heat.
   d. All heat engines receive heat from a high temperature reservoir and reject heat to a low temperature sink.

13 In which part of a heat engine does entropy change?
   a. Compressor
   b. Turbine
   c. Boiler
   d. All of the parts

14 Practical steam turbines operate in the
   a. Carnot cycle.
   b. Bryton cycle.
   c. Rankine cycle.
   d. Joule cycle.

15 The efficiency of a heat engine depends on
   a. the fuel used to boil water.
   b. the ratio of higher and lower temperatures.
   c. flow rate of the working fluid.
   d. average temperature of the working fluid.

16 In conventional thermal power plants heat is produced by
   a. electric power.
   b. mechanical power.
   c. steam circulation.
   d. combustion.

17 Which type of thermal power plants cause the most greenhouse gas emissions?
   a. Coal fired
   b. Natural gas fired
   c. Nuclear
   d. Geothermal

Answers: 1-a, 2-b, 3-d, 4-a, 5-c, 6-d, 7-a, 8-d, 9-b, 10-c, 11-b, 12-d, 13-c, 14-c, 15-b, 16-d, 17-a.

Research Topics and Problems

Research and Discussion Topics

1 In a white paper, discuss the benefits and drawbacks of shifting electric generation from coal to natural gas.

2 Compare the environmental impacts of water-cooled versus air-cooled thermal power plants.

3 Research emerging technologies to improve the efficiency of thermal power plants.

4 Research currently available technologies to reduce greenhouse gas emissions from fossil fuel burning power plants.

5 Discuss the impacts of thermal power plants on freshwater sources.

Problems

1 What is the efficiency of a thermal power plant that uses 10,000 Btu heat/kWh of generated electricity?

2 Water circulating in a steam turbine is heated to 350 °C and cooled during the thermodynamic process to 25 °C. What is the maximum efficiency of this turbine in percent?

3 Heat rate of a thermal power plant is 9500 Btu/kWh. What is its efficiency in percent?

4 Efficiency of a thermal power plant is 34%. What is its heat rate in Btu/kWh?

5 A 200-MW coal-burning power plant operates at 35% efficiency at full load. Given the heat content of the coal used in this power plant is 21 million Btu/ton, what is the daily coal consumption in tons?

6 A 150-MW coal-burning power plant consumes 1500 tons of coal per day when it operates at full load. Given the energy content of the coal used in this power plant is 19 million Btu/ton, what is the heat rate of this power generation plant?

7 A power plant delivers 250-MW using a steam turbine. Steam goes into the turbine at 300 °C and condensed water leaves the turbine at 40 °C. Assuming that the turbine operates as an ideal Carnot engine and the losses in the electric generator are negligible, determine the power loss in MW.

8 A steam engine converts 1500 million Btu input energy obtained by burning a fuel into mechanical energy. Steam goes into the turbine at 250 °C and condensed water leaves the turbine at 30 °C. Assuming that the turbine operates as an ideal Carnot engine determine the mechanical energy output in MWh.

9 An 80-MW steam turbine receives steam at 400 °C from the boiler. The working fluid leaves the condenser at 25 °C, after releasing excess heat to a river that flows at a rate of 45 m^3/s. Assuming thatm the engine operates on an ideal Carnot cycle, calculate the temperature rise of the river water downstream from the turbine. (Density of the river water: 0.9; specific heat of the river water: 4.18 kJ/kg.)

10 A 50-MW steam turbine operates at an efficiency of 38% and releases excess heat to cooling water that flows at a rate of $8 \, m^3/s$. Calculate the temperature rise of the cooling water. (Specific volume of the cooling water: $1 \, dm^3/kg$; specific heat of the water: 4.18 kJ/kg.)

Recommended Web Sites

- American Journal of Physics: https://aapt.scitation.org/journal/ajp
- Engineering Toolbox: https://www.engineeringtoolbox.com
- NIST Chemistry WebBook: http://webbook.nist.gov
- Scientific American: https://www.scientificamerican.com

References

Cengel, A.Y. and Boles, A.M. (2006). *Thermodynamics – An Engineering Approach*, 5e. New York, NY: McGraw-Hill Higher Education.

DOE/EIA MER, 2017. Monthly Energy Review, s.l.: s.n.

EIA, 2013. *Electric Power Annual 2013*, Washington, DC: s.n.

EIA, 2019. *Glossary*. [Online] Available at: https://www.eia.gov/tools/glossary [Accessed 15 January 2019].

Fenn, B.J. (1982). *Engines, Energy, and Entropy*. San Francisco: W. H. Freeman and Company.

Lemmon, E. W., McLinden, O., and Friend, D. G., 2015. *Thermophysical Properties of Fluid Systems*. [Online] Available at: http://webbook.nist.gov [Accessed 9 2015].

Patel, R.C. and Karamchandani, C.J. (1996). *Elements of Heat Engines*, 16e. Gandhi Nagar Gruh Vadodara: Acharya Publications.

Rossini, F.D., Wagman, D.D., Evans, W.H. et al. (1952). *Selected Values of Chemical Thermodynamic Properties*. Washington, DC: US: Government Printing Office.

Sonmtag, E.R., Borgnakke, C., and Van Wylen, J.G. (1998). *Fundamentals of Thermodynamics*, 5e. New York, NY: Wiley.

8

Hydropower

Canadian and American hydroelectric power plants in Niagara Falls. The Sir Adam Beck Hydroelectric Generating Station in Ontario, Canada (left) was opened in 1922, and the Robert Moses Niagara Hydroelectric Power Station located in Lewiston, NY, was opened in 1961. Both stations are pumped storage hydroelectric power plants pumping water up in two separate lakes at night, one in Canada and one in the US, generating electricity during peak hours using the water stored in the lakes.

Energy for Sustainable Society: From Resources to Users, First Edition. Oguz A. Soysal and Hilkat S. Soysal.
© 2020 John Wiley & Sons Ltd. Published 2020 by John Wiley & Sons Ltd.

8.1 Introduction

Power of flowing water has been harnessed since ancient civilizations used them to drive sawmills, grain mills, metal forging apparatus, and mining equipment. Waterwheels were common in many parts of the world, and early waterwheel designs were either undershot or overshot types and were quite inefficient. The first submerged vertical axis turbine, invented in 1832 by French engineer Benoît Fourneyron, was a breakthrough that increased efficiency.

Hydropower was used to generate electricity for the first time on September 30, 1882 on the Fox River in Appleton, Wisconsin; the facility was developed by Appleton Paper manufacturer H.J. Rogers. The earlier generation units built in New York City by Thomas Edison were using steam power to drive DC generators, while the Appleton plant used the natural hydropower of the Fox River.

In 1895, Nikola Tesla and George Westinghouse built the first larger scale hydroelectric power plant in Niagara Falls to generate electricity for Buffalo, NY and New York City. This plant was the first facility using AC generators invented by Tesla with 100-kV overhead lines for long-distance transmission of electric power.

Hydropower currently provides about 2.5% of the world's total primary energy supply (IEA 2018). In 2017, the installed capacity of hydroelectric power plants in the World was 1.15 TW (IRENA 2019). The same year, hydroelectric plants around the world generated 4,060 TWh electricity to supply 16% of the world's electric energy need (BP 2018). Hydropower is a clean and renewable primary energy source. Development of a hydroelectric power plant is a multi-purpose project that involves building a dam and installing large turbines and generators. In addition to generating electricity, hydroelectric plants serve the region by regulating water flow of rivers, flood prevention, and storage of irrigation water. Dam reservoirs make artificial lakes that often provide touristic and recreational activities.

Conventional hydroelectric generation is the oldest technology used to transfer the energy of rivers to distant locations. Hydropower is a proven and predictable resource. Efficiency of hydroelectric generation can reach as high as 90%; the highest efficiency that can be obtained among all energy sources.

8.2 Basic Concepts of Hydrodynamics

Hydrodynamics is a branch of fluid mechanics that study and analyze forces, power, and energy associated with moving liquids, particularly water. Liquids are practically non-compressible fluids, meaning that the change of volume with pressure and temperature is negligible.

8.2.1 Density and Specific Mass

Density is defined as mass per unit volume. In the international unit system (SI) it is expressed in kg/m^3. The density of liquids depends more strongly on temperature than pressure.

$$\rho = \frac{m}{V} \tag{8.1}$$

To simplify calculations across different unit systems, it is sometimes more convenient to use the dimensionless quantity "relative density" or "specific gravity," defined as the ratio of the density of a substance to the density of a well-known standard material, usually pure water. At $4\,°C$, the density of pure water is $1000\,kg/m^3$. The specific gravity of natural water containing minerals and other materials is greater than unity.

The reciprocal of density is the specific volume, which is defined as volume per unit mass (m^3/kg). Specific weight is the weight of a unit volume, and includes the gravitational acceleration $g = 9.81$.

$$\gamma = g \cdot \rho \,[N/m^3] \tag{8.2}$$

8.2.2 Pressure

The pressure of a liquid is the normal force exerted per unit area.

$$P = \frac{F}{A} \tag{8.3}$$

In the SI, the unit of pressure is N/m^2, which is called a "pascal" (Pa). Because one pascal is a too small for pressures encountered in practice, the kilopascal (kPa) and megapascal (MPa) are more commonly used. Other frequently used international units of pressure are bar, atmosphere, and kgf/cm^2.

> $1\,bar = 10^5\,Pa = 100kPa = 0.1MPa$
>
> $1\,atm = 101,325Pa = 1.01325bars$
>
> $1\,kgf/cm^2 = 9.807 \times 10^4\,Pa = 0.9807\,bar = 0.9679atm \tag{8.4}$

In the imperial unit system, pressure is usually expressed in pound-force per square inch (psi). One atmosphere is equivalent to 14.696-psi.

The actual pressure is measured relative to absolute vacuum and named absolute pressure. Most pressure-measuring devices are calibrated to read zero at normal atmospheric pressure. In most practical hydrodynamic calculations, the atmospheric pressure is excluded from the absolute (actual) pressure.

The pressure difference between two points in a homogeneous and static liquid is proportional to the difference of their depth from the free liquid surface.

$$\Delta P = \frac{mg}{A} = \frac{\rho g V}{A} = \rho g(h_2 - h_1) \tag{8.5}$$

In this expression A represents the area and $V = Ah$ is the volume of liquid above that area.

8.2.3 Flow Rate

Flow rate is the amount of liquid flowing in unit time through a section of a confined path, such as stream, channel, watercourse, duct, or pipe. The amount can be volume or mass. The mass flow rate is expressed in kg/s or ton/s in SI; in the British (imperial) unit system (IS), the unit is pounds per minute (lb/min). The volumetric flow rate is measured in l/s or m^3/s (SI), and in IS the common unit is gallons per minute (gal/min).

$$Q_m = \frac{dm}{dt} \quad Q_v = \frac{dv}{dt} \tag{8.6}$$

The velocity of the liquid may be different at different points of the flow. To avoid confusion with the volume, we will represent the velocity with the letter U. The average velocity of the flow is expressed as the average of the normal velocity U_n (m/s) across the entire cross-sectional area A_c.

$$U_{avg} = \frac{1}{A_c} \int_{A_c} U_n dA_c \tag{8.7}$$

For simplicity, we will drop the subscript avg and denote the average velocity in the direction of flow by U. For an incompressible flow of fluid having uniform density, we can write

$$Q_m = \rho U A_c \tag{8.8}$$

8.2.4 Conservation of Mass in Steady Liquid Flow

In fluid mechanics, the "conservation of mass" or "mass continuity" principle in a control volume can be expressed as "the net mass transfer to or from a control volume is equal to the net change (increase or decrease) in the total mass Δm_{cv} within the control volume during a time interval Δt" (Yunus and Cimbala 2014). Mathematical expression of the conservation of mass principle is:

$$m_{in} - m_{out} = \Delta m_{cv} \tag{8.9}$$

In the special case of a liquid flow through a conduit such as pipe, tube, or hose that doesn't change shape with pressure, the incoming mass must be equal to the outgoing mass, thus the mass flow rate is constant.

$$\frac{dm_{in}}{dt} = \frac{dm_{out}}{dt} \tag{8.10}$$

In other words, steady flow means that the total rate of mass entering a control volume is equal to the total rate of mass leaving it. In engineering systems, many elements have one input and one output. Therefore, the mass flow rate at the inlet is the same as the mass flow at the outlet. Furthermore, if the fluid is incompressible, which is the case for liquids, the volume flow rate is also the same at the inlet and outlet of the control volume.

8.3 Bernoulli's Principle

Daniel Bernoulli, a Swiss mathematician, stated an important principle of fluid dynamics in his book "Hydrodynamica" published in 1738, when he was working in St. Petersburg, Russia. Bernoulli's principle states that "the sum of the kinetic, potential, and flow energies of a fluid particle is constant along a streamline during steady flow when the compressibility and frictional effects are negligible" (Yunus and Cimbala 2014). In 1755 his colleague Leonhard Euler derived the mathematical expression of Bernoulli's principle. The expression can be written for unit mass in the following form.

$$\frac{1}{2}u^2 + gh + \frac{p}{\rho} = \text{constant} \tag{8.11}$$

In this equation, the first term corresponds to the kinetic energy of unit mass moving with a velocity u, the second term corresponds to the potential energy of unit mass at a point located at the elevation h. The third term is the flow energy related to the pressure p at the same point and specific mass of the liquid ρ.

Bernoulli's equation can be derived by using the conservation of momentum or conservation of energy principles. It is, however, valid in regions of steady, incompressible flow where net frictional forces due to the viscosity are negligible. Although approximate, this relationship is a useful tool in fluid mechanics calculations.

Different situations that occur in a liquid system can be analyzed by using the Bernoulli equation under given conditions. The value of the constant can be evaluated at any point on the streamline where the pressure, density, velocity, and elevation are known. The Bernoulli equation can also be written for any two points on the same streamline as

$$\frac{V_1^{\,2}}{2} + gh_1 + \frac{P_1}{\rho} = \frac{V_2^{\,2}}{2} + gh_2 + \frac{P_2}{\rho} \tag{8.12}$$

8.4 Euler's Turbomachine Equation

Figure 8.1 shows a simplified sketch of a turbine runner. The incoming water enters the runner with the absolute velocity V_1 at an angle β_1 with the tangent to circle of radius r_1. The runner blades are designed in such a way that when the turbine operates around its rated values, the water remains tangent to their surface at any point. Therefore, the water exits the blades with an angle β_2 with the tangent to the circle of radius r_2.

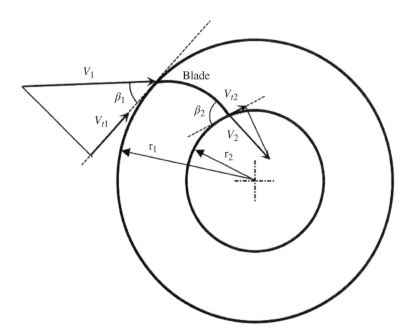

Figure 8.1 Vector diagram for a turbomachine.

Net torque on the rotor shaft is equal to the rate of change of angular momentum of the fluid. Since the water enters and exits the runner with angles, the tangential components of the velocity vectors must be considered. If the specific mass of water is ρ and the volume flow rate is Q, the torque can be expressed as

$$T = \rho Q(r_1 V_{t1} - r_2 V_{t2}) \tag{8.13}$$

$$V_{t1} = V_1 \cos \beta_1$$

$$V_{t2} = V_2 \cos \beta_2 \tag{8.14}$$

Substituting Eq. (8.14) into Eq. (8.13), the net torque on the shaft is obtained as:

$$T = \rho Q(r_1 V_1 \cos \beta_1 - r_2 V_2 \cos \beta_2) \tag{8.15}$$

Equation (8.15) is known as "Euler's turbomachine equation" and it applies to both turbines and pumps. The only difference between a pump and a turbine is the direction of the energy flow. A turbine is designed to convert the hydraulic energy of water into mechanical energy delivered by the shaft. A pump performs the reverse function, converting the mechanical energy input to the shaft into hydraulic energy output. In the case of a pump, the direction of the vectors V_1 and V_2 must be reversed. The power delivered to the shaft of a turbine rotating at the angular velocity ω is

$$P = \omega T = \omega \rho Q(r_1 V_1 \cos \beta_1 - r_2 V_2 \cos \beta_2) \tag{8.16}$$

The maximum torque is achieved for $\cos \beta_2 = 0$ or $\beta_2 = 90°$. Therefore, the maximum power expression is reduced to

$$P_{max} = \omega \rho Q r_1 V_1 \cos \beta_1 \tag{8.17}$$

Euler's turbomachine equation can be used to analyze the performance of most turbomachines.

8.5 Hydraulic Turbines

Turbomachines are energy converters that transfer energy between a moving fluid and a rotating shaft. Rotational pumps and compressors add energy to a fluid by increasing its pressure. Turbines, on the other hand, extract energy from a fluid to rotate a shaft. When the working fluid is water, turbomachines are called hydraulic pumps or turbines. Some types of hydraulic turbomachines can operate in two ways, either as a pump or turbine. In this section, our focus is hydraulic turbines or hydro-turbines.

Hydraulic turbines convert the potential energy associated with a difference in water elevation into mechanical work, which drives an electric generator. A hydraulic turbine has a stationary part with a water inlet and a rotating part turning a shaft. The rotating part of a hydro turbine is called the "runner;" basically a wheel mounted on the shaft which converts the kinetic energy of water to mechanical torque according Euler's equation [Eq. (8.15)].

Performance of different types of water turbines depends on the hydraulic head, flow rate, and runner velocity. Hydraulic head is defined as the elevation difference between upstream and downstream levels of the water supply (see Figure 8.8). The choice of turbine type depends on hydraulic conditions of the power plant site.

According to the principle of operation, turbines are grouped as impulse or reaction turbines. Kinetic energy of the flowing water can be transferred to the runner in two ways; by impulse or reaction. In an impulse turbine, the available potential energy of water is converted into kinetic energy before entering the runner, and the available power is extracted from the flow approximately at atmospheric pressure. In impulse turbines a water jet coming out of a nozzle hits the runner blades. In a reaction turbine, the pressure and velocity of water decrease as it flows through the runner blades. Reaction turbines do not have a nozzle, instead the runner enclosed in a casing rotates due to the exchange of momentum between the flowing water and runner blades.

Turbines installed in hydroelectric power plants today are either impulse or reaction turbines. Commonly used types in modern hydroelectric stations are known as Pelton, Francis, and Kaplan turbines, named after their inventors. A Pelton turbine is a typical example of impulse turbines, while Francis and Kaplan turbines are reaction turbines (Andrews and Jelley 2007). Figure 8.2 illustrates the runners used in each turbine types in photos of small-scale models.

Figure 8.2 Turbine runner types. (Models displayed in the visitor center of Grand Coulee Dam, Washington.)

In most conventional hydroelectric generation facilities, the turbine shaft is vertical, directly coupled to the generator. Generally, the rotation speed of hydroelectric turbine-generator systems is lower (in the order of 100/min) than steam and gas turbines. Generators used in hydroelectric plants have a larger number of poles to generate electric power at 50 or 60 rpm shaft rotation speed.

8.5.1 Pelton Turbine

The Pelton turbine is an improved version of the waterwheels used for centuries to drive grain-grinding mills and sawmills. Pelton turbines are classified as impulse turbines because a water jet hitting buckets turns the wheel-type runner. Inventor Lester Pelton was born and raised in Erie County, Ohio. He moved to California during the Gold Rush and he was inspired by waterwheels while he was working as a millwright and carpenter. Pelton did not find gold, but invented the most efficient type of impulse turbine, and patented his invention in 1878. Figure 8.3 shows (a) the original drawing from US patent 233692 and (b) installation of a Pelton turbine in the Walchensee Power Plant, Germany (Wikimedia Commons 2006).

The Pelton wheel has a certain number of specially shaped cups (also called buckets) mounted on the periphery. A water jet coming from a nozzle placed at an optimal angle hits the splitter ridge of the cup and turns back nearly 180°. A portion of the outer edge of each cup is cut out to allow the water jet to pass through and reach the next cup aligned with the jet direction. The maximum power of a Pelton turbine can be calculated using the Euler Turbomachine equation given in Eq. (8.15). Suppose that the water jet strikes the

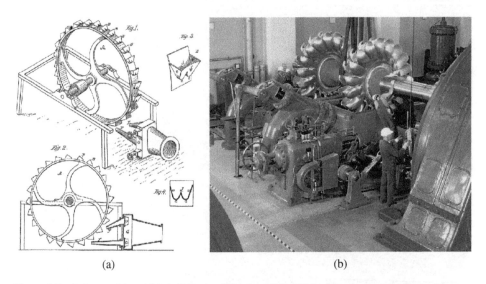

(a) (b)

Figure 8.3 Pelton turbine. (a) L.A. Pelton – US patent 233 692 (public domain). (b) Assembly of a Pelton turbine in the Walchensee Power Plant, Germany. Source: Courtesy of Voith Siemens Hydro Power.

bucket in the tangential direction to the wheel ($\beta_1 = 0$) and is reflected 180° without losing its speed v_j ($\beta_2 = 180°$). The cup advances with the velocity v_c. The size of the cup and the diameter of the water jet are negligible compared to the diameter of the wheel. Therefore, we can assume that $r_1 = r_2 = r$. Given the specific mass of injected water ρ and the flow rate Q, Euler's equation [Eq. (8.15)] can be modified to obtain the torque acting on the wheel.

$$T = 2\rho Q(v_j - v_c)r \tag{8.18}$$

Theoretical maximum power of the turbine is $P = \omega T$. Since $\omega = v_c/r$, the expression of theoretical maximum turbine power is simplified to:

$$P = \omega T = 2\omega r\rho Q(v_j - v_c) = 2v_c\rho Q(v_j - v_c) \tag{8.19}$$

Note that if the wheel is not turning, or $v_c = 0$, the power is zero. When the wheel turns at the same speed as the water jet, or $v_c = v_j$, the power is again zero. Between zero and the speed of the water jet, there is an optimal bucket velocity for which the shaft power is maximum. We can find this optimal velocity by taking the derivative of the power expression in respect to v_c and equating to zero.

$$\frac{dP}{dv_c} = \frac{d}{dt}[2\rho Q(v_j v_c - v_c^2)] = 2\rho Q(v_j - 2v_c) = 0 \tag{8.20}$$

Equation (8.20) yields $v_c = \frac{1}{2}v_j$. In other words, the Pelton turbine delivers the maximum shaft power when its runner turns with the tangential speed half of the velocity of the water jet. Maximum power expression is

$$P_{max} = \frac{1}{2}\rho Q v_j^2 \tag{8.21}$$

Maximum power occurs when the kinetic energy of the water jet is completely transferred to the shaft. In real situations, if the direction of the water jet is tangent to the wheel the reflected water would hit the backside of the neighbor bucket. Pelton turbines used in power plants are designed in such a way that the angle of the water jet can be adjusted to change the output power [see Figure 8.3(b)].

8.5.2 Francis Turbine

In 1848 in Lowell, MA, American engineer James B. Francis developed the most versatile turbine used in hydroelectric power plants. The Francis Turbine produces more power compared to the same size impulse turbines (for example the Pelton turbine) in a wide range of hydraulic heads. In a Francis turbine the runner is submerged in a volute-type casing like a curved funnel with decreasing cross-sectional area. A schematic view of the Francis turbine and generator unit at the Grand Coulee power station is shown in Figure 8.4. Water enters from the larger diameter intake with higher kinetic energy and exits from the center where the pipe diameter is smaller. While water travels through the volute, the kinetic energy is transferred to the runner blades by reaction force and it decreases as it approaches the discharge. Operation of the Francis turbine is the opposite of a centrifugal pump. Since energy transfer is based on reaction force, Francis turbines are classified as

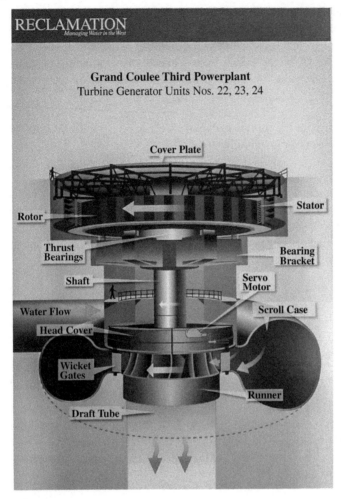

Figure 8.4 Schematic view of the Francis turbine and generator installed at Grand Coulee dam, Washington, US. Source: photo of a display at the Grand Coulee Dam Visitor Center.

reaction turbines. Figure 8.5 shows the runner of a Francis Turbine removed from Robert Moses Hydroelectric Power Station at Niagara Falls, NY.

Water enters the volute tangentially and is directed by a set of fixed guide vanes (stay vanes), to adjustable vanes called "wicket gates," which direct the water toward the runner blades. Wicket gates are a series of doors that surround the turbine runner. They change water flow similar to vertical window blinds that control the light. As the angle of the wicket gates change the flow rate of water passing through the runner blades changes; consequently, the power output can be quickly adjusted. Unlike the impulse turbines, water fills the entire casing before leaving the turbine. The rotational momentum of the running water is transferred to the runner. The Euler turbine equation [Eq. (8.15) is used to calculate the power of a Francis turbine for given blade shapes and water flow rate.

Figure 8.5 A Francis turbine runner displayed at the Robert Moses Niagara Hydroelectric Power Station visitor center.

Figure 8.6 Wicket gates control mechanism during maintenance at Bonneville Dam, Oregon.

The power output of a Francis turbine is controlled by adjusting the direction of wicket gates controlled by hydraulic actuators that change the angle of water entering the runner. Figure 8.6 shows the wicket gate assembly of a Francis turbine during maintenance work.

8.5.3 Kaplan Turbine

The Kaplan turbine, developed by Austrian engineer Victor Kaplan in 1913, is a propeller-type reaction turbine with adjustable blades. The number of blades is between three and eight, much fewer than Francis turbines.

Figure 8.7 Kaplan turbine runner displayed at North Bonneville dam visitor center.

A Kaplan turbine is similar to a boat propeller; it may be enclosed in a case or submerged in streaming water like an underwater counterpart of a wind turbine. The power output of a Kaplan turbine may be controlled by changing the pitch of the blades, but in an enclosed design, adjustable angle wicket gates can be used to control the output power. Some Kaplan turbines have double controls whereby we can change the blade pitch and angle of the wicket gates, and there are also turbine propellers with fixed blades. Figure 8.7 shows a Kaplan turbine propeller, which was used at North Bonneville Dam in Oregon, currently displayed at the visitor center.

Kaplan turbines are more suitable for tidal power generation plants and run-of-river (ROR) type hydroelectric generation plants. In the Bonneville, Dalles, and John Day hydroelectric power stations located on the Columbia River near Portland, OR, all turbines are Francis-type with adjustable wicket gates (USACE 2013).

8.6 Hydroelectric Generation

The power of water can be harnessed in many ways. In a conventional hydroelectric station, a dam blocks the downstream end of the reservoir and a penstock (floodgate) conveys water to a turbine, which drives a generator. Turbines, generators, switchgear, and auxiliary equipment are installed in the powerhouse. The turbine-generator system converts the potential energy of the water stored in the dam into electricity. Step-up transformers next to the powerhouse increase the output voltage of generators to the high-voltage level suitable for long-distance transmission.

Power available at a hydroelectric station can be obtained by applying the Bernoulli principle. Water particles at different locations have different energy levels. Because potential energy is a relative entity, we will select an arbitrary reference plane to define different energy levels. Height of the water surface of the reservoir relative to the turbine input is h. At the free surface of the water, the pressure P is the atmospheric pressure and the velocity V is zero. According to the Bernoulli principle, the total energy per unit mass at the input of the turbine is equal to the total energy at the water surface. By taking the atmospheric pressure reference, the Bernoulli equation [Eq. (8.12) becomes

$$\frac{V^2}{2} + gh + \frac{P}{\rho} = \frac{W_t}{m} \tag{8.22}$$

The mass flowing per unit time is the volumetric flow rate multiplied by the specific mass ρ of the water flowing through the turbine. The maximum mechanical power of a hydraulic turbine is obtained by using the expression below, where g is the gravitational acceleration 9.81 m/s^2, H_{gross} is the "gross head," and Q represents the volume flow rate of the dam water with specific mass ρ.

$$P_{max} = \rho g Q H_{gross} \tag{8.23}$$

By definition, a gross head is the elevation difference between the reservoir surface and the surface of water stream exiting the dam. Because of the irreversible losses in the path of the water flow, the actual power output of a turbine is less than the value obtained from Eq. (8.23).

$$P = \frac{dW}{dt} = \rho g Q_v h \tag{8.24}$$

The Bernoulli principle does not consider the losses due to friction. In practice, the friction losses that occur while the flow reaches the turbine are expressed as a loss in potential energy, hence they are called "head losses" and expressed in meters. Therefore, for more accurate estimation of the input power, the head losses h_L should be deducted from the actual height of the reservoir surface.

$$H = h - h_L \tag{8.25}$$

In addition, the turbine and generator efficiencies must be included in the estimation of the electric power output of the hydroelectric power plant.

$$P_{elec} = \eta_T \eta_G (\rho g Q_v H) \tag{8.26}$$

The parameters of Eq. (8.26) are defined with their metric units:

P_{elec} : Electrical power output of the generator [W]

η_T : Turbine efficiency

η_G : Generator efficiency

ρ : Specific mass of the water [kg/m^3]

g : Gravitational acceleration [9.81 m/s^2]

Q_v : Volume flow rate [m^3/s]

H : Net head [m]

Table 8.1 Turbine suitability based on hydraulic head Gupta (2012).

Head class	Low head	Medium head	High head	Very high head
Range (m)	2–15	16–70	71–500	Above 500
Suitable turbine	Kaplan	Kaplan, Francis	Francis, Pelton	Pelton

8.7 Turbine Selection

Selection of a hydraulic turbine for a specific site depends on many factors including the hydraulic head, flow rate, power, and rotation speed. Hydraulic head is the principal parameter in choosing the turbine type. Pelton turbines are generally used for relatively higher heads from 200 up to 2000 m (Gupta 2012). Francis turbines are the common choice in a wide range of power plants with medium heads. Kaplan turbines are more suitable for lower heads (Table 8.1).

8.8 Hydroelectric Station Types

The amount of electric power that a hydroelectric plant can continuously deliver is called the generating capacity. Hydroelectric generation units are generally classified based on capacity as "large hydro" and "small hydro." While the classification based on size changes from country to country, conventional hydroelectric plants with an installed capacity of more than 30-MW are generally considered large. Smaller units may be medium size with a generation capacity between 100-kW and 30-MW, or micro-stations with a capacity of less than 100-kW. All hydropower facilities transform potential energy of water to rotational work by a hydraulic turbine coupled with an electric generator, which converts mechanical energy into electricity.

Hydroelectric generation facilities are often grouped into three main categories based on the type of water supply as storage (reservoir), pumped storage, and ROR hydropower plants.

Large conventional hydroelectric plants generally take water supply from a reservoir, artificial lake, pond, or an enlarged section of a river created by a dam. Conventional hydroelectric plants have relatively higher initial cost, but their typical economic life is long enough to recoup the cost, and the maintenance and operational costs are considerably lower compared to other power plants that make generated electricity cost competitive.

ROR stations are supplied from a river and may not have water storage at all, but a short-term water storage can regulate the hourly or daily variations of the water flow. A portion of the river is diverted to a channel or pipe to convey water to a hydraulic turbine, typically a Kaplan type. Such power stations can be constructed in a wide range of capacities, from a few kilowatts to more than one gigawatt. Because they do not require a large reservoir, they can be installed in a shorter time at relatively low cost. Larger conventional units supply the interconnected power network. The Bonneville power station on the Columbia River between Washington and Oregon in the US is an example of a large conventional ROR plant. The total installed capacity of the power station is 1242-MW. Small-scale ROR units can be installed at any location where a stream is available. They can supply either isolated users in remote areas or provide distributed generation for a large interconnected grid.

Pumped storage hydropower plants (PSH) are storage facilities rather than sources of energy. They have two reservoirs, one at a lower elevation and the other at a higher elevation. When electric energy demand peaks, the higher reservoir is drained through the turbines to the lower reservoir and generators supply the electric grid. When electric energy demand is lower, say during late night or early morning hours, energy flow is reversed. The generators operating as motors drive the turbines to pump water up to the higher reservoir. In this operation mode, excess electric energy of the grid is stored as potential energy of water.

In-stream technologies have also been developed to capture hydrokinetic energy from existing barrages, canals, waterfalls, free flowing rivers, currents, and tides. Hydrokinetic energy is generally converted into electric energy by generators coupled with submerged turbines, similar to a propeller or wind turbine.

8.9 Dam Structures

Dams are built across a river or stream for various purposes including freshwater storage, irrigation, regulation of seasonal water flow variations, and flood control. Dams also

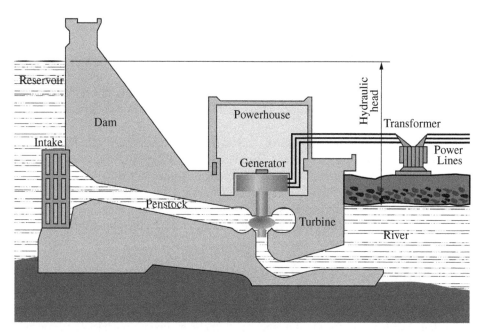

Figure 8.8 A conventional hydroelectric power plant. Source: Modified from public domain graphic produced by Tennessee Valley Authority.

Figure 8.9 Grand Coulee dam and power station. Source: © Soysal.

improve navigation of commercial transportation barges and provide a lake for recreational activities such as swimming, boating, and fishing.

Dams are one of the essential parts of conventional hydroelectric facilities. They form the reservoir and control the water flow to the turbines (see Figure 8.8). Intake structure is protected by a screen that prevents the passage of rubble, rocks, and soil carried by the river to the turbines. Tunnels and penstock direct water from the intake to the turbines. The tailwater tunnel discharges the water released from the turbine to the river, while a spillway controls the maximum water level in the reservoir and permits the passage of floodwater.

Dams are grouped into several classes based on their profile and building materials. The most common dam building materials are concrete, earth, and rock. A decision on the type of dam depends on the topography, available materials, and river properties (Jackson & Brown 2019).

Concrete gravity dams are constructed across a broad valley as a straight line. They resist the horizontal thrust of water by their own weight. All the dams constructed in the Columbia River Basin are concrete gravity type (Figure 8.9).

Arch dams are thick structures that curve upstream and join the steep slopes of a narrow valley. The water thrust is transferred either directly or through concrete abutments to the valley sides. Since the strength relies on the valley sides only favorable sites with solid rock formations are suitable for arch dams. Many of the large reservoir dams are arch gravity type.

One of the typical examples is the Glen Canyon Dam shown in Figure 8.10. The dam constructed on the Colorado River created Lake Powell near Page, Arizona. Total nameplate capacity of eight generation units powered by Francis turbines is 1312 MW (US Bureau of Reclamation 2016). Earth-fill dams may have a core filled with clay or concrete. Rock-fill dams are constructed by dumping loose rocks that are firmly compacted to form an embankment. Such dams are sealed by a watertight concrete coating on the upstream slope.

8.10 Strengths and Challenges of Hydroelectric Power Plants

Hydroelectric power plants have several advantages, which make them suitable for interconnected electric networks. On the other hand, there are major drawbacks and concerns in the development of large hydroelectric projects (Gupta 2012).

Figure 8.10 Glen Canyon dam and power plant. Source: © Soysal.

Benefits of hydroelectric power are now summarized.

- *Hydro power is renewable and clean:* Hydroelectric power plants do not produce greenhouse gases, volatile organic compounds (VOC), toxic emissions, and microscopic particulate matter (PM). In addition, there is no waste disposal problem like coal fired power plants and nuclear power plants.
- *Fuel transportation and storage:* Hydroelectric plants do not require continuous fuel transportation like coal fired plants. They have natural energy storage in their reservoir.

- *Energy cost is more stable and predictable:* Hydroelectric power plants do not have a fuel cost; the price of the generated electricity is based on the initial capital cost and maintenance cost. The operational costs of hydroelectric plants are lower compared to fossil fuel fired and nuclear power plants. Therefore, the cost of energy is not subject to market conditions like oil and natural gas.
- *Hydropower development is a multipurpose project:* A dam constructed for a hydropower project also serves as flood control, irrigation, and drinking water storage. Artificial lakes created by a dam improve the landscape and generate recreational and touristic benefits.
- *Reliability and energy security:* Hydroelectric power plants are relatively reliable and less dependent on external sources. Although some hydroelectric power plants are built on international rivers, availability of water is not as dependent as oil and natural gas.
- *Output power control:* Hydro turbines can start and stop quickly. Their mechanical power can be controlled easier than steam and gas turbines. The efficiency of modem hydraulic turbines is high in a range of loads. Hence, hydroelectric power plants can follow the variations of electric demand. They increase the controllability of the fuel mix in an interconnected electric system.
- *Economic life:* The life expectancy of hydropower equipment is about 50 years or more; longer than thermal or nuclear power plant equipment.

On the other hand, hydroelectric projects present the following challenges and concerns.

- *Capital cost:* The initial investment needed to build a hydroelectric power plant is considerably very high. Hydroelectric power plant projects usually require multinational consortiums for financing and development.
- *Preparation time:* Detailed feasibility analysis to develop a hydroelectric power plant involves multidisciplinary work including civil, mechanical, electrical, geological engineering, and economic analysis.
- *Location:* Hydroelectric power plants are usually located at remote places hard to access. The cost of transportation of heavy equipment and materials is usually very high.
- *Power generation regime:* The power output of hydroelectric power plants depends on natural parameters such as water flow in the river, rainfall, quantity of water available, and air temperature which evaporates water in the reservoir.
- *International relations:* Many large hydroelectric power plants share international rivers and use of water for electric generation is regulated by international treaties.
- *Social impacts:* Large hydroelectric projects require flooding of a populated area. Heritage, archeologic sites, historical monuments, and residences are affected of the project. Residences and monuments have to be moved to another location before construction. The planning of many hydroelectric projects takes longer time because of resolutions of public reactions.

8.11 Chapter Review

This chapter outlines fundamental principles and practical applications relevant to hydroelectric generation.

Bernoulli's Principle states that "the sum of the kinetic, potential, and flow energies of a fluid particle is constant along a streamline during steady flow when the compressibility and frictional effects are negligible."

Turbomachines are energy converters that transfer energy between a moving fluid and a rotating shaft. Euler's turbomachine equation is used to analyze the performance of most turbines and centrifugal pumps. The only difference between a centrifugal pump and a turbine is the direction of the energy flow.

Hydraulic turbines convert the potential energy associated with a difference in water elevation into mechanical work, which drives an electric generator.

Performance of different types of water turbines depend on the hydraulic head, flow rate, and runner velocity. Hydraulic head is defined as the elevation difference between upstream and downstream levels of the water supply.

According to the principles of operation, turbines are grouped as impulse or reaction turbines. In modern hydroelectric stations, Pelton, Francis, and Kaplan turbines are the most common turbine types.

The Pelton turbine is an impulse turbine where a water jet coming from a nozzle placed at an optimal angle hits the splitter ridge of a specially shaped cup. A Pelton turbine delivers the maximum shaft power when its runner turns with the tangential speed at half of the velocity of the water jet.

In a Francis turbine the runner is submerged in a volute-type casing like a curved funnel with decreasing cross-sectional area. Water enters the volute tangentially and is directed by a set of fixed guide vanes (stay vanes), to adjustable vanes called "wicket gates," which direct the water toward the runner blades.

The Kaplan turbine has a propeller-type runner similar to a boat propeller. A Kaplan turbine may be controlled by changing the pitch of the blades and/or the angle of wicket gates.

In a conventional hydroelectric station, a dam blocks the downstream end of the reservoir and a penstock conveys water to a turbine, which drives a generator.

The hydraulic head is the principal parameter in choosing the turbine type. Pelton turbines are generally used for relatively higher heads, Francis turbines are the common choice in a wide range of power plants with medium heads, and Kaplan turbines are more suitable for lower heads.

Hydroelectric generation facilities are often grouped in three main categories based on the type of water supply, as storage (reservoir), pumped storage, and ROR hydropower plants.

Large conventional hydroelectric plants generally take water supply from a reservoir, artificial lake, pond, or an enlarged section of a river created by a dam.

ROR stations are supplied from a river and may not have a water storage at all, but a short-term water storage can regulate the hourly or daily variations of the water flow.

PSH have two reservoirs, one at a lower and the other at higher elevation. At peak demand hours, the higher reservoir is drained through the turbines to the lower reservoir and generators supply the electric grid. When electric energy demand is lower, energy flow is reversed.

Dams are built across a river or stream for various purposes including freshwater storage, irrigation, regulation of seasonal water flow variations, and flood control.

Dams are grouped into several classes based on their profile and building materials. The most common dam building materials are concrete, earth, and rock.

Concrete gravity dams are constructed across a broad valley as a straight line. Arch dams are thick structures that are curved upstream and join steep slopes of a narrow valley.

Review Quiz

1 Which form of energy is not considered in the Bernoulli's equation?
   a. Kinetic energy
   b. Potential energy
   c. Fluid flow energy
   d. Thermal energy

2 Hydraulic turbines convert
   a. electric energy into potential energy of water.
   b. potential energy of water into mechanical work.
   c. mechanical work into kinetic energy.
   d. electrical energy to mechanical work.

3 Hydraulic head is
   a. the elevation difference between upstream and downstream levels of a turbine's water supply.
   b. the height of a dam.
   c. the elevation of the source of a river from sea level.
   d. the distance of the turbine intake below water surface.

4 Which one below is an impulse turbine?
   a. Francis
   b. Kaplan
   c. Pelton
   d. None of these

5 A Francis turbine is
   a. an impulse turbine.
   b. a reaction turbine.
   c. both an impulse and reaction turbine.
   d. either an impulse or reaction turbine depending on operating conditions.

6 Pelton turbines are more suitable for
   a. lower hydraulic head.
   b. lower speed.
   c. flow rate.
   d. higher hydraulic head.

7 Wicket gates are used for
   a. control of a reaction turbine's power.

b. control of water jet in an impulse turbine.
c. control of turbine speed.
d. security of a hydroelectric power plant.

8 Francis turbines are more suitable for
a. low and medium hydraulic head.
b. medium and high hydraulic head.
c. extremely high hydraulic head.
d. any hydraulic head.

9 Kaplan turbines are mostly used in
a. hydroelectric power plants supplied from a lake.
b. hydroelectric power plants supplied from ROR dams.
c. hydroelectric power plants installed at arch dams.
d. hydroelectric power plants installed at a higher elevation.

10 Arch dams are
a. thick structures curved upstream and joining steep slopes of a narrow valley.
b. long structures in a wide valley.
c. low structures joining steep slopes of a narrow valley.
d. arc-shaped structures across a river.

Answers: 1-d, 2-b, 3-a, 4-c, 5-b, 6-d, 7-a, 8-b, 9-b, 10-a.

Research Topics and Problems

Research and Discussion Topics

1 Discuss the major issues regarding hydroelectric power plant development on major rivers.

2 Research new technologies developed to harness the power of small water streams.

3 Research the benefits and challenges of pumped storage hydroelectric power plants.

4 Discuss the challenges of hydroelectric power generation on cross-border rivers.

5 What are the major impacts of hydropower development on local and regional economy?

Problems

1 A hydroelectric power plant will be installed 60 m below the water level of a reservoir. Determine the maximum fluid power in kW available to the turbine for 8 m³/s flow rate.

2 A hydraulic turbine located 85 m below the water level of a lake is supplied 3000 l/s water. The turbine operates at an efficiency of 85%. What is the mechanical power produced by the turbine in kW? (Specific mass of the water is 980 kg/m^3.)

3 A 5-MW hydroelectric power plant has a 100-m net head and is supplied 4 m^3/s water when it produces the rated power. Determine the overall efficiency of the power plant in percent. (Specific mass of water is 1000-kg/m^3.)

4 A hydraulic turbine operating with an efficiency of 80% takes water from a lake with the water level 100 m above the intake pipe. Determine the required water flow rate in m^3/s to obtain 10 MW mechanical power on the turbine shaft.

5 A power plant is to be built to generate 40-MW electric power from a reservoir. Given the water flow of 50 m^3/s, turbine efficiency of 76%, and generator efficiency of 90%, calculate the required vertical distance between the water level and the turbine intake pipe.

Recommended Web Sites

- US Army Corps of Engineers Digital Library: https://cdm16021.contentdm.oclc.org/digital
- US Army Corps of Engineers (USACE) publications: https://www.publications.usace.army.mil
- US Bureau of Reclamation (USBR): https://www.usbr.gov
- National Hydropower Association (NHA): http://www.hydro.org

References

Andrews, J. and Jelley, N. (2007). *Energy Science*. Oxford, UK: Oxford University Press.

BP (2018). *BP Statistical Review of World Energy 2018*. London: BP.

Gupta, M. K., 2012. *Power Plant Engineering*. s.l.: s.n.

IEA (2018). *Renewables Information: An Overview*. Paris: International Energy Agency.

IRENA (2019). *Renewable Capacity Statistics 2019*. Abu Dhabi: Iternational Renewable Energy Agency (IRENA).

Jackson, D. C. and Brown, J. G., 2019. *technology/dam-engineering/*. [Online] Available at: https://www.britannica.com/technology/ [Accessed 6 July 2019].

US Bureau of Reclamation, 2016. *Hydropower Program*. [Online] Available at: https://www.usbr.gov/power/edu/history.html [Accessed 17 July 2019].

USACE, 2013. *Digital Library*. [Online] Available at: https://cdm16021.contentdm.oclc.org/digital/collection/p16021coll11/id/2891 [Accessed 7 October 2019].

Wikimedia Commons, 2006. *Walchenseewerk Pelton*. [Online] Available at: https://commons.wikimedia.org/wiki/File:Walchenseewerk_Pelton_120.jpg [Accessed 7 October 2019].

Yunus, C.A. and Cimbala, J.M. (2014). *Fluid Mechanics*. s.l.:New York, NY: McGraw-Hill.

9

Wind Energy Systems

View of the first offshore wind farm in the United States from South of Block Island, RI. The 30-MW, five-turbine Block Island Wind Farm began commercial operations in December 2016. The project was developed by Deepwater Wind. (For more information visit: http://dwwind.com/project/block-island-wind-farm/ Accessed on 10/22/2019).

Energy for Sustainable Society: From Resources to Users, First Edition. Oguz A. Soysal and Hilkat S. Soysal.
© 2020 John Wiley & Sons Ltd. Published 2020 by John Wiley & Sons Ltd.

9.1 Introduction

Wind is a clean, sustainable, and renewable source of energy. Wind power has been an essential driving force of the economic development since early civilizations. For thousands of years, sailboats carried voyagers and merchandise on rivers, lakes, seas, and oceans. Windmills powered water pumps, grain-grinding mills, and various mechanical equipment throughout the centuries; Antique communities used windmills to pump water. In the eleventh century, grain-grinding vertical-axis windmills supported the agriculture in Persia and the Middle East. In the medieval age, crusaders and merchants traveling to Asian countries brought the idea of converting wind energy into rotational mechanical work to Europe. More advanced mechanisms using gear trains allowed development of horizontal axis windmills. By the thirteenth century, Dutch engineers used windmills to pump water out of areas below sea level. The windmills used to drain wetlands transformed swamps and marshlands into fields suitable for agriculture and other development. European countries adapted windmills for advanced industrial applications such as cutting and shaping wood, metals, or stones. The first immigrants to the New World began using windmills in the nineteenth century for agricultural irrigation.

The development of the steam engine and industrialization led to a gradual decline in the use of windmills for commercial applications because coal and petroleum provided more controllable and portable energy. The idea of using wind power to generate electricity dates back to 1887, however wind-powered electric generation gained its momentum during the oil crisis of the 1970s. Since the 1990s the share of wind power in electric generation has been continuously increasing around the world due to the concerns regarding depletion of fossil fuels, energy security, and climate change.

Since wind is a free source of energy operational costs are negligible, and there is no fuel transportation and waste disposal cost, wind powered generation facilities operate at full capacity as long as wind is available, and electric power supply is needed. Wind powered generation is easy to control; commercial wind farms can be stopped at any time if additional energy is not needed to supply demand.

The installed capacity of practical wind turbines range from a few hundred watts to megawatts. Small-scale wind turbines that supply grid-tie or stand-alone loads range from a few watts to kilowatts rated power. Commercial and community-size wind generation units have a capacity several hundred megawatts. Utility scale wind farms are installed on mountain ridges, plains, and offshore marine locations.

Individual small wind turbines installed by residential and commercial consumers are usually connected to the grid through bidirectional electric meters. Electricity generated by such wind turbines is primarily used to supply the individual consumers, but any excess electric power supplies the electric grid and is credited to the consumer through net-metering agreements.

Wind powered generation systems may be used to supply a standalone load, such as an island, an isolated agricultural or recreational facility, an off-grid building, or electronic equipment at a remote location. In a standalone application, electric generation must be backed up by some form of energy storage, Battery bank, water tank, or thermal storage are among the common energy storage systems for standalone wind generation. Wind powered generation units connected to an interconnected power grid without a backup energy storage present similar integration challenges as solar powered electric generation.

9.2 Sources of Wind

Since wind is a free and nondepletable energy source, it is essential to understand the sources and properties of wind for a realistic assessment of its potential for electric generation. Wind is a consequence of solar energy reaching the Earth's surface. Uneven heating effects of the sun create local temperature differences that change air pressure and density, leading to natural air flow. In addition to the temperature gradients, Earth's rotation and terrain conditions affect the direction and turbulence of the resulting wind.

Global wind patterns are the result of the temperature gradient due to the maximum solar irradiance that changes with altitude from Ecuador to the North and South poles. Climatic differences and the non-uniform surface of the Earth modify the global wind patterns on continental scales. Local winds depend on uneven heating of the surface, proportion of land and water masses, thermal properties of the ground cover, shading or reflection of the sunlight, and geographic features of the terrain. At the tropical regions, the warmer air rises, resulting in lower pressure at the lower parts of the atmosphere. On the other hand, arctic regions are continuously cooler than the other parts of the Earth. The higher pressure at the poles forces air to move toward the equator. However, rotation of the Earth causes deflection of the wind direction from east to west around the equator by the Coriolis force. Prevailing surface winds blowing from the east in the tropical regions are called "trade winds" because they have been used for centuries by captains of sailing ships to cross the oceans. In the northern hemisphere, trade winds blow from the northeast, while in the southern hemisphere, they blow from southeast, as shown in Figure 9.1.

Around the thirtieth parallel, the airflow becomes unstable and the wind direction is reversed. Between 30° and 60° latitudes, the prevailing winds move from the southwest to northeast. In practice the surface friction, eddy motions, and seasonal variations affect the prevailing winds. Three distinct wind patterns are observed in each hemisphere. The wind belt in the tropical regions is called a Hadley cell, the mid-latitude belt is a Ferrell cell, and the third wind pattern seen in the Arctic and Antarctic regions is the Polar cell.

Local winds develop mainly because of temperature differences between different regions. Prevailing global and regional winds as well as weather activities may reinforce or weaken local winds depending on their relative directions. From midday to sunset, mountains may be warmer than the valley depending on the exposure to the sunlight.

Figure 9.1 Global wind patterns.
Source: Simplified by the authors from Encyclopedia Britannica (n.d.).

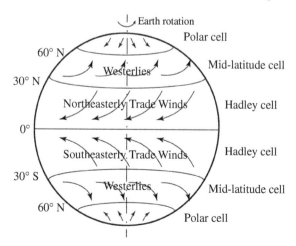

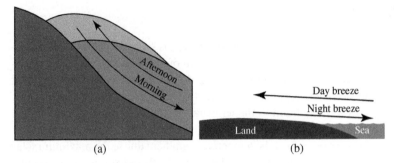

Figure 9.2 Local wind patterns; (a) mountain-valley wind, (b) seashore wind.

Cooler air from lower elevations rises to the higher elevations. After sunset, higher elevations in the mountains get cooler, hence air flows from higher elevations to the valley. Figure 9.2(a) illustrates the mountain-valley wind pattern.

Coastal regions are often windy because daily and seasonal differences between land and water temperatures cause local winds. Air temperature over a lake, sea, or ocean remains relatively steady because water has high heat capacity (3900–$4200\,\mathrm{J \cdot kg^{-1} \cdot K^{-1}}$) and both conduction and convection heat transfer from deeper layers regulate the temperature at the water surface. Soil, sand, and rocks have lower heat capacity (190–$350\,\mathrm{J \cdot kg^{-1} \cdot K^{-1}}$) and heat transfer from deeper layers occur only by conduction. Hence, the air temperature above water changes more slowly than the air temperature on the shore. During the day, a quicker increase of the land temperature causes air flow from the sea toward the land. In the evening and night hours, water remains warmer than the land, which gives rise to the night breeze from the land toward the sea. Figure 9.2(b) shows the coastal wind pattern; coastal wind patterns repeat over each 24-hour cycle, and in many regions and they are quite predictable.

Sirocco, Mistral, and Bora are among well-known seasonal wind patterns in Mediterranean regions, producing strong and sustained winds caused by the temperature differences between sea and land. The Sirocco originates from the warm, dry, tropical air in the Arabian or Sahara deserts, moves across the Mediterranean Sea and reaches the southeastern coasts of Europe. The Mistral occurs between the Bay of Biscay along the northwestern Atlantic coast of France and the Gulf of Genoa in the Mediterranean Sea. The air flowing through the Rhone Valley accelerates at the narrower regions and reaches its maximum speed in the south, where the Valley becomes wider. Other major seasonal wind patterns known as Monsoons develop in Asia, Africa, Central America, and Oceania due to the uneven heating of land and sea.

As the air stream rises over the smooth ridge of a mountain, compression of air layers accelerates the wind as shown in Figure 9.3(a). The acceleration rate depends on the shape of the ridge and direction of the prevailing wind. A desirable acceleration occurs on ridges between 6° and 16°. Steeper slopes greater than 27° are not favorable because of the turbulence (Mathew 2006). The acceleration effect is higher when the prevailing wind is perpendicular to the ridge. Triangular and smooth rounded top mountains increase the wind speed more effectively. This is why the Appalachian Mountains on the east coast of the United States are desirable locations for mountain ridge wind farms.

Mountain passes also enhance the prevailing wind mainly due to the Venturi effect, named after Italian physicist Giovanni Battista Venturi (1746–1822). As the air flows through the notches in the mountain, barriers increase the wind velocity. Width, length,

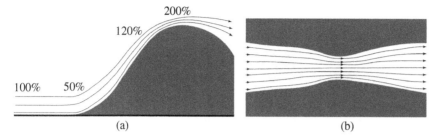

Figure 9.3 Wind acceleration effects; (a) mountain ridge acceleration, (b) Venturi effect.

and slope of the mountain pass affect the acceleration. The Venturi effect illustrated in Figure 9.3(b) is more significant in cities where wind is forced to pass between tall buildings, valleys, and mountain passes in the direction of the prevailing wind. Gorges in the wind direction between high hills are usually favorable sites for wind farm development.

Broad flat lands, prairies, steppes, and grasslands between regions with significant climate differences are exposed to strong winds suitable for windfarm development. For example, the cold air in Canada flowing to the warmer Gulf of Mexico offers great opportunity for wind power development in the Great Planes region of the United States.

Wind patterns often result from complex interactions of various physical phenomena. For example, on North America's Pacific coast, the ocean currents bring cold water to the shore, which is not far from desert areas where temperature is extremely high during the day. The mountain ranges parallel to the coastline have passes in the east–west wind direction. Air flow funneling through the mountain passes results in strong and steady wind during the day. Such a pattern presents a great opportunity for wind farm development since the electric demand also peaks during the day due to air conditioning loads.

9.3 Wind Shear

Wind speed and direction depend on the friction caused by the ground surface and shape of the objects around which wind flows. Wind shear is variation in wind velocity over a relatively short distance in the atmosphere. Such variations may be in any direction and wind flow may be turbulent. As the distance increases, the effect of such disturbances become less significant and wind flow tends to be more uniform.

Wind turbines are designed to turn to the wind direction by a natural force or mechanism called a yaw control system. Since yaw control can follow relatively slower directions of the wind flow, turbulences decrease wind turbine performance. Since wind turbines are designed to convert energy of the wind flowing in horizontal direction, vertical wind shear is more relevant in the assessment of a wind resource for electric generation. In theory, wind speed on the ground is zero because of frictions. Wind speed increases logarithmically by height. If the wind speed is known at a reference height H_r, wind speed at another height H can be estimated by using the expression below.

$$V(H) = V(H_r)\frac{\ln\left[\frac{H}{z_0}\right]}{\ln\left[\frac{H_r}{z_0}\right]} \tag{9.1}$$

Table 9.1 Surface roughness lengths and wind shear exponent for various terrains.

Land	Surface roughness length z_0 (m)	Wind shear exponent α
Ice	0.00001	0.07
Snow on flat ground	0.0001	0.09
Calm sea	0.0001	0.09
Coast with onshore winds	0.001	0.11
Snow-covered crop stubble	0.002	0.12
Cut grass	0.007	0.14
Short-grass prairie	0.02	0.16
Crops, tall-grass prairie	0.05	0.19
Hedges	0.085	0.21
Scattered tree and hedges	0.15	0.24
Trees, hedges, a few buildings	0.3	0.29
Suburbs	0.4	0.31
Woodlands	1.0	0.43

Source: Gipe (2004).

In the expression above z_0 represents surface roughness length, which characterizes the obstruction of the vegetation, hills, buildings, and other barriers to the wind flow. Surface roughness length of selected terrain types are listed in Table 9.1 (Gipe 2004). Plots in Figure 9.4 are generated using Eq. (9.1) to illustrate the vertical wind shear effect by surface roughness length. Smoother surfaces of ice, snow, and calm water cause smaller resistance to the air flow; hence, the wind speed change is relatively smaller above the reference height. Turbulence caused by trees, buildings, and woodlands is reduced as the height increases, then the ratio of the wind speed to the known value at the reference height increases drastically with height.

Variation of wind speed by height can be alternatively estimated by the exponential function below. The parameter α is named wind shear exponent. Typical values of the wind shear exponents for selected terrain types are listed in Table 9.1 (Gipe 2004).

$$V(H) = V(H_r)\left[\frac{H}{H_r}\right]^{\alpha} \tag{9.2}$$

Wind shear is an important factor in the selection of a wind turbine tower height. Even a small change of wind speed is dramatically enhanced since the output power of wind turbines is proportional to the third power of wind speed.

Weather stations usually measure wind speed at a 10-m height from ground. For wind resource assessment and estimation of wind turbine output power generation at a

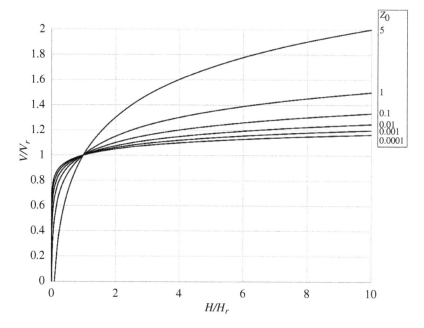

Figure 9.4 Normalized vertical wind shear by surface roughness.

prospective site, wind speed data may be collected at elevations as high as 100 m. Increasing turbine tower heights significantly increase the energy generation capacity of wind farms near woodlands and suburbs.

9.4 Wind Regimes

Random changes of wind speed in a broad time spectrum describe wind regimes. Short time turbulence effects have significant impacts on the performance of wind turbines. Especially wide variations in a short time-scale in the order of seconds or minutes, such as wind gusts, may create risky stresses on the blades and other moving turbines parts.

Hourly and daily variations are important for integration of large amounts of wind-power into an interconnected electric grid supplied from various types of generation units. At many locations such variations of wind speed are less predictable than solar irradiance. The prediction of short-term wind variations associated with weather conditions is more complicated and requires weather forecast information. At some locations, like the Pacific coast of the US, daily variations may be quite regular because of the geographic characteristics and may present advantages in balancing electric supply and demand. Electric utilities take special precautions to minimize the effect of large swings of power supply to the grid.

Seasonal variation of the average wind speed is generally fairly predictable. At certain locations, change of wind speed during the day may be also predictable based on the physical conditions that cause the air flow.

9.4.1 Site Wind Profile

Prediction of long-term changes of average wind speed is important in feasibility analysis for potential sites for wind farm development and forecast of resource availability for planning of future energy systems.

To assess the wind profile of a prospective wind turbine site, wind velocity and direction data must be collected over a reasonably long period. Wind data is collected by a datalogger receiving wind speed data from anemometers and wind direction information from wind vanes. Data collection instruments are mounted at known heights on poles located at one or more locations. In practice, 10, 50, or 100 m are common heights for anemometer poles or towers. Wind data is generally collected at a fixed sampling period (usually 10-minutes) over an adequate time interval, typically one year or several years. The collected wind speed data can be projected to the actual turbine height using Eqs. (9.1) or (9.2).

Yearly average wind velocity is calculated by summing up all the values measured over one year at a constant sampling period and dividing the total by the number of collected values as in Eq. (9.3).

$$V_{av} = \frac{1}{n} \sum_{i=1}^{n} V_i \tag{9.3}$$

Average wind velocity may be misleading for wind power estimation at a given site. Wind turbines essentially convert kinetic energy of the moving air into rotational mechanical power. Considering a constant and unidirectional wind without turbulences, the volume of air that flows through the rotor in one second is the product of the area swept by the rotor blades A_T and the wind speed V. For a specific mass of air, the mass of air passing through the blade swept area is ρA_T. The kinetic energy of the wind in unit time yields the power $m = \rho A_T V$.

$$P_W = \frac{1}{2} m V^2 = \frac{1}{2} \rho A_T V^3 \tag{9.4}$$

Note that P_w is the maximum power available from the wind, not the output power of the turbine. Equation (9.4) is independent of the type and structure of the wind turbine. For assessment of the wind potential at a site, it is more meaningful to calculate the average of the third power of the wind speed since the available energy can be scaled up for any blade swept area.

$$P_{av} = \frac{1}{n} \sum_{i=1}^{n} V^3 \tag{9.5}$$

Average wind speed at a location may change from one year to another, even over longer periods of decades or more. Such variations depend on the climatic fluctuations or global climate change. Long-term predictions and projections of average wind speed provide useful information to assess the resource potential in a prospective site for wind power development.

Rapid variations of the wind speed affect the performance of wind turbines due to the moment of inertia of the blades and the delay of the rotor axis control to follow wind direction. Probability distribution of wind speed must be also considered in the site assessment.

Standard deviation is a measure of the variability of wind speed. In statistical analysis, standard deviation is the deviation of individual values from the average value expressed in Eq. (9.6) for wind speed data. Lower values of standard deviation obviously indicate more uniform wind regime.

$$\sigma_V = \sqrt{\dfrac{\sum\limits_{i=1}^{n}(V_i - V_{av})^2}{n}} \tag{9.6}$$

For better assessment of the electric generation potential at a prospective turbine site, wind speed data is often grouped by values and presented in the form of frequency distribution. Such distribution gives the information about the number of hours for which the wind occurs within a specific speed range. Figure 9.5(a) shows the frequency of occurrence of wind speed recorded over one year in Western Maryland at a potential wind turbine

(a)

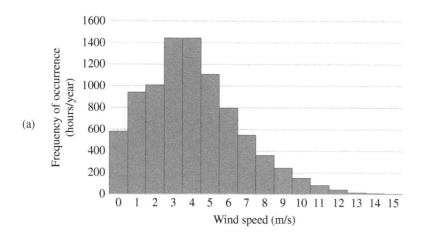

(b)

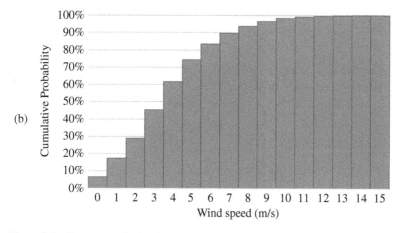

Figure 9.5 Frequency distribution of 10-minute average wind speed the authors collected over one year near Oakland, Maryland.

site. Each recorded wind speed value is the average over a10-minute interval. The graph indicates that the occurrence of 3 and 4 m/s wind speeds have the highest probability.

Figure 9.5(b) shows the cumulative probability distribution, which indicates how many hours per year a wind speed equal or lower than V occurs. The cumulative distribution is obtained by adding up the occurrences from zero to the given wind speed. For example, at the site where data was collected, the probability of a wind speed of 7 m/s or lower is 90% over one-year period.

Describing the wind profile of a potential turbine site by a continuous probability function has several benefits. Frequency distribution of wind speed at a site can be approximated by a skewed exponential probability density function. Weibull and Rayleigh distribution functions are the most common representations of wind speed distributions. Standard mathematical representation simplifies the comparison of different sites and simplifies the turbine selection. Manufacturers often characterize turbine performance for standard wind speed distribution using a Weibull or Rayleigh probability density function. Different turbine types can be realistically compared using mathematical distribution functions. Furthermore, frequency density and cumulative probability functions are helpful in estimating the electric generation potential based on data collected on site over a shorter period.

9.4.2 Weibull Distribution

A more general probability distribution that can fit to most wind profiles is the Weibull distribution, named after Swedish mathematician Waloddi Weibull, who described its properties around 1951. The Weibull probability density function is described by Eq. (9.7).

$$f(V) = \frac{k}{c}\left(\frac{V}{c}\right)^{k-1} e^{-(V/c)^k} \tag{9.7}$$

where k is shape factor and c is scale factor. A probability density function indicates the occurrence probability of a specific wind velocity V. Cumulative probability at which a wind speed equal or below a certain value is described by Eq. (9.8) for Weibull distribution.

$$F(V) = \int_0^\infty f(V)dV = 1 - e^{-\left(V/c\right)^k} \tag{9.8}$$

The cumulative distribution function can be used for estimation of the amount of time for which the wind is in a given velocity interval. Probability (Pb) of wind blowing within an interval is the difference of cumulative probabilities corresponding to the boundary velocities V_1 and V_2.

$$Pb_{(V_1 \leq V \leq V_2)} = F(V_2) - F(V_1) = e^{-\left(V_1/c\right)^k} - e^{-\left(V_2/c\right)^k} \tag{9.9}$$

Weibull density and cumulative probability functions are plotted in Figure 9.6 for a scale factor of 8 m/s and selected shape factors between two and six. As the form factor k increases, the maximum probability of wind velocities in a narrower range increase. Consequently, the cumulative frequency of wind velocities above the scale factor increases while the occurrence probability of velocities below the scale factor decreases.

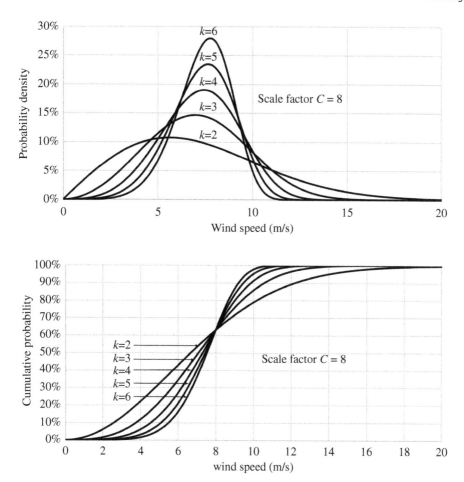

Figure 9.6 Weibull density and cumulative frequency functions.

Average wind speed over the period T for Weibull distribution is calculated by integrating the product of the wind speed and the probability distribution function, Eq. (9.7).

$$V_{av} = \frac{1}{T}\int_0^\infty \left[V\frac{k}{c}\left(\frac{V}{c}\right)^{k-1} e^{-\left(V/c\right)^k} \right] dV \qquad (9.10)$$

This equation can be arranged in the form of standard gamma function (Mathew 2006, pp. 68–69).

$$\Gamma(n) = \int_0^\infty e^{-x}x^{n-1} dx \qquad (9.11)$$

Hence, the average can be expressed as

$$V_{av} = c\Gamma\left[1 + \frac{1}{k}\right] \qquad (9.12)$$

Table 9.2 Average wind velocity for values of shape factor for unity scale factor.

		Decimal part								
		0	0.1	0.2	0.3	0.4	0.5	0.6	0.7	0.8	0.9
Integer part	1	1.000	0.965	0.941	0.924	0.911	0.903	0.897	0.892	0.889	0.887
	2	0.886	0.886	0.886	0.886	0.886	0.887	0.888	0.889	0.890	0.892
	3	0.893	0.894	0.896	0.897	0.898	0.900	0.901	0.902	0.904	0.905
	4	0.906	0.908	0.909	0.910	0.911	0.913	0.914	0.915	0.916	0.917
	5	0.918	0.919	0.920	0.921	0.922	0.923	0.924	0.925	0.926	0.927
	6	0.928	0.929	0.929	0.930	0.931	0.932	0.933	0.933	0.934	0.935
	7	0.935	0.936	0.937	0.937	0.938	0.939	0.939	0.940	0.941	0.941
	8	0.942	0.942	0.943	0.943	0.944	0.944	0.945	0.945	0.946	0.946
	9	0.947	0.947	0.948	0.948	0.949	0.949	0.950	0.950	0.951	0.951

Most spreadsheets, MATLAB®, and other computation software have built-in functions to calculate gamma functions. Table 9.2 shows average wind velocities calculated for Weibull distribution with unity scale factor c and shape factor varying from 1 to 9.9.

9.4.3 Rayleigh Distribution

In order to use the Weibull distribution, wind velocity data must have been collected over an adequately long period. Especially in preliminary assessment of a site, such data may not be available. A special case of the Weibull distribution for the shape factor of two is known as Rayleigh distribution. Substituting $k = 2$ in Eq. (9.12), the average wind velocity is obtained as

$$V_{av} = c\Gamma\left(\frac{3}{2}\right) = c\frac{\sqrt{\pi}}{2} \tag{9.13}$$

Hence, the scale factor for Rayleigh distribution is

$$c = \frac{2V_m}{\sqrt{\pi}} \tag{9.14}$$

Substituting this scale factor in Eq. (9.7), the expression of Rayleigh probability density function can be written as

$$f(V) = \frac{\pi}{2}\frac{V}{V_{av}^2}e^{-\left[\frac{\pi}{4}\left(\frac{V}{V_{av}}\right)^2\right]} \tag{9.15}$$

By similar substitution, the cumulative distribution is expressed as

$$F(V) = 1 - e^{-\left[\frac{\pi}{4}\left(\frac{V}{V_{av}}\right)^2\right]} \tag{9.16}$$

Rayleigh distribution has the advantage of using only one parameter, which is the average (mean) wind velocity given or estimated over a certain timeframe, for example daily, monthly, or yearly. The probability of wind above a velocity V_x can be estimated by the exponential expression in Eq. (9.17).

$$Pb_{(V>V_x)} = 1 - \left[1 - e^{-\left[\pi/4 \left(V_x/V_{av} \right)^2 \right]} \right] = e^{-\left[\pi/4 \left(V_x/V_{av} \right)^2 \right]} \tag{9.17}$$

9.5 Wind Turbine Types

A variety of turbine designs have been developed and tested throughout the evolution of the wind power industry. The Dutch windmill and multiblade American windmill shown in Figure 9.7 are examples of early windmill types used for grain grinding and water pumping in rural areas. Wind turbines are usually classified based on the axis on which their blades rotate. Two main groups are horizontal axis wind turbines (HAWT) and vertical axis wind turbines (VAWT). Whereas various types of vertical axis wind turbine structures as sketched in Figure 9.8 have been developed and tested, today HAWT dominate the markets of all sizes of wind turbines. Figure 9.9 shows one, two, and three blade HAWT types.

In practical terminology, a "rotor" is the set of blades attached to a central piece called the hub, that rotates the shaft of a generator. In general, the term "wind turbine" refers to the whole electromechanical energy conversion system that includes the rotor, generator, mechanical transmission, and electronic controls. The nacelle of a wind turbine is the enclosure that contains the generator, mechanical transmission, switchgear, and electronic circuits.

Dutch windmill   American windmill

Figure 9.7 Traditional European and American windmills.

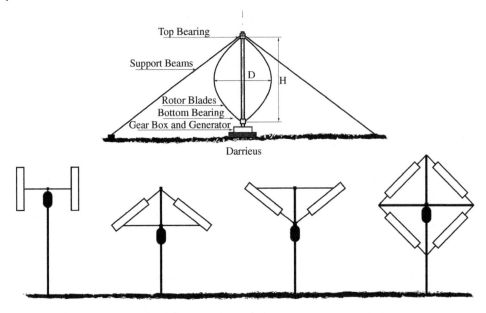

Figure 9.8 Vertical axis wind turbine (VAWT) types.

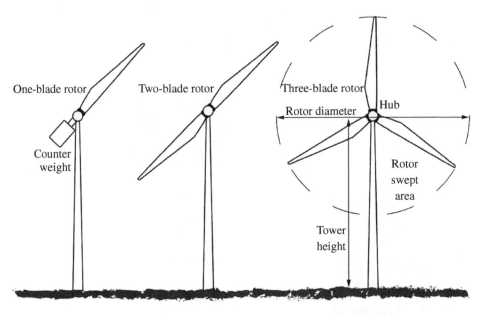

Figure 9.9 Three types of horizontal axis wind turbines (HAWT) used for electric generation.

Wind turbines are characterized by the orientation of the rotor axis. "Vertical axis wind turbines" (VAWT) have a rotor spinning around an axis vertical to the ground, and they have the advantage of producing maximum power for all wind directions. Larger VAWT units are always ground mounted, while smaller units have been developed for installation on the top of towers or tall buildings. The rotor of "horizontal axis wind turbines" (HAWT) rotates around an axis in horizontal orientation. Such turbines are always mounted on the top of a tall tower. By increasing the tower height HAWT can capture higher speed wind with less turbulence. Unlike VAWT, they must be is equipped with some mechanism to turn the axis in the wind direction. Small wind turbines are designed to turn the rotor axis to the wind direction naturally by aerodynamic forces. Commercial-size turbines have a motorized control mechanism to turn the rotor axis in the wind direction. Because they are mounted on a tall tower, the footprint of a HAWT at ground level is smaller and the land around the tower base can be used for other purposes.

While the operation of all wind turbines is based on the same principle, each type has benefits and drawbacks. Most of the commercial wind turbines are currently manufactured with horizontal axis, with three blade rotors. Small wind turbines up to 100-kW designed for residential or isolated low power applications differ significantly from larger units intended for commercial-scale power generation in the range from several kilowatts to megawatts.

9.5.1 Maximum Turbine Power and Torque

Power theoretically available in the wind was described in Section 9.4.1 by Eq. (9.4), copied below for reading convenience.

$$P_w = \frac{1}{2}\rho A V^3 \tag{9.18}$$

In this expression A is the area swept by the blades and V is the wind speed representing the air density in kg/mρ^3, which depend on the elevation of the site and temperature. In most practical cases, air density may be taken as 1.225 kg/m^3. If a wind turbine converts all kinetic energy available in the flowing air, the wind speed should drop to zero while passing through the blades. Since this is not the case in practice, a wind turbine with a given blade-swept area can convert only a fraction of the energy available in wind into mechanical work.

Since HAWT are more common in practice, we will analyze the power developed by a rotor turning in a vertical plane, as shown in Figure 9.10. We will assume that the static air pressures are the same on sections 1 and 2, far enough from the front and back of the rotor. In addition, the air stream will be considered steady, and frictions and wake around the rotor will be neglected. As the cross-sectional area of the air stream changes, the speed of the incoming (upstream) wind V is reduced to V_d far behind (downstream) the rotor. The turbine rotor area is A_T and the wind speed crossing the rotor section is V_T.

Bernoulli's equation can be applied to the steady frictionless incompressible flow shown in Figure 9.10.

$$p + \frac{1}{2}\rho V^2 = p_1 + \frac{1}{2}\rho V_T^2 \tag{9.19}$$

$$p + \frac{1}{2}\rho V_d^2 = p_2 + \frac{1}{2}\rho V_T^2 \tag{9.20}$$

Subtracting both sides of Eqs. (9.19) and (9.20) we obtain:

$$p_1 - p_2 = \frac{1}{2}\rho(V^2 - V_d^2) \tag{9.21}$$

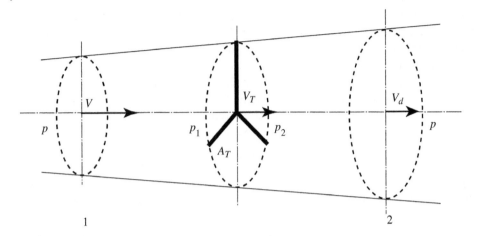

Figure 9.10 Idealized airflow through a wind turbine rotor.

The thrust is the product of the pressure difference before and after the rotor and the area.

$$F_T = (p_1 - p_2)A_T = \frac{1}{2}\rho A_T(V^2 - V_d^2) \tag{9.22}$$

At the same time, thrust can also be expressed in terms of the difference of momentum before and after the rotor:

$$F_T = m(V - V_d) = \rho A_T V_T(V - V_d) \tag{9.23}$$

Comparing Eqs. (9.22) and (9.23) we obtain

$$V_T = \frac{V + V_d}{2} \tag{9.24}$$

We see that the speed of wind across the rotor is different from the speed of wind approaching the turbine. This is because the blades convert part of the kinetic energy of the approaching wind to shaft power. The ratio of the wind speed reduction by the rotor to the speed of undisturbed wind is called the "axial induction factor" and expressed in Eq. (9.25).

$$a = \frac{V - V_T}{V} \tag{9.25}$$

The speed of air stream crossing the rotor and far after passing the rotor can be expressed in terms of incoming wind speed and the axial induction factor. Using Eqs. (9.24) and (9.25) we can write

$$V_T = (1 - a)V \tag{9.26}$$

$$V_d = (1 - 2a)V \tag{9.27}$$

The kinetic energy converted to shaft power by the rotor is the difference of the kinetic energy of the incoming wind at speed V_1 and the kinetic energy left to the outgoing wind V_2.

$$P_T = \frac{1}{2}\rho A_T V_T(V^2 - V_d^2) \tag{9.28}$$

Substituting V_T and V_d in Eq. (9.28) the power output can be expressed as

$$P_T = \frac{1}{2}\rho A_T V^3 [4a(1-a)^2] \tag{9.29}$$

Comparing Eqs. (9.4) and (9.29), we define the power coefficient as the ratio of the power developed by the turbine to the power of the wind.

$$C_P = \frac{P_T}{P_w} = 4a(1-a)^2 \tag{9.30}$$

To determine the axial induction factor that produces the maximum power, we can differentiate the power coefficient and equate to zero. This operation gives $a = 1/3$. Substituting this optimal value into Eq. (9.30) we obtain

$$C_{p,\max} = \frac{16}{27} \tag{9.31}$$

The maximum power that a wind turbine can deliver is known as Betz Law, named after the German physicist Albert Betz. The shaft power of a wind turbine cannot exceed the Betz limit.

$$P_{T,\max} = \frac{16}{27} \times \frac{1}{2}\rho A_T V^3 \tag{9.32}$$

9.5.2 Performance Coefficients

The performance of a wind turbine rotor can be assessed using the wind speed, rotor swept area, and the actual output power of the turbine. These parameters also help in design calculations.

One of the indicators used to describe the performance of a wind turbine is the power coefficient. In practice, the output power of a wind turbine is always smaller than the Betz limit. Power coefficient of a wind turbine is defined as the ratio of the actual output power to the power available from the kinetic energy of wind.

$$C_P = \frac{P_T}{P_w} = \frac{2P_T}{\rho A_T V^3} \tag{9.33}$$

If the total kinetic energy of the wind were converted to rotational power, the maximum force exerted on the rotor would be

$$F = \frac{P_w}{V} \tag{9.34}$$

Since torque is the product of the force and radius, the maximum torque available to a rotor of radius R is obtained as

$$T_w = P_w \frac{R}{V} = \frac{1}{2}\rho A_T V^2 R \tag{9.35}$$

The actual torque developed on the rotor is less than the maximum torque T_w. The torque constant is defined as the ratio of the torque developed on the rotor to the maximum torque.

$$C_T = \frac{T_T}{T_w} = \frac{2T_T}{\rho A_T V^2 R} \tag{9.36}$$

Remembering that rotational mechanical power is the product of torque and the angular velocity, the turbine torque is found by dividing the power developed on the turbine shaft to the angular velocity of the rotor.

$$T_T = \frac{P_T}{\omega} = P_T \frac{60}{2\pi N} \tag{9.37}$$

In Eq. (9.37) N is the rotor speed in revolutions per minute (rpm). The ratio of the power coefficient to the torque coefficient provides another performance factor called tip speed ratio.

$$\lambda = \frac{C_P}{C_T} = \frac{R\omega}{V} \tag{9.38}$$

Tip speed ratio is an important criterion used for comparison of different wind turbine designs and selection of the suitable turbine for a particular application. For example, a wind turbine used for water pumping requires different tip speed ratio compared to a turbine used for electric generation. Wind turbines used for electric generation achieve the maximum efficiency at a particular tip speed ratio.

Example 9.1 The rotor diameter of a utility-size wind turbine rate 2 MW at 26 mph wind speed is 90 m. Determine the following considering the specific mass of air 1.225 kg/m³.

a) Power available to the rotor at 26-mph wind speed.
b) Maximum power according to the Betz limit.
c) Power coefficient for the rated turbine output.
d) Maximum torque at 26-mph wind speed when the rotor turns at 15 rpm.
e) Torque coefficient.
f) Tip speed ratio for the operating conditions determined above.

Solution

a) Given 90-m rotor diameter, 26-mph wind speed, and 1.224-kg/m³ specific mass of air, the power available from wind is calculated below.

$$P_w = \frac{1}{2}\rho AV^3 = \frac{1}{2} \times 1.225 \times \left(\frac{\pi \times 90^2}{4}\right) \times \left(26 \times \frac{1,609}{3600}\right)^3 \times 10^{-6} = 6.110 \text{ MW}$$

b) According to Betz Law, a wind turbine can develop 16/27 times the power available in the wind:

$$P_{T,max} = \frac{16}{27}P_w = \frac{16}{27} \times 6.110 = 3.620 \text{ MW}$$

c) Since the rated power at 26-mph wind is 2 MW, the power coefficient of this turbine is:

$$C_p = \frac{P_T}{P_w} = \frac{2}{6.110} = 0.327$$

d) Maximum torque that can develop on the shaft at 26 mph wind speed is

$$T_w = P_w \frac{R}{V} = 6.110 \times 10^6 \times \frac{90}{26 \times \left(\frac{1609}{3600}\right)} = 47.321 \times 10^6 \text{ Nm}$$

e) When the rotor turns at 15 rpm, the torque developed on the rotor shaft at rated power is

$$T_T = \frac{P_T}{\omega} = 2 \times 10^6 \frac{60}{2\pi \times 15} = 1.273 \times 10^6 \text{ Nm}$$

Then, the torque coefficient is obtained below using Eq. (9.36).

$$C_T = \frac{T_T}{T_w} = \frac{1.273}{47.321} = 0.027$$

f) Tip speed ratio is obtained as

$$\lambda = \frac{R\omega}{V} = \frac{90 \times (^{2\pi 15}/_{60})}{26 \times (^{1609}/_{3600})} = 12.16$$

Note that the ratio of the power coefficient to the torque coefficient obtained in parts (c) and (d) confirms this result.

$$\lambda = \frac{C_p}{C_T} = \frac{0.327}{0.027} = 12.11$$

9.5.3 Blade Aerodynamics

Blades of a wind turbine convert the energy of the air stream to rotational motion. According to the Betz law, the maximum power that the rotor can harness is limited to approximately 59% of the total power the wind delivers. The shape and size of turbine blades determine how much the turbine output can approach this limit. The typical power coefficient of a modern wind turbine lays around 30% (see Example 9.1).

To analyze the aerodynamic behavior of wind turbine blades, consider a small blade element located at a distance r from the rotor axis (Figure 9.11). Wind turbine blades have a cross-section similar to airfoils used in airplane wings. Airplane wings are designed to keep the airplane in the air with a lift force due to the airflow around them. The purpose of turbine blades, however, is to create rotational force on the rotor plane to produce torque and power on the turbine shaft. Figure 9.12 illustrates the wind forces acting on the airfoil at the cross section of the blade element.

9.5.3.1 Pitch Angle

Remember that according to Betz theory, the theoretical maximum power occurs for the axial induction factor $a = 1/3$. The velocity V of the incoming wind far from the rotor is reduced to two-thirds of its value as it reaches the rotor blades.

$$V_T = (1 - a)V = \frac{2}{3}V \tag{9.39}$$

When the rotor spins at an angular velocity ω, an observer moving with the blade element feels a tangential wind component that blows with the speed of ωr in the opposite direction to the rotor movement. The resultant w of the axial and tangential wind vectors produces aerodynamic forces on the blade element. The angle α between w and the chord line is called "angle of attack" and the angle β between the chord line and the rotor plane is called "pitch angle." The sum of α and β gives the flow angle represented by ϕ.

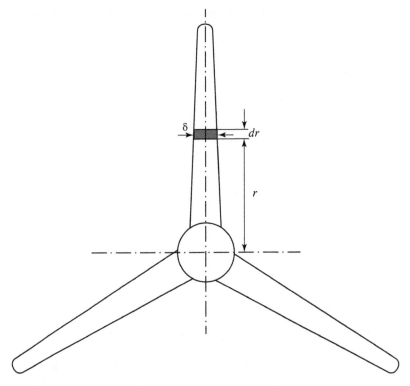

Figure 9.11 Infinitesimal blade section.

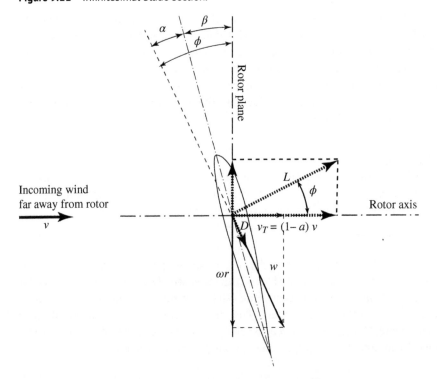

Figure 9.12 Forces developed on the airfoil of a wind turbine blade.

Equation (9.40) for the flow angle can be obtained from Figure 9.12 using trigonometric relationships.

$$\tan \phi = \frac{(1-a)V}{\omega r} = \frac{2}{3} \frac{V}{\omega R} \tag{9.40}$$

We can determine the optimal flow angle for the blade element by substituting $\omega = \lambda V/R$ from the expression of wind speed ratio in (9.38).

$$\tan \phi = \frac{2}{3} \frac{V}{\left(\lambda \frac{V}{R}\right) r} = \frac{2R}{3\lambda r} \tag{9.41}$$

$$\phi = \arctan \left[\frac{2R}{3\lambda r}\right] \tag{9.42}$$

For a given angle of attack, the pitch angle of the blade element can be found by subtracting the angle of attack from the flow angle.

$$\beta = \phi - \alpha \tag{9.43}$$

Equation (9.42) shows that the flow angle is a function of tip speed ratio and distance of the section to the rotor axis. The tip speed ratio is the ratio of power coefficient to the torque coefficient, therefore λ should be selected to achieve the desired turbine performance. As λr increases, the flow angle decreases. The sections along the blade must be therefore twisted to adjust the airfoil orientation to the optimal angle of attack.

9.5.3.2 Lift and Drag Forces

Suppose a hypothetical situation where the wind hits the blade element in the normal direction at a velocity w without any deflection. The force corresponding to the kinetic energy of the air hitting the area A_b of the blade element would be

$$F_n = \frac{1}{w} \left[\frac{1}{2}\rho A_b w^3\right] = \frac{1}{2}\rho bcw^3 \tag{9.44}$$

In reality, the airflow w hits the airfoil with an angle of attack as shown in Figure 9.12 and produces a *lift* force L perpendicular to its direction and a *drag force* D in the same direction due to the aerodynamic shape of the airfoil. In practice, dimensionless design parameters C_L and C_D known as lift and drag coefficients are used to calculate the lift and draft forces for a given airfoil.

$$L = C_L F_n = C_L \frac{1}{2}\rho bcw^2 \tag{9.45}$$

$$D = C_D F_n = C_D \frac{1}{2}\rho bcw^2 \tag{9.46}$$

Lift and drag coefficients are functions of the angle of attack. Most wind turbines require higher lift and lower drag forces for optimal energy conversion. Therefore, the blades are designed such that the angle of attack yields the maximum C_L/C_D ratio. As an example, the C_L/C_D ratio for the airfoil S805A is plotted in Figure 9.13 as a function of the angle of attack. For this airfoil, the optimal angle of attack is approximately 6.5°.

The same airfoil can be used for different blade sizes. Aerodynamic forces that develop on the actual size of a blade element are scaled using the Reynolds number, which is defined as the ratio of momentum forces to viscous forces. Therefore, the lift and drag coefficients

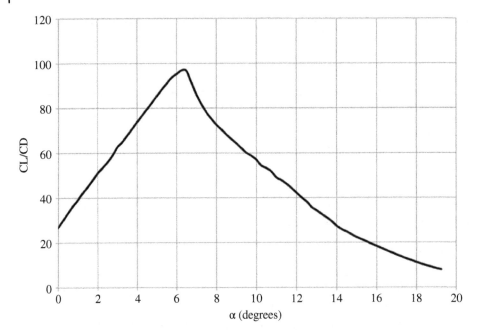

Figure 9.13 Lift-to-drag ratio for the airfoil S805A. Source: www.airfoiltools.com.

for a given airfoil change with the chord length and physical properties of the air crossing the turbine blades. For the chord length c and wind speed w, the Reynolds number can be calculated using Eq. (9.47).

$$R_e = \frac{\rho c w}{\mu} = \frac{c w}{\nu} \tag{9.47}$$

In Eq. (9.47), ρ is the density and μ is the dynamic viscosity of air. Kinematic viscosity defined as $\nu = \mu/\rho$ is substituted in the second part of the equation.

Many airfoil shapes developed for airplanes can also be used for wind turbine blades. Characteristics of airfoils developed by the National Advisory Committee for Aeronautics (NACA) are available in many official publications. The National Wind Technology Center (NWTC) has developed airfoils particularly suitable for HAWT. Detailed data for these airfoils are published at the NWTC Information Portal http://wind.nrel.gov (Tangler and Somers 1995). Specifications and wind tunnel tests of a large number of low-speed airfoils are summarized in Williamson et al. (2012).

The lift and drag coefficients can be determined by wind tunnel tests or computational fluid dynamics methods. Computational tools and online calculators are available for analysis of airfoil characteristics. A freely accessible web site (AirfoilTool.com 2019) contains a large airfoil library and implements the software named XFOIL described in Drela (1989, 2013) to compute and plot design characteristics of airfoils. Figure 9.14 shows the plots of the lift and drag coefficients for the airfoil S805A developed by the US National Renewable Energy Laboratory (NREL) for medium-size wind turbines. The C_L/C_D ratio plotted in Figure 9.13 reaches the maximum at the angle of attack of 6.5°.

Figure 9.14 Lift and drag coefficients for the airfoil S805A. Source: www.airfoiltools.com.

9.5.3.3 Chord Length

The axial component of the wind force that acts on a blade element can be determined using the projections of the lift and drag forces on the rotor axis.

$$F_a = (C_L \cos \phi + C_D \sin \phi) \left[\frac{1}{2} \rho bc w^2 \right]$$
(9.48)

At the maximum power generation ($a = 1/3$), w can be written in terms of the incoming wind speed v.

$$w = \frac{(1 - a)v}{\sin \phi} = \frac{3}{2} \frac{v}{\sin \phi}$$
(9.49)

For a rotor that has n blades, the total axial force on the blade elements located at the same distance from the rotor axis can be rewritten as

$$F_{a,total} = n \frac{1}{2} \rho bc \left[\frac{2}{3} \frac{v}{\sin \phi} \right]^2 (C_L \cos \phi + C_D \sin \phi)$$
(9.50)

On the other hand, the axial force can be determined by applying Betz condition

$$F_{a,total} = 2\pi r \rho b \left(\frac{2}{3} v \right)^2$$
(9.51)

In practice, airfoils are designed in such a way that the drag coefficient is much smaller than the lift coefficient, therefore $C_D \sin \phi$ is negligible. The following expression can be obtained by equating Eqs. (9.50) and (9.51).

$$c = \frac{4\pi r}{nC_L} \tan \phi \cdot \sin \phi$$
(9.52)

Recall from Eq. (9.42) that

$$\tan \phi = \frac{2R}{3\lambda r} \tag{9.53}$$

The following expression is obtained for the chord length by substituting Eq. (9.53) into Eq. (9.52).

$$c = \frac{8}{3} \cdot \frac{\pi R}{nC_L \lambda} \sin \phi \tag{9.54}$$

It should be noted that several assumptions were made in this analysis. In addition to neglecting the axial component of the drag force, turbulences and wake behind the rotor were not considered. More accurate methods to determine the power coefficient and blade dimensions can be found in Mathew (2006) and Burton, Sharpe, Jenkins et al. (2001).

9.5.4 Blade Design

Given the rated power and wind speed, the blade length is obtained from the required rotor swept area. The setting angle at different sections of the blade must be determined to achieve maximum performance of the wind turbine at the rated wind speed and power. In the previous paragraph we saw that the width and setting angle of the blade sections must gradually change from the root to the tip.

Example 9.2 Design a three-blade rotor for a horizontal axis wind turbine with the specifications listed below.

Turbine specifications:

Rated wind speed	9 m/s
Rated power	2 kW
Rotor speed at the rated wind speed	325 RPM

Assume a power coefficient of 0.4 and overall generator efficiency of 90%. Use NREL's S822 airfoil recommended for small wind turbines of 2–20 kW rated power. Based on available data, for this airfoil the maximum C_L/C_D occurs at an angle of attack of 5.75°, which yields a lift coefficient of 0.8582. (Density of air at the turbine location is 1.224 kg/m³.)

Solution
First, we must determine the blade length:

$$P = \frac{1}{2}C_P \eta \rho A_T V^3 \quad \Rightarrow \quad A_T = \frac{2P}{C_P \eta \rho V^3} = 12.45$$

$$R = \sqrt{\frac{A_T}{\pi}} = 2 \text{ m}$$

The tip speed ratio can be calculated for given rated wind speed and rotor speed.

$$\lambda = \frac{\omega R}{V} = \frac{2\pi N}{60} \cdot \frac{R}{V} = \frac{2 \times \pi \times 325}{60 \times 9} = 7.52$$

The blade is partitioned into nine sections, starting from 20% of the rotor radius allowing space for the hub. The flow angle ϕ, chord length c, and setting angle β are computed for each section radius using Eqs. (9.53) and (9.54). The calculation results are shown in Table 9.3.

Table 9.3 Design summary for Example 9.2.

Section	r (m)	ϕ (°)	$\beta = \phi - \alpha$	c (m)
1	0.398	22.4	16.7	0.293
2	0.597	15.9	10.2	0.224
3	0.796	12.3	6.5	0.177
4	0.995	9.9	4.2	0.145
5	1.195	8.3	2.6	0.123
6	1.394	7.2	1.4	0.106
7	1.593	6.3	0.5	0.094
8	1.792	5.6	−0.2	0.083
9	1.991	5.0	−0.7	0.075

9.6 Wind-powered Electric Generation

The idea of using wind power for electric generation dates back to the late nineteenth century, but commercial wind power development has accelerated since the 1970s energy crisis. Today generation capacity of wind powered units range from a few watts to several hundred megawatts. The share of the wind power in the fuel mix of electric generation is continuously growing.

As discussed in previous chapters, electricity is the most flexible and convenient form of energy to deliver renewable sources to end-users. In modern energy systems, utility scale facilities deliver wind energy to consumers in the form of electricity. Individual users install small wind turbines on their properties to convert wind into electric power.

Wind is available everywhere. However, inadequate use of wind power may lead to failures, and public misconceptions that might damage proper use of such an abundant and environment-friendly resource. Therefore, it is important to understand the technical characteristics, control options, and grid integration of wind powered generation units for more effective use of the existing wind potential.

9.6.1 Turbine-Generator Characteristics

Output power of a wind turbine is proportional to the circular area swept by the blades and cube of the wind speed. Horizontal axis wind turbines (HAWT) have become standard in the wind power industry. While any number of blades can be used, three-blade rotors are the most common in practical wind turbines.

Output power characteristics of a wind powered electric generation unit depend on the airfoil design, pitch angle of the blades, and generator performance. Figure 9.15 depicts the output power characteristics of a typical wind turbine with pitch control. Below a certain threshold wind velocity turbine blades cannot produce sufficient torque to turn the rotor. Wind turbines start generating electricity above a minimum wind speed called cut-in speed, which is around 3–4 m/s (7–9 mph) for most commercial turbines.

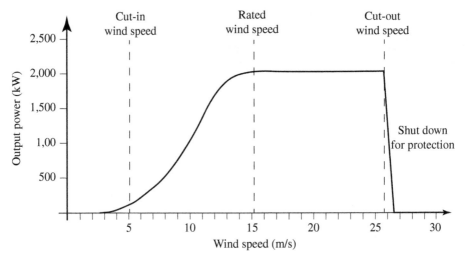

Figure 9.15 Typical output characteristic of a wind turbine-generator system with pitch control.

According to the Betz law, a wind turbine can convert up to 16/27 (59%) of the total kinetic energy available in the air flowing through the rotor into mechanical work. Practical wind turbines deliver less than this theoretical power because of aerodynamic limitations. Moreover, the output power delivered by the turbine-generator combination is reduced due to mechanical and electrical losses. The power that can be obtained from a practical turbine at a given wind speed is determined by multiplying the theoretical power available in the wind with the power coefficient C_p and the overall efficiency η of the conversion system. Above the cut-in speed, the power output of a wind turbine in its normal operation range is described by Eq. (9.55).

$$P = \eta C_p \left(\frac{1}{2}\rho A V^3 \right) \qquad (9.55)$$

Depending on the design, a typical wind turbine generates its rated power at a certain wind speed between 11 and 17 m/s (27–33 mph). The maximum electric power delivered by a practical wind turbine is constrained by the blade design, mechanical limitations of the turbine and electrical specifications of the generator. In addition, extremely high winds may create destructive forces and tensions on the mechanical structure. A wind turbine can operate safely up to a cut-out speed of 20–30 m/s (45–67 mph) because of mechanical and electrical limitations.

9.6.2 Output Power Control

Wind turbines are designed to limit electric generation to their maximum level and prevent the rotor from spinning at extremely high speed. There are two main methods to control the output power of a wind turbine commonly known as *pitch control* and *stall control*.

9.6.2.1 Pitch Control
Most of the commercial-size wind turbines are equipped with a pitch control system that continuously adjusts the blade angles to keep the power output at maximum level. Between

the cut-in wind speed and the rated power, pitch control adjusts the blade angles such that maximum wind power is converted into electricity. After the conversion system has reached its rated power, the pitch control mechanism sets the blade angle to maintain the power output constant as long as the wind speed is in a range that allows the safe operation of the turbine. If the wind speed exceeds the safety limits, a brake system blocks the shaft rotation and the pitch control mechanism turns the blades such that the torque becomes minimum (see Figure 9.12). Output power characteristic of a conversion system with pitch control is shown in Figure 9.15.

9.6.2.2 Stall Control

When the wind speed exceeds a certain level, air flow over the blades starts to create turbulence on the upper side of the airfoil, causing a drop in the lift force that reduces the produced torque. Airfoil can be designed such that the power developed on the rotor decreases above a certain wind speed. Reducing the power output naturally by airfoil shape at high wind speeds is known as stall control. In stall control the output power of the conversion system may reach a maximum, then decreases. If the wind speed exceeds the range that allows mechanically safe operation limits, a braking system prevents the shaft from turning idle. Since there is no mechanism to change the blade angle, the rotor must withstand the thrust forces caused by the high wind (Burton, Sharpe, Jenkins et al. 2001).

Stall control requires advanced blade design techniques including computational fluid dynamics simulation and wind tunnel tests. Stall controlled turbines are therefore cheaper and more reliable since there is no pitch control mechanism that increases the manufacturing and maintenance costs. Smaller wind turbines use stall control and have either an electromagnetic brake system or a furling mechanism to turn the rotor parallel to the wind direction.

9.6.3 Generator Types

The generators used in wind powered electric generation systems are either the synchronous or asynchronous type. The structure and operation principle of these generators are discussed Chapter 6 on electric energy systems. Both generator types have a similar three-phase stator winding arrangement, but different rotor structures. In synchronous generators a magnetic field produced by either permanent magnets or coils attached to the rotor induce voltages on the stator windings. As explained in Chapter 6, the magnitude of the generated voltage depends on the magnetic flux density, number of turns of the stator windings, and magnetic properties (magnetization or "B–H" characteristic) of the magnetic circuit. In asynchronous generators, stator windings induce currents created on the rotor conductors which create a rotating magnetic field. Hence, asynchronous generators are also known as induction generators. In normal operation, the magnitude of the AC voltage induced on the stator windings can be considered proportional to the rotation speed ω and the field current i_f.

$$E = k_f i_f \omega \tag{9.56}$$

The frequency of the output voltage is directly proportional to the rotation speed and inverse of the number of poles. Given rotation speed n in revolutions per minute (rpm) and the total number of poles p, the synchronous speed is found as

$$f_s = \frac{p \cdot n}{120} \tag{9.57}$$

As with any type of electric machine, both synchronous and asynchronous machines allow bidirectional energy flow. In synchronous operation with a strong grid supplied from other power plants, the terminal voltage and frequency can be considered constant. Synchronous machines must always operate at the synchronous speed and the corresponding frequency. Induction (asynchronous) generators, on the other hand, can achieve energy conversion when they operate at a speed different than the synchronous speed.

9.6.3.1 Synchronous Generators

Almost all generators used in conventional electric generation units driven by heat engines are of the synchronous type. In the early years of wind turbine development, considerable efforts were made to use synchronous generators similar to conventional units. Two major issues challenged the use of synchronous generators in wind turbines. First, the rotating field windings must be supplied DC to produce the magnetic field. The second challenge is direct coupling of synchronous generators with the turbine rotor, since the changes of the wind speed causes proportional variations of the voltage and frequency of the output power.

Most of the larger generators use a slipring-brush assembly to supply rotating field windings from a stationary external circuit. Rotating contacts increase the manufacturing cost and require more maintenance. Brushless synchronous generators developed to eliminate moving contacts use a self-excitation generator mounted on the shaft and a solid-state rectifier to supply the field windings. Development of strong rare-earth magnets allowed the use of permanent magnets in smaller wind turbine generators, which significantly reduced the cost and maintenance need in smaller turbines.

A synchronous machine driven by a wind turbine and connected to an electric grid produces both active and reactive power, depending on the mechanical power on its shaft and magnetic field produced by the rotor windings. Since the single synchronous generator cannot change the frequency of the interconnected grid, the rotation speed is strictly defined by the grid frequency. Active power that the generator delivers to the grid depends directly on the mechanical power driving its shaft. Similarly, since the single generator cannot change the voltage of the interconnected grid, the output voltage of the generator is the same as the grid voltage at the connection point. Changing the magnetic field which induces a proportional emf on the stator windings will change the active power absorbed from or delivered to the grid. Therefore, synchronous generators that have rotor windings offer the benefit of controlling both the active and reactive power generation. In permanent magnet generators only the active power can be controlled.

Direct coupling of the turbine rotor and generator creates both electrical and mechanical issues for synchronous generators used in wind powered energy conversion units. Output voltage and frequency of a synchronous generator are proportional to the rotation speed. Turbine blades can be designed for fixed speed operation, but the turbine achieves the maximum efficiency at one particular tip speed ratio $\lambda = R\omega/V$, which depends on the wind velocity and angular speed of the rotor.

Fixed shaft speed allows the synchronous generator to produce constant frequency and voltage. However, fixed speed turbine operation is sub-optimal for wind speeds other than the speed corresponding to the optimal tip-speed ratio. The rotor speed is then selected to yield maximum power in a range of wind speed, but the captured energy is less than the energy that could be captured with the optimal tip speed ratio.

One option to increase the efficiency is operating the turbine at two separate fixed speeds so that the tip speed ratio remains closer to the optimal value. In that case, either two separate generators with the same number of poles must be installed coupled with the rotor shaft via a gearbox with two output shafts turning at different speeds, or a single generator with the number of poles that can be changed depending on the speed. Both options clearly increase the cost.

The tip speed ratio can be maintained at the value that corresponds to the maximum efficiency by increasing the rotation speed proportionally to the wind speed. In variable speed operation, a solid-state electronic converter is used to transform the variable voltage and variable frequency AC produced by the generator into constant frequency and voltage on the output terminals to match the utility standards. Such converters can also control the active and reactive power flow from the generator to the grid depending on the changing operating conditions.

Wind turbines up to about 10 kW have permanent magnet three-phase generators designed to operate at variable speed. Small wind turbines are generally designed for either battery charging or grid-tie operation supplying a low voltage residential or small commercial loads. Grid-tie turbines have a computer-controlled converter to adjust the output voltage and frequency in real time to the values of the utility grid. The three-phase AC power with variable frequency and variable voltage is first converted to DC, then inverted to a single-phase AC with voltage and frequency values exactly matching the utility grid at all times for synchronous operation. Battery charging turbines are connected to a battery bank through a charge controller. The charge controller rectifies the AC output of the generator and adjusts the battery charging current according to the characteristics of the battery type.

9.6.3.2 Asynchronous (Induction) Generators

Asynchronous machines can perform energy conversion when the rotor speed is different than the synchronous speed ($n_s = 120f/p$). In motor operation of an asynchronous generator the parameter slip defined in Eq. (9.58) is positive. If the machine is forced to turn above the synchronous speed, the asynchronous machines operates as a generator.

$$s = \frac{n_s - n}{n_s}$$

$0 < s < 1$: motor operation

$s > 1$: generator operation $\hspace{3cm}$ (9.58)

Asynchronous (induction) generators have become more common in commercial-size wind turbines because of their simple structure, lower cost, and lower maintenance requirement.

In fixed speed operation, an asynchronous generator may be directly connected to the grid if the shaft speed is maintained equal to or larger than the synchronous speed corresponding to the grid frequency. Active power delivered to the grid increases as the rotation speed increases. An induction generator requires reactive power to produce a magnetic field necessary for induction. In standalone operation, this reactive power is supplied to the induction generator by a capacitor bank connected to the stator terminals. When the induction generator supplies an interconnected network with other generators the reactive power may be absorbed from the grid, but its direction and magnitude cannot be controlled as in the case of synchronous machines.

A doubly fed induction generator, which allows variable speed operation, consists of an induction generator with externally accessible rotor windings and a controlled electronic converter that supplies the rotor windings to produce a magnetic field. Such an arrangement has benefits of controlling both the active and reactive power during variable speed operations.

Advancements in power electronics and availability of high voltage and high current solid-state devices reduced the cost of converters to control grid-connected induction generators. Most of the utility-size modern wind turbines are equipped with induction generators and computer-controlled converters to maximize energy production.

9.6.3.3   Stand-Alone Operation

In stand-alone operations, the frequency of the generated voltage is proportional to the rotation speed. Since the wind speed changes continuously, if the turbine is designed to operate at variable speed, both the frequency and voltage of the generated power are variable. The turbine can be designed to operate at a nearly constant speed. In stand-alone operation an electronic converter is needed to regulate the voltage and frequency. Some small-size wind turbines are designed for battery charging. In these generators, the AC output voltage is rectified and controlled for suitable battery charging modes. In a stand-alone operation dump resistors may be used to dissipate the generated power at no-load operation to avoid acceleration of the system when the electric load is turned off while there is wind power turning the blades.

In larger units, the turbine power is transferred to a synchronous or asynchronous generator via a gear box to increase the shaft speed to the level of the synchronous speed. With a turbine designed for constant speed operation, the turbine speed is self-regulated by the power transferred to the grid. If the turbine operates at variable speed, an AC-to-AC converter is needed to maintain the voltage and frequency at the utility standards to ensure proper operation of the supplied electric load. In addition, a capacitor bank must be installed to supply the reactive power required for creation of the magnetic field. The capacitor bank is also required for self-excitation of the induction generator at start up.

9.6.3.4   Grid Connected Operation

In grid-connected operations, the output voltage and frequency are always equal to the grid voltage and frequency. If the generator is directly connected to the grid, the turbine must drive the generator at fixed speed. A gear box increases the turbine speed to the level of the synchronous speed corresponding to the utility frequency.

Since wind speed changes randomly, if not controlled, active and reactive power delivered to the grid will also change randomly. Such random changes may cause voltage and frequency fluctuations decreasing the power quality of the grid. In some extreme situations, excessive power added to or subtracted from the grid may cause a mismatch between supply and demand leading to electromechanical oscillations.

9.6.4 Grid Integration of Wind Powered Generation

Wind powered electric generation has no fuel cost, and the maintenance requirement and running cost including technical services, fuel transportation, and waste disposal is negligible compared to fossil fuel burning and nuclear power plants. Therefore, it is advantageous to operate the grid-connected wind conversion facilities at full capacity as much as possible. In fact, some countries mandate utility companies to purchase all energy generated from wind to allow wind farms operate at higher capacity factor. The major downside of wind-powered generation is its lower predictability than all other energy sources.

Clearly, an interconnected electric network cannot depend mainly on wind power. In modern energy systems with well-balanced fuel mix and adequate storage facilities, grid penetration issues are resolved by coordination of power generation.

9.7 Energy Output Estimation

Energy output of a wind turbine over a period is estimated considering the wind profile of the turbine site and power characteristics of the wind turbine. Since the kinetic energy available in wind is proportional to the third power of wind speed, a simple monthly or yearly wind speed average is often misleading. As indicated earlier, average of the cube of wind speed may be a better indicator for assessment of the wind potential. However, each wind turbine available on the market has different power characteristic and efficiency versus wind speed.

If wind speed data recorded with fixed intervals over a year is available, energy output of a particular wind turbine can be estimated quite realistically. Suppose that n values of wind speed have been recorded over one year at 10 minute or 1 hour intervals. If $P(V)$ is the function representing the power output of the given wind turbine, then the total energy output over a year is

$$E = \frac{8760}{N} \sum_{i=1}^{N} P(V_i) \tag{9.59}$$

In this expression 8760 is the number of hours in a year. Daily or monthly energy output can be estimated similarly, but one year is usually a more reasonable period for economic analysis and estimation of the break-even point for the initial investment.

Example 9.3 A 10-kW wind turbine is considered for a residential site. The 10-minute average wind speed data has been collected at the site over a one-year period. We would like to estimate the annual electric generation of the considered turbine, arranged as a probability density distribution as shown in the first two columns of Table 9.4. The power characteristic of the considered wind turbine is entered in column 3 from the data sheet provided by the manufacturer. Estimate the annual energy output of this turbine.

Solution
A spreadsheet similar to the one shown in Table 9.4 can be used for the requested estimation. Power output of the turbine for each wind speed is entered in column 2. Wind speed data recorded on-site is arranged in frequency distribution using the FREQUENCY function

Table 9.4 Data for Example 9.3.

1	2	3	4	5
Wind speed (m/s)	Turbine output power (kW)	Probability density (%)	Frequency of occurrence (h/yr)	Energy generation (kWh/yr)
0	0	0	0	0
1	0	3.08	270	0
2	0.14	5.87	514	423
3	0.43	8.15	714	2499
4	0.88	9.74	853	7306
5	1.51	10.57	926	14,785
6	2.35	10.68	936	23,496
7	3.43	10.17	891	31,094
8	4.8	9.20	806	35,566
9	6.42	7.93	695	35,393
10	8.21	6.55	574	30,859
11	10.02	5.19	455	23,642
12	11.37	3.95	346	15,560
13	11.76	2.90	254	8647
14	12.06	2.05	179	4423
15	12.14	1.39	122	2065
16	12.15	0.92	80	893
17	12.1	0.58	51	358
18	11.92	0.36	31	132
19	11.44	0.21	18	45
			Total:	237,185

of MS Excel® and copied in column 3. The frequency of occurrence of each wind speed is obtained by multiplying the probability density with 8760, the number of hours in one year (column 4). Contribution of each wind speed in the annual energy production is calculated in column 5 by multiplying the frequency of occurrence of each wind speed (column 4) and the turbine power output corresponding to that wind speed (column 2).

Total annual generation is obtained as 237,185 kW by adding all values in column 6. The estimated value is approximate because the hourly average wind speed is used in the probability distribution. The graphs shown in Figure 9.16 illustrate the wind speed distribution, turbine characteristic, and contribution of each wind speed in the annual generation.

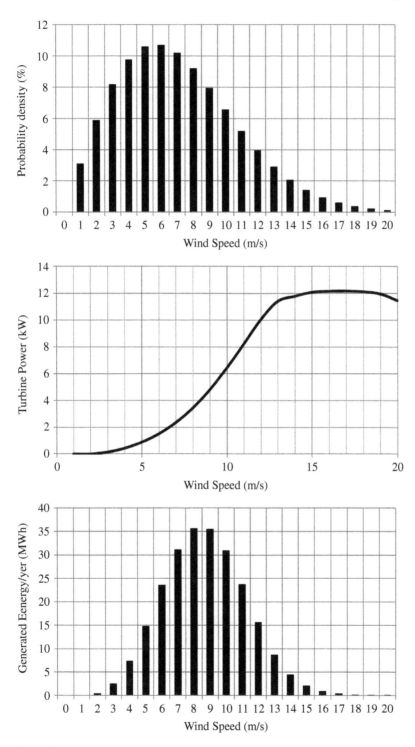

Figure 9.16 Graphical illustration for Example 9.3.

9.8 Chapter Review

This chapter presents technical aspects of wind power. Various wind patterns, analysis of wind regimes, and assessment of the wind resources are described.

Wind is a clean, sustainable, and renewable source of energy. Currently, wind power is delivered to end-users mainly in the form of electricity.

Global wind patterns are the consequence of temperature difference by latitude and rotation of the Earth. Local winds mainly result from pressure differences due to uneven heating effects on the Earth's surface.

Wind speed changes with height due to wind shear. Barriers and obstacles on the ground cause turbulence.

Wind patterns can be mathematically represented by Weibull or Raleigh probability density functions. Weibull distribution is a skewed exponential probability function with a shape factor and a scale factor. Rayleigh distribution is a special case of Weibull distribution for shape factor 2.

Commercial wind farms supply the interconnected grid. Grid-tie mid-size wind turbines supply communities and commercial loads. Individual small wind turbines installed by residential and commercial consumers are mostly connected to the grid through bidirectional electric meters. Wind powered generation systems may be also used to supply a standalone load, such as an island, an isolated agricultural or recreational facility, an off-grid building, or electronic equipment at a remote location.

Wind turbines convert kinetic energy of the air flowing through the blades into rotational mechanical work. In practice, turbines are classified as vertical or horizontal axis. The most common type is horizontal axis wind turbines (HAWT) with three blades.

Kinetic energy available in the air is proportional to the third power (cube) of the wind speed, blade swept area, and specific mass of the air. Maximum power that a wind turbine can deliver is 59% of the power available in the wind, known as the Betz limit.

Wind turbines are characterized by a power constant and torque constant. The ratio of these constants give the tip speed ratio, which is a useful indicator of the turbine performance. A wind turbine delivers maximum power at a particular tip speed ratio.

Turbine blades can be designed for fixed speed, two speed, or variable speed operation.

Wind powered generation units use either synchronous or asynchronous generators. Both have similar stator-holding three-phase windings. In synchronous generators either permanent magnets or field windings mounted on the rotor create the magnetic flux necessary to induce voltages on the stator windings. Asynchronous generators have conductor bars or coils on the rotor. The rotating magnetic field of the stator windings induce currents on the rotor conductors, which creates the magnetic flux in the air gap. Asynchronous machines are also known as induction machines.

Utility scale wind turbines mostly use induction generators designed for grid connected operation. Doubly fed asynchronous generators can operate at variable speeds, and the active and reactive power can be controlled through an electronic converter.

Smaller wind turbines operate at variable speeds and generate variable voltage and variable frequency three-phase AC power with a permanent magnet synchronous generator. Small wind turbines are designed either for grid tie operation or battery charging. In either type, an electronic circuit interfaces the generator terminals to the output for intended purposes.

Review Quiz

1   Wind power is mostly supplied to end-users in the form of
   a. heat.
   b. electricity.
   c. steam.
   d. pressured air.

2   Global wind patterns are a result of
   a. temperature change with latitude and Earth rotation.
   b. temperature differences between oceans and land.
   c. temperature change with altitude.
   d. air pressure change due to uneven heating of Earth's surface.

3   Local wind patterns are a result of
   a. temperature change with latitude and Earth rotation.
   b. temperature differences between oceans and land.
   c. air pressure change due to uneven heating of Earth's surface.
   d. temperature change with altitude.

4   Wind speed increases with height
   a. linearly.
   b. as a quadratic function of the height.
   c. as the cubic function of the height.
   d. as the power of the wind shear exponent.

5   Power available in wind is
   a. proportional to wind speed.
   b. proportional to the cube of the wind speed.
   c. a quadratic function of the wind speed.
   d. an exponential function of the wind speed.

6   What kind of energy is converted to mechanical work by a wind turbine?
   a. Kinetic energy
   b. Potential energy
   c. Thermal energy
   d. Electric energy

7   The power coefficient of a wind turbine is
   a. the ratio of the turbine power to the angular velocity of the rotor.
   b. the ratio of the power to the torque times angular velocity.

c. the ratio of the power extracted by the turbine to the total power contained in the wind.

d. the ratio of the shaft power to the electric power of a wind turbine

8 Which value of the axial induction factor produces the maximum power in a wind turbine?

a. 0.33

b. 0.50

c. 0.59

d. 1.00

9 What is the Betz limit?

a. Maximum height of a wind turbine

b. Maximum number of blades that a wind turbine can have

c. Maximum power coefficient that a wind turbine can have

d. Maximum energy that a wind turbine can generate in one hour

10 The ratio of the wind speed reduction by the rotor to the speed of undisturbed wind is called the

a. axial induction factor.

b. power factor.

c. tip speed ratio.

d. torque coefficient.

11 What is the maximum power coefficient for a wind turbine?

a. 0.25

b. 0.33

c. 0.59

d. 0.75

12 Which one of the following parameters has the largest effect on the power output of a wind turbine?

a. Wind speed

b. Specific mass of air

c. Blade swept area

d. Number of blades

13 The maximum electric power that a wind turbine can deliver is

a. limited to about 59% of the power available in wind.

b. equal to the power available in wind.

c. more than 90% of the power available in wind.

d. unlimited.

14 Tip speed ratio of a wind turbine is

a. the ratio of the torque coefficient to the power coefficient.

b. the ratio of the power coefficient to the torque coefficient.

c. the ratio of the tip speed of the blades to the torque developed on the rotor.

d. the ratio of the angular velocity of the rotor to the torque wind speed.

15 Suppose that the rotor speed increases while the power remains constant. How would the torque change?

a. Decreases

b. Increases

c. Remains constant

d. Fluctuates

16 Which ones of the following is a wind turbine type?

a. Pelton

b. Darrieus

c. Kaplan

d. Carnot

17 Rotor of a wind turbine refers to

a. the blade assembly and hub.

b. the rotating part where the blades are attached.

c. the shaft and mechanical transmission.

d. all rotating parts of the turbine-generator combination.

18 Most utility-size wind turbines utilize

a. DC generators.

b. permanent magnet AC generators.

c. asynchronous generators.

d. shunt generators.

19 Small-size wind turbines mostly utilize

a. permanent magnet AC generators.

b. single-phase AC generators.

c. asynchronous generators.

d. DC generators.

20 A doubly-fed induction generator is

a. a synchronous generator with two field windings.

b. an induction generator with two-phase windings.

c. an induction generator with a converter that produces one three-phase and one single-phase output.

d. an induction generator with a converter that supplies the rotor windings.

Answers: 1-b, 2-a, 3-c, 4-d, 5-b, 6-a, 7-c, 8-a, 9-c, 10-a, 11-c, 12-a, 13-a, 14-b 15-a, 16-b, 17-a, 18-c, 19-a, 20-d.

Research Topics and Problems

Research and Discussion Topics

1 Draft a white paper to discuss the practical advantages of using induction generators in wind turbines.

2 Research the current technologies used in utility-scale wind turbines.

3 What type of converters are used in grid-tie small wind turbines?

4 Find the technical specifications of two utility-size wind turbines and compare their strengths and drawbacks.

5 Find the technical specifications of two residential-size wind turbines and compare their strengths and drawbacks.

Problems

1 An anemometer placed on a 10-m pole measures 5 m/s wind speed. Estimate the wind speed at a 25-m turbine height assuming that the roughness length 0.15 m.

2 At 10-mph wind speed a turbine generates 1.5 kW power. What would the output power be if the wind speed increases to 20 mph (assume that stall does not occur at that speed)?

3 The blades of a three-blade horizontal axis wind turbine are 5 m long. What is the power delivered by the air stream to the blades for the wind speed of 8 m/s? (Specific mass of air is 1.224 kg/m³.)

4 The rotor blades of a horizontal axis wind turbine are 35 m long. Assuming that all losses are negligible, determine the maximum power in kW this turbine can generate for the wind speed of 10 m/s at the hub height? (Specific mass of air is 1.224 kg/m³.)

5 The rotor diameter of a horizontal axis wind turbine is 80 m. For a 10-m/s wind speed at the hub height the output power is measured 80 kW? What is the power coefficient of this wind turbine?

Recommended Web Sites

- American Wind Energy Association: https://www.awea.org
- NREL NWTC Information Portal: https://nwtc.nrel.gov
- XFOIL Subsonic Airfoil Development System: http://web.mit.edu/drela/Public/web/xfoil
- Airfoil tools: http://airfoiltools.com.

References

AirfoilTool.com, 2019. *Airfoil tools.* [Online] Available at: http://airfoiltools.com [Accessed 10 10 October 2019].

Encyclopedia Britannica, n.d. *Encyclopedia Britannica.* [Online] Available at: https://www .britannica.com/science/atmospheric-circulation#/media/1/41463/107938 [Accessed 22 October 2019].

Burton, T., Sharpe, D., Jenkins, N., and Bossany, E. (2001). *Wind Energy Handbook.* New York, NY: Wiley.

Drela, M. (1989). XFOIL: An analysis and design system for low Reynolds number airfoils. In: *In: Mueller T.J. (ed) Low Reynolds Number Aerodynamics. Lecture Notes in Engineering, Vol 54, pp 1–12.* Berlin, Heidelberg: Springer.

Drela, M., 2013. *XFOIL.* [Online] Available at: http://web.mit.edu/drela/Public/web/xfoil [Accessed 27 November 2015].

Gipe, P. (2004). *Wind Power – Renewable Energy for Home, Farm, and Business.* White River Junction, VT: Chelsea Green.

Mathew, S. (2006). *Wind Energy – Fundamentals, Resource Analysis, and Economics.* Berlin, Heidelberg, New York: Springer.

Tangler, J.L. and Somers, D.M. (1995). *NREL Airfoil Families for HAWTs.* Golden: NREL.

Williamson, G.A., McGranahan, B.D., Broughton, B.A. et al. (2012). *Summary of Low-Speed Airfoil Data.* Urbana-Champaign (IL): Dept. of Aerospace Engineering, University of Illinois at Urbana-Champaign.

10

Solar Energy Systems

Two common technologies to capture solar energy are Photovoltaic (PV) conversion and concentrated solar power (CSP) generation systems.

Top: A utility-size PV generation facility near Hagerstown, Maryland, with 25-MW installed capacity.

Bottom: Crescent Dunes Concentrated Solar Power (CSP) plant between Reno and Las Vegas in Nevada. The facility has 110-MW capacity and generates electric power day and night by storing thermal energy in molten salt.

Energy for Sustainable Society: From Resources to Users, First Edition. Oguz A. Soysal and Hilkat S. Soysal.
© 2020 John Wiley & Sons Ltd. Published 2020 by John Wiley & Sons Ltd.

10.1 Introduction

Sunlight is the main source of energy on Earth. The idea of harnessing solar energy for practical applications goes back to prehistoric times. Lenses and concave mirrors were used by ancient civilizations to make glass from sand, melt metals, or simply initiate fire by concentrating sunlight. In the eighteenth century, solar furnaces were constructed using polished steel to melt iron, copper, and other metals. French scientist Antoine Lavoisier reached temperatures as high as 1750 °C by focusing sunlight through lenses (Kalogirou 2004).

Today, a broad range of practical applications use solar energy for either direct heating or electric generation. Modern architectural approaches have also evolved for natural heating and lighting of passive solar buildings. This chapter will focus on active conversion systems that capture solar energy by solar collectors and photovoltaic (PV) cells.

Based on recent surveys, approximately 54% of the solar energy conversion capacity installed worldwide is used for electric generation (see Figure 5.2). The larger fraction of electric generation is currently based on direct conversion of solar radiation into electric power by solid-state PV technologies. Progress of PV cell technologies has significantly reduced the cost and efficiency of PV generation systems, making them a favorite choice for distributed electric generation. On the other hand, concentrated solar power (CSP) plants are emerging in utility-scale electric generation. CSP facilities capture solar energy to produce steam and generate electricity using conventional turbine-generator units. With massive energy storage capability, CSP technologies present a potential alternative to fossil fuel burning electric generation.

About 46% of the total solar capacity in the world directly produces thermal energy for a broad range of applications. In the residential sector, solar thermal systems are mainly used for space, water, and pool heating. Agricultural applications include greenhouse heating, drying, and produce processing. Desalination, distillation, process heating, and metal melting are among major solar thermal applications in the industrial sector.

As discussed in Chapter 5, the running expenses of solar energy conversion systems are extremely low since there is no cost associated with fuel supply and waste disposal. The maintenance cost of PV generation systems is negligible compared to conventional fossil fuel burning and nuclear power plants. The maintenance cost of a fixed PV array is limited to dust or snow removal and seldom for the replacement of broken parts. Solar thermal systems have slightly higher maintenance costs due to the mechanical equipment for circulation of the heat transfer fluid. The economic life of a typical PV module exceeds 25 years. CSP systems have the highest running costs among all solar conversion technologies because of the circulation of the working fluid, turbines, and generators. However, their operational costs are still lower than conventional power plants since the fuel, transport, and waste disposal expenses are zero.

Solar energy is clean, globally available, abundant in many locations, and it does not vanish by conversion to other forms. Therefore, using the solar potential wisely is essential for the sustainable development of modern society. The largest fraction of the investment for solar energy development is the initial cost. Design mistakes generally result in considerable investment loss and disappointment, which leads to the loss of public confidence in solar power.

This chapter introduces fundamental concepts regarding solar radiation and trigonometric relations used for the estimation of solar energy that can be captured by a solar collector or PV module. The operation principles and types of solar collectors are described. A brief introduction to solid-state physics principles relevant to PV energy conversion continue with a comparison of major PV cell technologies. The chapter concludes with design methods and evaluation of solar energy conversion systems.

10.2 Solar Radiation

Solar energy is the result of fusion reactions on the Sun. As briefly explained in Chapter 4, at extremely high temperature and pressure in the plasma that forms the mass of the Sun, hydrogen nuclei fuse into a helium nucleus releasing energy due to the decrease of total mass. As a result, the Sun loses about 4.3 million tons of its mass every second and releases a power of 3.845×10^{26} W. This energy radiates in all directions through the universe. Energy of the Sun reaches the Earth through space in the form of radiation.

10.2.1 Solar Constant

The Sun can be considered as a perfect radiation emitter (or black body) at a temperature around 5500–5800 K. The average energy flux incident on a unit area perpendicular to the sunbeam is called irradiance. The mean solar irradiance per unit time per unit area on a surface perpendicular to the rays of the sunlight in the absence of the Earth's atmosphere is defined as *solar constant* and often represented by "SC" (Mecherikunnel and Richmond 1980). Early NASA estimates, based on terrestrial measurements made at different solar zenith angles and extrapolated to the top of the atmosphere suggested values between 1338 and 1368 W/m^2 (Messenger and Ventre 2004). NASA/ASTM adopted 1353 W/m^2, which is the average of estimated irradiances and has an uncertainty of $\pm 1.5\%$. Some authorities consider a value as high as 1377 W/m^2 for solar constant.

Extraterrestrial irradiance depends on several factors, including the distance of the Earth to the Sun that changes by the day of the year, and nuclear activities known as solar storms that cause darker or brighter spots on the Sun's surface. For a given day n starting on January 1st, the extraterrestrial irradiance I_{ET} can be approximated by Eq. (10.1) where SC represents the solar constant (Duffie and Beckman 2013).

$$I_{ET} = SC \left(\frac{r_0}{r}\right)^2 = SC \left(1 + 0.033 \cos \frac{360n}{365}\right) \tag{10.1}$$

As sunlight travels through the atmosphere part of its energy is absorbed by air molecules, water vapor, and particulate matter. Direct normal irradiance (DNI) is the solar radiant energy per unit time (radiant power) received on an area of one square meter orthogonal to the Sun's beam at a location on Earth. Also known as direct beam irradiance, DNI is generally measured in watts or kilowatts per square meter. The difference between the extraterrestrial irradiance (solar constant) and DNI is equal to the losses that occur as the beam passes through the atmosphere. Therefore, DNI changes during the day depending on the position of the Sun in the sky and the position of the Earth on its orbit. Figure 10.1 illustrates the sunbeam crossing the atmosphere at summer solstice (June 21st) and reaching the

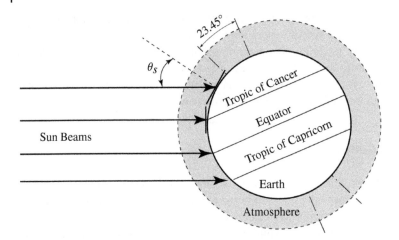

Figure 10.1 Sunbeams reaching the Earth on summer solstice (approx. June 21).

Tropic of Cancer at a 90° angle. At locations on the Northern Hemisphere, DNI is maximum at noon on the day of summer solstice, which is around June 21st, depending on the year.

The beams that are scattered in the atmosphere by clouds, fog, or haze cause diffuse horizontal irradiance (DHI) on a horizontal surface also expressed in kW/m². DHI does not depend on the position of the Sun during the day but it changes with the opacity of the air, cloudiness, precipitation, dust, smoke, fog, pollution, and other atmospheric factors. In measurement of the DHI the radiation emitted from the Sun disk (circumsolar radiation) is subtracted from the radiation diffused in the atmosphere.

Direct sunbeams reflected by the ground and surrounding surfaces may also reach a solar collector or PV module. The term "albedo" refers to the radiation that is reflected from ground-level surfaces to a point of interest. Albedo depends on the reflectance of the ground and other surfaces around the point of interest.

Global horizontal irradiance (GHI) is the total radiant power received from the Sun on a horizontal surface on Earth. It is calculated by adding DHI and the component of the DNI perpendicular to the horizontal surface. Given the angle of the incident sun beam with the line perpendicular to the surface (normal direction) θ_z the direct horizontal irradiance I_H can be calculated using Eq. (10.2).

$$\text{GHI} = \text{DNI} \times \cos(\theta_z) \quad [\text{W/m}^2] \tag{10.2}$$

In Eq. (10.2), DNI represents the incident beam irradiance (Direct Normal Irradiance) on the surface. For many locations around the world hourly solar data is available in databases and online calculators provided by national and international institutions such as National Renewable Energy Laboratory (NREL), National Oceanic and Atmospheric Administration (NOAA), and European Council Science Hub (EU-PVGIS). Several agencies have developed online tools for estimation of solar energy potential based on satellite data, terrestrial data collections, and reanalysis (see suggested web sites at the end of this chapter).

Solar energy captured on a horizontal surface is the integral of the irradiance over a certain interval and is called irradiation or insolation, and generally expressed in Wh/m².

Typical integration periods are hour, day, or month. Insolation changes throughout the year with the length of daylight time. The average of daily irradiations over one year allows a rough estimation of the yearly total solar energy received at a location. Yearly total insolation depends on the latitude and clarity of the sky at a particular location.

The ratio of the GHI at a specific location to the extraterrestrial irradiance obtained from Eq. (10.1) is defined as a dimensionless parameter called the clearness index.

$$K_T = \frac{GHI}{G_E} \qquad (10.3)$$

In Eq. (10.3) K_T describes the average attenuation of solar radiation by the atmosphere at a site over a given period, typically a month.

10.2.2 Effect of Clear Atmosphere on Solar Radiation

Solar irradiance reaching the Earth's surface is highly variable depending on the distance the beam travels in the atmosphere, pressure, humidity, cloudiness, and air quality. The effect of a clear atmosphere on sunlight is characterized by the concept called *air mass*, which represents the relative length of the solar beam path through the atmosphere; air mass is unity, abbreviated as AM1, when the sun is at its highest position in the sky on a clear day at sea level and under normal physical conditions. At other times of the day air mass is the inverse of the cosine of zenith angle ($1/\cos\theta_z$), Extraterrestrial air mass denoted by AM0 is relevant for satellite applications. AM1.5 is considered as the typical irradiance on Earth's surface on a clear day corresponding to the maximum solar irradiance of 1 kW/m^2.

Figure 10.2 shows the solar radiation spectrum for direct light at the top of Earth's atmosphere and at sea level. The sun produces light with a distribution similar to what would be expected from a 5778 K (5505 °C) blackbody, which is approximately the sun's surface temperature. These curves are based on the American Society for Testing and Materials (ASTM) Terrestrial Reference Spectra, which are standards adopted by the PV industry to ensure consistent test conditions, and are similar to the light that could be expected in North America. Regions for ultraviolet, visible, and infrared light are indicated on the figure.

The solar irradiance received at ground level may be higher than 1 kW/m^2 due to reflections (like on snow-covered ground on a clear day). Throughout the year the available maximum irradiance at a location varies considerably due to the rotation of the Earth, weather conditions, and air pollution.

10.2.3 Solar Geometry

Since Copernicus' (1473–1543) discovery, it is well known that Earth rotates around the sun. The Earth's orbit is elliptic but close to a circle, such that it can be considered a circle for solar calculations. The rotation axis of the Earth is tilted by 23.45° to its orbital plane, called an ecliptic plane and shown in Figure 10.3. The incident angle of the sunlight on a horizontal surface on Earth changes throughout the year, resulting in seasonal changes. Approximately on June 21 the beams of the sun are perpendicular to 23.45° North latitude, known as Tropic of Cancer (see Figure 10.1). Around December 21 the beams are perpendicular to the Tropic of Capricorn at 23.45° south latitude. Twice a year, day and night durations

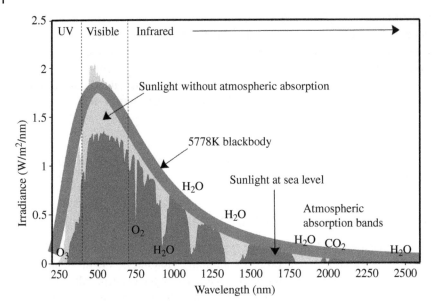

Figure 10.2 Effects of the atmosphere on the solar spectrum based on ASTM. Source: © Wikimedia Commons share alike. Available at https://commons.wikimedia.org/wiki/File:Solar_spectrum_en.svg#filelinks Accessed on October 20, 2019.

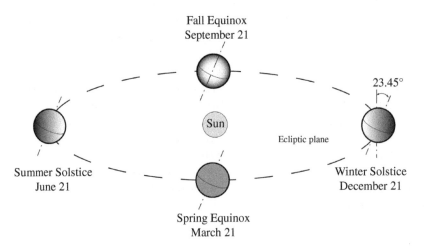

Figure 10.3 Rotation of Earth around the Sun (winter and summer solstices for the Northern Hemisphere).

are equal on the *vernal equinox* (March 21) and the *autumnal equinox* (September 21). At both equinoxes the beams of the Sun are perpendicular to the equator.

The solar declination angle, defined as the angle between the equatorial plane and the line drawn from the center of the Earth to the center of the Sun, changes throughout the year between −23.45° and +23.45°. Exact values of declination, which varies from year to year, are calculated and published in astronomical publications such as The American

Ephemeris and Nautical Almanac. The simple trigonometric expression below where the spring equinox is assumed to occur on day 81 (March 1st) is a sufficient approximation for solar energy calculations. In Eq. (10.4) n denotes the number of the day starting from January 1st and declination angle is in degrees.

$$\delta = 23.45 \sin \left[\frac{360}{365} (n - 81) \right] \tag{10.4}$$

Equation (10.4) is approximate because Earth completes one rotation around the sun in 365.25 days. The calendar is adjusted every four years (leap year) on February 29 which shifts the vernal equinox one day.

As previously noted, the solar declination angle δ changes throughout the year between −23.45° and +23.45°. It is zero at the vernal equinox (20/21 March) and autumnal equinox (22/23 September). In summer solstice (21/22 June) the declination is +23.45° and in winter solstice it is −23.45° (21/22 December).

While Figure 10.3 illustrates the rotation of Earth around the Sun, it is not convenient for solar energy calculations. To estimate solar energy that can be captured on a surface, we need to determine the position of the Sun in respect with the Earth. For solar energy calculations, it is common to consider the Sun moving relative to a solar module on Earth as shown in Figure 10.4.

In the Northern Hemisphere, true south is the orientation where the sun reaches its highest point at solar noon. Since a magnetic compass needle aligns with the magnetic field of the Earth, magnetic north and south indicated by a compass are slightly different than true south. The deviation of magnetic south from true south is called magnetic declination and changes by location. Magnetic declination at a certain location also slightly changes from one year to another but this change is negligible in practice. The compass reading must be corrected considering the magnetic declination at the location. Maps and calculators to find magnetic declination at a given location, such as (NOAA 2019), are available online.

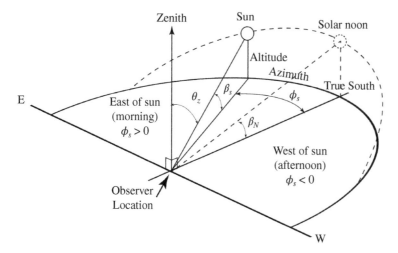

Figure 10.4 Solar angles relative to an inclined surface on the Northern Hemisphere (in Southern hemisphere, true south must be replaced by true north).

The Sun's apparent location east or west of true south is called the azimuth. Assuming that Earth completes one 360° revolution around its axis in 24 hours, the sun appears to move $\omega = \frac{(TLT-12.00)}{h} \times 15°$ degrees in one hour. The direction perpendicular to the horizontal plane at the observation point is the zenith. Zenith angle is the angle θ_z that complements β_s to 90°.

$$\theta_z = 90 - \beta_s \tag{10.5}$$

The angle β_s between the line defined by the sunlight beam and the horizontal plane is called the elevation or altitude of the sun.

10.2.4 Solar Time

Clock time is standardized to synchronize activities in a region and simplify the conversion between local standard times. The natural true time is indicated by the location of the Sun relative to an observer on the Earth. In solar energy calculations the position of the Sun and associated angles are determined according to the true solar time rather than the time clocks indicate.

True (or apparent) solar time is measured by the diurnal motion of the Sun. The time of day is expressed in terms of the hour-angle of the Sun westward from the meridian. Since the motion of Earth on its orbit is ecliptic and non-uniform, the rate of motion of the Sun in hour-angle is not constant. Hence, the solar noon is a function of the day throughout the year. Changes over longer periods than one year are negligible for solar calculations. Mean solar time is measured by the motion of a fictitious body called the mean Sim, which is supposed to move uniformly in the celestial equator. Mean solar time is uniform and regular in time; however, it cannot be determined by direct observation, but can be calculated indirectly by correcting the observed position of the Sun using the "Equation of Time," which gives the difference in hour-angle between the true Sun and the mean Sun (US Naval Observatory 2019). Deviation from the median solar noon for the day n can be calculated in minutes using the approximate equation of time below (Masters 2004).

$$E = 9.87 \sin 2B - 7.53 \cos B - 1.5 \sin B \text{ [minutes]}$$
$$B = \frac{360}{364}(n - 81) \text{ [degrees]} \tag{10.6}$$

At solar noon the Sun reaches the highest altitude β_N. At the latitude λ, the elevation of the Sun at noon is:

$$\beta_N = 90° - \lambda + \delta \tag{10.7}$$

The location of the Sun relative to an observer on Earth depends on the declination angle, latitude of the location, and time of the day. Solar time is generally expressed in terms of the deviation from the solar noon.

Since in one-hour Earth rotates $360/24 = 15°$, the angular deviation of the sun from solar noon (hour-angle) is calculated by multiplying the difference of the local true time and solar noon (both in hours) by 15°.

$$\gamma = \Delta T_s \times 15° \tag{10.8}$$

Table 10.1 Local time meridians for US standard time zones.

Time zone	Local time meridian (°)
Eastern	75
Central	90
Mountain	105
Pacific	120
Eastern Alaska	135
Alaska and Hawaii	150

In Eq. (10.8) ΔT_s is the difference of the true local time from solar noon. In the morning the difference is positive, in the afternoon it is negative. Since Earth rotates 15° per hour, rotation of one degree corresponds to φ minutes. Local longitude is obtained from GPS (global positioning system) coordinates. The local time meridian is identified from the standard time zone of the location. At Greenwich, UK both the local time meridian and local longitude are obviously zero. Local time meridians for the US standard time zones are given in Table 10.1.

$$T_s = T_c + 4 \times \text{(local time meridian − local longitude)} + E \qquad (10.9)$$

Once the true solar time is known at a given latitude λ the apparent position of the Sun for an observer on Earth can be defined by solar elevation and solar azimuth. Solar altitude (elevation) angle β is found from

$$\sin \beta_s = \cos \lambda \cos \delta \cos \gamma + \sin \lambda \sin \delta \qquad (10.10)$$

Solar azimuth ϕ_s is calculated as

$$\sin \phi_s = \frac{\cos \delta \cdot \sin \gamma}{\cos \beta_s} \qquad (10.11)$$

During spring and summer in the early morning and late afternoon, the Sun's azimuth may be more than 90° away from the true south. We can determine whether the azimuth is greater or less than 90° by using the equation below:

$$\text{If } \cos H \geq \frac{\tan \delta}{\tan \lambda} \quad \Rightarrow \quad |\phi_s| \leq 90° \quad \text{otherwise } |\phi_s| > 90°$$

10.2.5 Incident Solar Radiation on a Collecting Surface

Consider a surface inclined at an angle α relative to the horizontal plane as shown in Figure 10.5. The incidence angle θ_m between the line perpendicular (normal) to this surface and the direct solar beam can be calculated in terms of the solar azimuth, module azimuth, inclination angle, and elevation of the Sun by the trigonometric expression below.

$$\cos \theta_c = \cos \beta_s \cos(\phi_s - \phi_c) \sin \alpha + \sin \beta_s \cos \alpha \qquad (10.12)$$

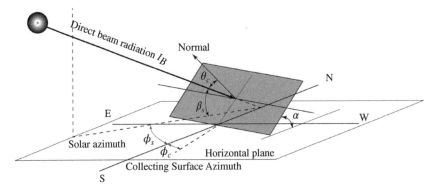

Figure 10.5 Incident solar radiation on an inclined surface.

The parameters of this expression are defined as:

θ_c: angle between direct beam and normal line to the collecting surface
β_s: elevation of the Sun
ϕ_s: azimuth of the Sun
ϕ_c: azimuth of the collecting surface
α: inclination of the collecting surface from the horizontal plane

The direct beam radiation incident on the module is found by multiplying the DNI at the location with the cosine of the angle of incidence expressed in Eq. (10.12).

$$I_{Bc} = I_B \cos \theta_c \tag{10.13}$$

CSP facilities and solar thermal units using parabolic collectors focus the beam radiation on a heat transfer element, hence diffusion and reflected radiation are not involved in such structures. However, non-concentrating solar collectors and PV cells collect the total solar radiation coming to their surface. Therefore, diffuse radiation I_{Dm} and reflected radiation I_{Rm} captured by the collecting element must be added to the direct beam radiation I_{Bm} to estimate the harnessed energy.

$$I_c = I_{Bc} + I_{Dc} + I_{Rc} \tag{10.14}$$

10.2.6 Estimation of Total Irradiance on an Inclined Surface

Total irradiance captured on a solar collector or PV module is the sum of direct beam, and diffused and reflected irradiances. If the total horizontal irradiance at the location is known, normal irradiance on the α-degree inclined surface can be found by substituting I_h in the trigonometric expression, Eq. (10.30), with the total value.

The most realistic estimation is based on long-term (at least one year) solar irradiance data collected on site. Solar irradiance is measured by using a calibrated pyranometer and a data logger. In practice, a typical data-sampling interval is 10-minutes. Most data loggers can record the average, standard deviation, maximum, and minimum values over the sampling interval. Long-term on-site data collection obviously requires specialized equipment and post-processing of the recorded data.

If a public database is available for the location, hourly irradiance data may be downloaded for use in Eqs. (10.12) through (10.14). For locations where reliable data is not available, direct beam radiation, diffuse radiation, and reflected radiation may be estimated as explained in the next section.

10.2.6.1 Estimation of Direct-Beam Radiation

Direct-beam radiation striking a surface is a function of the extraterrestrial radiation on the top of the atmosphere, and attenuation of the beam passing through the atmosphere. The American Society of Heating, Refrigerating, and Air Conditioning Engineers (ASHRAE) uses the empirical exponential decay model developed by (Threkeld and Jordan 1958). Based on collected data for a moderately dusty atmosphere with atmospheric water content typical in the US and other similar regions, the direct beam radiation can be estimated as

$$I_B = Ae^{-km} \qquad (10.15)$$

where A is the apparent extraterrestrial flux in W/m^2, and k is a dimensionless factor called optical depth. Given the solar elevation angle β_s, air mass ratio is calculated as

$$m = \frac{1}{\sin \beta_s} \qquad (10.16)$$

From the values suggested in ASHRAE (1993) the A and k values can be calculated using the approximations in Eqs. (10.17) and (10.18) (Masters 2004, p. 412).

$$A = 1160 + 75 \sin \frac{360(n - 275)}{365} \quad W/m^2 \qquad (10.17)$$

$$k = 0.174 + 0.035 \sin \frac{360(n - 100)}{365} \qquad (10.18)$$

In both equations n is the day number starting from January 1st.

10.2.6.2 Estimation of Diffuse Radiation

Diffuse radiation depends on many factors including moisture and solid particles in the atmosphere. It is generally assumed to arrive to a collecting surface from all directions with equal intensity. The model adopted by ASHRAE suggests that diffuse radiation is proportional to the direct-beam radiation I_B independent of the position of the Sun.

$$I_{Dh} = CI_B$$

$$C = 0.095 + 0.04 \sin \frac{360(n - 100)}{365} \qquad (10.19)$$

An inclined surface is exposed to a part of the diffuse radiation depending on the angle it is facing the sky. Considering that a surface perpendicular to the horizontal ground faces only half of the sky, total diffuse radiation captured by a surface inclined by α-degree is approximated as:

$$I_{Dc} = I_{Dh}\frac{1}{2}(1 + \cos \alpha) \qquad (10.20)$$

10.2.6.3 Reflected Radiation

Reflected radiation depends on the reflectance ρ of the ground and various reflective surfaces around a collecting surface. Ground reflectance ranges from 0.1 (dark colored pavement and roofs) to 0.8 (snow covered ground). For ordinary grass and vegetation, it is around 0.2. While on some bright days with snow-covered ground the reflected radiation may be considerably high, in general its effect on the total captured radiation is negligible in most cases.

Since both the direct and diffuse radiation bounce from reflective surfaces, the total reflected radiation can be approximated with similar assumptions to the diffuse radiation as:

$$I_{Rc} = \rho(I_{Bh} + I_{Dh})\frac{1 - \cos\alpha}{2} \qquad (10.21)$$

10.2.7 Solar Array Orientation

Both solar collectors and PV modules assembled in arrays must be oriented to capture the maximum solar energy. In the Northern hemisphere solar arrays oriented to true south can capture maximum direct radiation if the tilt angle is selected properly. In the Southern hemisphere, the best orientation is obviously true north. True south and true north differ from the magnetic south and north depending on the location. Westerly and easterly directions reduce the captured energy and diffuse radiation is practically the same for all directions.

In order to capture the maximum direct radiation, the array surface must always be perpendicular to the solar rays. In practice, solar arrays are either installed on the roof of a building, a pole, or on the ground. Roof-mounted arrays are limited to the orientation and inclination of the roof. Pole- or ground-mounted arrays can be inclined at a particular angle to optimize the energy capture over the entire year or certain seasons.

Two-axis solar trackers continuously follow both the azimuth and solar elevation angles. One-axis trackers follow only daily variations of the azimuth angle. Solar trackers maximize the energy output over one day and one year for an additional initial cost. In earlier years of the PV industry solar trackers were more effective to reduce the breakeven time of the initial investment. Dropping costs of PV modules per unit installed power has reduced the demand for solar trackers, since increasing the number of modules is often less expensive than using a solar tracker. Solar trackers may reduce the snow cover but at the same time add mechanical parts that increase the risk of failure and maintenance costs.

Fixed arrays may be optimized for maximum energy capture over the year or a certain part of the year. A fixed solar array with a tilt angle approximately equal to the latitude of the location captures the maximum total energy over the entire year. Since in summer the Sun is higher, energy capture may be maximized by decreasing the tilt angle 15°. The energy output can be maximized in winter months by increasing the tilt angle 15° above the latitude. Some fixed arrays may be designed to allow manual adjustment of the tilt angle for seasonal changes.

10.3 Solar Thermal Energy Conversion

10.3.1 Solar Collector Types

Modern solar collectors can be grouped in two categories as non-concentrating and concentrating collectors. Non-concentrating collectors simply absorb the solar radiation captured on their surface. Three essential types of solar collectors are flat panel, evacuated tube, and compound parabolic collectors (CPC). Concentrating collectors focus solar radiation on a smaller area using optical elements such as mirrors, lenses, curved reflectors, or their combinations to reach higher temperature levels. All types of solar collectors may be installed on a fixed array or mounted on solar trackers to maximize the captured energy.

Flat panel collectors (FPC) have an absorbing surface and metal tubes enclosed in a thermally insulated casing. Solar radiation passing through the transparent cover heats the dark colored plate called an absorber. Tubing where the heat transfer fluid circulates is either an integral part of the absorber or rigidly attached to it for best heat conduction. The fluid may be air or liquid. Water containing a certain amount of glycol to prevent freezing is a typical heat transfer fluid for FPC systems. The transparent cover has high transmittance for the solar radiation but low transmittance (practically zero) for long wave thermal radiation (red and infrared) emitted by the absorber. The cover also insulates the absorber and tubing by reducing the convection losses by restraining the stagnant air layer in the casing. Transparent cover is usually glazed and tempered glass to provide mechanical strength to withstand hail, ice, and falling objects. The back side of the FPC is insulated from the surroundings. Conventional flat plate collectors perform better in sunny and warm climates. Their performance is greatly reduced in cold, cloudy, and windy days because heat losses are larger at lower ambient temperatures.

Evacuated tube collectors (ETC) have a collector enclosed in a vacuum-sealed transparent tube similar to fluorescent light bulbs. The external tube has a diameter of a few inches. A smaller diameter metal pipe is attached to collector fins covered with a special layer of coating for better absorption of radiant energy. The small amount of fluid in the metal pipe is not circulated to the outside, instead it undergoes an evaporating-condensing cycle, and the vapor moving to the heat sink releases the latent heat as it condenses. Methanol is a commonly used fluid. A bank of individual ETCs are assembled on a module where they exchange produced heat to the external heat transfer fluid that circulates through the heating system.

ETC has several advantages over FPC particularly in colder climates. The vacuum envelope reduces the convection and conduction heat losses. ETC can operate at higher temperatures. In addition, individual tubes may be replaced, or the thermal system can be expanded as needed.

CPC reflect the incident sunlight to the absorber. This type of collector is composed of two parabolic sides facing each other. Incident sunlight coming from different directions within certain angle limits is directed to the absorber by internal reflections. Thus, seasonal and daily changes of solar orientation have a smaller effect on their conversion efficiency. CPC can be combined with ETC to increase the efficiency.

Solar collectors are installed facing the equator with a fixed angle inclination from the horizontal plane for optimal exposure to the direct solar irradiance. For an inclination angle equal to the latitude, a solar panel delivers maximum energy over the year. Because the Sun is higher in summer, an inclination of below the latitude (typically –15°) results in higher energy output in summer months. On the other hand, an inclination above the latitude (typically +15°) delivers more energy in winter months when sun is lower.

Concentrating solar collectors focus sunlight reflected from a larger surface on a small area to increase the temperature and reduce the losses. Extremely high temperature levels reached by concentrated solar collectors allow melting metals or the production of high temperature steam or gases to power heat engines for electric generation. Higher temperatures reached by using a molten metal or salt as the working fluid increases the thermodynamic efficiency of steam or gas turbines. Concentrated collectors have higher thermal efficiency because the heat losses are smaller on the absorber. Larger reflecting surface area can be obtained by installing a large number of mirrors mounted on individual solar trackers (see Case Study, Crescent Dunes CSP generation plant).

10.3.2 Solar Collector Performance and Efficiency

Under steady-state conditions, the useful heat delivered by a solar collector is equal to the radiant heat reaching its surface heat minus total loss. We write the expression of the useful energy delivered by the collector per second in Eq. (10.22) where the total mass flow of the heat transfer fluid in the collector is m_f (kg/s), its specific heat is c_f, temperature of the entering fluid is T_i, and temperature of the fluid exiting the collector is T_o.

$$q_u = m_f c_f [T_o - T_i] \tag{10.22}$$

If the incident solar irradiance is G_t (W/m^2), then the energy available for conversion to heat on the solar panel with surface area A_c is $q_c = G_t A_c$. Hence, the efficiency can be expressed as in Eq. (10.23).

$$\eta_c = \frac{q_u}{q_c} = \frac{m_f c_f}{q_s A_c}[T_o - T_i] \tag{10.23}$$

Performance of a solar panel can be analyzed using structural data and physical properties of its elements. Absorptivity, transmittance, and area of the absorber are essential parameters to calculate the total useful energy that a collector can absorb from available solar irradiance.

Absorptivity is the fraction of the total solar energy that the collector can absorb. Given the absorber transmittance $\alpha\tau$, total energy absorbed by a collector with total area A_c will be $A_c G_t \tau\alpha$. On the other hand, if the heat transfer loss coefficient of the collector U_L (W/m^2 °C) is known, for average plate temperature T_p and ambient temperature T_a, the total loss can be calculated as $A_c U_L(T_p - T_L)$. Equation (10.24) describes the useful energy delivered by a flat plate collector.

$$q_u = A_c[G_t \tau\alpha - U_L(T_p - T_a)] \tag{10.24}$$

Temperature of the heat transfer fluid that enters the panel T_i is more useful in practice than the average plate temperature. By using a correction factor defined as heat removal factor F_R the average plate temperature can be replaced with T_i. F_R depends on the fluid type and flow rate.

$$q_u = A_c F_R[G_t \tau\alpha - U_L(T_i - T_a)] \tag{10.25}$$

Dividing both sides by $G_t A_c$ the efficiency expression is obtained in Eq. (10.26).

$$\eta = F_R \tau\alpha - \frac{F_R U_L}{G_t}(T_i - T_a) \tag{10.26}$$

For incident angles from the normal to the panel surface below 35°, the product of absorptivity and absorber transmittance $\alpha\tau$ is approximately constant (Kalogirou 2004). Assuming that the incident angle is small enough and U_L remains constant, Eq. (10.26) is linear with a negative slope $-F_R U_L/G_t$.

In reality, the heat loss coefficient is a function of the difference between the collector inlet and ambient temperatures. In addition, energy conversion in evacuated tube panels (ETP) involves both convection and conduction heat transfer as well as the latent heat of the working fluid. The coefficients of the quadratic function, Eq. (10.27), can be estimated experimentally for all types of solar panels. ASHRAE Standard 93:1986 is one of the information sources for testing and performance evaluation of flat-plate and concentrating collectors.

Panel efficiency can be approximated using quadratic function as:

$$\eta = F_R \tau \alpha - \frac{c_1}{G_t}(T_i - T_a) - \frac{c_2}{G_t}(T_i - T_a)^2 \tag{10.27}$$

If we define $c_o = F_R \tau \alpha$ and $x = T_i - T_a$, Eq. (10.28) can be used to calculate the efficiency of a non-concentrating solar collector.

$$\eta = c_o - \frac{c_1}{G_t}x - \frac{c_2}{G_t}x^2 \tag{10.28}$$

For concentrating collectors, expression of efficiency is modified by including the optical efficiency η_o, concentration factor C, and beam irradiance G_b.

$$\eta = F_R \eta_o - \frac{c_1}{CG_b}(T_i - T_a) - \frac{c_2}{CG_b}(T_i - T_a)^2 \tag{10.29}$$

By denoting $k_o = F_R \eta_o$ $k_1 = \frac{c_1}{C}$ $k_2 = \frac{c_2}{C}$ $y = (T_i - T_a)$, the efficiency of a concentrating solar collector can be expressed as:

$$\eta = k_o - \frac{k_1}{G_b}y - \frac{k_2}{G_b}y^2 \tag{10.30}$$

10.4 Photovoltaic Energy Conversion

As the name implies, PV energy conversion transforms radiant energy of light directly into electricity. PV generation is based on energy exchanges between light and electrons in the crystalline structure of a semiconductor junction, thus it does not involve any moving pieces. Energy conversion occurs in PV cells. In practice several PV cells are connected in series and parallel to form a PV module with a certain voltage, current, and power ratings.

10.4.1 Structure of Silicon Crystal

Several semiconductor materials such as silicon (Si), cadmium telluride (CdTe), and gallium arsenide (GaAs) are currently used in the fabrication of PV cells. Polymer and organic PV cell technologies are also emerging. Resistivity of a pure semiconductor material is much higher than conductors but smaller than insulators. The resistivity changes drastically as an impurity is added. Properties of various semiconductors used in PV cell fabrication are presented in detail in Markvart and Castaner (2003). In this text we will limit our discussion to silicon, which is the most commonly used material in PV industry.

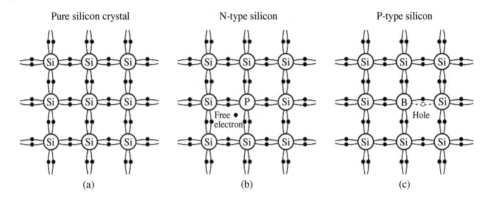

Pure silicon crystal   N-type silicon   P-type silicon

(a)   (b)   (c)

Figure 10.6   Schematic diagram of (a) pure, (b) N-type, and (c) P-type silicon crystal structures.

Silicon has four valence electrons, which hold the atoms together. The crystal structure of silicon has a cubic structure where each atom is tied to the neighboring atoms with the four valence electrons. Figure 10.6(a) is a simplified two-dimensional diagram of the pure silicon crystal.

If pure silicon is deliberately contaminated by a trace amount of phosphorus, which has five valence electrons, the crystal will contain an excess of free electrons. In solid-state physics terminology, the word *doping* refers to such intentional contamination processes. A silicon crystal that has excess electrons is called N-type. The free electron, if stimulated by an electric field, can move from one atom to another creating electric current. Crystal structure of an N-type silicon crystal is illustrated in Figure 10.6(b).

Pure silicon crystal doped with an element that has three valence electrons, such as boron, becomes P-type because it lacks electrons to form the even lattice structure of the pure crystal. As shown in Figure 10.6(c), a boron atom shares two electrons with three of its neighbor silicon atoms and only one electron with the fourth. The term *hole* signifies for a missing electron in solid-state electronics.

10.4.2   Operation of a PV Cell

Neither a pure, N-type, or P-type silicon crystal alone has a practical use in electronics. However, a junction of N- and P-type crystals forms a diode used in many electronic circuits. An N-type crystal sandwiched between two P-types or a P-type sandwiched between two N-types form a *bipolar junction transistor* (BJT), the pioneer of solid-state electronic elements. Various combinations of P–N junctions developed since the invention of the BJT have specific uses in electronics devices.

A PV cell is essentially a P–N junction and has many common features with semiconductor diodes. Structurally, however, a PV cell is made to absorb sunlight and produce electric current. Therefore, sunlight must penetrate into the crystal structure as much as possible, with minimal reflection. The surface area of a PV cell is much larger than a typical diode used in electronic circuits. Since one side of the cell is exposed to external light, it must be protected from contamination by a transparent film. PV cells are very delicate, thin elements. A PV module that combines series and parallel PV cells is enclosed in a sturdy aluminum frame and a durable highly transparent sheet.

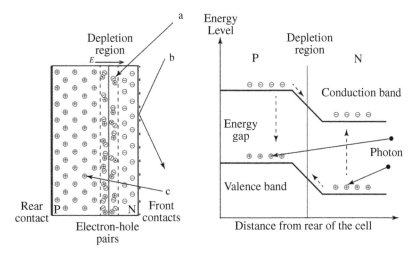

Figure 10.7 Photon penetration and energy levels in a P–N junction.

The front layer of the junction is a thin N-type crystal, the opposite layer is a relatively thicker P-type crystal. Rear side of the PV cell is covered by an aluminum sheet, which connects one end of the semiconductor to the external circuit. The top surface cannot be covered with a conductor layer since it would block sunlight. A grid of thin wires made from a good conductor, typically silver, is attached to the upper surface to conduct produced current while allowing sunlight to enter the junction. In addition, an anti-reflective coating reduces the reflection of the incident light.

Figure 10.7 illustrates schematically a P–N junction and corresponding energy levels. When a P–N junction is formed free electrons in the N region drift to the P region and combine with the holes to form electron-hole pairs around the junction. An electric field E is thus formed in the neutral area called the depletion zone. Free charge carriers cannot cross the depletion zone unless stimulated by an external energy source. A P–N junction supplied from an external source forms a diode. In semiconductor diodes a potential difference that creates an electric field large enough to overcome E can move electrons from the N region into the P region, which allows the flow of an electric current. A combination of P- and N-type semiconductors is often referred to as a P–N junction. A diode is forward biased when the voltage applied to the P–N junction is typically larger than 0.5–0.7 V depending on the material used. If the external potential difference is in the same direction as the internal field E, the depletion region becomes wider and current flow is blocked. In this case, the P–N junction is reverse biased.

In a PV cell, light entering the P–N junction stimulates charge carriers to create electric current if the junction is connected to an external circuit. Figure 10.7 illustrates three different cases for the light beam hitting the cell. The light beam marked "a" provides sufficient energy to move a charge carrier out of the depletion region, thus creating an electric current. The beam marked "b" is reflected on the cell surface and does not contribute to the energy conversion. The beam marked "c" enters the cell but does not have enough energy to overcome the electric field E. This type of beam can still form an electron-hole pair and produce heat in the cell.

In quantum theory, light is considered as a flux of particles named photons that carry radiant energy expressed as

$$E = h v = h\frac{c}{\lambda} \tag{10.31}$$

In the Eq. (10.31) h is the Planck constant (approx. $6.63 \cdot 10^{-34}$ J/s) and v is the frequency of the light wave corresponding to the photon.

Electrons shown in this matter are in one of several energy levels or *bands*. Valence electrons that are shared by adjoining atoms to hold the matter together are at a lower energy state known as the *valence band*. These electrons cannot move from one atom to another, therefore if all electrons are in the valence band, the substance cannot conduct electric current. Electrons in the higher energy level called a *conduction band* can move from one atom to another, thus conducting electricity. Electric conductors have many electrons in the conduction band while insulators have few or none. Doped semiconductor layers that form the junction have free electrons or holes separated by the depletion region. Valence and conduction bands are separated by an energy gap as shown on the right side of Figure 10.7. An energy gap is measured in electron volts ($1 \text{ eV} = 1.602 \cdot 10^{-19}$ J) and depends on the semiconductor material. Section 10.2 lists energy gaps of semiconductors commonly used in PV applications. If a photon entering the N-type layer of a PV cell has energy greater than the energy gap, it can promote (or excite) an electron from the valence band to the conduction band. The excess energy is converted to heat and dissipated to the surroundings as loss.

The spectral distribution of sunlight on the Earth's surface depends on the clarity of the air, elevation, and weather conditions. For maximum conversion of light into electric energy, the energy gap of the material used to make the PV cell must reasonably match the spectrum of the incident light.

Example 10.1 Compare the maximum wavelengths of the light spectrum that can produce electric current in a crystalline and amorphous silicon PV cell with the energy gaps given in Table 10.2.

Solution

Planck's equation will be used to calculate the photon energy as a function of wavelength. With $h = 6.63 \cdot 10^{-34}$ J/s and $c = 3 10^8$ m/s, we can write:

$$E_{photon} = \frac{hc}{\lambda} = \frac{(6.63 \cdot 10^{-34}) \times (3 \cdot 10^8)}{\lambda} = \frac{19.89 \cdot 10^{-26}}{\lambda} \text{ J}$$

Table 10.2 Energy gaps of some materials used in fabrication of PV cells.

Material	Energy gap (eV)
Crystalline Si	1.12
Amorphous Si	1.75
Cadmium Telluride (CdTe)	1.45
Gallium Arsenide	1.42

Source: (Markvart 2000).

Photon energy must be equal to the energy gap. Using $1eV = 1.602 \cdot 10^{-19}$ J to convert Joule to eV, we obtain the following maximum wavelengths for each material.

- Crystalline silicon:

$$\lambda_{max} = \frac{19.89 \cdot 10^{-26}}{1.602 \cdot 10^{-19} \times 1.12} = 1.109 \cdot 10^{-6} \text{ m} = 1109 \text{ nm}$$

- Amorphous silicon:

$$\lambda_{max} = \frac{19.89 \cdot 10^{-26}}{1.602 \cdot 10^{-19} \times 1.75} = 7.095 \cdot 10^{-7} \text{ m} = 709.5 \text{ nm}$$

We see that crystalline silicon PV cells can absorb a wider range in the solar irradiance spectrum compared to amorphous silicon PV cells, which can convert only the visible spectrum.

10.4.3 Output Characteristic and Delivered Power

As we explained, a PV cell is essentially a special type of diode which produces electric current from the incident light. Unlike the diodes used as electronic circuit elements, PV cells are active elements delivering energy. The output-voltage current characteristic of a PV cell is basically a diode characteristic shifted by the current created in the junction due to photon energy. The characteristic is reflected in respect to the voltage axis since the direction of the output current is reversed. The relationship between voltage V and the current I of an ideal PV cell is described by Shockley in Eq. (10.32).

$$I = I_{ph} - I_0 \left[e^{\frac{qV}{kT}} - 1 \right] \tag{10.32}$$

In this equation, I_{ph} represents the current created by photons entering the junction. The term subtracted from I_{ph} is the ideal diode equation where q is the charge of an electron $(1.602 \cdot 10^{-19})$, T is the absolute temperature, and k is the Boltzmann constant (approx. $1.38 \cdot 10^{-23}$ J/K). The diode reverse saturation current I_0 is typically extremely small compared to the short circuit current.

When the output terminals are short-circuited, the terminal voltage becomes zero and the current is approximately equal to the current produced by the photons entering the junction. I_{ph} is, therefore, approximately equal to the short circuit current of a PV cell, which can be easily measured by connecting an ammeter to the terminals when the cell is exposed to the sunlight. I_{ph} is a function of the irradiance incident to the cell. Usually this short circuit current is identified as 1 kW/m^2 solar irradiance orthogonal to the cell surface. Short circuit current of a PV cell is approximately proportional to the irradiance. If the short circuit current is known for an irradiance G_0, the short circuit current can be proportionally scaled for an irradiance G.

$$I_{sc}(G) = \frac{G}{G_0} I_{sc}(G_0) \tag{10.33}$$

In the open-circuit operation, no current is drawn from the cell. Open circuit voltage of an ideal cell can be obtained by equating the right side of Eq. (10.32) to zero. Since $I_{ph} \gg I_0$, the expression can be simplified as Eq. (10.34).

$$V_{oc} = \frac{kT}{q} \ln \left[1 + \frac{I_{ph}}{I_0} \right] \simeq \frac{kT}{q} \ln \left[\frac{I_{ph}}{I_0} \right] \tag{10.34}$$

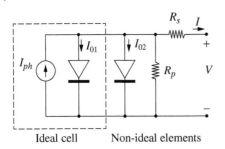

Figure 10.8 Equivalent circuit of a PV cell.

Ideal cell   Non-ideal elements

At first glance, the open circuit voltage seems to be directly proportional to absolute temperature. However, the reverse saturation current I_0 is highly temperature-dependent. Overall, open-circuit voltage will increase as the temperature decreases. For a typical cell, this decrease is about 0.5% per degree Celsius. Open-circuit voltage can be measured by simply connecting a voltmeter to the terminals of the cell exposed to sunlight. PV cells and modules are usually tested at 25 °C ambient temperature.

A practical PV cell can be represented by the equivalent circuit shown in Figure 10.8. The circuit enclosed with the dashed box corresponds to the ideal cell. Addition of a second diode, a parallel and a series resistance, reflect the non-ideal operating conditions. Series resistance represents the contact resistances of the front grid and back plate with the silicon layers. Surface effects at the cell edges result in an equivalent parallel resistance. Equation (10.35) is the modified expression of the open circuit voltage for a practical cell (Markvart and Castaner 2003).

$$I = I_{ph} - I_{01}\left[e^{\frac{V+IR_s}{kT}} - 1\right] - I_{02}\left[e^{\frac{V+IR_s}{2kT}} - 1\right] - \frac{V + IR_s}{R_p} \tag{10.35}$$

In practice, solar module specifications are given for standard test conditions (STC) defined with $1\,kW/m^2$ solar irradiance and 25 °C ambient temperature. Since the temperature of PV cells is generally higher when they operate at maximum power due to losses, STC may overestimate the actual performance. An alternative test standard called PTC (Photovoltaic Test Standard) is defined where the irradiance is again $1\,kW/m^2$ but the cell temperature (not ambient) is 25 °C. Cells are generally covered by an anti-reflective coating and a special laminate sheet to prevent degradation of the cell contacts. In addition, modules are enclosed in a metal frame, which provides mechanical strength. The PV cells mounted in a module are at a temperature different than the ambient temperature. Nominal operating cell temperature (NOCT) is defined as the temperature the cells would reach when they operate at open-circuit in an ambient temperature of 20 °C under solar irradiance of $G = 0.8\,kW/m^2$, and a wind speed less than $1\,m/s$. The cell temperature in degrees Celsius can be estimated using a linear approximation.

$$T_{cell} = T_{ambient} + \frac{NOCT - 20}{0.8}G \tag{10.36}$$

Voltage-current characteristic of an ideal PV cell is plotted in Figure (10.9). The curve is obtained using Eq. (10.32) for 25 °C cell temperature, $I_{sc} = 1A$ and $I_0 = 2\,nA$. Output power calculated as $P = VI$ is plotted on the same horizontal axis. The output power of a PV cell is zero at short- and open-circuit operations and has a maximum value for current and voltage values denoted by I_{mp} and V_{mp}.

Maximum power that a PV cell can deliver is the area of the rectangle defined by V_{mp} and I_{mp} lines. The ratio of the area of the maximum power rectangle to the area defined by I_{sc} and V_{oc} is defined as *fill factor*.

$$FF = \frac{V_{mp}I_{mp}}{V_{oc}I_{sc}} \tag{10.37}$$

The fill factor differs for various semiconductor materials and cell structures. A cell with a fill factor closer to unity produces more power for the same short circuit current and open circuit voltage. Therefore, the fill factor can be considered as a measure of the quality of a cell.

10.4.4 PV Technologies and Cell Efficiency

The performance of a PV cell depends on several factors including the solar spectrum that is absorbed, semiconductor material, junction structure, and internal resistances. PV cells are the building blocks of PV modules. Size and cost of the cell ultimately affects the overall cost of the PV generation unit. In practice, increasing the power delivered per cell area is desirable. One of the reasons is reducing the footprint of the generation system. The other reason is reducing the cost of structural materials that are needed to make a safe and reliable PV generation system.

Conversion efficiency of the first silicon cells developed in 1958 was about 11%. Thanks to the improvements in half a century, the conversion efficiency of commercial cells reached the level of 20%, while in the laboratory conditions up to 30% efficient cells have been tested.

Cost of unit delivered power is another factor in comparison of solar cells. The cost ratio of the first cells produced at the beginning of the PV industry was extremely high, above $1000/W. In the 1960s, PV cells were mostly used in spacecrafts, where solar power was the only energy source and cost was not a critical issue.

Various technologies have been developed to expand the region of the solar spectrum that can create current in the junction. As seen in Figure 10.2 and Example 10.1, the absorbed spectrum increases as the energy gap of the material decreases. Therefore, silicon crystal, which has the smallest energy gap among other listed materials can convert radiant energy over the wider region of the solar spectrum. In the quest to optimize the cost-effectiveness of PV products, compound materials such as gallium arsenide (GaAs), cadmium sulfide (CdS), and cadmium telluride (CdTe) were developed. Use of amorphous instead of crystalline silicon introduced new benefits and applications. Amorphous silicon absorbs a narrower range of the solar spectrum than crystalline silicon but absorbs light more effectively. Therefore, amorphous silicon cells can be thinner. Cells and modules made from amorphous silicon are also known as *thin film*. Although less efficient than crystalline cells, thin film technology allowed manufacturers of PV elements to deposit silicon film on a wide range of substrates to make rigid, flexible, transparent, curved, or fold-away modules.

New semiconductor materials and multi-junction cell technologies have improved the overall conversion efficiency by stacking two or more junctions with different energy gaps. The energy gap of amorphous silicon can be increased by inserting an alloy of intrinsic silicon with another element between the P- and N-type semiconductors. The resulting junction responds better to light at the blue end of the spectrum. In multi-junction cells, a wider energy gap junction is typically on the top absorbing higher-energy photons at the blue range of the spectrum, followed by other thin film junctions designed to absorb lower energy photons (Markvart and Castaner 2003).

Semiconductor materials and junction types are not the only ways to improve cell efficiency. Open-circuit voltage and short-circuit current affect the fill factor FF. Open-circuit voltage depends on the ratio of the photocurrent I_{ph} to the reverse saturation current I_0 as we see in Eq. (10.34). On the other hand, the short circuit current is approximately equal to the photocurrent. The fill factor can be maximized by reducing the reverse saturation current and optimizing the photocurrent. Reducing the series resistance and increasing the parallel resistance in the equivalent circuit shown in Figure 10.8 also significantly improves cell efficiency (Messenger and Ventre 2004).

10.5 PV Generation Systems

A PV unit is a solid-state generation system that consists of a PV array, electronic converters, switchgear, and protection elements. A PV generation unit may supply a standalone or grid-connected load. The installed capacity of a PV generation system ranges from a few watts to megawatts.

Standalone systems are typically used to supply buildings, facilities, or equipment at remote locations where a utility grid is unavailable. Portable PV units and pole-mounted standalone systems have installed power of a fraction of a kilowatt. Fixed standalone arrays mounted on the ground to supply a remote facility or building may have a capacity of several kilowatts.

Decreasing cost of PV modules and increasing energy prices motivated individuals to install grid connected (grid-tie) residential systems on their properties. Residential PV generation systems are generally mounted on the roof or a pole. The tilt angle of a roof-mounted solar array is fixed and its orientation and size are restricted by the architectural features of the building. Grid-connected (grid-tie) systems allow reducing the utility bill for a modest

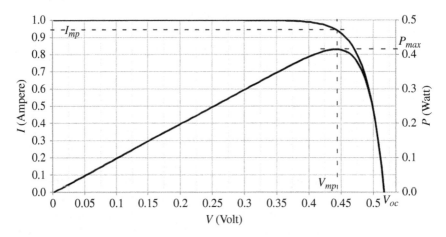

Figure 10.9 V-I characteristic and maximum power of an ideal PV cell.

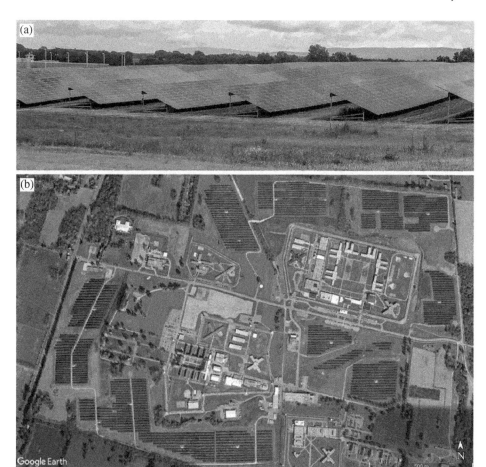

Figure 10.10 Solar farm near Hagerstown, Maryland, US. (a) Ground-mounted PV arrays. (b) Satellite view of the 1.1-MW capacity CES VMT solar farm operated by Consolidated Edison Solutions Inc. Source: Image captured from Google Earth on 6/29/2019.

initial investment. Simple grid-tie systems operate only when the utility power is available and then either supplement the energy received from the grid or supply the electric grid if excess energy is generated. A special net metering agreement is required for grid-tie genera-tion units. Battery backup adds the opportunity of electric generation when the grid power is lost and increases the reliability of electric supply in the building. Control of the energy penetration to the electric grid is an additional benefit of the battery backup small-scale PV generation units.

Utility-scale commercial solar generation stations are mounted on the ground. PV mod-ules are arranged in rows properly spaced to avoid self-shading. The land area covered by a typical solar generation unit is much larger than the footprint of a fossil fuel fired or nuclear-powered generation station. PV generation systems are modular and can be installed in a short time. The installation cost is lower compared to wind turbines since the structure does not need a deep foundation. Although PV generation does not cause pollution, the land used for utility scale solar farms are in general not usable for agricultural or other public purposes like wind farms. Abandoned strip mines, unused space around the highways, or restricted access safety and security areas may be suitable for commercial-size PV generation. Figure 10.10 shows a satellite and ground view of a 25-MW PV generation facility installed around the state correctional institution near Hagerstown, Maryland (US). The land covered by ground-mounted arrays cannot be used for commercial development or agriculture due to the security restrictions.

10.5.1 PV Generation System Configurations

The typical output voltage of a PV cell is in the order of 0.6–0.8 V, which is typical semiconductor diode voltage. PV modules, also called solar panels, combine PV cells (typically 60 cells) in a series and parallel configuration to generate an output voltage and power suitable for practical applications. PV modules are rated based on STC defined for $1 \, kW/m^2$ solar irradiance and 25 °C ambient temperature. The actual power output of a PV module changes in a broad range depending on the solar irradiance it receives.

A combination of PV modules forms a solar array with generation capacity equal to the total rated power of installed PV modules. Typical residential PV arrays have few kW generation capacity, depending on the generation goals of homeowners. Commercial PV arrays range from hundreds of kilowatts to several hundred megawatts. The actual electric generation of a PV array depends on many factors including the irradiation (insolation) at the location, tilt angle of the modules, orientation (azimuth), and shading.

Figure 10.11 shows schematic outlines of residential-scale standalone and grid-tie PV generation systems. Since PV array delivers direct current (DC) power at variable voltage and current depending on the solar radiation intensity, an electronic interface is required to adjust the output variables according to the load requirements. In a standalone system the interface is a charge controller to adjust the voltage and current based on the charging requirements of the battery type. If the system supplies alternating current (AC) load, an inverter must convert the DC voltage to AC at regulated voltage and frequency to standard values. In a grid-tie system, the inverter must precisely follow the voltage and frequency variations of the utility grid. In addition, utility companies have strict rules for the harmonic levels of the standalone systems. Inverters designed for grid-tie operation have a microcontroller that adjusts the voltage and frequency to match the grid values in real time.

Standalone units must have a battery storage since the generated power varies significantly with the captured solar radiation and no power is generated at night. In a grid-connected (grid-tie) system battery backup is optional. In many locations, the utility company may allow bidirectional energy flow with a net-metering contract. In such cases, any excess energy generated by the PV system supplies the grid and the value of that energy is credited to the consumer's utility bill.

Battery storage has benefits for both the consumer and the utility. At the consumer side, battery backup increases the reliability of electric supply and enhances energy security. From the utility standpoint, battery storage reduces the impacts of the variable electric generation on the grid performance. Smart-grid technologies monitor electric generation and

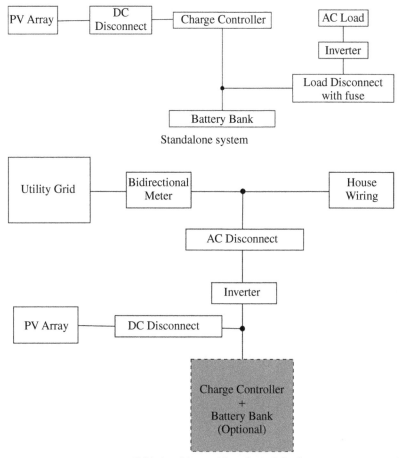

Figure 10.11 Residential scale PV generation system.

consumption throughout the distribution grid and adjust the energy flow to balance the supply and demand in real time.

Solar irradiance radiation data is available for many locations on Earth in official databases (NREL 2012).

10.6 Concentrated Solar Power

CSP plants generate electric energy indirectly from the solar heat. In a CSP plant sunlight is reflected by mirrors on a tower where a fluid is heated. The heat transfer fluid circulated through heat exchangers boils water and steam turbines drive electric generators. Such power plants can store solar energy as heat and continue electric generation at night.

Case Study 10.1

Crescent Dunes Concentrated Solar Power Plant (CSP)

Crescent Dunes CSP plant installed in Nevada (US) is the world's first utility-scale facility featuring an advanced molten salt power tower to generate electricity 24 hours a day using steam turbines. Figure 10.12 shows an aerial view of the 110-MW power plant.

Figure 10.12 Crescent Dunes 110-MW concentrated solar power (CSP) plant located in Nevada, US. Source: Courtesy of Solar Reserve (printed with permission).

Molten salt as both the heat transfer fluid and thermal energy storage medium provides efficient and cost-effective storage for solar energy. The tower height is 640 ft (195 m) total; with 540 ft (164.5 m) of concrete plus the 100-ft-tall receiver (30.5 m). The steam generation process is similar to the process used in conventional fossil fuel or nuclear power plants; the only difference is that the source of heat is solar energy. CSP with storage provides controllable and reliable capacity to meet demands throughout the day and night.

10.7 Chapter Review

This chapter presents fundamentals of solar energy conversion systems. Solar energy reaching the sun is harnessed either for heating or electric generation.

Solar thermal panels convert solar radiation into heat. Common types of solar thermal converters are flat plate, evacuated tube (ETC), parabolic, and concentrated solar collectors. Flat plate and ETC are mostly used for residential and commercial space or water heating. Concentrated solar collectors use optical elements to focus sunlight on a relatively small area to obtain high temperatures. Concentrated solar conversion systems are used for industrial applications including metal melting, steam generation, district heating, or electric generation.

Photovoltaic (PV) cells are solid-state elements that convert solar radiation directly into electric energy. PV cells are essentially a PN junction similar to a semiconductor diode but with a large surface exposed to sunlight. Silicon crystal, amorphous silicon, gallium arsenide, cadmium sulfide, and cadmium telluride are among semiconductor materials commonly used in PV cell fabrication. PV cells are combined in PV modules and PV modules are installed in series and parallel connections to form a solar array. With the decreasing cost of PV modules per watt, the installed power of PV generation systems has significantly increased during the last decades.

Review Quiz

1 Which one of the following makes the largest change on the short circuit current of a PV cell?
a. Temperature
b. Tilt angle
c. Latitude
d. Irradiance

2 Which one of the following makes the largest change on the open circuit voltage of a PV cell?
a. Irradiance
b. Temperature
c. Cell area
d. Angle of incidence

3 It is necessary to know the LOWEST possible temperatures at the site when sizing a PV system to prevent
a. over-voltage.
b. over-current.
c. under-voltage.
d. under-current.

4 In a PV inverter, what does the term MPPT mean?
a. Maximum power point tracking
b. Multiple power production technology
c. Maximum power processing technology
d. Minimum power penetration tracking

5 Which of the following factors is MOST LIKELY to affect irradiance?
   a. Longitude
   b. Latitude
   c. Module area
   d. Module tilt angle

6 Irradiance is a measure of solar power per unit of
   a. time.
   b. heat.
   c. distance.
   d. area.

7 Which cell technology has the HIGHEST efficiency?
   a. Single-crystal silicon
   b. Cadmium telluride
   c. Gallium arsenide
   d. Thin film (amorphous) silicon

8 In commonly available PV modules, what type of solar cell provides the most power in the smallest of space?
   a. Thin film
   b. Monocrystalline
   c. Gallium arsenide
   d. Cadmium telluride

9 Which of the following is the MOST CORRECT explanation of the PV effect?
   a. Heat from the sun gives electrons in silicon enough energy to escape their atomic bonds and flow, making an electric current.
   b. In a silicon wafer, "extra" electrons from atoms of phosphorus travel across the P–N junction to atoms of boron and fill in their "holes," creating an electric current.
   c. Particles of light (photons) penetrate into silicon, knocking loose electrons and giving them energy (voltage).
   d. Light causes the resistance in silicon to be reduced, thereby allowing electrons to escape from silicon atoms and flow.

10 What circuit arrangement is used to connect two PV modules to obtain a higher voltage?
   a. Series
   b. Parallel
   c. Delta
   d. Star

11 At a location with 40° latitude, which tilt angle will produce maximum energy in summer months?
   a. 40°

b. 55°

c. 25°

d. 60°

12 At a location with 35° latitude, which tilt angle will produce maximum energy in winter months?

a. 30°

b. 50°

c. 35°

d. 60°

13 Which conditions define Standard Test Conditions (STC)?

a. 1000 W output power and 25° cell temperature

b. 1000 W/m² and 25° ambient temperature

c. 1000 W/m² and 25° cell temperature

d. 800 W/m² and 25° cell temperature

14 Under what conditions can a PV module exceed its rated voltage output?

a. Hot, sunny summer day

b. Sunrise on a cold day

c. Cold, clear winter day

d. Sunset on a hot day

15 If the goal is to maximize the energy output during the winter months, what should the fixed array tilt angle be for a south-facing PV array?

a. Latitude angle +15°

b. Latitude angle –15°

c. Latitude angle

d. Latitude angle + azimuth

16 At which orientation a PV array produces maximum energy?

a. North

b. South

c. East

d. West

Answers: 1-d, 2-b, 3-a, 4-a, 5-b, 6-d, 7-a, 8-b, 9-c, 10-a, 11-c, 12-b, 13-c, 14-c, 15-a, 16-b.

Research Topics and Problems

Research and Discussion Topics

1 Compare economic and environmental impacts of replacing coal-burning power plants with PV generation.

2 Discuss the economic impacts of utility-scale PV development in populated areas.

3 What are the possible impacts of PV generation over energy independence of a country?

4 Discuss the potential benefits of concentrated solar power plants.

5 Hypothetically a PV generation plant installed on an area of 300 mi by 300 mi in a desert in the Western US can generate enough power to supply the whole country. Is such an idea feasible?

Problems

1 60 cells with the V-I curve shown in Figure 10.9 are connected in series in a PV module. Determine the voltage, current, and the power that this module can deliver when it operates at the maximum power.

2 The nameplate of a PV module gives the values below. Design an array of 1500 W generation capacity to be used with an inverter that has an input voltage range between 100 and 200 V without exceeding 600-V open circuit voltage.

3 At a location with latitude 40° north, solar irradiance is 800-W/m² at a certain time on a horizontal surface with 180° azimuth. Determine the irradiance on a 30° inclined surface with 160° azimuth.

4 A PV module has the specifications listed below. Calculate the total capacity of an array formed by connecting two series strings of five modules in parallel (total ten modules).
- Open circuit voltage: 56 V
- Short circuit current: 5 A
- Maximum power voltage: 52 V
- Maximum power current: 4–5 A

5 At a location, the average yearly solar irradiance is 4.5-kWh/day. Estimate the yearly electric energy generation of an array with the inclination equal to the latitude and total power capacity of 10 kW.

Recommended Web Sites

- American Solar Energy Society (ASES): www.ases.org
- National Oceanic and Atmospheric Administration (NOAA) Magnetic Declination: https://www.ngdc.noaa.gov/geomag/declination.shtml
- National Oceanic and Atmospheric Administration (NOAA) Solar Calculator: https://www.esrl.noaa.gov/gmd/grad/solcalc
- National Renewable Energy (NREL) Solar Research: https://www.nrel.gov/solar
- NREL National Solar Resource Database (NSRDB) International Database: https://nsrdb.nrel.gov/international-datasets

- NREL National Solar Resource Database (NSRDB) viewer: https://maps.nrel.gov/nsrdb-viewer
- NREL PVWatts Calculator: https://pvwatts.nrel.gov
- RETScreen Clean Energy Management Software by Natural Resources Canada (Retscreen Expert): https://www.nrcan.gc.ca/energy/retscreen/7465
- Solar Energy Industries Association (SEIA): www.seia.org

References

ASHRAE (1993). *ASHRAE Handbook: Fundamental*, 1993, ed. Atlanta, Georgia: American Society of Heating, Refrigeration, and Air Conditioning Engineers.

Duffie, J.A. and Beckman, W.A. (2013). *Solar Engineering of Thermal Processes*. Hoboken, NJ: Wiley.

Kalogirou, S.A. (2004). Solar thermal collectors and applications. *Progress in Energy and Combustion Science* 30: 231–295.

Markvart, T. (ed.) (2000). *Solar Electricity*, 2e. Chichester (West Sussex), UK: Wiley.

Markvart, T. and Castaner, L. (eds.) (2003). *Practical Handbook of Photovoltaics*. Amsterdam, Netherlands: Elsevier.

Masters, G.M. (2004). *Renewable and Efficient Electric Power Systems*. Hoboken, NJ: Wiley.

Mecherikunnel, A.T. and Richmond, J.C. (1980). *Spectral Distribution of Solar Radiation*. Greenbelt, MD: NASA.

Messenger, R.A. and Ventre, G. (2004). *Photovoltaic Systems Engineering*, 2e. Boca Raton, FL: CRC Press.

NOAA, 2019. *Magnetic Declination*. [Online] Available at: https://www.ngdc.noaa.gov/geomag/declination.shtml [Accessed 30 October 2019].

NREL, 2012. *Geospacial Data Science - Solar Data*. [Online] Available at: https://www.nrel.gov/gis/data-solar.html [Accessed 20 October 2019].

Threkeld, J.L. and Jordan, R.C. (1958). Direct solar radiation available on clear days. *Transactions - American Society of Heating Refrigeration and Air-Conditioning Engineers* 64: 45.

US Naval Observatory (2019). *The American Ephemeris and Nautical Almanac*. Washington DC: US Government Printing Office.

11

Energy Security

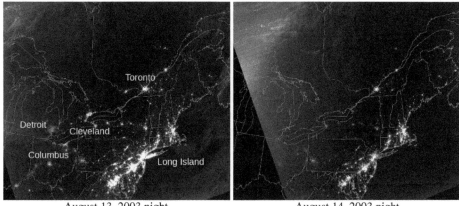

| August 13, 2003 night | August 14, 2003 night |

Photo: NASA Earth Observatory (public domain). Image courtesy of Chris Elvidge, US Air Force.

Satellite photo showing the US during the Northeastern blackout on August 13–14, 2003. A power line short circuit in Ohio was not reported to system operations control room due to a software bug. The cascaded failures caused power outages in the states of Ohio, Michigan, Pennsylvania, New York, Vermont, Massachusetts, Connecticut, New Jersey, and the Canadian province of Ontario. The outage affected an estimated 50 million people and 61,800 MW of electric load. The blackout began a few minutes after 4:00 p.m. Eastern Daylight Time (16:00 EDT), and power was not restored for four days in some parts of the United States. Parts of Ontario suffered rolling blackouts for more than a week before full power was restored. Estimates of total costs in the United States range between $4 billion and $10 billion (US dollars). In Canada, gross domestic product was down 0.7% in August, there was a net loss of 18.9 million work hours, and manufacturing shipments in Ontario were down $2.3 billion (Canadian dollars).

Source: U.S.-Canada Power System Outage Task Force Final Report (U.S.-Canada Power System Outage Task Force 2004).

Energy for Sustainable Society: From Resources to Users, First Edition. Oguz A. Soysal and Hilkat S. Soysal.
© 2020 John Wiley & Sons Ltd. Published 2020 by John Wiley & Sons Ltd.

11.1 Introduction

The term "energy security" was initially used in early the 1970s during the energy crisis, which particularly hit countries relying on foreign oil. The initial impacts of the oil crisis showed the vital importance of energy, especially for industrialized countries. Unprecedented energy shortages affected the economy, social life, public security, health services, and other essential services in many countries that rely on petroleum. The oil crisis triggered a rapid increase in prices of oil products, long lines at gas stations, shortages of basic products and services, and frequent power outages to conserve energy became usual during the energy crisis. Many Western countries started to change their energy policies, searching new sources and more efficient technologies to reduce their dependency on foreign petroleum.

The scope of energy security has changed over time covering a broader range of threats that might affect energy supply over a short or long term. Energy security is often seen as an "umbrella term" that covers many concerns linking energy, economic growth, and political power (Spagnol 2013).

As emphasized in previous chapters, energy is the most essential commodity for many sectors. Public security, emergency services, healthcare, national defense, and data transfer are among critical infrastructures that rely on uninterrupted energy supply. In extreme cases, energy interruptions and shortages significantly impact the economy and society. Unreliable supplies of essential goods and lack of public services deteriorates the welfare and quality of life.

The International Energy Agency (IEA) defines energy security as "uninterrupted availability of energy sources at an affordable price" (IEA 2014). In modern societies, all essential services, industrial productivity, transportation, commercial, and financial activities depend on continuous energy supply. Interruption of communication and computer networks can instantly cripple commerce at all levels. Even a few hours of electric power outage result in the interruption of important social and economic activities causing huge financial losses. If the electric outage lasts several days, refrigerated food stocks begin to perish, and food, water, and fuel shortages start. Energy security also requires a stable energy market and access of the users to affordable energy sources.

Supply-demand balance is the main factor that determines energy prices. A shortage of the energy supply affects not only the energy cost but the prices of all products and services. Economic stability and development of countries are therefore strongly linked to long-term energy security. A continuous and reliable supply of adequate amounts of energy is achieved by reducing the risk of energy deficits due to unavailability of resources, sudden fluctuations in supply-demand balance, or a disruption of the supply chain. Power system security can be also described as "the ability of the system to continuously fulfill its function against possible adverse situations" (Fulli 2016).

This chapter covers different aspects of energy security. Threats, targets, and mitigation are the main issues discussed in the following sections, with recent examples. Long-term energy security depends on future availability of energy sources. Vulnerability of the electric grid and critical infrastructure created the idea of energy independence, which has evolved to energy interdependence of nations. Concerns regarding climate change and deterioration of natural resources shape the development of energy systems, linking energy security to global sustainability.

11.2 Aspects of Energy Security

Different groups around the world view energy security from different perspectives. Consumers and energy-intensive industries are generally more concerned about disruptions of energy services and they desire reasonably priced energy to continue their functions. Energy exporting countries see energy products as a source of revenue and political power. Multinational oil companies profit from resource explorations and development of oil recovery facilities. Developing countries prioritize improvement of their energy supply capacity for economic stability and growth. Developed countries worry more about continuity and long-term availability of energy supply since any interruption or shortage can cause significant loss of production, which would propagate through the domestic and international markets.

Figure 11.1 illustrates the interactions of threats, targets, and mitigation measures. The duration and impacts of various events span the timeline ranging from fractions of a second

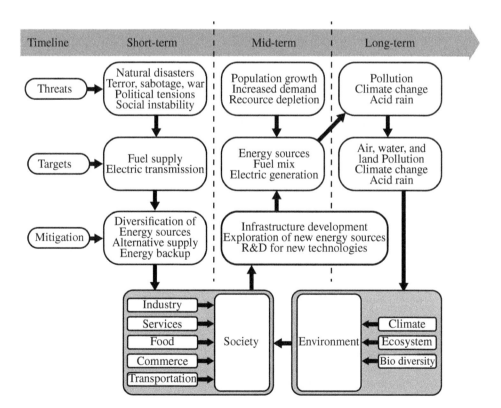

Figure 11.1 Energy security flowchart.

to generations. In this diagram, short-term refers to events up to a year, mid-term ranges from one year to several decades, and long-term covers up to several generations. The timeline may be further subdivided into shorter sections depending on the scope of analysis.

11.2.1 Types of Energy Security Concerns

Energy security concerns can be categorized based on the cause and duration of the possible interruptions of energy supply. Regular energy supply may be threatened by diverse factors and disrupted over a significantly wide range of time intervals from seconds to months, even years.

Primary sources reach consumers through transportation networks, electric transmission and distribution grids, and fuel delivery stations, which constitute a "supply chain." Even with abundant primary sources readily available, natural or artificial disasters can disrupt the supply chain by disabling fuel processing facilities, generation plants, fuel transport, or electric transmission. Hurricanes, major snowstorms, earthquakes, tsunamis, heat waves, and lightning strikes are typical natural disasters that affect continuous energy supply to a broad area. Artificial disasters are intentionally-created situations to disable energy supply. Terrorist attacks, strikes, embargos, economic sanctions, and a variety of other circumstances may create national or regional energy shortages.

Natural disasters and severe weather conditions have always presented the risk of energy supply discontinuities. Electric systems are designed to operate at a certain reliability level considering a reasonable risk of failure. Similarly, fuel supply also depends on the reliability of the transportation system and availability of alternative routes to deliver fuel to consumers. Several major hurricanes such as Katrina, Sandy, and Rita led to an integrated energy crisis in recent years, disrupting oil, natural gas, and electric supply to a large area.

Following the September 11, 2001 attacks in New York, Washington DC, and Maryland, terror became a major concern for energy security. Impacts of the energy crisis rippled through global economy with shortages of industrial products and significant price fluctuations.

Electric power is particularly important for continuity of critical functions in a modern society. Telecommunication networks, computers, industrial controls, security equipment, airport controls, and hospital emergency rooms are among mission-critical systems that rely on electric power. Supervisory control and data acquisition (SCADA) systems used in energy systems rely on continuous electric supply. Consequently, failure of the electric supply network also affects the process, transport, and delivery of other energy sources. The electric power grid is significantly vulnerable to a variety of natural or man-made disasters. Terrorist attacks, and computer viruses, spam, and other malware are among possible man-made disasters. Another threat for energy security is an "electromagnetic pulse" which may be caused by solar storms or detonation of a nuclear weapon at a high altitude.

Many catastrophic blackouts affected large populations worldwide for an extended time in recent years. Blackouts of Southern Brazil in 1999, Java-Bali in 2005, Northeast America in 2003, and Puerto Rico in 2018 were among historic electric outages affecting millions for several days. Most of the large-scale power outages result from cascading high voltage transmission line failures. Failure of a main feeder in a power grid may lead to failures of other transmission lines, which eventually shuts down a power grid supplying a large area.

As briefly summarized on page 437, the power grid in Northeastern America experienced one of the largest blackouts on August 14, 2003. About 50 million consumers lost power in Ohio, Michigan, New York, Pennsylvania, New Jersey, Connecticut, Massachusetts, Vermont, and Ontario. Some areas of the United States remained without electric power for four days and parts of Ontario experienced rolling blackouts for up to two weeks due to a generation capacity shortage. The major cause of the blackout was a computer failure, which prevented the system operators' awareness of a transmission line short circuit with overgrown trees (NERC 2004). In this event, cascading failures led to system instability that finally shut down power generation plants supplying a large region in the Northeastern US and Canada.

11.2.2 Short-term Energy Security

Short-term energy security is concerned with energy interruptions ranging from seconds to days. Electric faults, road damages, fires, and similar incidents can disable energy supply in a region. The geographic area affected from a short-term power outage or fuel supply interruption varies in a wide range from a small residential district to a highly populated metropolitan area, and even can propagate to a larger region.

Most of short-term threats are probabilistic and difficult to prevent. A resilient energy system must be able to recover quickly from disruptions of the supply chain. Fuel reserves and alternative supply paths increase the system resilience. Electric transmission and fuel supply infrastructure are more vulnerable to short-term threats. Developing backup power generation to supply critical equipment reduce the vulnerability of the supply network to short-term threats.

Natural causes and disruptive events created by people are the major causes that threaten energy availability in a relatively short time. Generally, the electric transmission grid and fuel supplies are vulnerable to short-term threats. Diversification of fuel supply, backup energy in the form of a fuel stock, electric storage, or pumped storage hydroelectric facilities reduce the risk of short-term interruptions.

Earthquakes, tsunamis, hurricanes, snowstorms, floods, and tornadoes cause electric outages from a few hours to several days. Electromagnetic pulses (EMP) caused by solar storms or detonation of a nuclear bomb at a high altitude can disable an electric network covering a large area. Refineries and oil processing facilities, oil and natural gas pipelines, large power plants, and electric transmission lines are vulnerable to terrorist attacks. In war conditions, such infrastructures become strategic targets. Not only physical attacks, but political tensions and extreme social conflicts also may impact the continuity of energy supply. Since the 1970s, withholding energy sources has become a political and economic leverage for countries. The oil crisis in 1974 and subsequent oil embargos are examples of such political and economic stresses. Energy harnessed from cross-border rivers may cause international conflicts that can even escalate to wars. Conversely, obstruction of water flow by an upstream country may impact electric generation in downstream countries resulting in electricity shortage for several months.

National defense, public safety, emergency health services, data storage, and processing centers are critical infrastructures that may be most affected by short-term energy interruptions. Backup electric generation units with easily accessible fuel stocks, battery backup, and other forms of energy storage can ensure continuity of electric supply.

Electric utilities feed a populated area from a regional interconnection via several transmission lines such that if one of the lines breaks down, the others continue to supply electricity. In electric power systems, the term "fault" signifies the failure of a transmission line or equipment such as a generator, transformer, and switchgear. Faults mainly occur either because of a short circuit or broken conductor.

11.2.3 Mid-term Energy Security

Mid-term energy security requires continuous and reliable energy supply in a time range of years to decades. Planning and development of energy production and distribution systems is essential to ensure adequate energy supply while energy demand is increasing. If the energy infrastructure development does not match the increasing demand, energy shortages may arise in the medium-term.

Energy infrastructure must be continuously maintained, renewed, improved, and expanded. Lack of maintenance may compromise mid-term energy security. Older technologies are often less efficient. Replacing fuel production and electric generation facilities reduces losses and improves the quality of energy sources. Research and development for new energy technologies increase mid-term energy security.

Development of large facilities like refineries, pipelines, coal processing plants, or electric generation plants takes many years. Major steps of an energy infrastructure include planning and site acquisition, permitting, financing, site preparation, construction, testing, and commissioning. The speed of development usually depends on social interactions and public acceptance. Today society is much more interconnected than before, and social media makes easier to share opinions, concerns, and reactions. Obtaining approval for a new facility takes typically several years. It is, therefore, reasonable to estimate a timeframe of 10–20 years for development of a large energy facility from the initial idea to becoming fully operational.

Energy planners strive to continually expand the energy infrastructure by considering the present energy consumption, growth rate, and estimated timeline of projects. In addition, decreased capacity and retirement of older facilities are also considered to ensure mid-term energy security.

The energy system of a country or region is supplied by a variety of primary sources. While some of these sources may be available internally, others must be imported from foreign markets. Political tensions and instability may restrict the availability of imported energy sources. Diversification of the primary sources and providers reduces the risk of their unavailability due to a possible social, political, or economic instability (Yergin 2006).

11.2.4 Long-term Energy Security

Population growth and a gradual increase of energy demand are long-term risk factors. Depletion of resources affects the production of usable energy sources throughout several decades. Climate change has possible effects on the availability of renewable energy sources in a period covering several centuries.

Proactive management of primary sources and planning of the energy system to meet the growing needs of the society and economy reduces the risk of energy shortages in the future. Sustainability of both nature and humanity require optimal use of resources with minimal effect on the environment.

Energy security is a complex issue that has technical and non-technical aspects. On the technical side, explorations of new fossil fuel reserves, research and development of cost-effective technologies for extraction and processing of energy sources, use of renewable energy sources, and managing the efficient end-use of energy sources are major factors. On the non-technical side, political and social issues that may affect extraction of energy sources and failure to make timely investments for development of energy processing facilities are among key considerations.

The biggest risk for the future availability of energy sources is depletion of fossil fuel reserves. The fact is that long-term energy security relies on exploration of new reserves as well as substitution of depleting fossil fuels with new alternative energy sources. This highlights the importance of research and development (R&D) in the energy field. Economic incentives established by government policies and funding sources motivate such R&D efforts.

11.2.5 Energy Security Indicators

Energy security can be assessed based on availability, accessibility, affordability, and acceptability of energy sources (Kruyt et al. 2009).

Availability of fossil fuels and nuclear energy relates to the physical existence of reserves that can be extracted by existing technologies. Availability of renewable sources of energy is constrained by geography, land use, and population density. Transformation of primary sources into usable forms of energy and delivery to consumers require adequate infrastructure. Availability of energy sources depends not only on the physical existence, but also the ability of the economy to make them available to the society.

In the context of security of supply (SOS), accessibility refers to geopolitical elements. Access to some energy sources may be subject to international trade restrictions. Political tensions, regional instability, and wars may limit the accessibility to certain vital sources, especially petroleum and natural gas. Most of the major rivers are shared by two or more countries. Use of the hydro-potential upstream for energy generation may conflict with the use of water for agriculture in a country downstream.

Affordability of energy sources depends on their cost to end-users. Initial investment to develop infrastructure, operational cost of facilities, and cost of transportation are main components of the total energy cost. In addition, energy prices are affected by market conditions, supply-demand balance, and economic and political tensions. As the coal, oil, and gas reserves near the surface are exhausted, the cost of extraction from deeper reserves increase. The cost of fuel obtained from deep-water wells and unconventional reserves is always higher compared to conventional reserves. In fact, some of the coal mines, oil, and natural gas fields may be abandoned as they become too expensive in the competitive market.

Acceptability of energy sources involves environmental and societal elements. Environmental concerns and public reactions may delay or even prevent exploitation of some energy sources. Development of electric generation stations, refineries, pipelines, and even extra-high-voltage electric transmission lines depend on public acceptance. Particularly, hydropower development requires flooding a large area, and therefore relocation of a population. Beyond the cost of relocation, disappearing cultural heritage and historically sensitive lands cause intense public reactions that may result in significant cost increase or even abandonment of some projects.

Four dimensions of energy security are interrelated. For example, depletion of limited sources affects the fuel prices, and consequently affordability. Restricted accessibility to sources due to political tensions also result in significant fluctuations of energy prices. Public reactions may prevent development of facilities or infrastructure to use abundant renewable energy sources.

Several criteria are used to assess and compare the energy security of countries. Some of the indicators commonly used in energy statistics and reports to evaluate specific aspects of energy security are listed below (Kruyt et al. 2009).

- Resource estimates: Quantity and likelihood of availability of fossil fuel resources.
- Reserve to production ratio (R/P): The ratio of economically feasible reserves to the production at a country or global level.
- Diversity indexes: Shares of diverse fuels in the total primary energy supply (TPES) or share of suppliers in import.
- Market concentration: Shares of the producers in the market.
- Import dependence: Import prices of energy carriers.
- Net energy import dependency: Import prices and shares of fuels in TPES.
- Political stability: Political risk ratings.
- Oil price: Price of imported crude oil.
- Carbon footprint: The amount of carbon emissions of the TPES.
- Energy intensity: TPES per gross domestic product.
- Energy use per capita: The ratio of TPES to the population.
- Share of oil in transport sector.
- Share of transport sector in total oil consumption.

Aggregated indicators are a combination of selected simple indicators. One of the aggregated indicators used by the IEA is related to the physical unavailability of fuels in markets where prices are regulated. Another index used by IEA assesses the price risks due to shares of producers and suppliers in the market considering political stability factors. Other aggregated indexes such as oil vulnerability index (OVI), Shannon index, and supply-demand index are described in Kruyt et al. (2009).

11.3   Cost of Electric Outages

Uninterrupted supply of electric power becomes increasingly more important as the society relies more on high technology for communication, production, and social life. A study conducted in 2001 for the Electric Power Research Institute (EPRI), surveyed a large number of establishments grouped in three sectors; digital economy, continuous process manufacturing, and fabrication and essential services (Lineweber and McNulty 2001). According to the study, the combined total cost of electric power outages to the US economy across all three sectors is estimated between $104 and $164 billion per year. Figure 11.2 breaks out the net cost of a single power loss by duration averaged across all establishments that participated

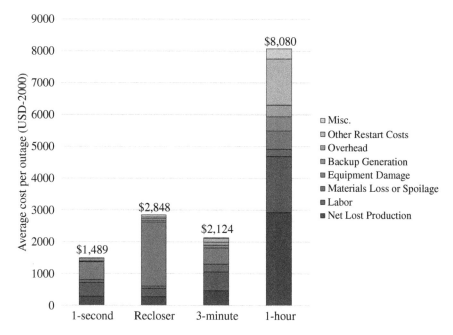

Figure 11.2 Total annual costs of outages in the US by sectors. Source: Lineweber and McNulty (2001, Table 2-1).

in the survey (Lineweber and McNulty 2001, Appendix D, Table D-1). The cost of a brief interruption that lasts around one second is obviously the lowest, but a combination of brief outages consisting of a one-second outage followed, a few seconds later, by another one-second outage, called a *recloser event*, costs more than a single one-second interruption. Equipment damage is more dominant in a recloser event. Interruptions in the order of one hour have the highest cost due to loss of production, labor, and the cost of restarting the operations. Net production and labor losses, backup generation, and overhead costs may be prorated for power losses of several hours.

Digital economy, which includes data storage, and processing operations, information technology (IT), and telecommunications require uninterrupted power supplies. Even a short period of interruption results mainly in disruption of services. Biotechnology, Internet providers, and financial institutions are among the industries that rely on digital services.

In continuous process manufacturing facilities raw materials are continuously fed to the production line, usually at a high temperature. Such manufacturing plants suffer mostly from loss of production and material waste when electric power becomes unavailable for a few hours. Some examples are metal (steel, aluminum, copper, etc.), paper, plastic, rubber, chemical process, and glass industries.

Fabrication and essential services include all manufacturing factories; electric, gas, water, utilities; pipelines, railroads, and electric public transportation services. In this sector, power outages cause loss of production, labor, and services.

The total annual cost of electric outages for each sector previously described are shown in Figure 11.3 (Lineweber and McNulty 2001, Figure 3-6). The cost of an electric power outage

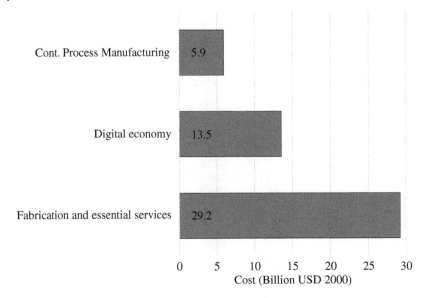

Figure 11.3 Average cost of electric outages by duration. Source: Lineweber and McNulty (2001).

has many implications such as loss of production, wasted or spoiled materials, equipment damage, backup generation, overhead, and payment for employee time or labor during the interruption of operations. In addition, backup generation and restarting the manufacturing process are also part of these costs. The cost savings during outages such as unused materials, savings on energy bills, and unpaid labor were insignificant. Longer power outages that last several days have hidden impact to the regional businesses that lose production in a competitive market economy.

Large-scale electric outages usually have significant social impacts. In addition to the loss of income for hourly working employees, darkness and lack of security, alarms, and monitoring systems increase crimes. Our society is now more connected than ever; cell phones, networked computers, Internet access, email, and social media are inevitable parts of our everyday life. An interruption of the electric power would disable all telecommunication including cell phones, information exchanges, and means of data transfer. Even if a building were heated by natural gas or petroleum, the ignition, circulation, and ventilation systems would not work without electricity. Inadequate heating may result in public health issues and consequently a loss of productivity. Elevators that stop working because of an extended blackout can cause public frustration and panic in high-rise buildings. Water supply also depends on electric power because of the filtration, water treatment, and pumping systems.

The comfort level desired for a modern lifestyle cannot be maintained without a continuous flow of energy. Shortage or unavailability of the energy supply results in a degradation of the quality of life and loss of production. Depletion of fossil fuel resources at a continuously growing rate, climate change, and environmental impacts of energy systems have emphasized the importance of energy related issues. Especially over the last decades, increasing energy prices, reliability of energy supplies, security threats, and natural disasters that caused extended power outages have motivated decision makers to create action plans for development of more reliable and sustainable energy supply. In addition, many consumers have considered developing individual on-site electric generation systems to supplement utility supply, secure backup power, or provide completely utility-independent sustainable energy.

11.4 Resource Availability

Increase of fuel consumption is usually described as a percentage of the currently used amount. If a quantity grows at a rate α proportional to the present amount q, its growth can be expressed by the differential equation (11.1).

$$\frac{dq}{dt} = \alpha q \tag{11.1}$$

In solving Eq. (11.1), we obtain the exponential growth function shown in Eq. (11.2), where Q_0 is the initial amount.

$$q(t) = Q_0 e^{\alpha t} \tag{11.2}$$

The initial quantity reaches a certain amount Q at the time t given in Eq. (11.3).

$$t = \frac{1}{\alpha} \ln\left(\frac{Q}{Q_0}\right) \tag{11.3}$$

The time interval at which the quantity doubles its value at a certain time has a particular importance in projecting the future amounts of this quantity. By setting $Q = 2Q_0$ in Eq. (11.3), the expression of the doubling time is obtained as in Eq. (11.4).

$$T_D = \frac{\ln(2)}{\alpha} = \frac{0.693}{\alpha} \tag{11.4}$$

Usually the growth rate is expressed in percentage of the present amount. The doubling time can be approximately calculated by dividing 70 with the given percent growth rate.

$$T_D = \frac{70}{A\%} \tag{11.5}$$

Equation (11.5) shows the change of doubling time with the growth rate. For example, if the consumption of a fuel increases 7% per year, it will double in approximately 10 years (see Figure 11.4). At present, the fossil fuel consumption is growing about 2% worldwide. This means that the consumption will double roughly in 35 years if the consumption rate remains constant at this value. Note that with the current growth rate, the consumption of fossil fuels would increase approximately four times during the average lifetime of a person.

Understanding the properties of exponential growth is extremely important in the use of finite resources. Suppose that a fuel has been consumed at a fixed rate since the beginning of time. The total consumption before a given instant t_1 can be calculated as shown in Eq. (11.6).

$$Q_1 = \int_{-\infty}^{t_1} q(t)dt = \int_{-\infty}^{t_1} (Q_0 e^{\alpha t})dt = \frac{Q_0}{\alpha} e^{\alpha t_1} \tag{11.6}$$

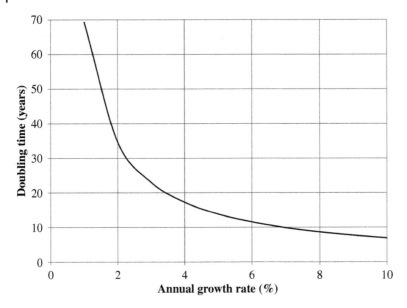

Figure 11.4 Doubling time as function of the growth rate.

In the same way, we can calculate the consumption over one doubling time starting at t_1.

$$Q_2 = \int_{t_1}^{T_D} (Q_0 e^{\alpha t}) dt = \frac{Q_0}{\alpha} (e^{\alpha T_D} - 1) e^{\alpha t_1} \tag{11.7}$$

Since $e^{\alpha T_D} = e^{\alpha \frac{\ln 2}{\alpha}} = 2$, we can see that

$$Q_2 = \frac{Q_0}{\alpha} (2 - 1) e^{\alpha t_1} = Q_1 \tag{11.8}$$

This result shows that the consumption during one doubling time is equal to the total amount consumed before it. For example, in the 1960s the consumption rate of fossil fuels was in the order of 7%. At that time, the consumption over a decade was equal to the total consumption in the history. Although the consumption of fossil fuels has considerably decreased worldwide, if the current growth rate of 2% remains the same during the next 35 years, more fossil fuels than what has been consumed thus far history would be depleted.

Another important concept to understand is the time left to consume an existing reserve completely. Suppose that the estimated total amount of proved fuel reserve is Q_R. If this fuel is consumed at a fixed rate until it vanishes, we can write Eq. (11.7), where T_E is the expiration time and Q_0 is the amount used at present ($t = 0$).

$$Q_R = \int_0^{T_E} (Q_0 e^{\alpha t}) dt = \frac{Q_0}{\alpha} (e^{\alpha T_E} - 1) \tag{11.9}$$

We can obtain the expiration time as shown in Eq. (11.10).

$$T_E = \frac{1}{\alpha} \ln \left(\frac{\alpha Q_R}{Q_0} + 1 \right) \tag{11.10}$$

Obviously, the consumption rate may not remain constant and new resources may be discovered using new survey methods. In addition, resources that are not considered feasible at present may become feasible in the future. Statistics given at a certain time should be interpreted by keeping these factors in mind.

Future availability of reserves can be estimated using a linear or exponential growth projection. In linear projection, the current production is assumed to remain the same in the future. In exponential projection, the production grows at a certain rate, which is usually expressed as the percentage of the production. Clearly, linear projection is a special case of exponential projection where the growth rate is zero. This growth rate can be estimated by using exponential regression of the production history. Depletion of fossil fuel reserves is usually estimated based on certain assumptions.

Let's suppose that the production of a fossil fuel at present is P_0 and the estimated growth of the production is β. The total amount Q_X extracted in an interval Δt would be:

$$Q_X = \int_0^{\Delta t} P_0 e^{\beta t} dt = \frac{P_0}{\beta}(e^{\beta \cdot \Delta t} - 1) \tag{11.11}$$

If the present reserve is Q_R, then Δt years later the remaining reserve would be

$$Q_{\Delta t} = Q_R - Q_X = Q_R - \frac{P_0}{\beta}(e^{\beta \cdot \Delta t} - 1) \tag{11.12}$$

Three scenarios are displayed in Figure 11.5 for depletion of fossil fuels in the years following 2013. The graphs assume that existing discovered reserves remain constant. These estimations are based on data published in the BP Statistical Review of World Energy 2014 (BP 2014). In the first scenario, where the production remains the same as in 2013, the depletion curve is a straight line with a slope defined by the "Reserve/Production ratio" (R/P). The second scenario assumes that the production will grow at the rate calculated from 2012 and 2013 productions. The third scenario uses the growth rate obtained from exponential regression of the production history.

On one hand, emerging technologies are revealing new reserves of fossil fuels and more economic ways of production, which were unknown or not feasible in the past. On the other hand, reclassification of proved reserves due to changing economic conditions shows an increase of fuel reserves in energy statistics. This virtual increase should not be misleading; we must always keep in mind that the total amount of available fossil fuel resources is continuously decreasing. If the current fuel mix and consumption scenarios remain the same, it is certain that the fossil fuel resources will vanish eventually.

11.5 Energy Interdependence

The concept of energy independence was developed after the oil crisis as developed countries were trying to reduce their dependence on foreign oil. To become energy independent, a country must exploit its domestic energy resources. While such effort stimulates exploration of internally available reserves and development of national infrastructures, it may also lead to overuse of limited deposits. Increased use of available renewables and improving energy efficiency is a more efficient way of reducing energy dependence rather than depleting limited fossil fuel reserves. However, with current technologies an energy system

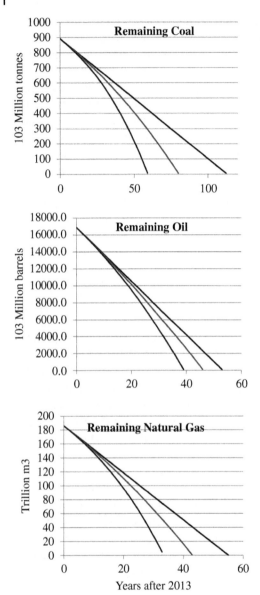

Figure 11.5 Fossil fuel reserves that would remain after the end of 2013 for three different growth rate of production scenarios. Source: BP Statistical Review of World Energy (2014).

cannot rely on intermittent renewable sources for continuous and uninterrupted energy supply. In addition, current transport systems depend mainly on oil products. Therefore, rather than energy independence, the concept of energy interdependence is more relevant from energy security and sustainable development perspectives.

The energy mix supplying a country is composed of a variety of sources. While many of these sources are locally available, some of them must be imported from other countries. Energy trade between nations may be bidirectional for more cost-effective use of fuels. Today all fossil fuels are extensively traded in international markets. Particularly, liquid fuels and gas are more frequently traded because of their lower transportation cost per unit energy compared to coal. Electric power is another way to trade energy sources that are not cost-efficient to transport like lignite, low quality coal, biomass, and municipal solid waste.

All countries around the world depend on each other to supply their energy needs. Few countries produce more energy than they consume. Whereas these countries can supply all their energy from internal resources, they rely on the income from fuels they export. Especially the crude oil market is one of the factors that determine the global economy. Projections of energy use, economic decisions, and investments depend heavily on the oil prices and availability.

While stable and reliable energy trades allow more efficient use of natural resources, energy independence is usually a concern for national autonomy. Energy dependence shows the extent to which a country relies upon imports in order to meet its energy needs. Energy dependency rate can be defined as the proportion of net imported energy that a country needs to supply all energy consumption (IEA 2014). Net imported energy is the difference between energy imports and exports. The gross energy consumption is the quantity of energy necessary to supply all energy requirements, including losses, stocks, and fuel supplied to international ships. A negative dependency ratio indicates that a country exports more energy than imported, if any. If this ratio exceeds 100%, then energy is being stored during the considered period. Equation (11.13) is used to calculate the energy dependency rate.

$$\text{Import dependency} = \frac{\text{Imported energy}}{\text{Energy demand}} \tag{11.13}$$

Beside the overall energy dependence, reliance on imports of certain types of fuel may be critical for a sustainable national economy. Energy shortages may arise if a fuel that is not produced from domestic resources cannot be either imported from external markets or substituted by other energy sources. Long-term energy shortages lead to an energy crisis.

Diversity of the fuel mix is important to reduce the risk of energy shortages in the case of unavailability of one of the fuels. In addition, diversifying the suppliers of critical fuels increases the reliability of fuel supply.

For example, the transportation sector uses mainly petroleum products. Unavailability of oil and refined oil products affects all sectors of the economy either directly or indirectly. Consequently, worldwide the most critical commodity is petroleum. All kinds of travel, public services, and delivery of mail and parcels rely on oil supply. In addition, the transportation network enables industrial logistics to supply raw materials and parts as well as delivery of finished products to customers. Unavailability of oil products may quickly paralyze the public services and industry of a country.

International pipelines carry large quantities of crude oil, liquid fuels, and gas between countries. A political tension or change of energy policies may affect the fuel supply to a large region since portions of such major pipelines are controlled by different nations.

Coal transportation to long distances is costlier than oil and gas. Still, about 15% of coal is traded internationally between continents. International trade in steam coal is mainly dominated by Atlantic and Pacific markets. Western Europe, particularly the UK, Germany, and Spain form the Atlantic market. The Pacific market consists of developing and OECD Asian importers, notably Japan, Korea, and Taiwan. Major coal exporters are Indonesia, Australia, Russia, USA, Colombia, South Africa, and Canada (World Coal Association www.worldcoal.org).

Energy interdependence is not limited to fossil fuel trades. Large rivers powering large hydroelectric plants cross several countries. Dams constructed on an international river are sometimes subject to political negotiations. Storing hydropower in dams or changing the flow of a river may reduce both energy and water supply in another country.

Operation and maintenance of nuclear power plants requires high-level technical knowledge, enriched fuel, and special parts. If a country does not have the necessary knowledge, experienced staff, support technology, and fuel-processing infrastructure, then nuclear power would eventually create an energy dependence.

Information technology (IT), data collection, process control systems, and telecommunication rely on electric power. Electricity is generated using a combination of various fuels. If a particular fuel is dominant in electric generation, sectors depending on electric power indirectly depend on that fuel as well.

Electric networks span many countries in each continent. International electric connections allow more efficient use of resources and installed generation capacity. Particularly, time difference between distant regions reduces the overall load variations on a large network, increasing the capacity factor of the generation plants. At the same time, sources such as hydro and nuclear power that cannot be transported, are shared between countries through the electric network. Countries interconnected by an electric network are energy interdependent since electricity plays a vital role for any part of the national economies.

11.6 Chapter Review

This chapter is an overview of essential energy security issues.

Energy security is usually defined as uninterrupted availability of energy sources at an affordable price. Energy security concerns can be categorized based on the cause and duration of the possible interruptions of energy supply.

Natural disasters or artificially created disturbances may cause short-term interruptions of energy supply. Major natural causes that may disable energy supply include hurricanes, earthquakes, tsunamis, floods, snowstorms, extreme heat, and other disasters. Terrorist attacks, intentional actions of individuals, groups, or nations are among artificial disruptions.

Short-term energy security is associated with energy interruptions ranging from seconds to days resulting from electric faults, road damages, fires, and similar incidents, and can disable energy supply in a region.

Mid-term energy security is continuous and reliable energy supply in a time range of years to decades. Adequate planning and development of energy infrastructure increase mid-term energy security.

Long-term energy security is concerned with population growth, depletion of limited reserves, and environmental interactions of energy systems.

Energy security indicators include availability, accessibility, affordability, and acceptability of energy sources. The energy security of nations is assessed by comprehensive models that include available resources in the national territories, energy imports, energy demand, and consumption growth rates.

The cost of electric outages results in losses of production, labor, and material. In addition, electric power loss may cause equipment damages.

Energy consumption growing as a fraction of its value at a certain time results in a growth represented by an exponential function. Such increase is known as exponential growth that doubles in a time interval obtained approximately by dividing 70 to the growth rate in percent.

Future availability of reserves can be estimated using a linear or exponential projection. Linear projection is based on the production to consumption ratio. Exponential projection assumes a constant growth rate as percentage of the consumption at all times.

Energy independence relates to self-sustainability of a nation. It has technical and economic consequences regarding overuse of available resources and environmental impacts. Development of energy conversion facilities powered by renewable sources and improving energy efficiency enhance national energy independence.

The concept of energy interdependence is more relevant for global energy security and sustainable development.

Review Quiz

1 What is energy security?
   a. Uninterrupted availability of energy sources
   b. Uninterrupted availability of energy sources at an affordable price
   c. Availability of energy sources at an affordable price
   d. Protection of energy production and supply facilities

2 What is the supply chain?
   a. Fuel transportation networks
   b. Electric transmission and distribution grids,
   c. Fuel delivery stations
   d. All above

3 Which one below is not a natural disaster causing energy supply interruption?
   a. Hurricane
   b. Winter storm
   c. A short circuit between power lines causing a cascaded power outage
   d. Electromagnetic pulse caused by a solar storm

4 A tsunami may cause
   a. a short-term electric outage less than a minute.
   b. a long-term energy shortage.
   c. a medium-term electric outage.
   d. None of the above.

5 A nuclear weapon detonated at high altitude causes
   a. electromagnetic pulses that propagate through transmission lines.
   b. destruction of power line.
   c. destruction of generators.
   d. destruction of fuel transportation network.

6 Mid-term energy security relates to
   a. several days of power outage due to a snowstorm.
   b. several months of energy shortage due to a political embargo.
   c. inability of an energy system to supply the growing demand.
   d. unaffordable energy prices.

7 Which one of the following is a major factor for long-term energy security?
   a. A natural disaster
   b. A terrorist activity
   c. Issues regarding the maintenance of energy infrastructure
   d. Depletion of limited resources

8 Energy demand growing as a percentage of the current demand results in
   a. exponential growth.
   b. linear growth.
   c. parabolic growth.
   d. hyperbolic growth.

9 If electric consumption increases 7% every year, the generation capacity must
   a. remain the same.
   b. double every year.
   c. double every 10 years.
   d. double every seven years.

10 Energy import dependency index is
   a. greater if a country imports more energy than produced.
   b. greater if a country imports more energy per demand.
   c. smaller if a country imports less energy than produced.
   d. smaller if a country imports more energy per consumption.

Answers: 1-b, 2-d, 3-c, 4-c, 5-a, 6-c, 7-d, 8-a, 9-c, 10-b.

Research Topics and Problems

Research and Discussion Topics

1 Research the impacts of the hurricane Catrina on the energy supply in the area affected.

2 Discuss the possible impacts of severe weather conditions on the interconnected electric grid.

3 Draft a technical report on possible impacts of high-altitude EMP on electric network.

4 Compose a white paper on vulnerability of a wide-area interconnected grid to terrorist attacks and discuss possible mitigation approaches.

5 Research natural gas trade between European Union (EU) and Russian Federation and discuss its effects on energy security of separate nations.

Problems

1 The energy consumption in a region increases at a rate of 4% every year. Assuming that the rate remains constant, in how many years will the energy consumption quadruple?

2 The energy consumption in a region doubles in 20 years. What is the average growth rate?

3 An island has a fuel stock with 50-quad net energy value. This fuel is used in an electric power station operating at 40% overall efficiency. The current electric consumption of 200-GWh per year is projected to increase 2% per year. How long can the fuel stock supply this island without being replenished?

4 An island consumes 1.2-quad energy in one year. This energy is supplied by imported crude oil of energy content 5.8 million Btu per barrel. How many million barrels of crude oil must be imported to meet the energy requirement of this island in two months?

5 A certain amount of fuel contains 15×10^{12} Btu of energy, and it is converted into electricity in a power station having 35% overall efficiency. The average daily demand on the station is 50 MWh. In how many days will the fuel be totally consumed?

Recommended Web Sites

- European Commission, Joint Research Center: https://ses.jrc.ec.europa.eu
- International Energy Forum (IEF): https://www.ief.org
- International Energy Agency (IEA): https://www.iea.org/topics/energysecurity

References

BP, 2014. *Statistical Review of World Energy.* [Online] Available at: http://www.bp.com/statisticalreview [Accessed August 2015].

Fulli, G. (2016). *Electricity Security: Models and Methods for Supporting the Policy Decision Making in the European Union.* Torino: Politecnico di Torino, Department of Energy, Doctoral School.

IEA (2014). *Energy Supply Security: Emergency Response of IEA Countries 2014*. Paris: International Energy Agency.

Kruyt, B., van Vuuren, D.P., de Vries, H.J.M., and Groenenberg, H. (2009). Indicators for energy security. *Energy Policy* 37: 2166–2181.

Lineweber, D. and McNulty, S. (2001). *The Cost of Power Disturbances to Industrial & Digital Economy Companies*. Madison, WI: Primen.

NERC (2004). *Technical Analysis of the August 14, 2003, Blackout*. Princeton, NJ: NERC.

Spagnol, G. (2013). *Energy Security*. Brussels: IERI.

U.S.-Canada Power System Outage Task Force, 2004. Final Report August 14, 2003; *Blackout in the United States and Canada – Causes and Recommendations*, s.l.: s.n.

Yergin, D. (2006). Ensuring energy security. *Foreign Affairs* 85 (2): 69–82.

12

Energy and Sustainable Development

Statue of famous American folk song writer Woody Guthrie in front of the Grand Coulee Dam on Columbia River in Washington State, USA. Grand Coulee is a multipurpose project for regulation of river flow, flood control, electric generation, and irrigation water for farmlands. Three power stations with total nameplate capacity of 6809 MW is the largest power plant in the US. The pumped storage hydroelectric power plant supports economic development in the region by generating electric energy as well as providing water for agricultural irrigation.

Source: © H&O Soysal

Energy for Sustainable Society: From Resources to Users, First Edition. Oguz A. Soysal and Hilkat S. Soysal.
© 2020 John Wiley & Sons Ltd. Published 2020 by John Wiley & Sons Ltd.

12.1 Introduction

Current state and progress of communities rely on productive use of natural resources. Food and water supply, industrial production, construction of infrastructure, operation of all equipment, and transportation involve use of abundant energy. Deployment of energy sources, on the other hand, interacts with availability of fresh water, clean air, fertile soils, and biodiversity; vital elements for sustainable development.

Energy systems are not isolated from the rest of the world. In fact, energy is the essential part of nature. Energy systems use diverse natural resources as primary energy sources. Energy use, on the other hand, impacts natural resources in many ways. Evidently, consumption of conventional energy sources at the current rate depletes hydrocarbon deposits that have formed over millions of years in the earth. Meeting the largest fraction of the world's energy need by burning fuels pollute the air, water, and soil; all critical elements of life. Air pollution leads to acid rain and global warming, causing changes of the ecosystem and climate in long term.

Energy and water are reciprocally interdependent. Energy systems use water for production of primary sources as well as cooling fluid for thermoelectric power plants. A considerable amount of energy is needed to extract, treat, and convey water for irrigation, industrial processes, and public use. Farming, fishery, dairy products, and the food industry rely on continuous availability of fresh water and energy. In addition, land use to extract and process fossil fuels, building dams for hydroelectric generation and biofuel production competes with agricultural fields needed for food production.

This chapter discusses the interactions of energy, water, and food supply considering pollution and climate change issues. First, effects of energy systems on ground level air pollution, acid rain, and climate change will be reviewed. Then, interactions of energy, water, and food supply will be discussed. Consumption at the users' end will be tracked back to exploitation of natural resources to produce the supplied fuel mix. The last section will discuss the environmental benefits of distributed energy production from clean sources.

12.2 Sustainable Development Goals

Sustainable development of any community depends primarily on energy, water, and food supply. The World Commission on Environment and Development defined sustainable development as "*development that meets the needs of the present without compromising the ability of future generations to meet their own needs*" (WCED 1987, p. 41). In this definition, the term "needs" includes access to clean water, food, sanitation, services, and high-quality energy sources for all communities around the world, especially in poor regions.

Maintaining the environment's ability to meet the needs of the future generations is the key factor for sustainable development. Furthermore, the United Nations has set forth seventeen sustainable development goals (SDG) to be achieved by 2030. SDG-7 specifically targets "*access to affordable, reliable, sustainable, and modern energy.*" SDG-6 targets the universal access to affordable, safe, and clean drinking water, as well as sanitation for all individuals. Sustainable development goals SDG-6 and SDG-7 are closely interlinked. According to the World Health Organization, 2.1 billion persons around the world have

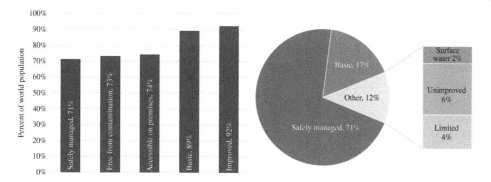

Figure 12.1 World population's access to drinking water. Source: WHO/UNICEF Joint Monitoring Programme for Water Supply, Sanitation and Hygiene (JMP); https://washdata.org/monitoring/drinking-water (accessed 8/14/2019).

limited access to safely managed drinking water. Many areas without direct access to water do not have access to modern energy systems either.

Domestic water services include availability, quality, and access to main sources used by households for drinking, cooking, personal hygiene, and other uses. In some remote areas, clean water is still extracted from wells or springs using buckets and hand pumps. In some urban areas, insufficient or interrupted water supply enforces individuals to carry water in containers and use limited amounts of water. Most limitations are associated with lack of continuous reliable energy supply to water services (Figure 12.1). Another target of SDG-6 is reducing the number of populations without managed sanitation services. Wastewater collection and treatment is closely linked to the availability of modern, reliable, and continuous energy services (SDG-7). A population of about 4.5 billion persons does not have safely managed basic sanitation (WHO/UNICEF 2019). Unreliable energy results in disruptions of adequate water supply, sanitary services, and treatment of wastewater, which all threaten the public health (WHO/UNICEF 2019).

Both access to water and energy impact food supply. SDG-2 targets ending hunger, achieving food security, and improved nutrition. Supply of irrigation water requires energy. In underdeveloped regions of the world, agriculture is challenging because of freshwater availability. In developing and developed countries water use for irrigation or energy has been debated for decades between agriculture and energy sectors. Energy use for food spans from farmlands to the consumers. While agriculture and fishing are not energy intensive, transforming agricultural produces into food, beverage, and food products is energy intensive. Transportation, refrigeration, and marketing of food products rely on continuous energy supply. Food processing, cooking, and refrigerated food storage constitute a considerable part of residential and commercial energy consumption.

All goals stated in sustainable development agenda of the United Nations directly or indirectly involve adequate supply of energy (UN 2015). Sustainability, reliability, and affordability of energy supply are central to access clean water, modern sanitation, food, and health services. Urban infrastructure, security, quality education, industrial development, and economic growth also depend on uninterrupted energy supply. Goals involving mitigation of climate change and pollution, protection of natural resources on land and water require cleaner energy conversion and efficient use of secondary sources.

12.3 Environmental Impacts of Energy Systems

Energy systems are major contributors of water, land, and air pollution. Fossil fuel and biomass burning power plants and all vehicles using combustion engines pollute air by releasing carbon dioxide, toxic gases, and microscopic particles. Coal burning power plants produce a huge amount of solid waste, which is typically discarded as landfill. Although some part of the solid waste that occurs from burning fossil fuels is used to make industrial materials as a side product, the remaining part pollutes the soil, ultimately reaching plants with ground water. Large power plants use water from rivers and lakes for cooling. If special precautions are not taken, coal-fired power stations may release toxic compositions containing mercury, arsenic, and lead that are absorbed especially by fish and seafood. Heat discharged to the environment changes the ecosystem. Among all human activities, energy production and use are the main causes of gas emissions that lead to the alarming phenomena known as acid rain and global warming.

12.3.1 Ground Level Air Pollution

Air pollution that occurs at the lowest level of the atmosphere is specified as ground-level air pollution. Major components of the ground-level air pollution are toxic gases such as carbon monoxide, sulfur dioxide, nitrogen oxides, ammonia (NH_3) compounds, microscopic particles called particulate matter (PM), and volatile organic compounds (VOC).

Carbon dioxide is not considered an air pollutant since it is also produced by the respiration of humans and animals and recycled naturally by plants through the photosynthesis process. But excessive amounts of carbon dioxide released from fuel combustion that the vegetation cannot handle reduce the air quality and cause respiratory problems for humans and animals. In addition, carbon dioxide emission produced by burning fuels leads to other environmental issues that will be discussed in the following sections.

In large metropolitan areas and industrial regions, extreme ground-level air pollution builds up a dark yellow or brown haze under certain weather conditions. This combination of smoke and fog, commonly called "smog," contains ground-level ozone (O_3) and a number of chemicals, including toxic gases, carbon dioxide (CO_2), soot, dust, and VOC such as benzene, butane, and other hydrocarbons. Smog has several long-term adverse health effects on people living in metropolitan and industrial areas. However, sometimes wind can carry smog to distant smaller cities and rural locations (EIA 2016).

Carbon monoxide (CO) is formed from incomplete combustion that occurs in wildfires and burning of low-quality fuels and biomass in traditional stoves. It is a colorless, odorless, and poisonous gas that reduces the ability of red blood cells to supply oxygen to the brain and other organs. Low levels of carbon monoxide slow down body functions, reflexes, and may impair judgment. It more severely affects pregnant women and people with heart diseases. High levels of carbon monoxide, especially in closed spaces, may be lethal.

According to the World Health Organization (WHO), in 2012 nearly three million persons worldwide died prematurely for reasons attributable to outdoor ambient air pollution. Lung diseases including cancer, heart disease, and stroke were among the lethal consequences of air pollution (WHO 2016, p. 44, Table 5). With the current increase of energy supply, the International Energy Agency (IEA) projects this number to reach 4.5 million by 2040.

Energy-related human activities are main sources of ground level air pollution. Fuel combustion for electric power generation and industrial heating, industrial processes, transportation vehicles, ineffective use of low-quality fuels and conventional biomass in less developed regions are major sources of human-caused air pollution.

12.3.2 Acid Rain

All fossil fuels contain a small amount of sulfur, which reacts with oxygen during combustion to form sulfur dioxide (SO_2). In addition, various nitrogen oxides represented as NO_x (nitric oxide NO, nitrogen dioxide NO_2, nitrous oxide N_2O, and others) form as fossil fuels burn. Oxidation compounds of sulfur and nitrogen react with water vapor and other chemicals in the atmosphere in the presence of sunlight to form sulfuric and nitric acids. The acidic mixture in the atmosphere dissolves in clouds and fog, then eventually comes down to the ground with raindrops, resulting in the phenomenon called "acid rain." While the soil can neutralize some part of the acid, the ground water can still be acidic in regions where power plants and heavy industry using low quality coal and oil products pollute the air excessively. Acidic ground water changes the ecosystem in rivers and lakes and damages the plants, ultimately deteriorating forests and vegetation.

12.3.3 Greenhouse Effect and Climate Change

Transparent substances transmit the visible range of the light spectrum at different rates depending on the wavelength. A surface heats by absorbing the sunlight and emits longer wavelength infrared (IR) radiation which is partially blocked by the atmosphere. For instance, we all have experienced that on a sunny and cold winter day a closed space behind glass window warms up. This is because the window glass allows the visible part of sunlight but stops most of the IR reflected from heated interior surfaces. Increased heating caused by this phenomenon is known as "greenhouse effect" because it is used in agricultural greenhouses to grow plants in cooler weather.

Certain gases such as water vapor, carbon dioxide, methane, and others grouped as greenhouse gases (GHG) change the rate at which the atmosphere transmits the IR radiation reflected from Earth's surface that warms up by sunlight during the day. The natural greenhouse effect regulates the temperatures on Earth allowing normal seasonal changes. However, the modified greenhouse effect caused by a combination of gases released from human activities changes the thermal balance of the Earth causing local increase or decrease of temperature.

In addition to the greenhouse effect, the ozone layer at higher levels of the atmosphere (stratosphere) protects the Earth from the ultraviolet (UV) radiations coming from the Sun. It was discovered that certain gases such as sulfur hexafluoride (SF_6), hydrofluorocarbons (HFCs), and perfluorocarbons (PFCs) destroy the ozone layer at certain parts of the stratosphere. The combined consequences of the greenhouse effect and depletion of the ozone layer lead to a gradual temperature change in the long term, mostly perceived as "global warming." Regional changes of average temperatures result in climate change. At the 21st convention of United Nations Framework on Climate Change (UNFCC), 196 countries reached a consensus to limit the global temperature rise 2 °C above the pre-industrial era temperature levels by the year 2100.

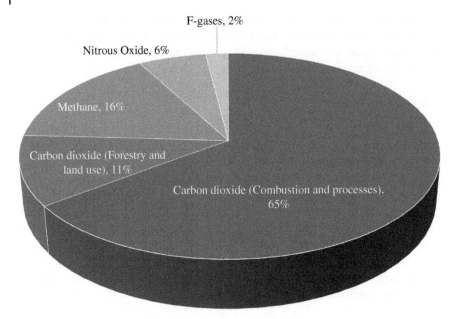

Figure 12.2 Composition of global greenhouse gas emission by mass. Source: IPCC (2014).

Carbon dioxide, nitrous oxide (N_2O), methane, and fluorinated gases (F-gases) are the main contributors to the greenhouse effect. Figure 12.2 shows the components of the most significant greenhouse gases in the global emission (IPCC 2014). Carbon dioxide and nitrous oxide are the dominant greenhouse gases produced by combustion of fossil fuels, biomass, and solid waste. Fluorinated gases are mostly released during industrial processes. Four sub-categories of fluorinated gases are hydrofluorocarbons (HFCs), perfluorocarbons (PFCs), sulfur hexafluoride (SF_6), and nitrogen trifluoride (NF_3).

Combustion of fossil fuels releases considerable amounts of carbon dioxide and water vapor resulting from the oxidation of hydrocarbons. Water vapor concentration in the atmosphere is mostly balanced by natural precipitation. Forests and vegetation offset some part of the CO_2 concentration in the atmosphere by photosynthesis. However, human activities release much more CO_2 than the natural cycle can remove. The excess CO_2 contributes to the climate change by increasing the greenhouse effect.

Methane is the largest byproduct of natural gas, therefore it is a valuable fuel. Although in small amounts, natural gas pipelines, and gas and oil processing facilities may leak methane to the atmosphere. A larger amount of methane is discharged naturally from coal mines and oil wells. If the discharged methane is not used on-site or commercialized as a byproduct, it is freely burned (flared) rather than released into the atmosphere to prevent its greenhouse effect. Methane also occurs naturally from the digestion process of livestock (mostly cows), fertilizers, organic waste, sewages, and swamps. Volcanic eruptions, geysers, mines, and wildfires also release a mixture of carbon monoxide, carbon dioxide, sulfur oxides, and methane to the atmosphere.

Human activities have significantly increased the GHG concentration in the atmosphere over more than 150 years, since the industrial revolution. Fossil fuel fired electric

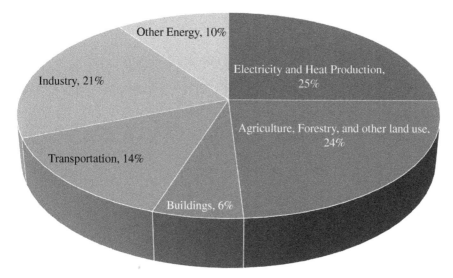

Figure 12.3 Sources of greenhouse gas emission. Source: IPCC (2014).

Table 12.1 Global warming potential of greenhouse gases.

Species	Chemical formula	Lifetime (yr)	Global Warming Potential (GWP)		
			20 yr	100 yr	500 yr
CO_2	CO_2	variable	1	1	1
Methane	CH_4	12 ± 3	56	21	6.5
Nitrous oxide	N_2O	120	280	310	170
Fluorinated gases	$C_xH_yF_z$	Variable	460–16,300	140–23,900	42–34,900

Source: *Climate Change 1995, The Science of Climate Change: Summary for Policymakers and Technical Summary of the Working Group I Report,* (IPCC 1996, p. 22).

generation, transportation vehicles powered by internal combustion engines, heavy construction and farming equipment, industrial heat production, chemical processes, and building heating systems are major sources of human made GHG emissions. Figure 12.3 shows the share of human activities in global GHG emission.

The global warming potential (GWP) of greenhouse gases are expressed relative to the equivalent amount of CO_2 in the atmosphere that would produce the same greenhouse effect. GWP of CO_2 is unity. Although the concentration of methane, nitrous oxide, and fluorinated gases are much lower than carbon dioxide, their lifetime in the atmosphere is longer, and their greenhouse effect is larger. The GWP of methane and nitrous gases released from energy-related applications are listed in Table 12.1 (IPCC 1996). The largest amount of fluorinated gases is emitted from industrial processes rather than energy conversion.

The average greenhouse gas emissions of selected fuels are listed in Table 12.2 based on the values compiled by the US EPA Greenhouse Gas Reporting Program. Although the

Table 12.2 Greenhouse gas emissions from combustion of selected fuels used in USA.

Fuel type	CO_2 kg/GJ	CH_4 g/GJ	N_2O g/GJ
Anthracite coal	98.3	10.4	1.5
Bituminous coal	88.4	10.4	1.5
Sub-bituminous coal	92.1	10.4	1.5
Lignite coal	92.6	10.4	1.5
Coal coke	107.7	10.4	1.5
Municipal solid waste	86.0	30.3	4.0
Petroleum coke (solid)	97.1	30.3	4.0
Plastics	71.1	30.3	4.0
Tires	81.5	30.3	4.0
Agricultural byproducts	112.0	30.3	4.0
Peat	106.0	30.3	4.0
Solid byproducts	100.0	30.3	4.0
Wood and wood residuals	88.9	6.8	3.4
Natural gas	50.3	0.9	0.1
Landfill gas	49.4	3.0	0.6
Other biomass gases	49.4	3.0	0.6
Biodiesel (100%)	70.0	1.0	0.1
Ethanol (100%)	64.9	1.0	0.1
Rendered animal fat	67.4	1.0	0.1
Vegetable oil	77.3	1.0	0.1

Source: US EPA Greenhouse Gas Reporting Program web site (accessed on 7/25/2019) https://www.epa.gov/ghgreporting.

amounts of methane and nitrous oxide are much lower than the amount of carbon dioxide, their greenhouse effects are higher. To obtain the GWP of each gas, the values should be multiplied with the corresponding GWP given in Table 12.1.

Example 12.1 Determine the total CO_2 equivalent of greenhouse gas emission per kWh output energy for an electric power plant burning bituminous coal electricity operating at 35% overall efficiency. Consider 100-year horizon for GWP calculation.

Solution
The GWP of gases per GJ released from burning bituminous coal is found using the values from Table 12.2 and Table 12.1 for 100-year horizon.

$$GWP_{fuel} = 88.4 + \frac{10.4 \times 21 + 1.5 \times 310}{1000} = 89.1 \text{ kg } CO_2 \text{ equivalent}$$

Table 12.3 Emission factors for end-use of common secondary sources.

Source	CO₂ equivalent	Unit
Electric generation (US National Average)	0.76	kg/kWh
Commercial boiler		
Bituminous coal	2929.00	kg/ton
Lignite	2437.00	kg/ton
Natural gas	2.42	kg/1000 m³
Residential fuel oil	3595.00	kg/1000 l
Distillate fuel oil	3222.00	kg/1000 l
LPG	1927.00	kg/1000 l
Stationary reciprocating engine		
Gasoline	3139.00	kg/1000 l
Distillate fuel oil	3212.00	kg/1000-l
Natural gas	2.64	kg/1000 m³
Small turbine		
Distillate fuel oil	3222.00	kg/1000 l
Natural gas	2.45	kg/1000 m³

Source: Deru and Torcellini (2007, p. 12, Tables 9 and 11).

To generate 1-kWh output energy, the power plant operating at 35% efficiency uses

$$W_{in} = \frac{1000 \times 3600}{0.35} \times 10^{-6} = 10.286 \text{ GJ}$$

input energy in gigajoule. The GWP of gas emission per kWh is:

$$GWP_{plant} = 10.286 \times 89.1 = 916.5 \text{ kg } CO_2 \text{ equivalent.} \therefore$$

The amount of greenhouse gas emissions depend on the fuel mix and technology. Table 12.3 shows the estimated average emission factors for common secondary sources delivered to end users. Emission factors for electricity depend on the fuel mix used for generation. Since electric generation stations powered from various primary sources supply an interconnected electric grid and regional interconnections exchange electric power, it is more reasonable to consider the national average emission. Local fuel mix and resulting emissions in the US can be obtained from the EPA Power Profiler web site (EPA 2019b). Table 12.3 also shows emission factors of various fuels burned in commercial boilers, reciprocating engines, and small turbines. The given values include the precombustion emissions that occur during extraction, processing, transformation, and transportation of fuels.

12.3.4 Carbon Footprint of Consumers

Carbon footprint generally refers to the total GWP of all gas emissions caused by energy related activities. Carbon footprint is associated with energy consumption rather than production, and it is expressed in tons of CO_2 equivalent gas emission per year. It is not limited to the on-site direct carbon dioxide emission from burning fuels but also includes the greenhouse effects of all gases released throughout the supply chain of fuels, electricity, and all other energy carriers delivered to users.

Calculation of the carbon footprint requires comprehensive data including GWP of all fuels used on-site including greenhouse gas emissions that occur during their production, processing, transformation, and delivery. Several institutions have developed spreadsheets and online tools to help consumers estimate their carbon footprint. For example, the Energy Profiler® developed by the US EPA gives the fuel mix used for electric generation in an interconnection and the GWP per unit electricity delivered to consumers. The Carbon Footprint Protocol of the World Resources Institute calculates the carbon footprint of a company, city, country, and other entities. Simple calculations can be made by using GWP of common fuels given in Table 12.3.

The Sankey diagram in Figure 12.4 illustrates carbon emissions produced by energy use in the US in 2014 (LLNL-DOE 2015). In the flowchart, only the emissions attributed to their physical sources are considered. Since emissions resulting from combustion of fuels in residential, commercial, industrial, and transportation sectors are not shown, the diagram does not realistically represent the carbon footprint of the country. However, it shows the areas where opportunities exist to reduce the carbon footprint of the country.

Electric generation and transportation are the major sources of carbon emissions. Average greenhouse gas emissions and total CO_2 equivalent emissions of electric generation plants in the US are listed in Table 12.4. Fossil fuel burning power plants are clearly the major sources of greenhouse gases, and replacing such power plants with renewables would significantly avoid carbon emissions.

The US national average of greenhouse gas emission is 760 kg/MWh (Table 12.3). Although regional grids are connected to each other, the energy exchanges to balance load flow do not have a significant effect on local emission rates. Users may approximate the carbon dioxide equivalent emissions due to the consumption of grid electricity on site considering the fuel mix of the utility that supplies their region. Transmission and distribution losses must be added to the on-site consumption to obtain the total actual electricity delivered from generation plants. In the US, transmission losses range between 4.79% and 5.63%, with the national average of 4.95%. The Power Profiler online calculator (https://www.epa.gov/energy/power-profiler#/) is a useful tool to estimate the contribution of consumers in the US to the greenhouse gas emissions by using grid-supplied electricity. On-site electric generation from wind and/or solar energy is an effective way to reduce the end-users carbon footprint.

The transportation sector mainly relies on petroleum derivative liquids. Electric powered public transportation, hybrid vehicles (HVs), and electric vehicles (EVs) reduce both dependence on petroleum and the carbon footprint. However, electric public transportation and EVs still indirectly cause emissions through electric generation. Therefore, reducing the emissions created by electric generation while increasing the share of electric power in transportation would reduce the carbon footprint of countries more successfully.

Estimated U.S. Carbon Emissions in 2014: ~5,410 Million Metric Tons

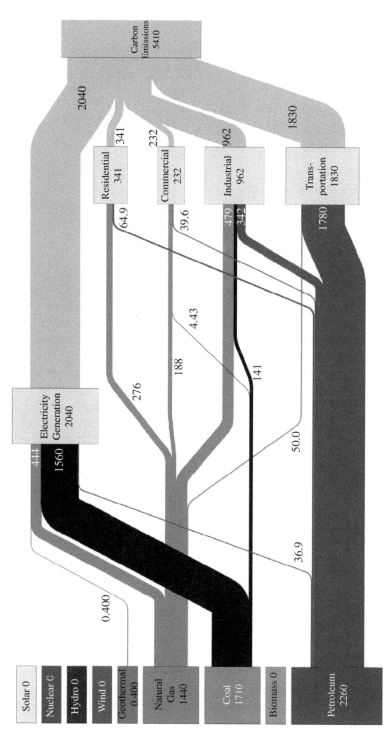

Figure 12.4 Carbon emissions from energy usage of sectors in the USA. Source: LLNL 2015. Data is based on DOE/EIA-0035(2015-03), March, 2015. Prepared by Lawrence Livermore National Laboratory and the Department of Energy.

Table 12.4 Average greenhouse gas emissions of power plants in the US by primary energy source.

	CO_2 (kg/MWh)	CH_4 (kg/GWh)	N_2O (kg/GWh)	CO_2 equivalent emission rate (kg/MWh)
Coal	1278.4	140.6	21.0	1285.1
Gas	1156.2	20.4	2.4	1133.0
Oil	1391.3	56.7	11.1	1392.0
Geothermal	43.2	0.0	0.0	43.2
Biomass	113.1	175.7	23.8	122.3

Source: US EPA eGrid database 2016 (EPA 2019a).

12.4 Energy, Water, and Food Interactions

Water, food, and energy are crucial necessities for human life, well-being, and economic growth of the society. On one hand, water is essential for production of fuels and cooling of power plants. On the other, energy use is vital for a range of water processes, including extraction, treatment, desalination, purification, and transport of fresh water. These two necessities are strongly interrelated since energy conversion needs water, and treatment and supply of fresh water needs energy.

Both water and energy are the major elements of the modern food supply chain. Agriculture relies on adequate water for irrigation and fuel to operate farming equipment. Processing of agricultural produces and meat needs clean freshwater and reliable electric supply. Continuity and reliability of electric supply is critical for refrigerated food storage. Transportation and distribution of food products use fuels.

Water pollution is inevitable during production and transformation of secondary energy sources. Agricultural pesticides and chemicals used in the food industry ultimately drain to freshwater sources. In addition, greenhouse gases released from production and use of energy is the major cause for climate change, which would affect the water and food supply in long term.

The hybrid Sankey diagram prepared by the U.S. Department of Energy's Water-Energy Technology Team (Figure 12.5) illustrates the interactions of energy and water flows in the US from sources to consumer sectors. Energy values are estimated for 2011; water values are based on available data from 2005 to 2011 (DOE 2014). The diagram clearly shows how water and energy flows are interdependent. A significant amount of water is used for cooling of thermoelectric generation units. The largest fraction of the cooling water is discharged to the ground surface and some part is drained to the oceans. While the total amount of water remains constant, energy production uses about half of the water available from all sources. Agriculture, which is the main source for the food industry, uses about one-third of the total available water mainly for irrigation. On the other hand, public water supply and wastewater treatment services depend on electric energy. Although the water and energy usages are different for each country, the interactions shown in Figure 12.5 is a typical example for the developed part of the world.

2011 Estimated U.S. Energy-Water Flow Diagram

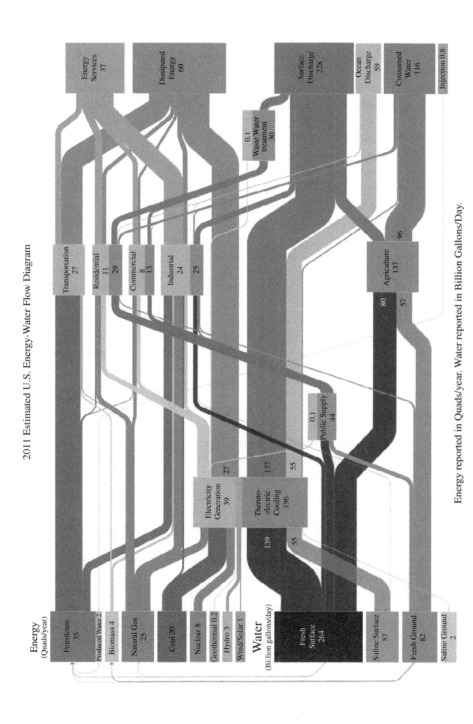

Energy reported in Quads/year. Water reported in Billion Gallons/Day.

Figure 12.5 Water and energy interdependency in the USA. Source: DOE 2014, p. 10, Figure 2.1.

12.4.1 Water Sources

About 71% of the Earth's surface is water-covered. Oceans hold approximately 96.5% of all the water on earth. Water also exists in the air as vapor, and in the form of clouds, fog, and humidity. Freshwater sources are grouped as surface and ground water. Water accumulated in rivers, streams, reservoirs, and lakes is surface water. Snow, icecaps, and glaciers are freshwater reserves in solid form. Water penetrated in the ground as soil moisture or accumulated below the land surface in the soil is called groundwater. Surface and groundwater sources are regenerated via the natural cycle.

Aquifers are large, deep layers of sand and gravel that store or transmit groundwater. Some aquifers contain water from the ice age. The recharging rate of aquifers is extremely slow compared to a human lifetime; therefore, they are considered non-renewable water sources. Major aquifers located under the Midwest and the high plains regions in the US provide ground water resources for several states. According to USGS, the water levels in aquifers is being rapidly depleted by growing irrigation and contaminated by fertilizers and herbicides used in agriculture.

Freshwater containing small amounts of dissolved chemicals may be used for irrigation, public services, cooling, and certain industrial processes. Most freshwater sources must be disinfected, filtered, and purified to be used as drinking water. Non-freshwater (brown water) resources include saltwater, urban or industrial wastewater, and agricultural drainage water. Such water sources contain minerals, bacteria, algae, and other contaminants above levels acceptable for specific end-uses. Brown water can however, be purified to supplement freshwater resources. All water treatment facilities consume energy in the form of heat or electricity.

12.4.2 Water Use for Energy

Worldwide, energy systems withdraw about 10% of all available water mainly for power plant operations, as well as production of fossil fuels and biofuels (IEA 2016). In the US, nearly half of all available water is used for energy as seen in Figure 12.5 (DOE 2014).

According to the IEA World Energy Outlook 2016, 397 billion cubic meters (bcm) of water was used in 2014 worldwide for energy related purposes (IEA 2009, p. 13). Figure 12.6 shows the global water withdrawals and consumptions for primary energy production and power generation. Power generation units withdraw considerably more water (350 bcm) than primary energy production facilities (47 bcm). Generation stations, however, return a large part of the withdrawn water to the same source while a considerable amount of water used for primary energy production is smaller, and is evaporated or transferred to another source. Evaporated water will eventually return to the ground as precipitation but not necessarily to the same location, therefore energy-related operations may cause water shortage in regions where water availability is limited.

Extraction and processing of primary energy sources worldwide used 47 billion cubic meters of water in 2014. About 35% of this water returned directly to the ground while 65% evaporated or drained to other sources. Water withdrawal for primary energy production varies over a wide range depending on the type of fuel and operation. Among primary sources, biofuels consume the largest portion of water, followed by coal and oil production.

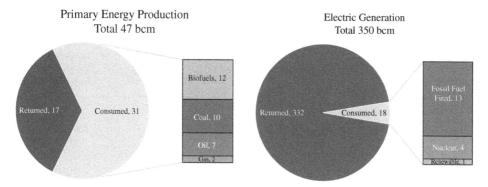

Figure 12.6 Global water withdrawals for primary energy production and electric generation in 2014. Source: IEA (2016, p. 13, Table 1).

Water is mainly needed for cultivation of biomass feedstock. The water input comes from precipitation and irrigation. Part of the water returns to the ground, and part is evaporated or absorbed by the plants. The amount of irrigation water depends on the soil characteristics and climate. For example, ethanol corn grown in the northern plains of the US needs more groundwater irrigation than the East Central region. In regions where significant irrigation is needed, biofuel farmers may have to pump water out from aquifers, which are not sustainable water resources. In addition to irrigation, biofuel production requires considerable amounts of water for grinding, liquefaction, separation, and steam drying.

Ethanol is one of the most popular biofuels mixed with gasoline. Ethanol can be produced from various plants such as sugarcane, soybean, corn, and cellulosic fibers. Cellulosic ethanol production consumes 1.9–5.4 gal of water for each gallon of ethanol produced, but if switchgrass and forest wood residue is used as feedstock, no additional water is used for irrigation. Corn ethanol production is, however, extremely water-intensive. Dry mill operations to produce corn ethanol consume about three gallons of water for one gallon of ethanol. Water consumption from corn farming to production of corn ethanol can be as high as 160 gal of water per gallon of ethanol produced. Table 12.5 shows the volumes of water consumed to produce one unit volume of corn ethanol by the US Department of Agriculture regions (Wu and Chiu 2011).

In coal production, water is used for mining as well as processing. In some coal mines, groundwater must be removed before production can begin. The highly contaminated water may be treated to be used as freshwater. Coal preparation and handling facilities use considerable amounts of water for separation of coal from soil and other minerals. Water is frequently sprinkled during surface mining and coal storage to prevent fires.

Conventional and unconventional oil drilling techniques use water for extraction. Primary oil recovery in conventional wells is based on bringing the oil, gas, and water mixture to the surface by natural pressure. As oil wells age, production from primary recovery declines and secondary recovery, also known as water flooding, is used to enhance extraction. In secondary recovery, separate wells are drilled to inject water into the formation. Whereas much of the injection water is recycled, fresh groundwater may be also needed. Enhanced oil recovery (EOR) is a used to improve oil production. One of the EOR methods

Table 12.5 Water consumption to produce one unit volume of corn ethanol in the US.

USDA region	Region 5 Iowa, Indiana, Illinois, Ohio, Missouri	Region 6 Minnesota, Wisconsin, Michigan	Region 7 North Dakota, South Dakota, Nebraska, Kansas
Share of US ethanol production (%)	50	15	23
Corn irrigation (groundwater)	12	19	224
Corn irrigation surface water	2	3	12
Ethanol production	3	3	3
Total without co-product allocation	17	25	239
Total with co-product allocation	11	17	160

Source: Wu and Chiu (2011, p. 34, Table 5).

is carbon dioxide injection to reduce surface tension. Steam, or micellar polymer injection is used to reduce the viscosity of the crude oil. Injection water requirements vary with recovery technology. While primary recovery needs about 0.21 gal freshwater to extract one gallon of crude oil, secondary recovery may need up to 8.6 gal of water per gallon of crude oil. Water intensive EOR technologies may use as much as 343 gal of water to extract one gallon of crude oil. Based on 2005 production data, 1171 million gallons per day was used in 2005 for EOR technologies (Wu and Chiu 2011, p. 45, Tables 8 and 9).

Hydraulic fracturing techniques used for unconventional oil and gas recovery require large amounts of water. Surface mining of oil sands and steam-assisted gravity drainage (SAGD) technologies used to reduce the viscosity of extra heavy oil are also water intensive. In addition to withdrawal, chemicals used for fracturing present a risk of ground water contamination.

About 70% of the electric power is generated in thermal power stations. Such generation units withdraw the largest fraction of water for cooling a working fluid circulated in heat engines or equipment heated because of the losses. Cooling is particularly an essential part of thermoelectric generation. Throughout the thermodynamic cycle, the condenser discharges part of the heat in the working fluid to a low temperature sink, which is mostly a river, lake, or sea inlet. A larger amount of the cooling water returns to the source while some evaporates or drains out to another source. Water evaporated during electric generation or drained to sea is consumed since it is not returned to the freshwater sources. Advanced air-cooling techniques and combined heat and power (CHP) generation systems have reduced the amount of water withdrawn and consumed by power generation. Renewables includes bioenergy, geothermal, concentrating solar power (CSP), solar photovoltaics (PV), and wind. PV and wind generation units do not use water. Bioenergy, geothermal, and CSP withdraw cooling water as other thermoelectric generation.

12.4.3 Energy Use for Water

Collection, treatment transfer, and distribution of freshwater rely on energy supply. Energy is mainly needed for extraction of groundwater and surface water, transfer from one location

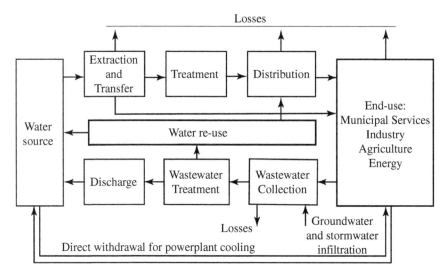

Figure 12.7 Energy-related segments of water processing.

to another, filtration, distillation, purification, and distribution of clean water to the users. Figure 12.7 illustrates the major processes from the water sources to end-users and from end-users back to sources. All steps of this block diagram involve some form of energy.

Water sources include groundwater, surface water, and salt water. Freshwater can be directly extracted from the ground or collected from surface water sources. In regions where freshwater sources are scarce, desalination facilities produce freshwater from saltwater drawn from seas and oceans. Water collection and desalination are the most energy intensive processes. Extraction of groundwater and transfer over long distances to treatment facilities uses electric energy to operate pump motors. Desalination processes involves evaporation, which is also energy intensive. Various fuels as well as renewables such as geothermal or solar energy can be used to produce heat needed for the desalination process.

Water treatment processes includes filtration, decontamination, purification, and softening. The treatment process is mostly electric powered. The water distribution system adjusts water pressure and conveys clean water to end-users. Some industrial processes may withdraw untreated water directly from extraction facilities or surface water without passing through the distribution network. Cooling water for industry and electric generation is typically withdrawn from the surface water or saltwater sources and returned to the same source. Some part of the cooling water is lost by evaporation.

Industrial and municipal wastewater treatment plants use electric energy to circulate water between storage pools and a wide variety of purification processes to remove contaminants. Wastewater may be collected and discharged directly by gravity or by forced circulation using pumps. Part of the treated water may be re-used, and the rest is returned to the sources.

Estimation of water related energy consumption is complicated because specific processing facilities use energy for many other purposes too. While some developed countries offer relatively realistic information about water-related energy consumption, estimation of energy used for global water supply is based on surveys and assumptions, and is often unreliable data.

Energy intensity for water supply depends on the type, location, and characteristics of water sources. Surface water carried by a river may be simply collected in a pond by gravity, but extraction of groundwater from wells depends on the depth and pump efficiency. Desalination technologies consume significant amounts of energy to remove dissolved salts from water. Desalination plants use energy-intensive distillation or reverse osmosis technologies, as distillation requires evaporation of large amounts of water. A reverse osmosis process pressurizes water to pass through semi-permeable membranes. The amount of energy consumption in all desalination methods depends on the salinity of the source water. For example, seawater desalination requires significantly more energy than slightly salty river water.

Topography, distances, and pumping efficiency are the main factors that affect energy used for water conveyance. Energy used in treatment processes depends on the quality of both the water source and the freshwater supplied to users. Wastewater treatment energy is also subject to the composition of the rejected water (CEC 2005). Table 12.6 shows energy intensities of main segments of the water cycle estimated for water treatment and pumping in California.

Based on the estimates of the IEA, the global energy use for water services is around 120 million tons of oil equivalent (Mtoe) per year. More than half of this energy is consumed in the form of electricity, totaling 820 TWh, which represents about 4% of the total global electricity consumption (IEA 2018). Figure 12.8 show the fraction of energy use for various activities related to freshwater supply and wastewater treatment in the world.

Globally, the largest amount of electricity is used for freshwater extraction. Wastewater collection and treatment plants consume about one-quarter of the electric power delivered for all water services. In developed countries, energy used for wastewater treatment can rise up to 40%. Approximately 20% of electricity is consumed for distribution and delivery

Table 12.6　Energy intensity of water treatment and pumping in California.

Water cycle segments	kWh/million gallon
Treatment	
Drinking water treatment	100–16,000
Wastewater treatment and distribution	1100–4600
Pumping	
Water supply/conveyance	0–14,000
Primary drinking water distribution	700–1200
Recycled water distribution	400–1200
Groundwater for agriculture	500–1500

Sources: CEC (2005) and CPUC (2010).

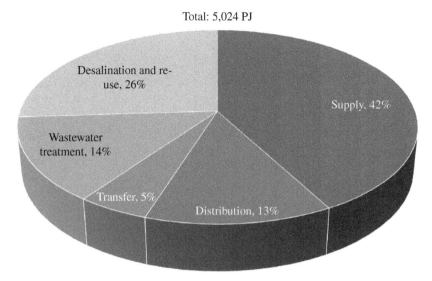

Total: 5,024 PJ

Figure 12.8 Global energy use for water services. Source: IEA (2018, p. 3, Figure 1.1).

of water to consumers. Although seawater desalination is one of the most energy-intensive processes, electric use is smaller because other primary sources such as solar and geothermal are used to evaporate saltwater for distillation.

12.4.4 Energy Invested for Energy

Extraction, process, and delivery of energy sources to end-users require a certain energy input; for example, drilling oil wells, opening mineshafts, and surface-mining use heavy equipment that consume gasoline or diesel fuel. Transport systems to carry extracted coal, oil, or gas to processing plants consume various energy sources. Mechanical or chemical processes to obtain better quality and usable fuels also consume energy. Agricultural equipment, machines, and vehicles to grow feedstock for the biofuel industry needs refined petroleum products (gasoline, diesel, kerosene, lubricants, etc.) for their operation. In addition, all kinds of fuels are transported to the end-users by trains, trucks, and ships, which also consume energy.

The indicator known as "Energy Return on Energy Invested" is the ratio of energy acquired from a fuel to the total energy consumed for its production, including processing, transportation, and delivery to consumers. This dimensionless ratio is represented by the acronyms EROEI, ERoEI, or simply EROI.

The concept of energy return on energy invested was originally defined in the 1970s, when the cost and future availability of oil became a concern. The expression below is the general definition of EROI:

$$EROI = \frac{\text{Energy acquired}}{\text{Energy used to get that energy}} \tag{12.1}$$

EROI should not be confused with conversion efficiency, which is the ratio of the output to the input energy. Efficiency is always less than unity (or 100%) since the output energy of

Table 12.7 $EROI_{st}$ for various energy resources used in the US.

Resource	$EROI_{st}$
Domestic oil and gas	11–18
Imported oil	18
Natural gas	10
Coal	80
Bitumen from tar sands	2–4
Shale oil	5
Nuclear	5–15
Hydropower	> 100
Wind turbines	18
Solar PV	6.6
Flat plate solar collectors	1.9
Concentrated solar collectors	1.6
Sugarcane based ethanol	0.8–10
Corn based ethanol	0.8–6
Biodiesel	1.3

Source: Murphy and Hall (2010).

a conversion device is smaller than the input energy due to losses. EROI, however, must be greater than unity to benefit from a resource. In other words, a resource is more beneficial if it can produce more energy than the energy used to make it available to consumers.

Estimation of the EROI is often challenging and depends on which point of the energy supply chain it is defining. From the point of production to the consumer, energy involved in transportation and distribution services indirectly affect the overall EROI.

$EROI_{st}$ is the standard EROI obtained as the ratio of the energy acquired at point of production to the sum of direct and indirect energy used to explore the source, reach the reserve, and extract that energy. It is estimated for the energy produced at the mine mouth, wellhead, or farm gate. Direct energy is the sum of all energy used on site to extract or produce the primary energy source. Indirect energy is the offsite energy to explore the sources, produce materials, manufacture equipment, build structures required for production, and support the production on site. Equation (12.2) is the most common EROI definition used to compare different primary sources. Table 12.7 shows estimated $EROI_{st}$ for common fuels.

$$EROI_{st} = \frac{\text{Energy acquired at the source}}{\text{Energy used to find and extract that energy}} \qquad (12.2)$$

$EROI_{pou}$ is the ratio of the energy acquired at the point of use to the total energy invested to find, reach, extract, process, and supply that energy to end-users. The denominator includes additional energy needed to deliver the resources to consumers. For example, the fuel consumed by trucks, trains, and ships to carry the fuel to the point where it is used are

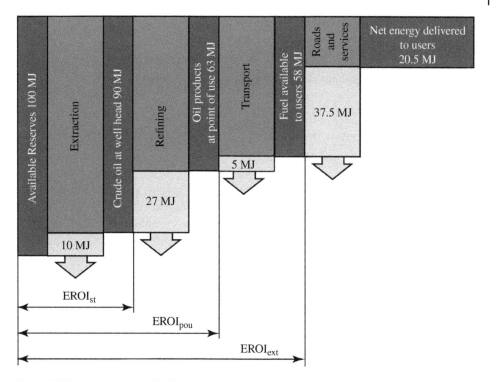

Figure 12.9 An example of EROI boundaries for petroleum. Source: Gagnon et al. 2009.

included in the calculation.

$$\text{EROI}_{\text{pou}} = \frac{\text{Energy supplied to society}}{\text{Energy used to get and deliver that energy}} \tag{12.3}$$

EROI_{ext} extends the EROI_{pou} by including additional energy required to enable the use this energy. For example, the energy used to maintain the transportation infrastructures such as roads, bridges, etc., that are necessary to use transportation fuels is considered in the estimation.

$$\text{EROI}_{\text{ext}} = \frac{\text{Energy supplied to society}}{\text{Energy consumed to get, deliver, and use that energy}} \tag{12.4}$$

EROI analysis is important for cost-benefit comparison of various resources, assessment of the resource feasibility, and evaluation of the commercial viability of a potential resource. However, selected boundaries must be consistent to make meaningful comparisons. Different definitions of EROI are illustrated in Figure 12.9

Energy production and consumption data are available at a great accuracy from databases provided by several institutions (e.g. US Energy Information Administration, International Energy Agency, United Nations, World Bank, etc.). Therefore, energy returned to society is usually determined quite easily. However, estimation of the energy consumed to produce the usable energy source requires reliable and accurate industrial, commercial, and agricultural data.

Example 12.2 Calculate standard, point-of-use, and extended EROI for petroleum using data given in Figure 12.9.

Solution
Applying the EROI definitions at the boundaries shown in Figure 12.9, we obtain:

$$EROI_{st} = \frac{90}{10} = 9$$

$$EROI_{st} = \frac{63}{10 + 27} = 1.70$$

$$EROI_{st} = \frac{58}{10 + 27 + 5} = 1.38$$

Note that one unit of energy consumed yields nine units of energy at the wellhead. At the point of use, EROI falls to 1.70 due to the energy used to transform crude oil to usable fuels. When the energy and services used to deliver the fuels to consumers are included, $EROI_{ext}$ drops to 1.38. In other words, to deliver fuel containing 1.38-unit energy to the consumers, 1 unit of energy is consumed to make that fuel available to end-users.

Studies show that a fuel cannot sustain without the support of other fuels if the EROI at the mine-mouth or farm-gate is less than 3 : 1 (Hall et al. 2013). Therefore, with current technologies, many alternative energy sources are not sustainable if not subsidized with fossil fuels.

EROI is not the only factor considered in social and economic decisions. For example, coal has higher EROI than natural gas; however, natural gas replaced coal in residential space heating because gas is cleaner, more cost efficient, and easier to transport. Global electric generation, on the other hand, still depends on low quality coal abundantly available in many regions of the world.

Biodiesel and corn-based ethanol have lowest EROI compared to other energy sources. In the United States, motor gasoline is blended with about 10% corn-based ethanol to meet the requirements of the 1990 Clean Air Act. In many cities with high smog levels the use of reformulated gasoline (RFG) has been required since 1995. The US Energy Information Administration (EIA) estimates that blending motor gasoline with 10% ethanol reduces the fuel efficiency by about 3%. Addition of corn-based ethanol to gasoline has been a source of public debate because of lower EROI, efficiency decrease, and land use. Similarly, lower EROI and environmental impacts rose public concerns about the use of oil sands, and shale gas (Table 12.7).

12.5 Energy Management

Energy management is coordination of planning, development, and operation of an energy system to supply energy needs of the society. Proactive and balanced management of production, conversion, and distribution of energy sources is central for sustainable development of a society since all energy related activities are closely related to natural resources, freshwater availability, air quality, food supply, and global climate changes. Sensible energy management considers cost-effective and reliable operation of the energy system while preserving natural resources, protecting the environment, and observing ethical values. Energy management activities can be performed at various levels including resource management; coordination of fuel transportation; balancing electric generation, transmission, and load flow; facility management; and building energy management. To simplify, we can group such activities into two major groups as resource coordination and load-side energy management. All layers of energy management are obviously linked since energy system is a whole from resources to final consumption.

12.5.1 Resource Coordination

Energy resource coordination is proactive administration of all available energy sources to allow production of fuels, electricity, and other energy carriers necessary for operation and development of a society. Primary energy mix that feeds a country's energy system depends on many factors including resource availability, development goals, economic conditions, international agreements, and political decisions. Typically, national and local administrations develop policies to regulate long-term use of the available resources. In free democratic countries, public opinions, concerns, and reactions weigh in political preferences.

Utilization of energy sources in a country affects the global balance of natural resources, economic development, and climate change. The United Nations Framework Convention on Climate Change (UNFCCC), Organization for Economic Cooperation and Development (OECD), Organization of Petroleum Exporting Countries (OPEC), Intergovernmental Panel on Climate Change (IPCC), and other international organizations set common goals that constrain the exploitation of energy resources to protect mutual interests of the member countries.

The decision of OPEC members in 1974 to limit oil production that led to the world-wide oil crisis has been a crossroad in the evolution of the global energy industry. The 1997 Kyoto Protocol and subsequent Paris Agreement in 2017 signed by the leaders of the UNFCC member countries set international goals to reduce the risks and effects of climate change. The Paris Agreement significantly affects long-term energy policies and resource allocations around the world.

Large rivers are critical resources for energy, agriculture, and transportation. Interests of competing sectors are typically resolved through national regulations. However, many big rivers cross political boundaries of several countries. According to the United Nations Department of Economic and Social Affairs (UNDESA), 145 nations around the world have territories within international river basins. The timing and flow rate of water released by upstream users has crucial implications downstream. For example, agricultural users in a downstream country may need water for irrigation at the same time as an upstream country needs it for hydroelectric generation. Many countries sharing cross-border rivers have negotiated the use of water resources and signed agreements. In some situations, however, disputes over shared water resources may escalate to international conflicts. As of 2006, out of 145 agreements on transboundary water resources, 57 were on hydropower management (UNDP 2006, p. 222, Figure 6.2).

Significant financial investment is necessary for exploration of resources and development of infrastructure for extraction, processing, and distribution of energy sources. Historically, large multinational companies have dominated oil exploration and recovery projects. Coal mines, hydroelectric dams, and nuclear facilities are typically developed by consortiums of large multinational companies. Such complex projects need long-term commitments and significant support of governments. Consequently, resource coordination for sustainable development of a national energy system depends principally on socioeconomic factors and national policies rather than resource availability and possible technical options.

12.5.2 Supply-side Energy Management

The purpose of supply-side energy management is coordination of activities involving production and distribution of secondary energy sources to the consumers. Since continuous supply to consumers relies on the effectiveness of the distribution network, preferences have shifted from lower to higher energy-intensity fuels. Hence, throughout the evolution of energy systems, liquids replaced solid fuels and gases replaced liquids. In modern energy systems, liquid fuels derived from petroleum and liquefied petroleum gases (LPGs) are primarily used in the transportation sector and mobile applications, while natural gas and electricity are essential energy sources for residential and commercial sectors. Hydrogen is emerging as an energy carrier for both transportation and stationary applications.

Management of fuel supply relies on a transportation network from production facilities to retail dealers. Although fuels can be stocked for an extended time in physical storage facilities, fuel production ideally must follow the consumption. Therefore, supply-side energy management is mostly driven by the supply-demand balance, which affects energy prices. Energy cost, in turn, impacts the composition of the consumed energy mix.

Energy transfer in the form of electric power enables supply of a fuel mix including renewables, nuclear energy, municipal solid waste, and low quality fuels that are not cost-effectively supplied to consumers. Electricity can be converted to all other forms of energy needed for practical applications. From an energy management standpoint, the largest challenge of electric supply is direct storage of electric energy. Although the unit cost of batteries is decreasing with modern technologies, electric storage in large amounts is expensive. Therefore, electric power must be generated as much as it is consumed at any time.

Consumption of electric power depends on consumer activities. Industrial and commercial consumers draw more electric power during operation hours. Residential users consume more electricity during the evening hours for lighting and entertainment. Street and building lighting also need more electric power after dark. In an interconnection that combines diverse groups of users, time variations of the electric consumption become smoother. As an example, monthly and hourly variations of the electric demand in the state of New York are illustrated in Figure 12.10.

Separate regional grids are generally electrically linked to create larger interconnected network. In the continental US, three main interconnections comprise 10 regional grids (Figure 12.11). The US grid has interconnections with Canadian and Mexican grids, this North American interconnected network is managed by North American Electric Reliability Corporation (NERC). National electric grids of countries in Europe and Asia are as well interconnected. Individual interconnections may implement different load management strategies to optimize the operation of the electric supply system, but cross-boundary interconnections are managed based on common quality and reliability standards.

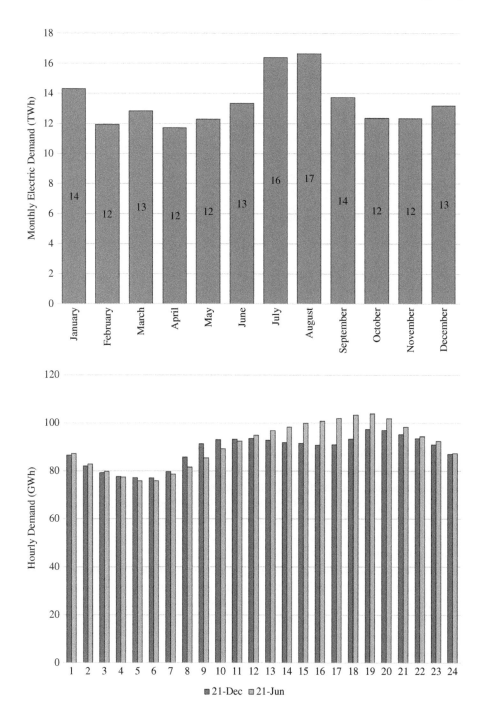

Figure 12.10 Monthly and daily variations of electricity demand in 2018 load in New York. Source: EIA, U.S. Electric System Operating Data, https://www.eia.gov/realtime_grid (accessed 8/13/19).

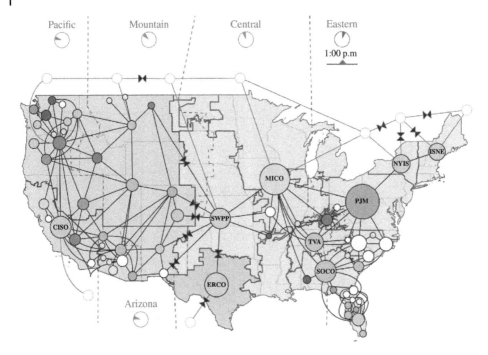

Figure 12.11 Electric interconnections in the continental United States. Source: Diagram provided by EIA website https://www.eia.gov/realtime_grid (accessed 8/29/2019).

Controllability of generation stations depends on the primary energy source used and conversion techniques. For example, the output power of gas-burning and hydroelectric stations can be controlled more easily than nuclear power plants as the demand changes. On the other hand, the output of wind and solar generation change continuously over a wide range with variable wind speed and solar irradiance.

For cost-effective operation of the electric supply system, larger generation facilities that have higher initial cost are loaded near their installed capacity as much as possible. Such units are called *base-load power plants*. Smaller units that are easier to control are operated at variable output power to supply load peaks.

To balance supply and demand variations, various types of electric generation facilities are interconnected to supply diverse groups of consumers spread over a wide geographic area.

Power outputs of generation stations are coordinated in such a way that the total electric generation equals the total consumption at any instant. In an interconnected network, energy flow between generation plants and consumers continuously changes. Pumped-storage hydroelectric plants enable bidirectional energy flow depending on the supply-demand balance. At late night, when electric consumption is lower, these facilities pump water to a higher pool to store excess electricity in the form of potential energy. At peak hours, water stored in the higher pool is used to generate electricity.

Electric power generation offers additional opportunities in coordination of energy flow from production to end-users. Primary sources that cannot be directly delivered to consumers and low-quality fuels that are not cost-effective for long distance transportation or unsuitable for use in industrial, commercial, or residential applications can be included in the fuel mix for electric generation. For example, long-distance transportation of lignite is not cost-effective because of its low heating value and relatively higher water content. In addition, it is not suitable for most industrial, commercial, or residential heating applications because of the higher amounts of ash, air pollution, and solid contaminants it produces. However, lignite can be used to generate electricity in a power plant installed close to the mine. The large amounts of ash and residues that cannot be recycled for industrial byproducts are typically transported back to the mine for landfill. Another example is coordination of electric generation and natural gas production to optimize the supply-demand balance of both energy sources. Most residential and commercial buildings use natural gas for heating in winter. In summer, however, electric consumption increases because of air conditioning use. Therefore, natural gas production facilities can be operated at higher capacity year-round if the excess natural gas is used for electric generation in summer to fulfill the increased air conditioning load.

The main objectives of supply-side energy management programs include optimizing the energy mix and providing consumers with reliable and affordable energy services. Since energy supply is a strategic issue for welfare and economic development of a nation, governments control energy supply by laws, standards, regulations, and codes. In many countries, state-owned establishments provide essential energy production and distribution services. In a monopolized economy, energy mix used for electric generation and energy prices are centrally determined by the administrative offices, leaving no options to consumers.

Starting from the 1980s, the concept known as "deregulation" has introduced competition into the electric power industry in several states of the US. In deregulated energy markets, energy prices vary among companies and consumers can choose their energy provider. Hence, consumer choices may eventually affect the composition of primary sources, methods used for electric generation, and supply-side energy management procedures.

12.5.3 Load-side Energy Management

Consumers have several options to meet their energy needs using various fuels, electric power, and other energy carriers. The types of secondary energy sources depend on the final application. For example, various fuels, electricity, or district steam can be used for heating; but light fixtures, household appliances, communication devices, and computers use electricity exclusively. The main objective of the load-side energy management is optimization of the energy consumption by selecting suitable energy sources and controlling the operation of devices to maintain the desired level of productivity and comfort. Conscious consumers also consider the impacts of their final energy use on the environment and natural resources.

Load-side energy management is mainly implemented in buildings and facilities. Improving energy efficiency is the first step to save energy. Well-insulated building envelopes, effective use of daylight, and passive solar heating significantly reduce the energy consumption in buildings (Soysal et al. 2016).

Consumers use a diverse combination of secondary sources to meet their energy needs; typical sources used in buildings are electricity, natural gas, fuel oil, LPG, and biomass

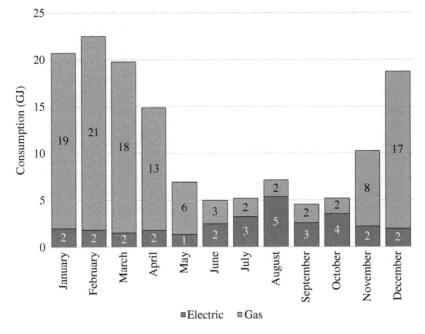

Figure 12.12 Monthly energy consumption of a typical residence in a moderate climate using natural gas for space and water heating, electricity for all other applications.

(wood pellets, firewood, etc.). In developed countries, electricity and natural gas are the main supply for residential consumers. In less developed and poor regions, conventional biomass, charcoal, and coal is extensively used. Use of low-quality fuels impact air quality, produce excessive solid waste, and lower the living standards.

Space and water heating, ventilation, air conditioning, cooking, refrigeration, washing, and powering various consumer electronics devices are among major end-use applications. The consumption of each source changes depending on the activities and outdoor temperature, showing short-term (hourly) and long-term (seasonal) variations. Figure 12.12 shows an example household in a moderate climate using only natural gas and electricity. Natural gas consumption is higher in winter and electricity consumption is higher in summer because of air conditioning. Seasonal variations of natural gas and electricity consumption of similar users vary by climate.

Energy profiles of consumers affect the management of energy sources. Beside the total energy need, energy flow rate is also an important parameter for reliable and continuous energy supply. Utilities try to forecast variations of demand in advance to plan daily and seasonal energy supply strategies. Government policies may impose particular ways and timing of nationwide consumption of certain energy sources. Implementation of daylight-saving time and restriction of driving private cars in certain days are examples of policies for regulation of energy usage.

Pricing strategies are effective ways to modify the consumers' energy demand profiles. Instead of charging a fixed-rate for electricity, utility companies may apply tiered pricing that increases the price rate when consumption exceeds certain thresholds. Time-of-use pricing changes the kilowatt-hour price according to days of the week and hours of the day depending on the peak demand times. In general, weekends are off-peak since businesses and some industrial facilities are not open. Peak hours of the workdays vary by seasons. In winter, electric demand is generally higher from 7 to 11 a.m. and from 5 to 7 p.m. In summer months, these time slots are mid-peak, and from 11 a.m. to 5 p.m. electric demands peak mainly because of air conditioning. During off-peak hours the price rate of electricity can be as low as half the rate charged during on-peak hours. In order to implement time-of-use rates, the utility company must install smart meters that record the time of energy use at the entry point of each customer.

During the energy crisis in the 1970s, the concept of energy conservation has evolved. To avoid confusion with the fundamental "conservation of energy" principle of physics, we will use the term "energy conservation" for various ways of avoiding unnecessary energy consumption. Energy conservation is basically reducing the energy consumption by avoiding certain applications. In extreme energy shortage periods, especially during the energy crisis, some governments restricted lighting of streets and public places, illumination of buildings during non-working hours, and decorative lights, and even applied scheduled electric outages up to several hours during peak-hours. Whereas such strategies may be short-term solutions to energy problems, they have adverse impacts on productivity, social life, education, health, public services, and security. Improving energy efficiency, rather than restricting energy usage, is a rational long-term solution for sustainable development.

The following basic principles can be applied to save energy in buildings and facilities:

- Operating equipment only when needed and as much as required
- Eliminating simultaneous use of heating and cooling equipment
- Setting temperatures at appropriate levels
- Using passive solar heating and lighting effectively
- Supplying heaters and coolers from the most efficient sources
- Minimizing energy losses by using energy-efficient equipment
- Operating power equipment around the design specifications.

Energy flow in buildings and facilities can be optimized by using dedicated feedback control systems. A typical building energy management control system (EMCS) collects energy-related information by temperature, humidity, occupancy, and light sensors installed at multiple points in the building. Computer software processes the information received from sensors and energy meters, then turns on or off HVAC (heating, ventilation, and air conditioning) units, lights, and electric devices to optimize energy consumption. Some of the equipment may have individual sensors and feedback control systems.

Equipment used in buildings can be grouped as critical, non-critical, and deferrable. For example, communication, data processing, and security systems are usually critical equipment and require uninterruptible power supply. Equipment that can tolerate short-term energy interruptions like refrigerators, home entertainment, water heaters, cooking ranges, and non-emergency lighting are non-critical. Appliances that can be used at a scheduled time like a clothes washer, dryer, and dishwasher are deferrable equipment. Classification of equipment depends on the activities supported by the supplied energy. For example, lighting is critical in an emergency and surgery rooms, and elevators may be critical equipment in a public high-rise building.

If electric prices change based on the monthly consumption (tiered pricing) or depending on the time of day (double tariff), EMCS may schedule the operation of some deferrable electric loads to minimize energy costs. A laundry washer and dryer, and dishwasher can be programmed to automatically operate during late-night hours when electricity is cheaper. Some customers can install electric heaters with thermal storage like hot water tanks or heavy concrete blocks. Heaters are operated when the electricity price is lower, and stored thermal energy can heat the building all day.

If on-site electric generation from renewable sources contributes to the energy purchased from a utility, deferrable equipment may be operated when the renewable source is available. In a warmer climate, solar generation can supply air conditioning in midday when the sun is higher, and AC is more needed. In an agricultural facility where wind power is used to generate electricity, water pumps may be operated during windy hours and water stored in a tank at high elevation can be used for irrigation as needed. If on-site solar and/or wind powered generation is backed up with some kind of energy storage, EMCS may include a forecast-based schedule to coordinate battery charging and discharging times with supply of deferrable loads (Soysal et al. 2019).

Load-side energy management is linked to the use of primary resources and protection of the environment. National and international institutions have developed programs to increase the public awareness about the importance of energy use. These programs also encourage energy efficiency at the consumer end. Some examples are certification programs developed by the US Green Building Council's Leadership in Energy and Environmental Design (LEED), Passive House Institute (PHI) based in Germany, Passive House Alliance US (PHAUS), and Energy Star program of the US EPA. Governments offer rebates, tax refunds, and various financial incentives to promote the use of energy-efficient appliances and building materials based on the certified improvements. The International Energy Agency publishes energy data and reports for researchers and policymakers, and the International Protocol for Climate Change (IPCC) provides detailed information about environmental impacts of energy production and consumption. Links to the web sites of these institutions are listed at the end of this chapter.

12.5.4   Site Energy and Source Energy

Energy used on-site by consumers is often referred to as *site energy*. Fuels, electricity, and other energy carriers delivered to end-users are produced by processing and transforming primary energy sources. Equipment used to extract or produce primary sources requires energy and some part of the primary sources are lost during screening, cleaning, and refining processes. In addition, trucks, trains, and various marine vessels that carry fuels from the production facilities to consumers also consume fuels. Electric generation from primary sources result in considerable conversion losses due to the thermodynamic cycle used in thermal generation and heat losses in generators. Electric transmission lines, transformers, and switchgear in the transmission system also dissipate part of the generated electricity in the form of heat. Consequently, significant amounts of source energy are used to produce clean and convenient secondary energy sources delivered to the end-users.

Consumers are usually more focused on the retail cost and convenience of delivered fuels and energy carriers as they try to optimize the composition of their fuel mix. Consumer preferences and efficiency of energy conversion equipment used on site, however, affect the overall efficiency, effectiveness, and sustainability of the whole energy system. Source energy is an equitable common unit to evaluate the energy performance of end users, particularly residential and commercial buildings.

Source energy factor (SEF), also known as source-site ratio is a dimensionless indicator that reflects the total quantity of primary energy sources used to deliver unit energy to consumers in some usable form. In other words, when a consumer uses a certain amount of fuel or electricity on site, SEF times the used energy is drawn from primary resources.

$$SEF = \frac{\text{Primary sources used to produce and deliver site energy}}{\text{Energy used on site}} \quad (12.5)$$

For fuels, source energy includes the energy used to extract, process, and transport the primary source to the consumer. Electricity is generated from a wide variety of fuels and sources using different types of conversion systems. The overall efficiency of electric generation systems depends on the fuel mix and the efficiency of technologies used to transform energy from one form to another in the conversion chain. Additional transmission losses occur until the generated electric power reaches the end users. SEF for the energy used on site is calculated using Eq. (12.5).

Electricity and natural gas are the most common energy sources for residential buildings. Coal, fuel oil, kerosene, propane, and LPG are other popular fuels used for space and water heating in individual buildings. Large commercial complexes, campuses, industrial plants, hospitals, and community services may also use energy carriers such as district steam, hot water, and chilled water for heating and cooling functions. In addition, some consumers supplement their fuel mix with on-site electric generation, solar heating, or ground source geothermal heating and cooling systems. Each energy source used or produced on site affects the source energy consumption in different proportions. Therefore, to compare the efficiencies of buildings it is necessary to convert all site energy sources into the corresponding source energy using their source-site ratios.

12.5.4.1 Direct Use of Fuels

Coal, natural gas, and petroleum products are common fuels delivered to consumers for space and water heating. In industrial and service facilities, diesel fuel may be used for power and emergency electric generation. Source energy corresponding to the derived fuels used on site includes the energy needed for extraction and processing as well as the losses that occur in transportation and distribution of the fuel.

The Energy Star standard created by the US EPA and Department of Energy in 1992 evaluates the energy performance of appliances, equipment, and buildings. Later, many countries including European Union, Canada, Australia, Japan, New Zealand, and Taiwan adopted similar programs to provide an equitable assessment of the energy performance of buildings. For the purposes of the Energy Star program, the EPA recommends the national source-site ratios shown in Table 12.9.

As part of the Energy Star US EPA has developed an online source called Portfolio Manager® to guide consumers in evaluating the energy performance of buildings, appliances, and energy conversion equipment. Table 12.9 shows the source-site ratios used for common fuels delivered to consumers in Energy Star Portfolio Manager (EPA 2018).

Table 12.8 Source energy factors for common fuels.

Fuel	Source energy factor	Higher heating value	
Anthracite coal	1.029	12,700 Btu/lb	29,539 kJ/kg
Bituminous coal	1.048	12,155 Btu/lb	28,270 kJ/kg
Subbituminous coal	1.066	8818 Btu/lb	20,509 kJ/kg
Lignite coal	1.102	6465 Btu/lb	15,038 kJ/kg
Natural gas	1.092	1010 Btu/ft³	37,631 kJ/m³
Residual fuel oil	1.191	149,500 Btu/gal	41,666 kJ/l
Distillate fuel oil	1.158	138,700 Btu/gal	38,656 kJ/l
Gasoline	1.187	100,000 Btu/gal	27,870 kJ/l
LPG	1.151	91,000 Btu/gal	25,362 kJ/l
Kerosene	1.205	135,000 Btu/gal	27,870 kJ/l

Source: Deru and Torcellini (2007).

Example 12.3 In a typical year, a single-family household purchases 2500 m³ of natural gas and 12000 kWh of electric energy from utility companies. Calculate the total site-energy used in this house and estimate the total source energy consumed to supply this site energy.

Solution
Using energy content data given in Table 12.8, we first determine the energy produced by burning natural gas.

$$Q_{gas} = 2{,}500 \times 37{,}631 \times 10^{-3} = 94{,}077 \text{ MJ}$$

Consumed electric energy is equivalent to

$$Q_{electric} = 12{,}000 \times 3{,}600 \times 10^{-3} = 43{,}200 \text{ MJ}$$

Total site energy is

$$Q_{site} = Q_{gas} + Q_{electric} = 94{,}077 + 43{,}200 = 137{,}277 \text{ MJ}$$

Source energy is estimated using the ratios for delivered fuels given in Table 12.9.

$$Q_{source} = 1.05 \times 94{,}077 + 2.80 \times 43{,}200 = 98{,}781 + 120{,}960 = 276{,}687 \text{ MJ}$$

Note that total source energy consumed to supply this house is about twice the actual energy used on site. Although the site-energy received from the electric grid is half of the energy obtained from natural gas, the source energy used to produce that electricity is 22% more than the source energy of the natural gas because of the conversion losses.

12.5.4.2 Use of Grid Electricity
In residential and commercial buildings about half of the energy is consumed is in the form of electricity. The largest portion of electric power delivered to consumers is generated at centralized generation plants.

Table 12.9 Average source-site ratios for fuels and energy carriers used on site.

Energy type	USA	Canada
Electricity (grid purchase)	2.80	1.96
Electricity (on-site solar or wind installation)	1.00	1.00
Natural gas	1.05	1.01
Fuel oil (1,2,4,5,6, diesel, kerosene)	1.01	1.01
Propane and liquid propane	1.01	1.04
Steam	1.20	1.33
Hot water	1.20	1.33
Chilled water	0.91	0.57
Wood	1.00	1.00
Coal/coke	1.00	1.00
Other	1.00	1.00

Source: EPA (2018).

Electricity generated at large centralized power plants is delivered to consumers via an electric grid over long distances through transformers, transmission lines, and cables. Losses that result from heating of transmission lines and transformers must also be considered in estimating the primary energy equivalent of electricity used on site.

An interconnected system supplying a country or region usually consists of several regional networks where different types of primary sources are used for electric generation. Energy converted to electricity at power plants reaches the consumers through transmission and distribution networks where W_t units of energy are lost. For grid electricity, the SEF is expressed in Eq. (12.6), where W_s is the total source energy consumed to generate W_e units of electric energy. In this equation, η_T represents the losses of the electric grid.

$$\text{SEF} = \frac{W_s}{W_e - W_t} = \frac{W_s}{W_e(1 - \eta_T)} \tag{12.6}$$

For consistency, the same unit must be used for all energy involved in the calculation. Energy data published in official databases can be used to estimate the SEF for a certain region or country.

Source-site ratio of electricity generated in a fossil fuel burning power station depends on the type and quality of the fuel and efficiency of the conversion technology used. An interconnected electric grid is supplied by diverse types of power stations using various primary sources and operating at different conversion efficiencies. The source-site ratio differs from one interconnection to another because of the fuel mix and types of the generation facilities. Every national or regional electric grid exchanges energy with other electric grids. For example, the continental United States is supplied by three separate interconnections with connections to each other, as well as cross-border interconnections with Canadian and Mexican grids. Electric grids continuously exchange electric power to balance the regional energy flow. Therefore, the fuels and efficiency of all generation units are used to obtain a more realistic source-site ratio for a large area interconnection.

Example 12.4 Based on Energy Information Administration, the US consumed 38.83 quad to generate electricity in 2018 (EIA 2019). Net electric generation was 14.25 quad and transmission losses were 0.94 quad. Determine total source energy used to generate 1-kWh site energy.

Solution
Substituting the given values in Eq. (12.6), we find:

$$SEF = \frac{38.83}{14.25 - 0.94} = 2.917$$

Therefore, one 1 kWh electric energy used on site corresponds to 2.917 kWh source energy. This is consistent with the source-site ratio of 2.80 suggested by EPA Portfolio Manager.

From the consumer perspective, electricity seems to be the most efficient energy carrier. In fact, electricity is generally cleaner, safer, more controllable, more accessible, and easier to be converted into various forms of energy than other fuels. However, as Table 12.8 shows, electricity drawn from the electric grid may consume more primary sources than other secondary sources used for the same purposes. Therefore, from the holistic perspective, using grid electricity for heating purposes is less efficient than direct use of other fuels. For heating and cooking purposes, the use of grid electricity should be reserved for small domestic appliances with heating components and to systems that cannot be directly supplied from other energy sources like lighting, ventilation, cooling, refrigeration, power equipment, audio-visual sets, computers, communication devices, control equipment, and security systems. Plug-in or battery-operated power equipment and tools are generally more efficient than their alternatives burning petroleum derivatives since some fraction of the grid electricity is generated from renewable sources and large conversion systems used in electric generation are more efficient than small engines.

Many industrial plants use grid electricity for process heating, metal melting, drying, and other thermal purposes mainly because of its controllability, local efficiency, simplicity, and cleanness. Such facilities consume considerably more primary sources compared to their counterparts directly burning fuels. Industrial heating applications are justifiable if CHP or renewable heat sources like solar or geothermal energy are not available or feasible.

12.5.4.3 On-site Electric Generation
Some consumers choose to generate electricity on site for several reasons. Hospitals, data centers, security services, communication centers, cell phone towers, airports, and military bases are critical infrastructures that require continuous energy supply. These facilities need to install emergency backup systems to generate their own energy when the grid power becomes unavailable. Diesel, gasoline, and natural gas are common fuels for such emergency generation systems. Renewable energy sources with battery backup can be also used for smaller loads. Industrial, commercial, and residential consumers may want to

install their own generation systems using solar arrays and wind turbines to reduce or off-set their energy bills. The number of net zero energy buildings has increased due to rising and volatile energy prices. Remote facilities such as remote homes, farm buildings, high-way rest areas, or recreation facilities may generate their own electricity because the utility grid is either unavailable or too expensive to access. Usually, solar and wind power with or without battery backup are economic choices for such end-users.

The SEF can be estimated based on the fuels or primary energy used to generate elec-tricity. If wind, solar, or small hydroelectric generation systems are used, the SEF is unity because the energy source is a renewable primary source and practically no transmission losses are involved. For generation systems that use non-renewable fuels, the efficiency of the generation system should be included in the estimation of the overall source-energy factor.

Example 12.5 Suppose that the household described in Example 12.4 generates 60% of the electricity from a PV array installed on site. How much primary source energy is saved?

Solution
Since the source-site ratio for on-site electric generation is unity, the new source energy becomes:

$$Q_{source} = 1.05 \times 94{,}077 + 0.6 \times 43{,}200 \times 1 + 0.4 \times 43{,}200 \times 2.80 = 173{,}085 \text{ MJ}$$

Therefore, $276{,}687 - 173{,}085 = 103{,}603$ MJ source energy is avoided by using on-site solar generation. This corresponds to a 37% of saving on the total source energy.

12.6 Chapter Review

This chapter discusses the economic development aspects of energy production, conver-sion, and use. Sustainable development requires meeting the current needs without com-promising the ability of future generations to meet their own needs. Basic needs of all com-munities around the world include access to clean freshwater, sanitary services, food supply, and adequate energy supply. While all these commodities are interrelated, reliable and continuous energy supply is central to the availability of others. However, energy-related activities are the major sources of ground-level air pollution, acid rain, and global climate change.

Significant amounts of water are used for energy production and processing. Most of that water is evaporated or heavily contaminated. Energy conversion uses water for cooling pur-poses. While the amount of water used for conversion is greater, the larger fraction returns to the same source.

Extraction, purification, and distribution of freshwater and wastewater management use energy; worldwide, the largest amount of electricity is used for freshwater collection. In developed countries, municipal and industrial wastewater treatment can consume as much energy as freshwater extraction.

The ratio of the total energy used for extraction, processing, conversion, and distribution to the energy used is represented by an indicator named EROI, which stands for "energy return on energy invested." The value of EROI depends on the boundary at which it is defined. Wellhead or farmgate EROI is the ratio of the energy acquired to the energy consumed to find and extract (or produce) the energy source. The denominator of point-of-use EROI includes in addition the energy used for fuel processing and delivery. Higher EROI is more desirable since a larger amount of energy is acquired by consuming smaller energy. Hydropower has the highest EROI value among all energy sources. Coal, oil, and natural gas have higher EROI than biofuels.

Coordination of planning, development, and operation of an energy system to supply the needed energy is known as energy management. Resource coordination, supply-side energy management, and load-side energy management are interlinked and assure effective use of resources while minimizing the environmental impacts.

Energy delivered to consumers is produced using a mix of various primary sources. Source-site ratio relates energy used on site to all primary sources consumed for production, conversion, and transportation of secondary sources supplied to end-users. Electricity has higher source-site ratio than fuels because of the conversion losses.

Carbon footprint of energy consumers includes the carbon dioxide equivalent of all greenhouse gases released throughout the energy supply chain from extraction and production of primary sources to the delivery of fuels, electricity, and all types of energy carriers to the point of use.

In conclusion, sustainable development of the humanity depends on sensible use of energy without exhausting the available resources and by preserving water and food supply, eco diversity, and global climate.

Review Quiz

1 Which organization established the sustainable development goals?
   a. OECD
   b. United Nations
   c. OPEC
   d. IEA

2 Which one of the gases below has higher global warming potential?
   a. Carbon dioxide
   b. Methane
   c. Nitrous oxide
   d. Carbon monoxide

3 In a thermoelectric power station water is mainly used for
   a. cooling the working fluid.
   b. cleaning the facility.
   c. fuel processing.
   d. fuel production.

4 Which process below consumes more energy than others?
   a. Freshwater extraction
   b. Water purification
   c. Wastewater treatment
   d. Water transport

5 Which gas below is not considered a toxic pollutant?
   a. Carbon monoxide
   b. Sulfur dioxide
   c. Nitrous oxide
   d. Carbon dioxide

6 Which type of electric power plant below has the highest emission rate?
   a. Nuclear
   b. Geothermal
   c. Coal fired
   d. Natural gas

7 In energy systems EROI means
   a. total energy used to produce and deliver an energy source.
   b. energy returned over the input of a conversion system.
   c. energy recovered on input energy.
   d. energy resources gained over financial investment.

8 A fuel is more preferable if it's EROI is
   a. larger.
   b. smaller.
   c. zero.
   d. negative.

9 Which energy source below has the highest source-site ratio?
   a. Coal
   b. Fuel oil
   c. Electricity
   d. Natural gas

10 Production of which secondary energy source uses the largest amount of water?
   a. Petroleum
   b. Nuclear energy
   c. Natural gas
   d. Electricity

Answers: 1-b, 2-c, 3-a, 4-c, 5-d, 6-c, 7-a, 8-a, 9-c, 10-d.

Research Topics and Problems

Research and Discussion Topics

1   Draft a paper on the energy use for purified water sold in PET bottles.

2   Estimate the total CO_2 equivalent greenhouse gas emission of your home over one year using the information from utility bills and additional fuels purchased.

3   Compile data from official websites and draft a report discussing the environmental and economic implications of adding ethanol to gasoline.

4   What are the environmental benefits and possible challenges of storing and purifying rainwater on site for domestic use?

5   What are the environmental implications of producing artificial snow in a large ski resort?

Problems

1   Efficiency of a 100-MW thermoelectric power station is 30% when it operates at full capacity. Compare the GWP of this power plant (a) when bituminous coal is burned as a primary source. and (b) if the station is converted to natural gas assuming that its efficiency remains the same.

2   A 50-gal residential water heater burns 250-Therm in one year (1 Therm = 105 Btu). If the same size electric water heater powered from the utility grid provides the same amount of hot water, how much source energy would each heater consume?

3   How much electric energy is consumed to fill a 3000-gal tank installed on a 150 ft tower? Assume 80% efficiency for the motor and pump.

4   A car has the average fuel consumption rate of 16 miles per gallon. (a)
a) How much carbon dioxide is released per mile if the fuel is regular gasoline? (b)
b) How much carbon dioxide is released per mile if the fuel contains 10% ethanol?

5   Estimate the amount of water used to supply a 100-W lightbulb for 10 hours in the US grid?

6   Given the emission rates of electric generation in a region, calculate the carbon footprint of a household that consumes 8000 kWh in one year. CO_2: 725 kg/MWh; SO_2: 1.7 kg/MWh; NO_x: 1.6 kg/MWh.

7   The fuel mix of an electric utility is given below. Determine the total carbon equivalent greenhouse gas emission per kWh electric usage. CO_2: 563 kg/MWh; SO_2: 0.54 kg/MWh; NO_x: 0.41 kg/MWh.

8 Calculate the yearly source-energy consumption for the residence with the energy consumption profile described in Figure 12.12.

9 A farmer uses electricity supplied from the utility grid to pump irrigation water from a 50-m deep well. Assuming 80% overall efficiency of the pumping system, calculate the energy intensity in kWh/l for 500 l per minute flow rate.

10 Estimate the carbon dioxide emission avoided in one year by using on-site solar energy for water heating in a residence where in average 40 l/day of water is heated 25 °C above ambient temperature.

11 Estimate the yearly carbon dioxide offset by using twenty solar landscaping lights from dusk to dawn instead of twenty 5 W lightbulbs supplied from grid electricity.

12 A family uses 12 000 kWh electric energy in one year. (a) Calculate the electric energy generated by the power plants in the region to supply this family considering 4.49% total transmission and distribution losses. (b) Given the fuel mix of the electric utility supplying the information below, estimate the carbon footprint of this family due to electric usage.
Regional fuel mix: Gas (16.7%); Coal (49.8%); Nuclear (27.6%); Hydro (0.9%); Wind (3.2%); Biomass (0.6%); Solar (0.1%); Oil (0.4%).

Recommended Web Sites

- EPA Energy Star: www.energystar.gov
- WRI Greenhouse Gas Protocol: https://ghgprotocol.org
- Intergovernmental Panel on Climate Change (IPCC): https://www.ipcc.ch
- International Energy Forum: https://www.ief.org
- NREL BeOpt: https://beopt.nrel.gov
- NREL Building Optimization Tool: https://beopt.nrel.gov
- Passive House Institute (PHI): https://passivehouse.com
- Passive House Institute US (PHIUS): https://www.phius.org
- United Nations Sustainable Development Goals (SDG): https://www.un.org/sustainabledevelopment
- United Nations Water: https://www.unwater.org
- US Environmental Protection Agency (EPA): https://www.epa.gov/energy
- US EPA Power Profiler: https://www.epa.gov/energy/power-profiler#/
- US Green Building Council: https://new.usgbc.org
- World Energy Council (WEC): https://www.worldenergy.org
- World Resources Institute (WRI): https://www.wri.org

References

CEC (2005). *California's Water-Energy Relationship*. Sacramento, CA: California Energy Commission.

CPUC (2010). *Embedded Energy in Water Studies Study 1: Statewide and Regional Water-Energy Relationship*. Sacramento, CA: California Public Utilities Commission Energy Division.

Deru, M. and Torcellini, P. (2007). *Source Energy and Emission Factors for Energy Use in Buildings*. Golden, CO: National Renewable Energy Lab.

DOE (2014). *The Water-Energy Nexus: Challenges and Opportunities*. Washington DC: US Department of Energy.

EIA (2016). *Energy and Air Pollution-World Energy Outlook Special Report*. Paris: Energy Information Agency.

EIA, 2019. *U.S. Energy Consumption by Source and Sector, 2018*. [Online] Available at: https://www.eia.gov/totalenergy/data/monthly/pdf/flow/electricity.pdf [Accessed 4 August 2019].

EPA, 2018. *Energy Star Portfolio Manager Technical Reference: Source Energy*. [Online] Available at: https://www.energystar.gov/buildings/tools-and-resources/portfolio-manager-technical-reference-source-energy [Accessed 4 August 2019].

EPA, 2019a. *Emissions & Generation Resource Integrated Database (eGRID)*. [Online] Available at: https://www.epa.gov/energy/emissions-generation-resource-integrated-database-egrid [Accessed 16 August 2019].

EPA, 2019b. *Power Profiler*. [Online] Available at: https://www.epa.gov/energy/power-profiler#/ [Accessed July 2019].

Gagnon, N., Hall, C.A.S.,.a., and Brinker, L. (2009). A preliminary investigation of energy return on energy investment for global oil and gas production. *Energies* (http://www.mdpi.com/journal/energies) 6.

Hall, C.A.S., Lambert, J.G., and Balogh, S.B. (2013). EROI of different fuels and the implications for society. *Energy Policy* (http://www.elsevier.com/locate/enpol) 10.

IEA (2009). *Bioenergy – A Sustainable and Reliable Energy Source,* Main Report. Paris: International Energy Agency.

IEA (2016). *Water Energy Nexus (Excerpt From World Energy Outlook 2016)*. Geneva, Switzerland: International Energy Agency.

IEA (2018). *Energy, Water, and the Sustainable Development Goals (Excerpt From World Energy Outlook 2018)*. Paris: International Energy Agency.

IPCC (1996). *Climate Change 1995, The Science of Climate Change: Summary for Policymakers and Technical Summary of Working Group I*. Cambridge, New York: Cambridge University Press.

IPCC (2014). *Renewable Energy Sources and Climate Change Mitigation*. New York, NY: Cambridge University Press.

LLNL-DOE, 2015. *Carbon Flow Charts*. [Online] Available at: https://flowcharts.llnl.gov/commodities/carbon [Accessed 15 August 2019].

Murphy, D.J. and Hall, C.A.S. (2010). *Year in Review – EROI or Energy Return on (Energy) Invested*. Annals of the New York Academy of Sciences.

Soysal, O.A., Soysal, H.S., and Guy, J. (2016). Evaluating the cost-benefit of an energy-efficient house. *Journal of the National Institute of Building Sciences*. February, Vol. 4(1): 16–20.

Soysal, O. A., Soysal, H. S., and Manto, C. L., 2019. *Method and instrumentation for sustainable energy load flow management system performing as resilient adaptive microgrid system*. USA, Patent No. 10,169,832.

UN, 2015. *Transforming Our World: The 2030 Agenda for Sustainable Development*. [Online] Available at: https://sustainabledevelopment.un.org [Accessed 14 August 2019].

UNDP (2006). *Human Development Report 2006 – Beyond Scarcity: Power, Poverty, and the Global Water Crisis*. New York: United Nations Development Programme.

WCED (1987). *Our Common Future*. Oxford: Oxford University Press.

WHO (2016). *Ambient Air Pollution: A Global Assessment of Exposure and Burden of Disease*. Geneva, Switzerland: World Health Organization.

WHO/UNICEF, 2019. *Joint Monitoring Programme (JMP) for Global Data on Water Supply, Sanitation and Hygiene (WASH)*. [Online] Available at: https://washdata.org/data [Accessed 14 August 2019].

Wu, M. and Chiu, Y. (2011). *Consumptive Water Use in the Production of Ethanol and Petroleum Gasoline – 2011 Update*. Chicago: Argonne National Laboratory.

A

Unit Conversion Factors

A.1 Metric Prefixes

Kilo: 10^3, Mega: 10^6, Giga: 10^9, Tera: 10^{12}, Peta: 10^{15}, Exa: 10^{18}

Units commonly used in energy statistics and calculations.

Dimension	Standard metric units (SI)	Imperial (IS) and non-metric units	Common conversion factors
Length			
		inch (in.)	1 in. = 0.0254 m 1 m = 39.370 in.
	meter (m)	feet (ft)	1 ft = 0.3048 m 1 m = 3.280 ft
		yard (yd)	1 yd = 0.9144 m 1 m = 1.094 yd
		mile (mi)	1 mi = 1609.344 m 1 km = 0.621 mi
Volume	cubic meter (m³)		1 m³ = 1000 l
	liter (l)		
		fluid ounce (fl oz)	1 l = 33.814 fl oz 1 fl oz = 0.0296 l
		imperial gallon (gal)	1 l = 0.2199 gal 1 gal = 4.546 l
	liter (l)	US gallon (gal)	1 l = 0.2642 gal 1 gal = 3.785 l
		barrel (bbl)	1 bbl = 42 gal (US) 1 bbl = 158.97 l
		cubic yard (cu yd)	1 m³ = 1.308 cu yd 1 cu yd = 0.765 m³
		cubic foot (cu ft)	1 m³ = 35.315 cu ft 1 cu ft = 0.0283 m³

Energy for Sustainable Society: From Resources to Users, First Edition. Oguz A. Soysal and Hilkat S. Soysal.
© 2020 John Wiley & Sons Ltd. Published 2020 by John Wiley & Sons Ltd.

Dimension	Standard metric units (SI)	Imperial (IS) and non-metric units	Common conversion factors
Mass			
	kilogram (kg) metric ton (t)		$1\,t = 1000\,kg$
		pound (lb)	$1\,kg = 2.205\,lb$ $1\,lb = 0.454\,kg$
		short ton (S/T)	$1\,S/T = 2000\,lb$ $1\,S/T = 907.185\,kg$
		long ton (L/T)	$1\,L/T = 2240\,lb$ $1\,L/T = 1016.047\,kg$
Speed/velocity			
	meter per second $(m\,s^{-1})$		$1\,m\,s^{-1} = 3.6\,km\,h^{-1}$ $1\,km\,h^{-1} = 0.2778\,m\,s^{-1}$
		kilometer per hour $(km\,h^{-1})$	
		mile per hour (mph)	$1\,m\,s^{-1} = 2.237\,mph$ $1\,mph = 0.447\,m\,s^{-1}$
		knot (kn)	$1\,m\,s^{-1} = 1.944\,kn$ $1\,kn = 0.514\,m\,s^{-1}$
Energy/work			
	Joule (J) Watt second (W s) Newton meter (N m)		$1\,J = 1\,W\,s$ $1\,N\,m = 1\,J$
		kilowatt hour	$kWh = 3600\,kJ$
		British Thermal Unit (Btu)	$1\,Btu = 1055.056\,J$ $1\,kWh = 3412\,Btu$
		quad	$1\,quad = 10^{15}\,Btu$
		calorie (cal)	$1\,cal = 4.184\,J$ $1\,J = 0.239\,cal$
		food calorie (Cal)	$1\,Cal = 1000\,cal$
		therm (thm)	$1\,thm = 99\,976.13\,Btu$
		ton of oil equivalent (toe)	$1\,toe = 41.868 \times 10^9\,J$ (or GJ) $1\,toe = 39.6831\,MBtu$ $1\,toe = 1.4286\,tce$ $1\,toe = 11.630\,MWh$[a]
		ton of coal equivalent (tce)	$1\,tce = 29.3076 \times 10^9\,J$ (or GJ) $1\,tce = 27.7781\,MBtu$ $1\,tce = 0.7\,toe$ $1\,tce = 8.141\,MWh$[a]

Dimension	Standard metric units (SI)	Imperial (IS) and non-metric units	Common conversion factors
Power			
	Watt (W)		
	Joule per second $(\mathrm{J\,s^{-1}})$		$1\,\mathrm{W} = 1\,\mathrm{J\,s^{-1}}$
		horsepower (HP)	$1\,\mathrm{HP} = 745.7\,\mathrm{W}$ $1\,\mathrm{HP}\ (\text{electric}) = 746\,\mathrm{W}$
Temperature			
	Celsius (C) Kelvin (K)		$^\circ\mathrm{K} = 273.5 + {}^\circ\mathrm{C}$
		Fahrenheit (F)	$^\circ\mathrm{F} = {}^\circ\mathrm{C} \times 9/5 + 32$ $^\circ\mathrm{C} = ({}^\circ\mathrm{F} - 32) \times 5/9$

a) Source: IEA Key World Energy Statistics 2017

Common unit abbreviations.

Abbreviation	Unit
Bcm	billion cubic meters
CCF	centum cubic feet (100 cubic feet)
Gcal	gigacalorie
GCV	gross calorific value
GW	gigawatt
GWh	gigawatt hour
kcal	kilocalorie
kg	kilogram
kJ	kilojoule
kWh	kilowatt hour
MBtu	million British thermal units
Mt	million tons
Mtoe	million tons of oil equivalent
MWh	megawatt hour
TJ	terajoule
toe	ton of oil equivalent
TWh	terawatt hour
USD	United States dollar

B

Calorific Values of Common Fuels

Gaseous fuels [at 0 °C (32 °F) 1 atm]

	Net calorific value (NCV) [1]			Gross calorific value (GCV) [2]			Density	
	Btu/cu ft	Btu/lb [3]	MJ/kg [4]	Btu/cu ft	Btu/lb [3]	MJ/kg [4]	g/cu ft	kg/m³
Natural gas	983	20 267	47.141	1 089	22 453	52.225	22.0	0.776 924
Hydrogen	290	51 682	120.21	343	61 127	142.18	2.55	0.090 053
Still gas (in refineries)	1 458	20 163	46.898	1 584	21 905	50.951	32.8	1.158 323

Liquid fuels

	Net calorific balue (NCV) [1]			Gross calorific value (GCV) [2]			Density	
	Btu/gal	Btu/lb [3]	MJ/kg [4]	Btu/gal	Btu/lb [3]	MJ/kg [4]	g/gal	kg/m³
Crude oil	129 670	18 352	42.686	138 350	19 580	45.543	3 205	705.015
Conventional gasoline	116 090	18 679	43.448	124 340	20 007	46.536	2 819	620.106
Reformulated (low-sulfur) gasoline	113 602	18 211	42.358	121 848	19 533	45.433	2 830	622.525
CA reformulated gasoline	113 927	18 272	42.500	122 174	19 595	45.577	2 828	622.085
U.S. conventional diesel	128 450	18 397	42.791	137 380	19 676	45.766	3 167	696.656
Low-sulfur diesel	129 488	18 320	42.612	138 490	19 594	45.575	3 206	705.235
Petroleum naphtha	116 920	19 320	44.938	125 080	20 669	48.075	2 745	603.828

Energy for Sustainable Society: From Resources to Users, First Edition. Oguz A. Soysal and Hilkat S. Soysal.
© 2020 John Wiley & Sons Ltd. Published 2020 by John Wiley & Sons Ltd.

	Net calorific balue (NCV) [1]			Gross calorific value (GCV) [2]			Density	
	Btu/gal	Btu/lb [3]	MJ/kg [4]	Btu/gal	Btu/lb [3]	MJ/kg [4]	g/gal	kg/m^3
NG-based FT naphtha	111 520	19 081	44.383	119 740	20 488	47.654	2 651	583.150
Residual oil	140 353	16 968	39.466	150 110	18 147	42.21	3 752	825.341
Methanol	57 250	8 639	20.094	65 200	9 838	22.884	3 006	661.241
Ethanol	76 330	11 587	26.952	84 530	12 832	29.847	2 988	657.281
Butanol	99 837	14 775	34.366	108 458	16 051	37.334	3 065	674.219
Acetone	83 127	12 721	29.589	89 511	13 698	31.862	2 964	652.002
E-diesel additives	116 090	18 679	43.448	124 340	20 007	46.536	2 819	620.106
Liquefied petroleum gas (LPG)	84 950	20 038	46.607	91 410	21 561	50.152	1 923	423.009
Liquefied natural gas (LNG)	74 720	20 908	48.632	84 820	23 734	55.206	1 621	356.577
Dimethyl ether (DME)	68 930	12 417	28.882	75 610	13 620	31.681	2 518	553.894
Dimethoxy methane (DMM)	72 200	10 061	23.402	79 197	11 036	25.67	3 255	716.014
Methyl ester (biodiesel, BD)	119 550	16 134	37.528	127 960	17 269	40.168	3 361	739.331
Fischer–Tropsch diesel (FTD)	123 670	18 593	43.247	130 030	19 549	45.471	3 017	663.660
Renewable diesel I (Super cetane)	117 059	18 729	43.563	125 294	20 047	46.628	2 835	623.625
Renewable diesel II (UOP-HDO)	122 887	18 908	43.979	130 817	20 128	46.817	2 948	648.482
Renewable gasoline	115 983	18 590	43.239	124 230	19 911	46.314	2 830	622.525
Liquid hydrogen	30 500	51 621	120.07	36 020	60 964	141.8	268	58.953
Methyl tertiary butyl ether (MTBE)	93 540	15 094	35.108	101 130	16 319	37.957	2 811	618.346
Ethyl tertiary butyl ether (ETBE)	96 720	15 613	36.315	104 530	16 873	39.247	2 810	618.126
Tertiary amyl methyl ether (TAME)	100 480	15 646	36.392	108 570	16 906	39.322	2 913	640.783
Butane	94 970	19 466	45.277	103 220	21 157	49.21	2 213	486.802
Isobutane	90 060	19 287	44.862	98 560	21 108	49.096	2 118	465.904
Isobutylene	95 720	19 271	44.824	103 010	20 739	48.238	2 253	495.601
Propane	84 250	19 904	46.296	91 420	21 597	50.235	1 920	422.349

Solid fuels

	Net calorific value (NCV) [1]			Gross calorific value (GCV) [2]		
	Btu/ton	Btu/lb [3]	MJ/kg [4]	Btu/ton	Btu/lb [3]	MJ/kg [4]
Coal (wet basis) [6]	19 546 300	9 773	22.732	20 608 570	10 304	23.968
Bituminous coal (wet basis) [7]	22 460 600	11 230	26.122	23 445 900	11 723	27.267
Coking coal (wet basis)	24 600 497	12 300	28.61	25 679 670	12 840	29.865
Farmed trees (dry basis)	16 811 000	8 406	19.551	17 703 170	8 852	20.589
Herbaceous biomass (dry basis)	14 797 555	7 399	17.209	15 582 870	7 791	18.123
Corn stover (dry basis)	14 075 990	7 038	16.37	14 974 460	7 487	17.415
Forest residue (dry basis)	13 243 490	6 622	15.402	14 164 160	7 082	16.473
Sugar cane bagasse	12 947 318	6 474	15.058	14 062 678	7 031	16.355
Petroleum coke	25 370 000	12 685	29.505	26 920 000	13 460	31.308

Source: Argonne National Laboratory, GREET, The Greenhouse Gases, Regulated Emissions, and Energy Use in Transportation Model; GREET 1.8d.1, Argonne, IL, released August 26, 2010. Available online at http://greet.es.anl.gov/ (accessed 12/2019).

Notes

[1] The Net Calorific Value (also known as lower heating value) of a fuel is defined as the amount of heat released by burning a specified quantity (initially at 25°C) and returning the temperature of the combustion products to 150°C, which assumes the latent heat of vaporization of water in the reaction products is not recovered. The NCV are the useful calorific values in boiler combustion plants and are frequently used in Europe.

[2] The Gross Calorific Value (also known as higher heating value) of a fuel is defined as the amount of heat released by a specified quantity (initially at 25°C) once it is combusted and the products have returned to a temperature of 25°C, which takes into account the latent heat of vaporization of water in the combustion products. The GCV are derived only under laboratory conditions, and are frequently used in the US for solid fuels.

[3] The heating values for gaseous fuels in units of Btu/lb are calculated based on the heating values in units of Btu/ft^3

[4] The heating values in units of MJ/kg, are converted from the heating values in units of Btu/lb.

[5] For solid fuels, the heating values in units of Btu/lb are converted from the heating values in units of Btu/ton.

[6] Coal characteristics assumed by GREET for electric power production.

[7] Coal characteristics assumed by GREET for hydrogen and Fischer-Tropsch diesel production.

C

Abbreviations and Acronyms

a.m.u.	Atomic mass unit
ABWR	Advanced Boiling Water Reactor
AC	Alternating current
ACAA	American Coal and Ash Association
ACR	Advanced CANDU reactor
AEO	Annual Energy Outlook (a publication of IEA)
AFC	Alkaline Fuel Cell
Ah	Ampere-hour
ALWR	Advanced Light Water Reactor
AM	Air Mass
API	American Petroleum Institute
ASHRAE	The American Society of Heating, Refrigerating and Air-Conditioning Engineers
ASTM	American Society for Testing and Materials
AWEA	American Wind Power Association
Bbl	Billion barrel
bcm	Billion cubic meter
BGR	Bundesanstalt für Geowissenschaften und Rohstoffe (Federal Institute for Geosciences and Natural Resources, Germany)
BJT	Bipolar Junction Transistor
BOP	Blowout preventer
BP	British Petroleum
Bq	Becquerel (SI of radioactivity)
BTL	Biomass-to-liquids
Btu	British thermal unit; quadrillion quad (1015 Btu)
BWR	Boiling Water Reactor
CAA	The Clean Air Act
CAL	Calorie, a unit of energy
CANDU	Canada Deuterium Uranium pressurized-heavy water reactor
CAPP	Canadian Association of Petroleum Producers
CBM	Coal bed methane
CCGT	Combined cycle gas turbine
CCP	Coal combustion products such as fly ash, bottom ash, boiler slag

Energy for Sustainable Society: From Resources to Users, First Edition. Oguz A. Soysal and Hilkat S. Soysal.
© 2020 John Wiley & Sons Ltd. Published 2020 by John Wiley & Sons Ltd.

cgs	Centimeter-gram-second system
CHP	Combined heat and power (also called cogeneration)
CHPP	Coal handling and preparation plant
CIS	Commonwealth of Independent States
COP21	21st yearly session of the Conference of the Parties (United Nations Climate Change Conference)
CPC	Compound Parabolic Collectors
CPUC	California Public Utilities Commission
CSP	Concentrated Solar Power
CSS	Cyclic Steam Stimulation; also known "steam soak" or "huff and puff"
CTL	Coal to liquid; coal liquefaction
DAPL	Dakota Access Pipeline
dB	Decibel
DC	Direct current
DEA	Diethanolamine
DHI	Diffuse Horizontal Irradiance
DMFC	Direct-Methanol Fuel Cell
DNI	Direct Normal Irradiance
DOE	US Department of Energy
DST	Daylight saving time
DTU	Danmarks Tekniske Universitat (Technical University of Denmark)
DWT	Dead-weight tonnage
EBR I	Experimental Breeding Reactor I
EGS	Enhanced Geothermal Systems
EHV	Extra-Large-High-Voltage
EIA	US Energy Information Administration
EJ	Exajoule (EJ) (SI unit of energy equal to 10^{18} J)
EMCS	Energy Management Control System
EMF	Electromotive Force
EMIS	Electromagnetic Isotope Separation
EOR	Enhanced Oil Recovery
EPA	Environmental Protection Energy
EPRI	Electric Power Research Institute
ERCB	The Energy Resources Conservation Board
EROI	Energy Returning on Energy Invested
ESMAP	The Energy Sector Management Assistance Program
ETC	Evacuated Tube Collector
ETP	Evacuated tube panels
EUI	Energy Use Intensity
EV	Electric Vehicle
FAEE	Fatty Acid Ethyl Ester
FAME	Fatty Acid Methyl Ester
FBC	Fluidized bed combustion
FBR	Sodium Cooled Fast Breeder Reactor
FC	Fixed carbon

FF	Fill factor
FGD	Flue-Gas Desulfurization
FPC	Flat panel collector
GCR	Gas-Cooled Reactor
GCV	Gross calorific value (also known as high heating value)
GDP	Gross domestic product
GFC	Gas-Cooled Fast Reactor
GHG	Greenhouse Gas
GHI	Global Horizontal Irradiance
GHP	Geothermal Heat Pump
GIF	OECD Nuclear Energy Agency Generation IV International Forum
GJ	Gigajoule
GTL	Gas-to-liquid
GW	Gigawatt
GWP	Global warming potential
HAWT	Horizontal axis wind turbine
hp	Horsepower
HV	High voltage; Hybrid vehicle
HVAC	Heating, Ventilation, and Air Conditioning
HVDC	High-voltage direct current
IAEA	International Atomic Energy Agency
IEA	OECD International Energy Agency
IEF	International Energy Forum
IGA	International Geothermal Association
IHP	International Hydropower Association
INES	International Nuclear and Radiological Event Scale
IPCC	Intergovernmental Protocol for Climate Change
IR	Infrared Radiation
IREA	International Renewable Energy Agency
IRENA	The International Renewable Energy Agency
IS	Imperial (British) unit system
KCL	Kirchhoff's Current Law
kV	Kilovolt
kVA	Kilovolt ampere
KVL	Kirchhoff's Voltage Law
LED	Light-emitting diode
LEED	Leadership in Energy and Environmental Design
LFR	Lead-Cooled Fast Reactor
LLNL	Lawrence Livermore National Laboratory
LNG	Liquified natural gas
LOCA	Loss-of-Coolant Accident, an accident scenario in nuclear reactors
LPG	Liquefied natural gas
LTO	Light Tight Oil
LV	Low voltage
LVRT	Low voltage ride through

LWGR	Light Water Graphite Reactor
MATS	Mercury and Air Toxics Standards
MCFC	Molten Carbonate Fuel Cell
MEA	Monoethanolamide
MJ	Mega Joule
MKSA	The system of units using based on meter, kilogram, second, Ampere
MMF	Magnetomotive Force
MPa	Mega Pascal
MPPT	Maximum Power Point Tracking
MSR	Molten-salt Reactor
Mt	Million tons
Mtoe	Million ton of oil equivalent
MV	Medium voltage
MW	Megawatt
MWA	Megawatt-ampere
MWh	Megawatt-hour
ACA	National Advisory Committee for Aeronautics
NASA	National Aeronautics and Space Administration
NASA GSFC	NASA Goddard Space Flight Center
NCV	Net calorific value (also known as lower heating value)
NEC	National Electric Code (US)
NEI	Nuclear Energy Institute
NEMA	National Electrical Manufacturers Association
NERC	North American Electric Reliability Corporation
NGDC	National Geophysical Data Center
NGL	Natural Gas Liquids
NGPL	Natural Gas Plant Liquids
N m	Newton meter
NOAA	National Oceanic and Atmospheric Administration
NOCT	Nominal Operating Cell Temperature
NRC	Nuclear Regulatory Commission
NREL	National Renewable Energy Laboratory (US)
NWTC	National Wind Technology Center (US)
OECD	Organization for Economic Co-operation and Development
OPEC	Organization of the Petroleum Exporting Countries
ORC	Organic Rankine Cycle
OVI	Oil vulnerability index
Pa s	Pascal second
PAFC	Phosphoric Acid Fuel Cell
PEM	Proton Exchange Membrane fuel cell
PHAUS	Passive House Alliance US
PHES	Pumped-storage Hydroelectric Energy Storage
PHI	Passive House Institute (Germany)
PHIUS	Passive House Institute (US)
PHP	Pumped-storage Hydroelectric Power stations

PHWR	Pressurized Heavy-Water Reactor
PM	Particulate matter
PPP	Purchasing power parity
PSH	Pumped-Storage Hydroelectric power station
psi	Pounds per square inch
PTC	PV Test Conditions
PV	Photovoltaic
PWR	Pressurized Water Reactor
quad	Quadrillion Btu, 10^{15}
R/P	Reserve to production ratio
REM/rem	Roentgen equivalent man (ionizing radiation)
REN21	Renewable Energy Policy Network for the 21st Century
RFG	Reformulated Gasoline
rms	Root-Mean-Square (effective value)
ROR	Run-of-river
RPM/rpm	Revolutions per minute
RPO	Recovery point objective
SAGD	Steam-Assisted Gravity Drainage
SC	Solar constant
SCADA	Supervisory Control and Data Acquisition
SCO	Synthetic Crude Oil
SCWR	Supercritical-water-cooled Reactor
SDG	Sustainable development goals
SEF	Source Energy Factor
SI	International Unit System
SiC	Silicon Carbide Fiber-Reinforced Matrix composites
SFT	Sodium-cooled fast reactor
SOFC	Solid Oxide Fuel Cell
SOS	Security of supply
STC	Standard Test Conditions
SVC	Switched Virtual Circuit
tce	Tons of coal equivalent
TFC	Total final energy consumption
TFEC	Total Final Energy Consumption
toe	Tons of oil equivalent
TPED	Total Primary Energy Demand
TPES	Total primary energy supply
TWh	Terawatt-hour
UNDESA	United Nations Department of Economic and Social Affairs
UNDP	United Nations Development Program
UNFC	United Nations Framework Classification
UNFCCC	United Nations Framework Convention on Climate Change
UNSCEAR	United Nations Scientific Committee on the Effects of Atomic Radiation
UPS	Uninterrupted Power Supply
USACE	United States Army Corps of Engineers

USD	United States dollar
USGS	US Geological Survey
UV	Ultraviolet
VA	Volt-ampere
VAr	Volt-ampere-reactive
VAWT	Vertical axis wind turbine
VDE	Verband der Elextrotechnik (Germany)
VHTR	Very-high temperature reactor
VM	Volatile matter
VOC	Volatile organic compounds
WCA	World Coal Association
WCED	The World Commission on Environment and Development
WCI	World Coal Institute
WEC	World Energy Council
WHO	World Health Organization
WRI	World Resources Institute

Glossary

Acid rain Precipitation containing harmful amounts of nitric and sulfuric acids formed primarily by sulfur dioxide and nitrogen oxides released into the atmosphere when fossil fuels are burned; also known as *acid precipitation* or *acid deposition*.

Active power The component of electric power that performs mechanical work or produces thermal energy (heat) in unit time, measured in watt or metric multiples of watt; also known as "real power." The term *active power* is used to distinguish it from Reactive power.

Adiabatic A reversible thermodynamic process that occurs without gain or loss of heat and without a change in entropy.

Airfoil A cross-section part or surface, such as a wing, propeller blade, or rudder, that interacts with a flow of air to provide stability, rotation, lift, or thrust.

Albedo (Astronomy) The fraction of incident electromagnetic radiation reflected by a surface, especially of a celestial body. (General Physics) The ratio of the intensity of light reflected from an object, such as a planet, to that of the light it receives from the sun.

Alternating current (AC) Electric current in which the direction of flow is reversed periodically.

Amorphous silicon A semiconductor material with a disordered, non-crystalline internal atomic arrangement, that can be deposited in thin-film layers (a few micrometers in thickness) to produce thin-film photovoltaic cells on glass, metal, or plastic substrates.

Ampere (A) or amp The standard unit for the electric current intensity.

Ampere-hour (Ah) The unit often used to quantify the energy storage capacity of a battery. Corresponds to the flow of one ampere of constant current for one hour.

Anticline A formation of stratified rock, raised up by folding, into a broad arch so that the strata slopes down on both sides from a common crest.

Apparent power The product of the effective (rms) values of the voltage across the terminals of an AC circuit, and the current flowing through the same terminals. It is measured in "volt-amperes" (VA) or metric multiples of VA.

Appliance A device or piece of equipment, typically powered by electricity, used to perform a particular energy-driven function.

Battery capacity The total number of ampere-hours that can be withdrawn from a fully charged cell or battery.

Energy for Sustainable Society: From Resources to Users, First Edition. Oguz A. Soysal and Hilkat S. Soysal.
© 2020 John Wiley & Sons Ltd. Published 2020 by John Wiley & Sons Ltd.

Battery cycle-life The number of cycles of full charge and discharge up to a specified depth that a cell or battery can undergo before failing to meet its specified capacity or efficiency performance criteria.

Battery self-discharge The rate at which a battery will lose its charge without delivering energy to an external load.

Battery state of charge Ratio of the amount of energy stored in a battery to the maximum energy stored at full charge, usually expressed in percentage.

Battery Two or more cells electrically connected for electrical energy storage.

Becquerel (Bq) An International System unit of radioactivity, equal to one nuclear decay or other nuclear transformation per second.

Beneficiation Crushing and separating ore to separate valuable substances from waste by any of variety of techniques.

Bitumen The constituents of coal specifically develop for industrial use such as asphalt, tar, mineral waxes, etc.

Black liquor A liquid residue formed during the pulping of wood to make paper; having a high concentration of lignin and capable of being used as a biomass fuel.

Blackbody A theoretically (hypothetical body) perfect absorber without reflection of the electromagnetic radiation incident on its surface.

Bloom Any coating similar in appearance, such as that on new coins or on rocks or minerals.

Byproduct Secondary or additional product resulting from the feedstock use of energy, or the processing of nonenergy materials.

Calandria A cylindrical vessel through which tubes pass in an evaporator, heat exchanger, or nuclear reactor.

Casing A metal pipe or tube used as a lining for a water, oil, or gas well.

Charge controller A device that controls the charging rate and/or state of charge for batteries.

Chord A straight line connecting the leading and trailing edges of an airfoil.

Coke The solid carbonaceous product obtained by destructive distillation of coal.

Collar Any of various rings used on like-devices to limit, guide, or secure a machine part.

Commutator A segmented metal cylinder or disc mounted on the armature shaft of an electric motor, generator, etc., used to make electrical contact with the rotating coils and ensure unidirectional current flow.

Conduction Transmission through a medium or passage, especially the transmission of electric charge or heat through a conducting medium without perceptible motion of the medium itself.

Convection A process of heat transfer through a gas or liquid by bulk motion of hotter material into a cooler region.

Cryogenic Relating to very low temperatures; requiring or suitable to very low-temperature storage.

Current at maximum power (Imp) The current at which maximum power is available from a module.

Dam A physical barrier constructed across a river or waterway to control the flow of or raise the level of water.

DC-to-DC converter A solid-state electronic circuit that changes the voltage and current levels of DC power.

Deep-cycle battery Type of battery that can be discharged to a large fraction of its capacity many times without diminishing its capacity.

Derate A decrease in the available capacity of an electric generating unit, commonly due to a system or equipment modification, environmental, or operational conditions.

Derrick A tall framework over an oil well, used to support boring equipment or hoist and lower lengths of pipe.

Dielectric A substance with extremely low electrical conductivity that practically does not conduct electric current.

Diode Electronic component that allows current flow in only one direction.

Direct current (DC) Electric current that flows in only one direction.

Dolomite A magnesia-rich sedimentary rock resembling limestone.

Dopant A substance such as boron, phosphorus, or arsenic, added in small amounts to a pure semiconductor material to alter its conductive properties for use in electronic components.

Dope (Electronics) To add impurities to (a semiconductor) to modify its properties.

Drag force A force on an airfoil proportional to the fluid velocity in the direction of motion.

Eddy current Electric current induced by an alternating magnetic field in a massive conductor, such as the core of an electromagnet, transformer, etc.; also called Foucault current.

Electric circuit A path for transmitting electric current.

Electric current The time rate of change of electric charge.

Electrical grid An integrated system for transmission and distribution of electric power, usually covering a large geographic area.

Electrolyte An ionized liquid in which current flows by migration of ions from one electrode of a battery to the other.

Electrolyte An ionized solution or molten substance that conducts electricity.

Endothermic Causing or characterized by absorption of heat.

Firedamp A combustible gas, mainly methane, often occurring in mines in association with bituminous coal; the explosive mixture of coalmine gases and air.

Fissile Property of a chemical element undergoing nuclear fission as a result of the impact of slow neutrons.

Fission A nuclear reaction in which a heavy atomic nucleus such as an isotope of uranium, plutonium, or thorium splits into two fragments of comparable but unequal mass and releases free neutrons and energy.

Flue gas The smoke in the uptake of boiler fire consisting mainly of carbon dioxide, carbon monoxide, nitrogen oxides, sulfur oxide, and particulate matters.

Furling A mechanism in some wind turbines that turns the rotor axis in a direction other than the wind flow to decrease the rotation speed of the blades.

Fusion A reaction in which two nuclei combine to a form a nucleus with the release of energy.

Gel-type battery Lead-acid battery in which the electrolyte is composed of a silica gel matrix.

Gigawatt (GW) Unit of power typically used in energy systems to describe the installed capacity or output power of generation facilities. One gigawatt is equal to one billion watts, one million kilowatts, or one thousand megawatts.

Grid-tie A PV, wind, or hydroelectric system that supplies power directly to the utility grid. Also called *grid-connected, grid-interactive, utility-intertie,* and other similarly descriptive terms.

Hybrid system A combination of electric generation units converting various sources of energy such as solar, wind, hydraulic potential, or fossil fuels into electricity.

Hydraulic fracturing/fracking The process of extracting oil or natural gas by injecting a mixture of water, sand (or gravel), and certain chemicals under high pressure into well holes in dense rock to create fractures such that the sand or gravel holds open, allowing the oil or gas to escape.

Hysteresis Phenomenon in which the magnetic flux density of a ferromagnetic material lags behind the change of external magnetic field.

Ingot A mass of metal cast in a convenient form for shaping, re-melting, or refining.

Insolation The amount of solar radiation reaching a given area; also called *irradiation*.

Inverter A solid-state electronic circuit that converts DC to AC power.

Irradiance Average solar radiant power density incident on a unit area perpendicular to the beam measured in $W\,m^{-2}$.

Irradiation Solar energy captured over a certain time interval on a unit horizontal surface; also called *insolation*.

Isentropic A thermodynamic process in which entropy remains constant.

Isobaric A thermodynamic process in which pressure remains constant.

Isothermal A thermodynamic process in which temperature remains constant.

Isotope One of two or more atoms having the same atomic number but different mass numbers.

Joule heating The process by which the passage of an electric current though a conductor produces heat; also known as *ohmic heating* and resistive heating.

Junction box An electrical box designed to be a safe enclosure in which to make proper electrical connections. On PV modules this is where PV strings are electrically connected.

Kerogen A fossilized material in shale and other sedimentary rock that yields oil upon heating.

Latent heat Heat which is absorbed or released by a substance during a change of state (fusion or vaporization) at constant temperature.

Lignocellulose A combination of lignin and cellulose that strengthens woody plant cells.

Maximum Power Point Tracker (MPPT) A controller that automatically changes the output voltage and current of a PV module or array such that the power delivered to the load is maximum at all times.

Mole (Chemistry) The base unit in the international unit system (SI) representing the amount of a substance that contains as many atoms, molecules, ions, or other elementary units as the number of atoms in 0.012 kilogram of carbon-12.

Monocrystalline cell A type of PV cell made from a single silicon crystal.

Nacelle The enclosure of a wind turbine that houses energy-conversion components including the shaft, generator, and gear train, coupled with the hub on which the rotor blades are attached.

Naphtha Any of several highly volatile, flammable liquid mixtures of hydrocarbons distilled from petroleum, coal, tar, and in making various chemicals.

Neutrino A stable electrically neutral elementary particle of an atom with spin of $\frac{1}{2}$ and an extremely small mass (close to zero) at rest.

ohm The unit of resistance to the flow of an electric current.

Open-circuit voltage (V_{oc}) The voltage across the terminals of an electric source or converter when no load current is drawn.

Orthogonal Relating to, consisting of, or involving right angles; perpendicular.

Pantograph A diamond-shaped frame, mounted on a train, tram, or trolley roof sliding on an overhead wire to conduct electrical current while the vehicle moves.

Particulate matter A small, discrete mass of solid or liquid matter that remains individually dispersed in air, gas, or liquid, usually considered to be a pollutant.

Pay zone A reservoir rock in which oil and gas are found in exploitable quantities.

Photovoltaic (PV) cell A semiconductor element that converts solar radiation directly to electric energy.

Piezoelectricity Electric charge accumulation in certain solid materials such as crystals and some ceramics in response to applied mechanical stress. Piezoelectric effect produces electric potential when a crystal is subjected to mechanical vibration, or changes the dimensions of a crystal when a voltage is applied.

Pinch-out Petroleum reservoir or a subsurface pool of hydrocarbons trapped by overlying rock formations with lower permeability.

Pitchblende A massive radioactive mineral also known as *uraninite*, containing various oxides of uranium and radium.

Polycrystalline cell A type of PV cell made from multiple silicon crystals; also called *multi-crystalline*.

Positron A subatomic particle with the same mass as an electron and a numerically equal but positive charge.

Power factor The ratio of the average active power of an AC circuit or system to the product of the effective values of the terminal voltage and current.

Proppant A solid material such as sand or ceramic used to facilitate unconventional oil extraction by keeping the fissures of a hydraulic fracture open.

Protactinium A rare, extremely toxic, radioactive, lustrous, metallic element.

PV array Any number of photovoltaic modules connected together to provide a single electrical output at a specified voltage.

PV cell The basic element of a photovoltaic module.

Radiation (Nuclear physics) The emission of energy as electromagnetic waves or as moving subatomic particles, especially high-energy particles which cause ionization and nuclear decay.

Rare-earth magnet Strong permanent magnets made from alloys of rare earth elements.

Reagent A substance used in a chemical reaction to detect, measure, examine, or produce another substance.

Scrubber A device or process for removing pollutants from smoke or gas produced by burning fuels with relatively high sulfur content.

Shale gas The natural gas that is extracted from shale by hydraulic fracturing (fracking).

Shale oil An unconventional oil, produced from oil shale rock fragments by pyrolysis, hydrogenation, or thermal dissolution; also called *tight oil*.

Shale A soft fine-grain laminated sedimentary rock formed from compressed mud or clay layers.

Short circuit current Current drawn from an electric circuit that contains sources when the output terminals are connected (short-circuited) to each other.

Sievert A derived SI unit of dose for the amount of ionizing radiation used as a measure of the biologic effect of low levels of ionizing radiation on the human body (symbol: Sv).

Solar tracker A system capable of rotating about one or two axes, to follow the sun throughout the day to maximize the power output of a PV array.

Specific heat The ratio of the amount heat needed to raise the temperature of a certain amount of a substance by one degree to the amount of heat needed to raise the temperature of the same amount of a reference substance, usually water, by one degree.

Thumper truck A seismic device that generates controlled seismic energy used to produce shock waves for seismic surveys.

Tight oil An unconventional oil, produced from oil shale rock fragments by pyrolysis, hydrogenation, or thermal dissolution; also called *shale oil*.

Tension-leg-platform (TLP) A vertically moored floating structure normally used for the offshore production of oil or gas.

Toroid A surface or a solid body with a hole in the middle like a doughnut formed by revolving a circle around an axis on the same plane at a distance greater than its radius.

Valence electron An electron at the outer shell of an atom that can participate in the formation of a chemical bond with other atoms if the outer shell is not closed.

Wellhead A component at the surface of an oil or gas well that provides the structural and pressure-containing interface for the drilling and production equipment.

Index

Energy for Sustainable Society: From Resources to Users, First Edition. Oguz A. Soysal and Hilkat S. Soysal.
© 2020 John Wiley & Sons Ltd. Published 2020 by John Wiley & Sons Ltd.